2D Semiconducting Materials for Electronic, Photonic, and Optoelectronic Devices

Two-dimensional semiconducting materials (2D-SCMs) are the subject of intensive study in the fields of photonics and optoelectronics because of their unusual optical, electrical, thermal, and mechanical properties. The main objective of *2D Semiconducting Materials for Electronic, Photonic, and Optoelectronic Devices* is to provide current state-of-the-art knowledge of two-dimensional semiconducting materials for various applications. Two-dimensional semiconducting materials are the basic building blocks for making photodiodes, light-emitting diodes, light-detecting devices, data storage, telecommunications, and energy-storage devices. When it comes to 2D-SCMs, electronic, photonic, and optoelectronic applications, as well as future plans for improving performance, no modern book covers as much ground. The planned book will fill such gaps by offering a comprehensive analysis of 2D-SCMs.

This book covers a range of advanced 2D materials, their fundamentals, and the chemistry for many emerging applications. All the chapters are covered by experts in these areas around the world, making this a suitable textbook for students and providing new guidelines to researchers and industries.

- Covers topics such as fundamentals and advanced knowledge of 2D-SCMs
- Provides details about the recent methods used for the synthesis, characterization, and applications of 2D-SCMs
- Covers the state-of-the-art development in 2D-SCMs and their emerging applications

This book provides directions to students, scientists, and researchers in semiconductors and related disciplines to help them better understand the physics, characteristics, and applications of 2D semiconductors.

2D Semiconducting Materials for Electronic, Photonic, and Optoelectronic Devices

Edited by
Anuj Kumar and Ram K. Gupta

CRC Press
Taylor & Francis Group
Boca Raton London New York

CRC Press is an imprint of the
Taylor & Francis Group, an **informa** business

Designed cover image: Shutterstock_1196754286

First edition published 2025
by CRC Press
2385 NW Executive Center Drive, Suite 320, Boca Raton FL 33431

and by CRC Press
4 Park Square, Milton Park, Abingdon, Oxon, OX14 4RN

CRC Press is an imprint of Taylor & Francis Group, LLC

ISBN: 978-1-032-57352-6 (hbk)
ISBN: 978-1-032-57452-3 (pbk)
ISBN: 978-1-003-43944-8 (ebk)

DOI: 10.1201/9781003439448

Typeset in Palatino
by Apex CoVantage, LLC

Dedicated to Dr. Shawn Naccarato, whose inspiration, innovation, and motivation have driven the exploration of semiconducting materials for advanced applications.

Contents

11. Logic Devices Based on 2D Semiconducting Materials 147

Muhammad Mubeen, Hasan Ubaid Ullah, and Yinghui Han

12. Memristors Based on 2D Semiconducting Materials 162

Zhengyue Li and Huarui Sun

20. Two-Dimensional (2D) Semiconductors for Solar to Hydrogen Fuel 282

Spandana Gonuguntla, Aparna Jamma, Bhavya Jaksani, and Ujjwal Pal

21. 2D Semiconductors for Next-Generation Thermoelectric Materials 296

Abu B. Siddique, Sergio O. Martinez and Mallar Ray

About the Editors

Dr. Anuj Kumar is working as Assistant Professor at GLA University, Mathura, India. His research focus is on designing catalytic materials for energy production and storage. He has published more than 160 articles, 5 books, and 50 book chapters in reputed peer-reviewed journals (like *Advanced Materials, Energy & Environmental Science, Materials Today, Applied Catalysis B: Environment and Energy, Chemical Engineering Journal, Journal of Materials Chemistry A, Coordination Chemistry Reviews, Chemistry of Materials, Applied Energy,* and *Small*); in addition to eight patents (two U.S. and six Indian), two Indian patents have granted by the Indian government. He has been awarded the Young Researcher Award 2020/2021/2022 by InSc and by Centre for Education Growth & Research. He is serving as Editor (*Nano LIFE*), Section Editor (*Organic Mini Reviews*), Advisory Board Member (*Carbon Energy*), Guest Editor, and Editorial Board Member for various journals. He is also working as Expert Consultant for research in various national and international universities.

Dr. Ram K. Gupta is Professor of Chemistry at Pittsburg State University (PSU), Kansas, USA. He is also serving as Associate Vice-President for Research and Support at PSU and Director of Research at the National Institute for Materials Advancement (NIMA). Dr. Gupta has been recently named by Stanford University as being among the top 2% of research scientists worldwide. Dr. Gupta's research spans a range of subjects critical to current and future societal needs including semiconducting materials and devices, bio-polymers, flame-retardant polymers, green energy production and storage using nano-structured materials and conducting polymers, electrocatalysts, optoelectronics and photovoltaics devices, organic–inorganic heterojunctions for sensors, nanomagnetism, biocompatible nanofibers for tissue regeneration, scaffold and antibacterial applications, and biodegradable metallic implants.

Dr. Gupta has published over 350 peer-reviewed journal articles, made over 450 national/international/regional presentations, chaired and organized multiple sessions at national/international meetings, wrote several book chapters (120+), edited over 50 books, and received several million dollars for research and educational activities from external agencies. He is also serving as Editor, Associate Editor, Guest Editor, and Editorial Board Member for various journals.

Contributors

Amit Agarwal
Indian Institute of Technology Kanpur
Kanpur, India

Mumtaz Ali
Department of Chemical Engineering
    Hanyang University
Seoul, Republic of Korea

Hassan Anwer
National University of Sciences and
    Technology
Islamabad, Pakistan

Zafar Arshad
National Textile University
Faisalabad, Pakistan

Somnath Bhowmick
Indian Institute of Technology Kanpur
Kanpur, India

Prakash Chandra
Pandit Deendayal Energy University
Gandhinagar, India

Yogesh Singh Chauhan
Indian Institute of Technology Kanpur
Kanpur, India

Li Chen
Anhui University
Hefei, China

Ghulam Dastgeer
Sejong University
Seoul, South Korea

Yash Desai
Pittsburg State University
Kansas, USA

Ayat N. El-Shazly
Central Metallurgical Research
    and Development Institute
Cairo, Egypt

Spandana Gonuguntla
CSIR-Indian Institute of Chemical
    Technology
Hyderabad, India
Academy of Scientific and
    Innovative Research
    (AcSIR)
Ghaziabad, India

Zheng Guo
Anhui University
Hefei, China

Ram K. Gupta
Pittsburg State University
Kansas, USA

Mahmoud Adel Hamza
Ain-Shams University
Cairo, Egypt
The University of Adelaide
Adelaide, Australia

Yinghui Han
University of Chinese Academy
    of Sciences
Beijing, China

Bhavya Jaksani
CSIR-Indian Institute of Chemical
    Technology
Hyderabad, India
Academy of Scientific and
    Innovative Research (AcSIR)
Ghaziabad, India

Aparna Jamma
CSIR-Indian Institute of Chemical
Technology
Hyderabad, India
Academy of Scientific and
Innovative Research
(AcSIR)
Ghaziabad, India

Eom Jonghwa
Sejong University
Seoul, South Korea

Daljit Kaur
DAV University
Jalandhar, India

Gurpreet Kaur
Lovely Professional University
Phagwara, India

Surinder Pal Kaur
Indian Institute of Technology Ropar
Rupnagar, India

Deok-kee Kim
Sejong University
Seoul, South Korea

Rajashree Konar
Bar-Ilan University
Ramat Gan, Israel

Prabhukrupa C. Kumar
ICT-IOC
Bhubaneswar, India

T. J. Dhilip Kumar
Indian Institute of Technology Ropar
Rupnagar, India

Anuj Kumar
GLA University
Mathura, India

Yang Li
Harbin Institute of Technology
Harbin, China

Zhengyue Li
Harbin Institute of
Technology
Shenzhen, China

Dilip K. Maiti
University of Calcutta
Kolkata, India

Sunil Kumar Baburao Mane
Khaja Bandanawaz University
Kalaburagi, India

Sergio O. Martinez
Tecnológico de Monterrey
Monterrey, Mexico

Neeraj Mehta
Banaras Hindu University
Varanasi, India

Mostafa Moslempoor
University of Isfahan
Isfahan, Iran

Mahmoud Moussa
The University of Adelaide
Adelaide, Australia
Beni-Suef University
Beni-Suef, Egypt

Muhammad Mubeen
University of Chinese Academy
of Sciences
Beijing, China

Ramakanta Naik
ICT-IOC
Bhubaneswar, India

Ateeb Naseer
Indian Institute of Technology
Kanpur
Kanpur, India

Gilbert Daniel Nessim
Bar-Ilan University
Ramat Gan, Israel

Sobia Nisar
Sejong University
Seoul, South Korea

Shiv Kumar Pal
Banaras Hindu University
Varanasi, India

Ujjwal Pal
CSIR-Indian Institute of Chemical
    Technology
Hyderabad, India
Academy of Scientific and Innovative
    Research (AcSIR)
Ghaziabad, India

Jing-Kai Qin
Harbin Institute of Technology
Harbin, China

Mamta Rani
DAV University
Jalandhar, India

Mallar Ray
Tecnológico de Monterrey
Monterrey, Mexico

Sudipta Ray
A.K. Mitra Institution for Girls
Kolkata, India

Laraib Sajjad
Sejong University
Seoul, South Korea

Mohammed Salah
Minia University
Minya, Egypt

Subrata Senapati
ICT-IOC
Bhubaneswar, India

Naghma Shaishta
Khaja Bandanawaz University
Kalaburagi, India

Esmaeil Sheibani
University of Isfahan
Isfahan, Iran

Abu B. Siddique
Tecnológico de Monterrey
Monterrey, Mexico

Manmeet Singh
IIT Kanpur
Kanpur, India

Ramesh Sivasamy
Amrita Vishwa Vidyapeetham
Karnataka, India

Marutheeswaran Srinivasan
Amrita Vishwa Vidyapeetham
Karnataka, India

Huarui Sun
Harbin Institute of Technology
Shenzhen, China

Konrad Szaciłowski
AGH University of Science
    and Technology
Kraków, Poland

Hasan Ubaid Ullah
University of Chinese Academy
    of Sciences
Beijing, China
University of Agriculture
Faisalabad, Pakistan

Edgar Mosquera Vargas
Universidad del Valle
Santiago de Cali, Colombia

Cheng-Yan Xu
Harbin Institute of Technology
Harbin, China

Liang Zhen
Harbin Institute of Technology
Harbin, China

1

Fundamentals of 2D Semiconducting Materials

Rajashree Konar* and Gilbert Daniel Nessim
*Correspondence: rajashreekonar@gmail.com

1.1 Introduction

The grapheme breakthrough in 2004 led to extensive research into the family of two-dimensional semiconducting materials (2D SCMs) [1]. The growing attention toward 2D semiconductors can be attributed to the challenges in achieving a substantial bandgap in graphene [1]. Despite its great properties, graphene lacks a bandgap that limits its targeted applications. 2D SCMs are emerging semiconductor materials that promise to overcome this limitation and offer new properties for specific applications [2]. Intriguingly, they have excellent performance even at the monolayer. The complex band structures and the heterostructures free of lattice-mismatch offer potential avenues for tailoring specific mechanisms to cater to the requirements of diverse electronic systems. Over hundreds of distinct 2D SCMs have been successfully isolated through experimentation, exhibiting bandgap values ranging from a few millielectronvolts to several electronvolts [2]. Additionally, several other semiconductors are expected to be extracted soon. Thanks to the wide range of available 2D materials, one can use a specific SCM for tailored applications.

Researchers have successfully produced single layers of different materials like boron nitride, silicon, boron, germanium, phosphorus, and transition metal dichalcogenides (TMDCs) [3]. Additionally, heterostructures of these 2D SCMs have been synthesized by simply integrating multiple layers of these specific materials [4]. Engineering 2D SCMs in the nanoscale dimensions provides unprecedented opportunities for advancing next-generation technologies. The confinement observed in reduced dimensionality systems gives rise to peculiar physical properties in 2D systems [5]. However, a more thorough comprehension of these properties is still required. Understanding and manipulating the physical properties of emerging 2D materials is essential for their effective use in modern electronic devices. So, these materials require precise control of their properties.

As previously stated, integrating multiple 2D crystals in a vertical stack presents numerous possibilities [6]. Heterostructures, held together by interlayer van der Waals (vdW) forces (the same forces that bond layered materials), provide a broader range of combinations than conventional growth techniques. The increasing variety of 2D crystals leads to a corresponding increase in the complexity of heterostructures that can be constructed accurately at the atomic level. The combination of various crystals in a stack yields notable synergistic effects. In the realm of first-order approximation, it is possible for charge redistribution to take place among the neighboring crystals of a stack and even those situated farther away [2]. Adjacent crystals can cause modifications in each other's structure. Furthermore, changes can be made by adjusting the placement of individual components relative to one another.

DOI: 10.1201/9781003439448-1

This chapter provides insight into the 2D SCM family through classification and understanding their crystal structures, affecting their characteristics, in addition to a few examples of reported applications in the relevant fields.

1.2 2D Semiconducting Materials Beyond Graphene

Graphene has been functioning as a stand-alone 2D material in the past decade [7]. Nonetheless, due to its relatively small bandgap, its utilization is restricted in numerous fields, as previously mentioned [7]. The surge in research on 2D materials such as graphene and its analogs directly results from their potential for various technological applications. This includes electronic devices, sensors, catalysts, energy conversion, and storage devices, which can fully exploit their remarkable electrical, optical, chemical, and thermal properties [1]. In addition to graphene, other 2D materials with diverse elemental compositions and crystallographic structures offer a wide range of material options with adjustable chemical and physical properties [1]. These materials are ideal for high-performance active components, such as electrode materials, and for electrochemical energy storage devices [8]. Various strategies have been consistently researched and created to expand the options for inorganic 2D materials. In recent years, a significant focus has occurred on 2D semiconductors or 2D SCMs such as 2D TMDCs, MXenes, hexagonal boron nitride (h-BN), phosphorene, black phosphorous (BP), germanene, and silicene [8]. The materials in question share a comparable structure to graphene, featuring a flat topology and extremely thin dimensions consisting of only one to a few atomic layers [9]. The atomic scale thickness of these materials leads to distinctive properties that can differ from those of their bulk lamellar counterparts [8]. The distinct electrical and electronic properties of electrons in the 2D plane are attributed to their quantum confinement [8]. In the following sections, we dedicate a thorough understanding of the 2D SCMs based on their categories.

1.2.1 2D Transition Metal Dichalcogenides

TMDCs are structured in an MX_2 pattern, where M represents a transition metal such as Mo, W, Re, and Nb, and X represents a chalcogen atom, which can be S, Se, or Te [10]. The material's properties can vary from semiconducting to metallic or superconducting, based on the combination of M and X [10]. The band structures of TMDCs undergo a notable transformation as the number of layers decreases from the bulk crystal [10]. This makes TMDCs an exceptional class of 2D material for diverse applications.

1.2.1.1 Crystal Structure and Electronic Structures Analysis of 2D TMDCs

Based on the above properties, it must be noted that the crystal structures of the 2D TMDCs are variable. The crystal structures and the number of d-electrons play a significant role in shaping the electronic configurations of 2D TMDCs [11]. The partially filled d bands of transition metals are considered critical as they determine their overall electronic behavior [11]. Bulk transition metal dichalcogenides (TMDCs) consist of monolayers stacked in an ordered manner and held together by the weak vdW interaction [11]. The material in question exhibits various polymorphs and tends to form distinct crystal structures, namely, the 1T, 2H, and 3R phases [10]. In the case of group IVB TMDCs and group VIB TMDCs,

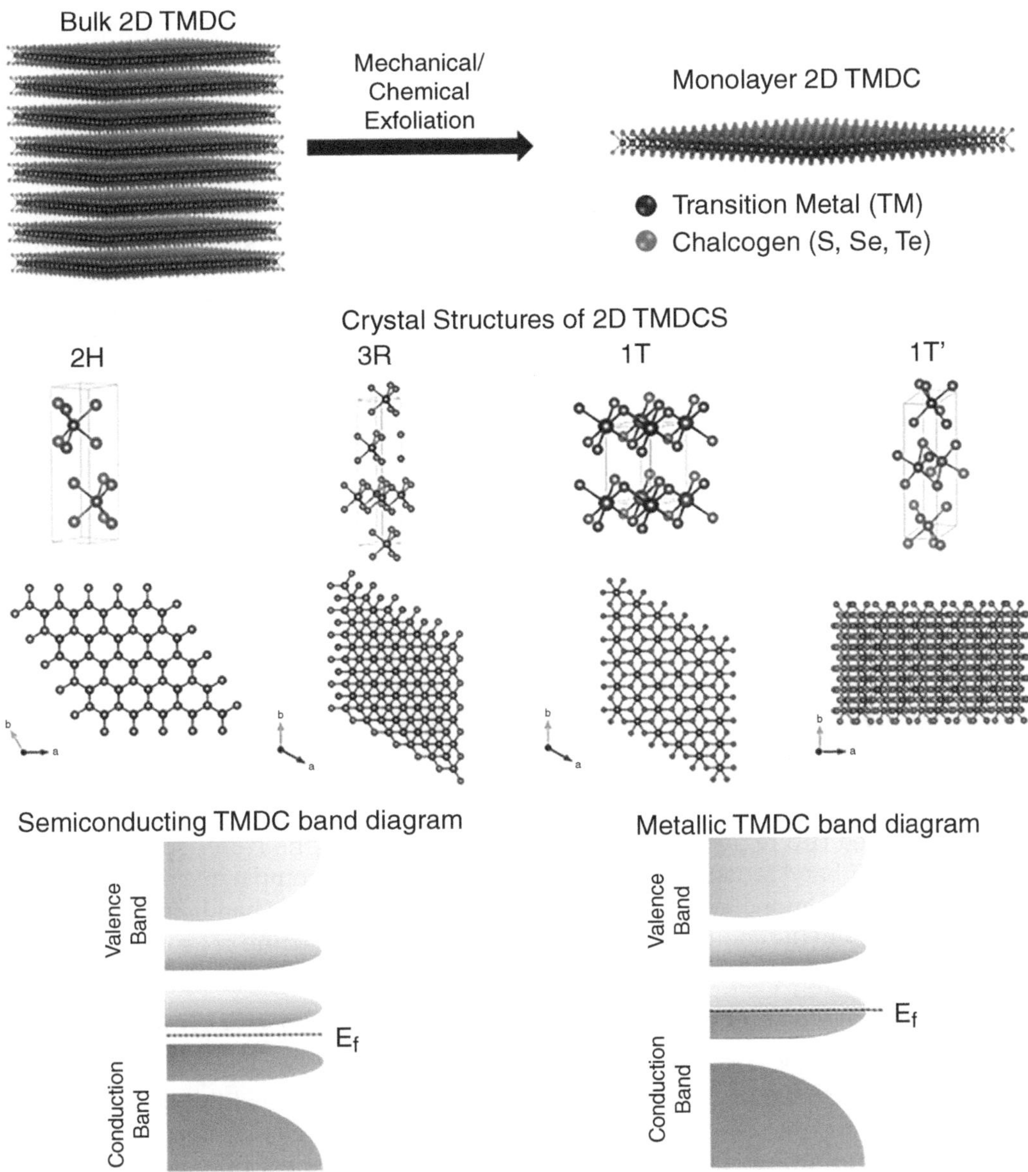

FIGURE 1.1
Schematic showing the differences between bulk and monolayer 2D TMDCs, the differences in crystal structures (2H, 3R, 1T, and 1T′, with respect to their coordinations), and band diagrams depicting differences between semiconducting and metallic TMDCs.

the 1T phase is more stable than the 2H phase when reversing the latter [12]. This could be due to the lower formation energies of group IVB TMDCs. **Figure 1.1** shows the difference between bulk and monolayer 2D TMDCs and the possible phases (such as 2H, 3R, 1T, 1T′ known as hexagonal, rhombohedral, and trigonal).

The various structural phases of these materials can also be analyzed based on the stacking orders of the three atomic planes (chalcogen–metal–chalcogen) that make up each layer

[13]. During the 2H phases, chalcogen atoms from different atomic planes occupy the same position and are aligned on top of each other perpendicularly to the layer. This results in an ABA stacking pattern [12]. The 1T phases showcase an arrangement of ABC stacking [14]. The thermodynamically stable phase can be either the 2H or 1T phase, depending on the combination of transition metal elements from groups IV, V, VI, VII, IX, or X and chalcogens (S, Se, or Te). However, the alternative structure can be frequently acquired as a metastable phase [2]. The current knowledge on the existence of the two phases, in either stable or metastable form, and on other properties of TMDCs is shown in **Figure 1.1**. In the case of five out of six chemically distinct bulk TMDCs that are created by group VI transition metals (metal = Mo or W; chalcogen = S, Se, or Te), the 2H phase remains thermodynamically stable, while the 1T phase can be achieved as a metastable phase [10, 14]. An exception is represented by WTe_2, for which the stable bulk phase at room temperature is the orthorhombic $1T_d$ phase [14]. It is also well known that WSe_2 crystallizes in the 2H phase compared to the 3R or 1T' phases (3R is produced only at high temperatures beyond 1000°C, and 1T' is made only using specific precursors in solvothermal syntheses) [14, 15]. On the other hand, WS_2 and $MoTe_2$ can be naturally obtained in the 2H, 3R, and 1T' phases, as these structures are thermodynamically stable in any synthesis route [10, 16]. The 2H variants are typically characterized as semiconducting among all the phases.

Other than the well-known 2D SCMs in the TMDC family (as shown in **Figure 1.1**, the band diagram schematics of the semimetals and semiconductors), there are many examples of other TMDCs relevant to optoelectronic device fabrications. For instance, 1T-phase group IVB TMDCs find wide usage in different device applications [17]. Among the known materials, $TiSe_2$, $TiTe_2$, $ZrTe_2$, and $HfTe_2$ exhibit semimetallic properties, characterized by a minor overlap between the upper edge of the p-orbital chalcogen valence band and the lower edge of the d-orbital transition metal conduction band [17].

The Zr and Hf TMDC families are all semiconductors with an indirect bandgap from the visible to infrared (IR) region [17, 18]. These crystals belong to the *P-3m1* space group, suggesting that their band structures from monolayer to bulk are comparable [17]. Even transitioning from bulk to monolayer, these materials retain their indirect bandgap. This feature is much different from the complex variations in the electronic structure of group VIB TMDCs.

1.2.1.2 *Fundamentals and Examples of Optoelectronic Properties in 2D TMDCs*

Research on 2D TMDCs reveals that the bulk MX_2 variants are categorized as indirect bandgap semiconductors, whereas their monolayer counterparts exhibit direct bandgaps that are highly desirable for optoelectronic applications [19]. The existence of the native bandgap provides an excellent current on/off ratio in field-effect transistors, which has been limited in 2D systems. Monolayer TMDC usually has a direct bandgap [19]. The valence band maximum (CBM) and valence band minimum (VBM) are located at a high-symmetric K point [19]. The valence band exhibits a clear split in its band structure because of spin-orbit coupling (SOC) [19]. Compared to MoX_2, the split in the valence band of WX_2 is larger [19]. The strength of SOC interactions is directly proportional to the order of W.

Xia et al. explain the following trends from their findings [19]:

(a) As the atomic number of element X increases from sulfur to tellurium, the CBM and VBM energies of MX_2 also increase accordingly. CBM is smaller than the VBM. Specifically, the VBM of $MoSe_2$ is 0.63 eV higher than molybdenum disulfide's (MoS_2), while its CBM is 0.37eV higher than MoS_2's. The only exception is WTe_2, whose CBM is slightly lower (by 0.06eV) than WSe_2 [1].

(b) In the common-X system, it has been observed that WX_2 exhibits higher CBM and VBM compared to MoX_2 [19]. This indicates a type-II band alignment in the lateral heterostructures of MoX_2–WX_2. VBM of WS_2 is observed to be 0.39 eV higher than that of MoS_2, while its CBM is 0.35eV higher [19]. 2D quantum well structures have been reported as an effective means to enhance LEDs' external quantum efficiency (EQE) based on TMDCs [19].

Recent studies indicate that single-photon emission is more prevalent in 2D materials with defects [20]. Using vdW heterostructures has enabled the demonstration of electrically driven single-photon emission from 2D TMDCs, like WSe_2 and WS_2 [9]. 2D TMDCs are promising gain materials for laser generation due to their high binding energy and lack of dangling bonds. A photonic crystal cavity was used to confine direct-gap excitons to the surface, resulting in an optical-pumped continuous-wave laser with ultralow thresholds [21].

Another critical property observed in 2D TMDC SCMs is the Raman scattering enhancement by introducing plasmonic nanostructures [21]. Most of the current research focuses on combining 2D semiconducting TMDCs with integrated plasmonic nanostructures. This approach offers strong light–matter interactions with great potential for developing high-performance optoelectronic devices.

The electronic properties of TMDCs and their layered structures are highly dependent on factors such as atomic composition, the number of layers, and the arrangement of the layers in terms of rotation or stacking orientation. Bulk MoS_2 is thinned to a monolayer transition from an indirect to a direct-gap semiconductor [22]. This significantly enhances exciton photoluminescence (PL) at room temperature due to the direct electron transition without any loss of momentum [22]. The enhancement of the quantum yield of PL is attributed to the significant increase in the excitonic electromagnetic decay rates through the surface plasmon resonant energy. The valleys in most monolayer TMDCs are caused by the broken inversion symmetry in combination with time-reversal symmetry, resulting in opposite electron spins at the energy-momentum dispersion. A plasmonic nanowire-WS_2 system was utilized to demonstrate the valley-dependent directional coupling of light. The coupling efficiency of the valley pseudospin in WS_2 to the transverse optical spin of the same handedness was 90% ± 1% [23].

In 2D TMDC superconducting materials, various nonlinear optical phenomena such as saturable absorption, second-harmonic generation (SHG), third-harmonic generation, and two-photon absorption have been observed [22]. In 2D TMDCs, the interaction length between the pump laser and light is typically very short due to their atomic thinness. Consequently, light-harmonic generation could be more efficient. In most TMDCs, the absence of inversion symmetry on the surface can lead to a robust SHG signal when subjected to intense optical pumps and plasmonic enhancement. A recent study showed that by using remotely excited surface plasmon polaritons, generating and controlling SHG in a single silver nanowire and monolayer MoS_2 hybrid system is possible [24]. This process leads to the emission of SHG around subwavelength waveguides, which are collimated axially but divergent transversely [24]. These findings have important implications for manipulating SHG emissions in nanoscale systems. A plasmonic metasurface enables the entanglement of light's phase and spin while enhancing and driving the nonlinear emission of 2D materials placed on top. Using a spin-related geometric phase Au metasurface, second-harmonic valley photons in monolayer WS_2 were coherently pumped by light, separated, and then directed to predetermined directions at room temperature [25].

1.2.2 Hexagonal Boron Nitride as Promising 2D SCM for Electronic and Optoelectronic Devices

Two-dimensional h-BN shares a lattice structure like graphene but with a 1.8% longer lattice constant [26]. It is a wide bandgap semiconductor with a bandgap of approximately 5.9 eV [26]. Its atomic structure consists of alternating boron and nitrogen atoms with a single-atom thickness. h-BN has superb chemical stability, mechanical robustness, high thermal conductivity, and is electrically insulating [26]. The utilization of 2D h-BN and its related composite systems is gaining attention as it presents a promising solution for various challenges that may arise in developing complex electronic devices and systems.

Various defects exist in 2D-h-BN nanosheets, such as edges, grain boundaries, dislocations, interstitial atoms, and vacancies [26]. Defects influence the electronic structure of 2D-h-BN nanosheets in their local atomic structure. These defects can modify the way the nanosheets function. Defects in 2D-h-BN nanosheets also include dopants. Dopant atoms such as carbon, hydrogen, metal, fluorine, and oxygen have been identified in 2D-h-BN nanosheets [26]. During chemical vapor deposition (CVD) or subsequent treatments, impurity dopants typically arise [27]. Introducing point defects in 2D-h-BN nanosheets leads to a deviation in the chemical bonding and coordination environment from the original lattice, affecting its structural and electronic properties [27]. So, the introduced point defects lead to variable alterations in the electronic structure. h-BN is also a well-known lubricant [26]. It is an effective charge leakage barrier layer in electronics due to its exceptional electrical insulation properties. The utilization of h-BN has extended to include applications in devices such as deep-ultraviolet (DUV) light-emitting and photodetectors [26]. The diverse mechanical, thermal, physical, and chemical properties exhibited by h-BN have made it an attractive candidate for an array of applications such as electronic, optoelectronic, and structural materials [26]. h-BN has been implemented in a variety of device applications. These include serving as encapsulation materials and dielectric substrates and being used as tunneling barriers [26]. **Figure 1.2a–f** shows the growth and stacking patterns of h-BN, high-resolution transmission electron microscopy images (both experimental and simulated) showing electron densities, synthesis mechanisms, and vertical transport in 6L h-BN [28].

h-BN is widely used as a substrate for graphene-related electronics. This helps to reduce disorderedness in graphene surfaces because h-BN has a smooth surface, almost free of charge traps and dangling bonds, a lattice constant nearly close to that of graphite, and a wide bandgap. For example, high-performance graphene devices were obtained using large-domain-sized 2D-h-BN sheets [29]. The graphene device on a substrate of h-BN sheets showed three times higher mobility compared to the same device without h-BN sheets [29]. Atmospheric pressure CVD was utilized to synthesize large-area h-BN, which was then integrated as a gate dielectric in graphene devices [29]. The graphene devices (transistors) exhibited a comparable performance with and without h-BN. The mobility of the graphene device (in the few thousand $cm^2/(V{\cdot}s)$ range) before and after the integration remained the same. Charge transfer across the h-BN layer in field-effect-transistor (FET) devices containing graphene has shown n-type characteristics while applying high stress to the device layers [29]. h-BN also finds application in electron tunneling devices as it is stable against high voltages and does not cause problems due to stress-induced leakage current, charge trapping or de-trapping, and premature breakdown [26]. Localized defects in 2D-h-BN monolayers and a few layers demonstrate a single-photon emission [26]. The active single-photon emitters at room temperature are perfect for 2D devices in quantum information applications and can be used in nanophotonic circuits on a chip.

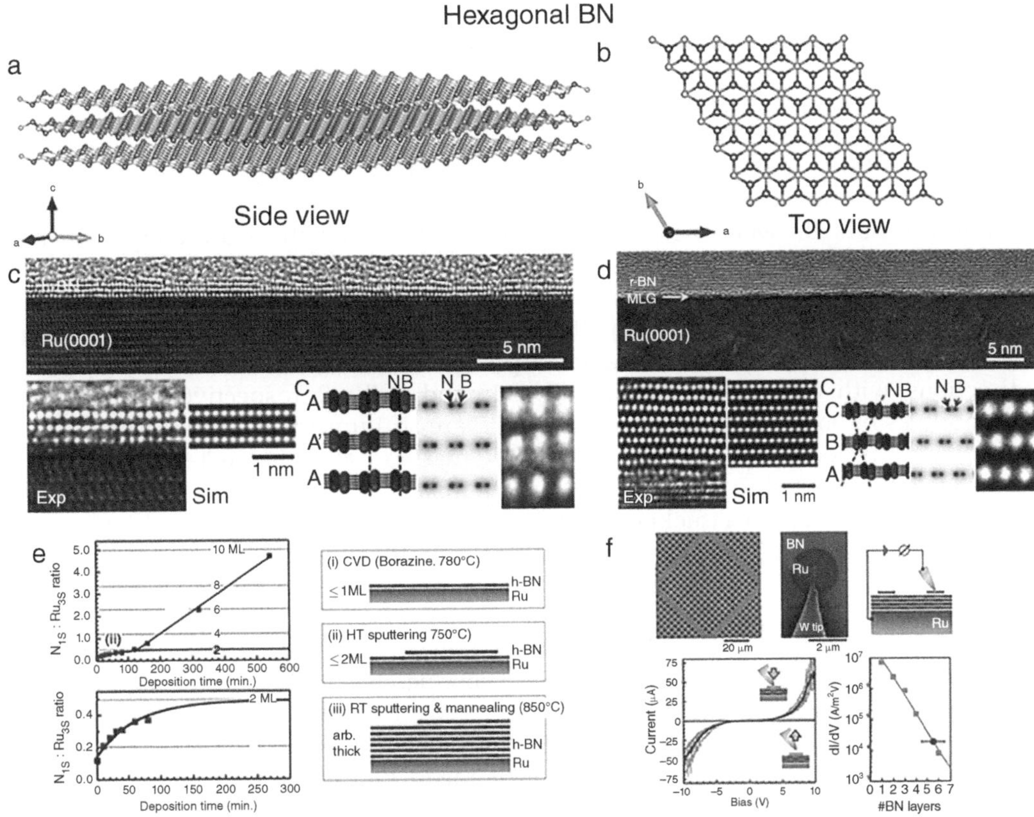

FIGURE 1.2
Crystal structures of h-BN from (a) side view and (b) top view; (c) A–A′–A stacking of h-BN on Ru (0001) as shown in the high-resolution transmission electron microscopy image of a 3L h-BN film (considering both experimental and simulated conditions containing the model of A–A′–A stacked h-BN with computed electron density map; (d) A–B–C stacked rhombohedral BN on graphene-covered Ru (0001) with 10 L film comparison, model of rhombohedral BN, and computed electron density map; (e) few-layer h-BN growth by chemical vapor deposition and reactive magnetron sputtering with side schematics; and (f) vertical transport through 6 L h-BN sandwiched between Ru electrodes. Adapted with permission [28]. Copyright (2013), American Chemical Society

1.2.3 Understanding the Role of BP in 2D SCM-Based Optoelectronic Devices

Among all the various phosphorus allotropes, BP stands out as the most stable and boasts exceptional properties like high carrier mobility and tunable bandgap [30]. These properties make BP a highly sought-after 2D-layered semiconductor for advanced next-generation optoelectronics.

The three common allotropes of phosphorous are white, red, and black [31]. In contrast to white and red phosphorus, BP is the most stable among all phosphorus allotropes. BP has an orthorhombic crystal structure where the P atoms are arranged in a honeycomb lattice due to the puckering of double layers [31]. Monolayer BP consists of sp^3 P atoms covalently bonded to three other adjacent P atoms with a bond length of 2.18 Å [30]. These P atoms have one lone pair of electrons and form a quadrangular pyramid structure. The arrangement of three P atoms in a layer exhibits an attractive physical property due to their orientation. Two atoms are positioned in the layer's plane at an angle of 98°, while the third atom is located between the layers at 103° [30]. This arrangement results in anisotropic properties that affect optical, mechanical, thermoelectric, and electrical conductance.

Compared to semiconducting TMDCs like MoS_2, BP has a stronger layer-to-layer interaction, resulting in a more pronounced dependence on its bandgap on thickness [30]. As the thickness of BP is reduced from more than ten (10) layers to a monolayer, its electronic bandgap varies from approximately 0.33 to 2.0 eV [30]. The effective masses of electrons and holes in bulk BP were measured experimentally along the x-direction (armchair), y-direction (zigzag), and z-direction (perpendicular to the 2D plane) [30]. The results showed that the effective masses were approximately 0.08 (0.08), 1.00 (0.65), and 0.13 (0.28) m0, respectively, where m0 represents the free electron mass [30]. In three to five-layered BP, it was predicted that electrons and holes possess similar effective masses in the x and y directions, with values of approximately 0.15 and 1.00 m0, respectively [30]. The carrier mobilities of both monolayer and bilayer BP are still uncertain despite knowledge of their intrinsic properties. For electron and hole masses/mobilities in 1–5-layer BP, more experiments are needed to understand its mechanisms. Among semiconducting layered TMDCs, BP thin-film with a thickness of 4 nm or greater boasts significantly higher carrier mobility than MoS_2 and other counterparts [30]. Ultrathin BP containing 4 nm thickness and six to eight layers, when placed between h-BN layers, displays remarkable hole mobility of approximately 5,000 cm^2 V^{-1} s^{-1} at room temperature. This makes it appropriate for advanced electronic applications. **Figure 1.3a-d** shows two examples of strain engineering in few-layered BP to show enhanced electrical conductance [32] and ultra-high photoresponses on frequency modulation [33].

The moderate bandgap of thin-film BP, approximately 0.33 eV, makes it an excellent candidate for photonics applications. Due to this bandgap, BP can strongly interact with light across a broad range of wavelengths, even in the challenging near-IR and mid-IR ranges [30]. This feature makes it an ideal material for use in various technical applications. Interestingly, BP can also be thinned down using plasma without significant degradation in electronic or photonic properties [31].

1.2.4　2D Semiconducting MXenes for Optoelectronics and Related Electronic Devices

MXenes are a diverse group of two-dimensional materials comprising transition metal carbides, nitrides, and carbonitrides with a chemical formula of $M_{n+1}X_nT_x$ (n = 1, 2, or 3) [34]. These materials feature early transition metals, such as Ti, Ta, Mo, Cr, Zr, Hf, and Nb, as well as carbon and/or nitrogen atoms [34]. The surface functional groups of MXenes are represented by T_x, which denotes various functional groups such as –O, –OH, –F, and –Cl. So far, more than 30 MXenes have been experimentally synthesized by selective acid etching and subsequent exfoliation, Lewis acidic etching, and CVD methods [34]. They have abundant and tunable electronic, optical, mechanical, biological, and chemical properties. Most MXenes have high conductivity; for example, $Ti_3C_2T_x$ has a conductivity of around ~104 S cm^{-1} [34]. Their exceptional hydrophilicity and high metallic conductivity make them ideal contact materials for electronic and optoelectronic devices, offering unique benefits.

MXenes are reported to be metallic or semimetallic. Very few semiconducting MXenes are known, some of which are Sc_2CO_2, Sc_2CF_2, $Sc_2C(OH)_2$, Ti_2CO_2, Zr_2CO_2, and Hf_2CO_2 [35]. The semiconducting MXenes exhibit band gap values between 0.24 and 1.8 eV, which makes them suitable for absorbing light in the visible to mid-IR region of the electromagnetic

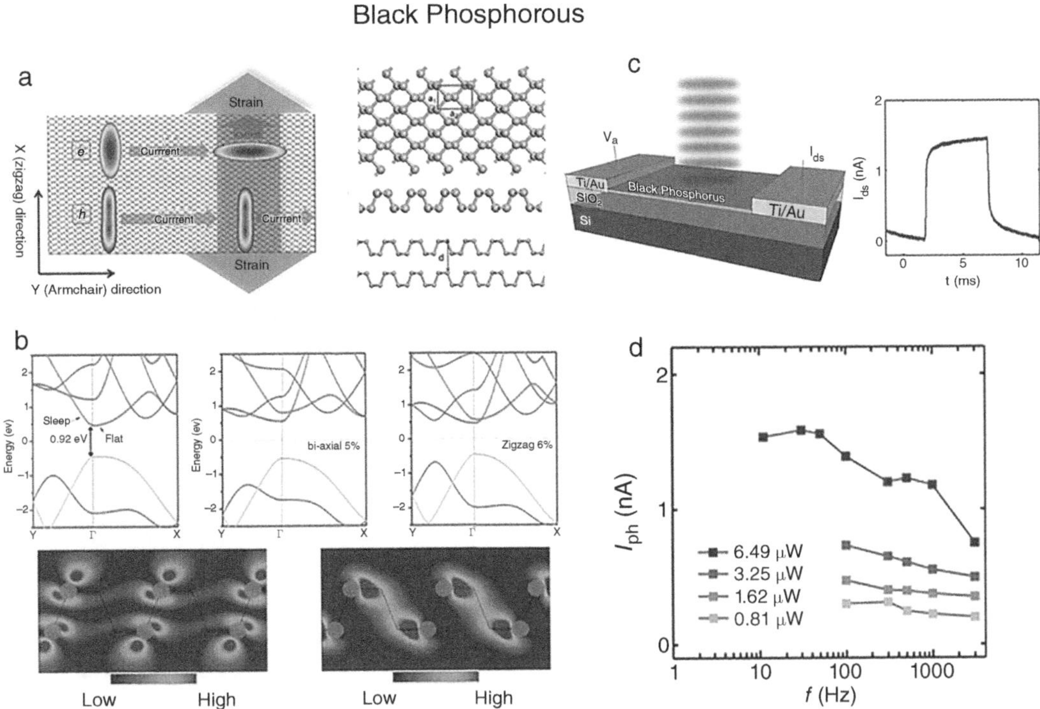

FIGURE 1.3

(a) Schematics showing how biaxial or uniaxial strain of about 4–6% can rotate the preferred conducting direction in a few-layer BP by 90°; (b) band structures of intrinsic monolayer phosphorene, phosphorene with a 5% biaxial strain, and phosphorene with a 6% zigzag uniaxial strain. The side view of the electronic wave function of the lowest-energy conduction band (the red-color band in panel a) at the Γ point. That of the second-lowest-energy conduction band at the Γ point; (c) photocurrent measured in one period of modulation of the light intensity (λ = 640 nm, Pd = 6.49 μW, Vds = 100 mV) in few-layered BP; (d) photocurrent as a function of the modulation frequency in few-layered BP for different incident power levels. Adapted with permission [32]. Copyright (2014), American Chemical Society. Adapted with permission [33]. Copyright (2014), American Chemical Society

spectrum [35]. In addition, the band structures of MXenes can be modified by surface functional groups, transition metal atoms, and the presence of biaxial strain [35]. Surface functional groups, such as –OH, –F, –Cl, and –O, can have an impact on the electronic structure of a pristine MXene system. This is because –OH, –F, and –Cl can accept one electron from metal atoms, while –O can accept two electrons. Studies utilizing first-principle calculations have revealed that the selective termination of MXenes with distinct surface groups can significantly impact the opening or closing of their band gap [35]. Ti_2C with F groups exhibits metallic properties, whereas the O-functionalization of Ti_2C, Zr_2C, and Hf_2C leads to a semiconducting behavior. Recent calculation studies have shown that the band gap of M_2CO_2 (M=Ti, Zr, Hf) MXenes increases with the atomic number of the metal due to their identical outer shell configurations [35]. Ti_2CO_2, Zr_2CO_2, and Hf_2CO_2 are believed to exhibit a shift in their band gap from an indirect to a direct one because of biaxial strain. **Figure 1.4a-c** shows the electrical and magnetotransport properties in a 2D semiconducting Ti_2CO_2 MXene [36]. The combined approach of Rode's iterative scheme with density functional theory (DFT)-based methods showed that the electron-doped

Semiconducting MXenes

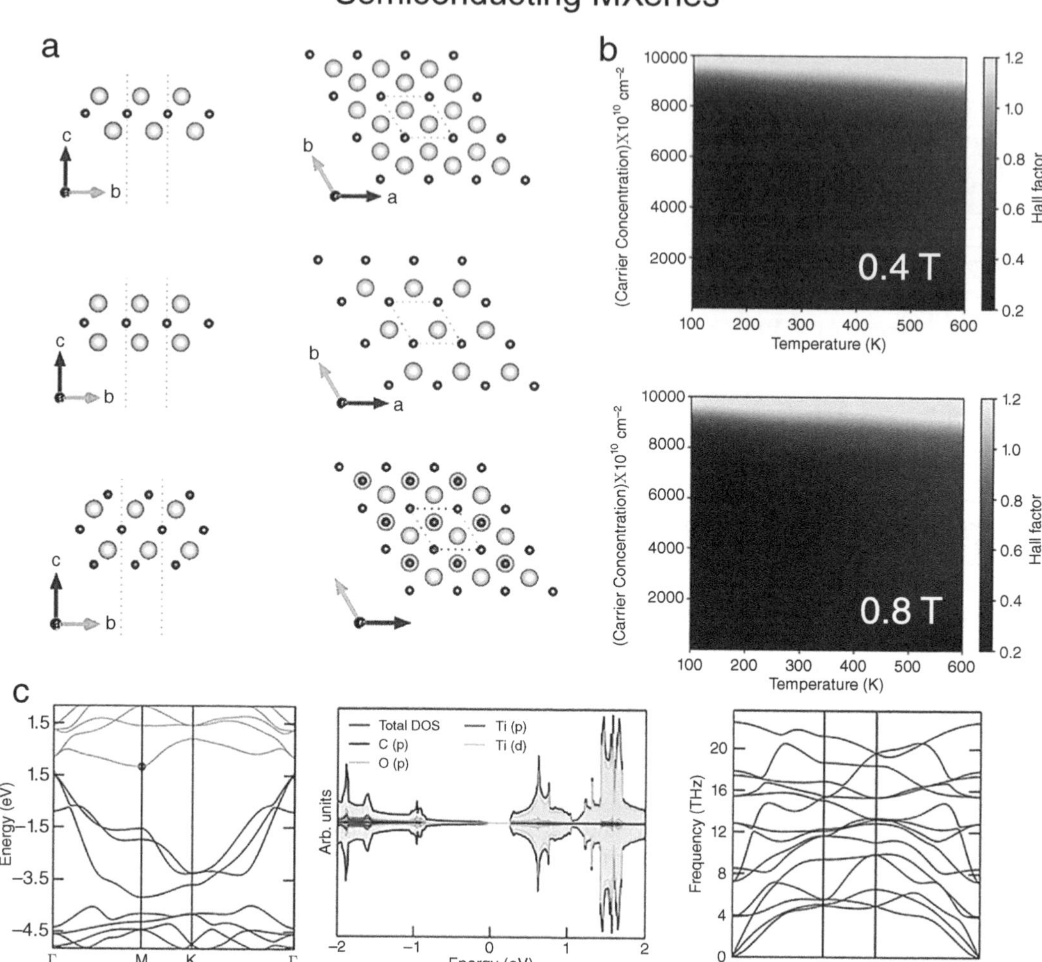

FIGURE 1.4

(a) Side view and top view of the 1T and 2H phases of Ti_2C and Ti_2CO_2; (b) hall scattering factor as a function of temperature and carrier concentration at 0.4 T and 0.8 T; (c) Band structure of Ti_2CO_2 with the VBM and CBM, respectively, and the projected density of states including phonon dispersion of Ti_2CO_2 along the high symmetry directions of the Brillouin zone. Adapted with permission [36]. Copyright (2022), Royal Society of Chemistry.

Ti_2CO_2 varies greatly with temperature and concentration, ranging from 0.2 to around 1.3 for weak magnetic fields [36].

The hydrophilic and highly conductive properties of MXenes make them an ideal material for use in photodetectors [34]. Additionally, their transparency and ease of preparation will make them popular in this application. These MXenes can serve as Ohmic or Schottky contacts for various semiconductors. This is due to their vast and adjustable work function range, making them ideal for photo-detecting applications such as Schottky diodes and metal–semiconductor–metal photodetector configurations. This is possible using Schottky contacts and photoconductor detectors that require Ohmic contacts.

1.3 Do Heterostructures of 2D SCMs Hold the Key to Unlocking New Potential in Optoelectronics and Related Device Applications?

In the past decade, 2D SCMs have been used in various applications, including ultrafast lasers, modulators, photodetectors, sensors, and frequency conversion devices [9]. However, many theoretical studies determined that their heterostructure formation can lead to better conceptual devices [37]. vdW heterostructures can lead to newer enhanced properties in the devices. 2D–2D interlayer interactions can provide better stimulation of the surfaces toward light and show superior electronic properties and better compatibility in the devices.

Graphene and semiconductor hybrid heterostructures can create advanced optoelectronic devices that overcome graphene's lack of bandgap. Graphene-based optoelectronic devices have been thoroughly investigated, and some have demonstrated competitive performance levels comparable to traditional semiconductor devices [7]. vdW heterostructures can be built using 2D SCMs due to their exceptional electronic, magnetic, and optical properties, eliminating the lattice-mismatch issue [37]. 2D SCMs can boost charge carrier mobilities, increase optical absorption, broaden spectrum response, reduce dark current, and improve response time [37].

The band alignment of vdW heterostructures should also be crucial in understanding the overall properties. They can be classified as [37]:

(i) *Type I (straddling type)*: The location of the conduction band minimum and CBM in materials is critical. For light-emitting applications, type-I band alignment is typically preferred.

(ii) *Type II (staggered type)*: The conduction band minimum and CBM exist in a device as separate components. In hybrid and sensitive photodetection applications, type-II band alignment is commonly utilized.

(iii) *Type III (broken-gap type)*: Type-III band alignment, also known as staggered band alignment, is like type-II band alignment but with a significantly lower conduction band minimum than the CBM. This configuration uses low-power optoelectronic devices and tunnel field-effect transistors.

So robust interlayer covalent bonding can be used to stack and create different heterostructures with desirable properties. Dean et al. employed mechanical stacking to produce vertical graphene devices on h-BN interface layers [38]. Devices placed on h-BN substrates demonstrated superior carrier mobility, minimal doping, low roughness, and enhanced chemical stability compared to other substrates like SiO_2 and Ge [38]. Li et al. reported a general growth technique for 2D vertical vdWs heterostructures by selectively patterning nucleation sites on monolayer or bilayer TMDCs and successively producing different 2D vdWs hetero series such as VSe_2/WSe_2, and $NbTe_2/WSe_2$ [39]. Xu et al. fabricated $MoSe_2$–WSe_2 lateral in-plane epitaxial heterostructures [6].

As evident, layered vdW heterostructures formed by stacking two 2D materials provide new opportunities for device design. **Figure 1.5a** shows the rationale for designing a $MoSe_2/WS_2$ structure for dye-sensitized solar cells (DSSCs) and hydrogen evolution reaction (HER) active electrodes [40]. Such an innovative microstructure can easily replace the more expensive Pt electrode. **Figure 1.5b** shows a 1D–2D Te-ReS_2 p–n junction with an ultrafast photoresponse of 5 ms at 180 A/W [41]. Other examples such as the ultrathin p-GaTe/n-MoS_2 vdW

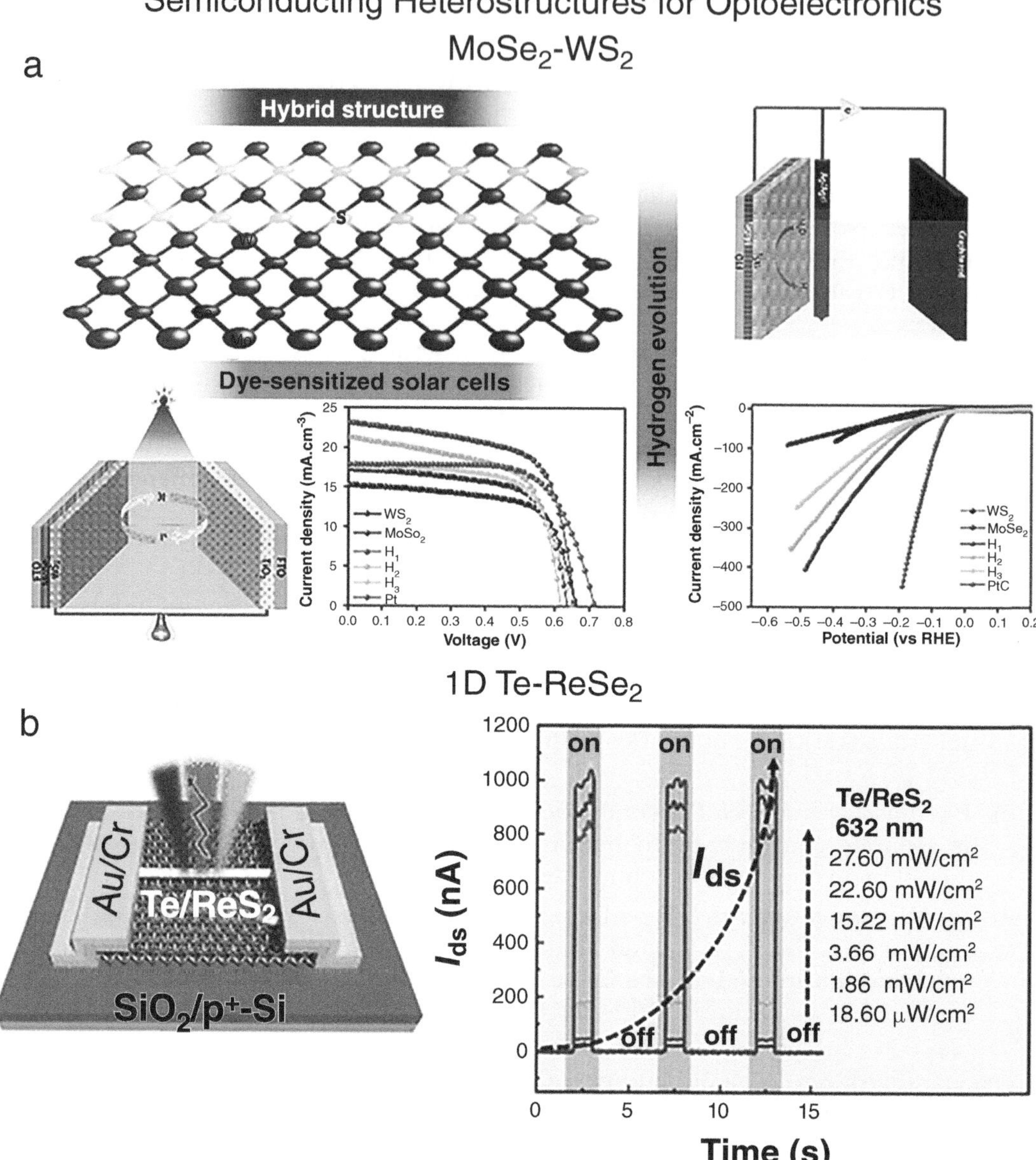

FIGURE 1.5

(a) Designing MoSe$_2$/WS$_2$ structures as a counterelectrode for DSSCs and an active electrode for hydrogen evolution to replace the nonabundant Pt; (b) high-performance one-dimensional (1D) p-type tellurium (Te) and 2D n-type ReS$_2$ p–n heterojunction with an ultrafast photoresponse (5 ms), high responsivity (180 A/W), and specific detectivity (109). Adapted with permission [40]. Copyright (2021), American Chemical Society. Adapted with permission [41], Copyright (2021), American Chemical Society.

heterostructure boast impressive figures, with a rectification ratio of 4105, EQE of 61.68%, and photoresponsivity of 21.83A/W [42]. Photodetectors based on vertical heterojunction arrays of MoS$_2$/WS$_2$ were fabricated, exhibiting a remarkable photoresponsivity of 2.3A/W upon excitation at 450 nm [43]. According to Wang et al., self-driven MoS$_2$/Si heterojunction photodetectors exhibit outstanding sensitivity to a wide range of wavelengths, spanning from visible to near-IR light [44]. They achieve exceptional detectivity levels, reaching up to approximately 10^{13} Jones, and offer an impressive response speed of roughly 3 ns [44]. A p–n diode with black phosphorus has been successfully developed using either homojunction black phosphorus or the combination of p-type black phosphorus and n-type MoS_2 vdW heterojunction [45]. A high-quality p–n junction is demonstrated by the strong gate tunable current-rectifying property of the p–n diode. The device exhibits substantial photoresponse to a 633 nm incident laser when employed as a photodiode [45].

1.4 Conclusion

We discussed progress in 2D semiconducting materials for electronic devices and explained their basic properties and principles. We also showed how they can produce better band alignment and photo responses. 2D TMDCs' polaritons enhance light–matter interactions but mostly at longer wavelengths, posing challenges for visible/near-IR optoelectronics. So we have suggested how tuning the band alignments by choosing the appropriate 2D semiconducting components can produce intense light–matter interactions among the layers. In the section on 2D semiconducting TMDCs, we have also explained how plasmonic nanostructures can be formed by simply functionalizing these 2D TMDCs using metallic nanoparticles. In the case of h-BN, we have considered its extremely wide bandgap that can sometimes render this material as an insulator. We have discussed the usage of 2D h-BN as a graphene substitute, which shows enhanced electron mobility and stability against high voltages. We have also discussed the role of BP and its thickness dependency toward its bandgap. We have also examined a few examples of semiconducting MXenes of the type M_2CO_2, which can have exciting developments if combined with other 2D SCMs. Most of the existing research has been conducted using intricate nanofabrication techniques rather than automated assembly methods and grinding the gap will result in significant progress in related areas.

References

[1] F. Wang, Z. Wang, C. Jiang, L. Yin, R. Cheng, X. Zhan, K. Xu, F. Wang, Y. Zhang, J. He, Progress on electronic and optoelectronic devices of 2D layered semiconducting materials, *Small*, 13 (2017) 1604298.

[2] T. Wei, Z. Han, X. Zhong, Q. Xiao, T. Liu, D. Xiang, Two dimensional semiconducting materials for ultimately scaled transistors, *iScience*, 25 (2022) 105160.

[3] X. Jing, Y. Illarionov, E. Yalon, P. Zhou, T. Grasser, Y. Shi, M. Lanza, Engineering field effect transistors with 2D semiconducting channels: Status and prospects, *Adv. Funct. Mater.*, 30 (2020) 1901971.

[4] Q. Zeng, Z. Liu, Novel optoelectronic devices: Transition-metal-dichalcogenide-based 2D heterostructures, *Adv. Electron. Mater.*, 4 (2018) 1700335.

[5] A. Moumen, R. Konar, D. Zappa, E. Teblum, G. D. Nessim, E. Comini, Room-temperature NO_2 sensing of CVD-modified WS_2-WSe_2 heterojunctions, *ACS Appl. Nano Mater.*, 6 (2023) 7323–7329.

[6] C. Huang, S. Wu, A. M. Sanchez, J. J. P. Peters, R. Beanland, J. S. Ross, P. Rivera, W. Yao, D. H. Cobden, X. Xu, Lateral heterojunctions within monolayer $MoSe_2$–WSe_2 semiconductors, *Nat. Mater.*, 13 (2014) 1096–1101.

[7] F. Bonaccorso, Z. Sun, T. Hasan, A. C. Ferrari, Graphene photonics and optoelectronics, *Nat. Photonics*, 4 (2010) 611–622.

[8] A. Chaves, J. G. Azadani, H. Alsalman, D. R. da Costa, R. Frisenda, A. J. Chaves, S. H. Song, Y. D. Kim, D. He, J. Zhou, A. Castellanos-Gomez, F. M. Peeters, Z. Liu, C. L. Hinkle, S.-H. Oh, P. D. Ye, S. J. Koester, Y. H. Lee, P. Avouris, X. Wang, T. Low, Bandgap engineering of two-dimensional semiconductor materials, *npj 2D Mater. Appl.*, 4 (2020) 29.

[9] W. Ahmad, A. K. Tareen, K. Khan, M. Khan, Q. Khan, Z. Wang, M. Maqbool, A review of the synthesis, fabrication, and recent advances in mixed dimensional heterostructures for opto-electronic devices applications, *Appl. Mater. Today*, 30 (2023) 101717.

[10] R. Konar, G. D. Nessim, A mini-review focusing on ambient-pressure chemical vapor deposition (AP-CVD) based synthesis of layered transition metal selenides for energy storage applications, *Mater. Adv.*, 3 (2022) 4471–4488.

[11] S. Ahmed, J. Yi, Two-dimensional transition metal dichalcogenides and their charge carrier mobilities in field-effect transistors, *Nano-Micro Lett.*, 9 (2017) 50.

[12] J. Xiao, M. Zhao, Y. Wang, X. Zhang, Excitons in atomically thin 2D semiconductors and their applications, *Nanophotonics*, 6 (2017) 1309–1328.

[13] S. Wang, X. Liu, P. Zhou, The road for 2D semiconductors in the silicon age, *Adv. Mater.*, 34 (2022) 2106886.

[14] M. Mohl, A. Rautio, G. A. Asres, M. Wasala, P. D. Patil, S. Talapatra, K. Kordas, 2D tungsten chalcogenides: Synthesis, properties and applications, *Adv. Mater. Interfaces*, 7 (2020) 2000002.

[15] R. Konar, Rosy, I. Perelshtein, E. Teblum, M. Telkhozhayeva, M. Tkachev, J. J. Richter, E. Cattaruzza, A. Pietropolli Charmet, P. Stoppa, M. Noked, G. D. Nessim, Scalable synthesis of few-layered 2D tungsten diselenide (2H-WSe$_2$) nanosheets directly grown on tungsten (W) foil using ambient-pressure chemical vapor deposition for reversible Li-ion storage, *ACS Omega*, 5 (2020) 19409–19421.

[16] B. Rajeswaran, R. Konar, S. Guddala, T. Sharabani, E. Teblum, Y. R. Tischler, G. D. Nessim, Nanostructure-free metal-dielectric stacks for Raman scattering enhancement and defect identification in CVD-grown tungsten di-sulfide (2H-WS$_2$) nanosheets, *J. Phys. Chem. C*, 126 (2022) 20511–20523.

[17] X. Wu, M. Sun, H. Yu, B. Huang, Z. L. Wang, Unraveling the unique response behaviors of ultrathin transition metal dichalcogenides to external excitations, *Nano Energy*, 115 (2023) 108721.

[18] K. W. Lau, C. Cocchi, C. Draxl, Electronic and optical excitations of two-dimensional ZrS$_2$ and HfS$_2$ and their Heterostructure, *Phys. Rev. Mater.*, 3 (2019) 074001.

[19] C. Xia, J. Li, Recent advances in optoelectronic properties and applications of two-dimensional metal chalcogenides, *J. Semicond.*, 35 (2016) 5 051001.

[20] R. Konar, B. Rajeswaran, A. Paul, E. Teblum, H. Aviv, I. Perelshtein, I. Grinberg, Y. R. Tischler, G. D. Nessim, CVD-assisted synthesis of 2D layered MoSe$_2$ on Mo foil and low-frequency Raman scattering of its exfoliated few-layer nanosheets on CaF$_2$ substrates, *ACS Omega*, 2022, 7, 4121–4134.

[21] S. Yan, X. Zhu, J. Dong, Y. Ding, S. Xiao, Nanophotonics, 2D materials integrated with metallic nanostructures: Fundamentals and optoelectronic applications, *Nanophotonics*, 9 (2020) 1877–1900.

[22] T. Mueller, E. Malic, Exciton physics and device application of two-dimensional transition metal dichalcogenide semiconductors, *npj 2D Mater. Appl.*, 2 (2018) 29.

[23] S. K. Chaubey, G. M. A, D. Paul, S. Tiwari, A. Rahman, G. V. P. Kumar, Directional emission from tungsten disulfide monolayer coupled to plasmonic nanowire-on-mirror cavity, *Adv. Photonics Res.*, 2 (2021) 2100002.

[24] Z. Li, C. Liu, X. Rong, Y. Luo, H. Cheng, L. Zheng, F. Lin, B. Shen, Y. Gong, S. Zhang, Z. Fang, Tailoring MoS$_2$ valley-polarized photoluminescence with super chiral near-field, *Adv. Mater.*, 30 (2018) 1–7.

[25] G. Hu, X. Hong, K. Wang, J. Wu, H.-X. Xu, W. Zhao, W. Liu, S. Zhang, F. Garcia-Vidal, B. Wang, P. Lu, C.-W. Qiu, Coherent steering of nonlinear chiral valley photons with a synthetic Au–WS$_2$ metasurface, *Nat. Photonics*, 13 (2019) 467–472.

[26] C. Maestre, B. Toury, P. Steyer, V. Garnier, C. Journet, Hexagonal boron nitride: A review on selfstanding crystals synthesis towards 2D nanosheets, *J. Phys. Mater.*, 4 (2021) 044018.

[27] H. Liu, C. Y. You, J. Li, P. R. Galligan, J. You, Z. Liu, Y. Cai, Z. Luo, Synthesis of hexagonal boron nitrides by chemical vapor deposition and their use as single photon emitters, *Nano Mater. Sci.*, 3 (2021) 291–312.

[28] P. Sutter, J. Lahiri, P. Zahl, B. Wang, E. Sutter, Scalable synthesis of uniform few-layer hexagonal boron nitride dielectric films, *Nano Lett.*, 13 (2013) 276–281.

[29] S. M. Kim, A. Hsu, M. H. Park, S. H. Chae, S. J. Yun, J. S. Lee, D.-H. Cho, W. Fang, C. Lee, T. Palacios, M. Dresselhaus, K. K. Kim, Y. H. Lee, J. Kong, Synthesis of large-area multilayer hexagonal boron nitride for high material performance, *Nat. Commun.*, 6 (2015) 8662.

[30] F. Xia, H. Wang, J. C. M. Hwang, A. H. C. Neto, L. Yang, Black phosphorus and its isoelectronic materials, *Nat. Rev. Phys.*, 1 (2019) 306–317.

[31] V. Eswaraiah, Q. Zeng, Y. Long, Z. Liu, Black phosphorus nanosheets: Synthesis, characterization and applications, *Small*, 12 (2016) 3480–3502.

[32] R. Fei, L. Yang, Strain-engineering the anisotropic electrical conductance of few-layer black phosphorus, *Nano Lett.*, 14 (2014) 2884–2889.

[33] M. Buscema, D. J. Groenendijk, S. I. Blanter, G. A. Steele, H. S. J. Van Der Zant, A. Castellanos-Gomez, Fast and broadband photoresponse of few-layer black phosphorus field-effect transistors, *Nano Lett.*, 14 (2014) 3347–3352.

[34] X. Zhang, J. Shao, C. Yan, R. Qin, Z. Lu, H. Geng, T. Xu, L. Ju, A review on optoelectronic device applications of 2D transition metal carbides and nitrides, *Mater. Des.*, 200 (2021) 109452.

[35] X. Bai, Y. Xue, K. Luo, K. Chen, Q. Huang, X.-H. Zha, S. Du, Two-dimensional half-metallic and semiconducting lanthanide-based MXenes, *ACS Omega*, 7 (2022) 40929–40940.

[36] A. K. Mandia, N. A. Koshi, B. Muralidharan, S.-C. Lee, S. Bhattacharjee, Electrical and magneto-transport in the 2D semiconducting MXene Ti_2CO_2, *J. Mater. Chem. C*, 10 (2022) 9062–9072.

[37] Z. Yang, J. Hao, Recent progress in 2D layered III–VI semiconductors and their heterostructures for optoelectronic device applications, *Adv. Mater. Technol.*, 4 (2019) 1900108.

[38] C. Dean, A. F. Young, L. Wang, I. Meric, G.-H. Lee, K. Watanabe, T. Taniguchi, K. Shepard, P. Kim, J. Hone, Graphene based heterostructures, *Solid State Commun.*, 152 (2012) 1275–1282.

[39] J. Li, X. Yang, Y. Liu, B. Huang, R. Wu, Z. Zhang, B. Zhao, H. Ma, W. Dang, Z. Wei, K. Wang, Z. Lin, X. Yan, M. Sun, B. Li, X. Pan, J. Luo, G. Zhang, Y. Liu, Y. Huang, X. Duan, X. Duan, General synthesis of two-dimensional van der Waals heterostructure arrays, *Nature*, 579 (2020) 368–374.

[40] D. Vikraman, S. Hussain, S. A. Patil, L. Truong, A. A. Arbab, S. H. Jeong, S.-H. Chun, J. Jung, H.-S. Kim, Engineering $MoSe_2$/WS_2 hybrids to replace the scarce platinum electrode for hydrogen evolution reactions and dye-sensitized solar cells, *ACS Appl. Mater. Interfaces*, 13 (2021) 5061–5072.

[41] J.-J. Tao, J. Jiang, S.-N. Zhao, Y. Zhang, X.-X. Li, X. Fang, P. Wang, W. Hu, Y. H. Lee, H.-L. Lu, D.-W. Zhang, Fabrication of 1D Te/2D ReS_2 mixed-dimensional van der Waals p-n heterojunction for high-performance phototransistor, *ACS Nano*, 15 (2021) 3241–3250.

[42] F. Wang, Z. Wang, K. Xu, F. Wang, Q. Wang, Y. Huang, L. Yin, J. He, Tunable $GaTe$-MoS_2 van der Waals p–n junctions with novel optoelectronic performance, *Nano Lett.*, 15 (2015) 7558–7566.

[43] Y. Xue, Y. Zhang, Y. Liu, H. Liu, J. Song, J. Sophia, J. Liu, Z. Xu, Q. Xu, Z. Wang, J. Zheng, Y. Liu, S. Li, Q. Bao, Scalable production of a few-layer MoS_2/WS_2 vertical heterojunction array and its application for photodetectors, *ACS Nano*, 10 (2016) 573–580.

[44] L. Wang, J. Jie, Z. Shao, Q. Zhang, X. Zhang, Y. Wang, Z. Sun, S.-T. Lee, MoS_2/Si heterojunction with vertically standing layered structure for ultrafast, high-detectivity, self-driven visible–near infrared photodetectors, *Adv. Funct. Mater.*, 25 (2015) 2910–2919.

[45] Y. Deng, Z. Luo, N. J. Conrad, H. Liu, Y. Gong, S. Najmaei, P. M. Ajayan, J. Lou, X. Xu, P. D. Ye, Black phosphorus–monolayer MoS_2 van der Waals heterojunction p–n diode, *ACS Nano*, 8 (2014) 8292–8299.

2

Synthesis and Characterization of 2D-Semiconducting Materials

Prabhukrupa C. Kumar, Subrata Senapati, and Ramakanta Naik*

**Correspondence: ramakanta.naik@gmail.com*

2.1 Introduction

Many essential technologies, including electronics, computers, communications, optoelectronics, and sensing, are built on semiconductors. Due to the advancement of discrete and integrated semiconductor devices and circuits, semiconductors are now an integral part of modern civilization [1, 2]. In recent years, material science has witnessed a remarkable surge in interest and innovation, driven by the pursuit of novel materials with extraordinary electronic and optoelectronic properties. Among these materials, 2D semiconducting materials (SCMs) have emerged as a captivating area of study, offering unprecedented opportunities for advancing electronic devices, photodetectors, sensors, and more [3, 4]. Unlike their bulk counterparts, 2D semiconductors possess unique physical and chemical properties driven by their atomically thin structure. This distinctive class of materials has garnered substantial attention due to its potential to revolutionize the landscape of modern electronics and photonics [5, 6]. This synthesis-focused exploration aims to shed light on the methods, challenges, and prospects in the synthesis of 2D SCMs, offering a comprehensive overview of the latest advancements in this field. As we delve deeper into this exciting realm, we will unravel the intricacies of producing 2D semiconductors with tailored properties, setting the stage for a new era of innovation and application in nanoelectronics and beyond [7].

Tunability of the semiconductor characteristics and miniaturization of the relevant devices to almost atomically thin dimensions are very desirable in today's electrical and optical technologies. While the renowned graphene set the stage for 2D materials research, a wealth of new possibilities has emerged with the discovery of a diverse range of 2D materials [8–10]. The intrinsic versatility and tunability of these 2D materials concerning their bandgap properties particularly make these materials fascinating [11]. This tunability is achieved through a lot of ingenious techniques and methodologies. Researchers have demonstrated the ability to finely tune the bandgap in these materials by manipulating factors such as layer thickness, heterostructure, strain engineering, chemical doping, alloying, intercalation, substrate engineering, and the application of external electric fields. Such precise control over the bandgap not only broadens the horizons of fundamental semiconductor physics but also opens a world of possibilities for tailored applications in next-generation electronic and optoelectronic devices. This

DOI: 10.1201/9781003439448-2

chapter will explore the cutting-edge strategies employed to synthesize these 2D SCMs with unprecedented potential.

In materials science, several synthesis methods have emerged as powerful tools for preparing these 2D SCMs [12]. One such method is chemical vapor deposition (CVD), which involves the controlled deposition of precursor molecules onto a substrate, followed by thermal annealing to achieve the desired 2D structure. Another method is mechanical exfoliation (ME), where bulk materials are mechanically peeled apart to obtain ultrathin layers. Additionally, solution-based methods like liquid-phase exfoliation and hydrothermal synthesis have shown potential in producing large quantities of 2D semiconductors. The importance of these synthesis methods lies in their ability to create materials with specific mechanical, optical, and electronic properties, opening avenues for applications in energy storage, optoelectronics, electronics, and sensing [13]. By understanding and harnessing these synthesis techniques, researchers can unlock the full potential of 2D SCMs and propel advancements in various technological fields [14].

2.2 Synthesis

The synthesis of 2D SCMs encompasses a rich landscape of techniques, each offering unique advantages and enabling precise control over material properties. CVD is one of the most prominent methods [15]. Liquid-phase exfoliation (LPE) is another crucial approach, where bulk crystals are exfoliated into 2D flakes. Molecular beam epitaxy (MBE) offers ultra-high vacuum conditions to synthesize 2D SCMs with remarkable purity and precision. Bottom-up techniques, such as solution-based and chemical synthesis, are gaining traction for their ability to craft custom-designed structures at the atomic level. These diverse synthesis methods are pivotal in not only producing 2D SCMs but also tailoring their electronic and optical properties, which facilitates the integration of 2D SCMs into a wide array of applications. The ability to select the most appropriate synthesis technique for a given application underscores the importance of understanding and harnessing these methods to unlock the full potential of 2D SCMs.

The discovery of remarkable properties in ultrathin 2D materials has spurred extensive research across a spectrum of scientific fields, including physics, chemistry, biology, materials science, and medicine. Yet, the precise synthesis of these materials with specific thickness and lateral dimensions has proven challenging due to factors like anisotropic crystal growth and chemical solid bonds within the crystal structure. Each synthetic approach has its own set of advantages and limitations when it comes to producing various types of materials. These methods can be classified into top-down and bottom-up approaches based on how nanostructures are generated. The distinction between these classifications lies in the processes used to craft these structures. The bottom-up approach is about preparing nanoscale materials from atomic or molecular precursors, allowing them to react, grow, or self-assemble into more intricate configurations. In contrast, the top-down approach involves precisely removing material from larger or bulk solids to shape nanoscale structures [16]. In this part of the chapter, we will discuss these synthesis methods in detail. The synthesis procedures are categorized and represented in **Figure 2.1.**

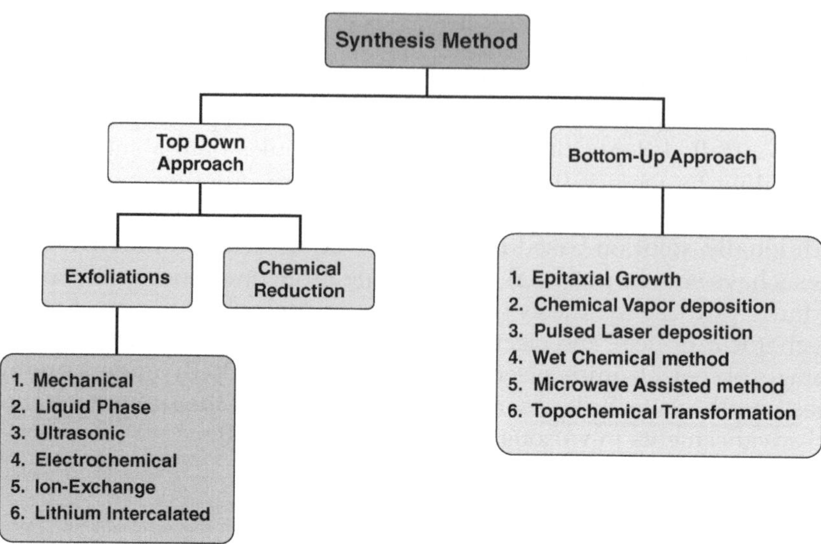

FIGURE 2.1
Schematic representation of different types of synthesis procedures.

2.2.1 Top-Down Approach

In the top-down approach, the preparation of 2D SCMs involves the controlled reduction or exfoliation of bulk or layered materials to produce ultrathin 2D nanosheets. This method is particularly valuable when dealing with materials naturally occurring in bulk or layered forms. The top-down approach offers excellent control over the size, shape, and thickness of 2D SCMs [17]. It is crucial for materials that are challenging to synthesize using bottom-up methods or when precise patterning is required. However, it may introduce defects and limited scalability compared to bottom-up techniques. The choice between top-down and bottom-up approaches depends on the specific material and desired properties for a given application. A combination of both methods is often employed to achieve the desired results. Two techniques that come under this approach are exfoliation and chemical reduction. There are different types of exfoliation strategies used to prepare the 2D SCMs: ME, LPE, ultrasonic exfoliation (UE), electrochemical exfoliation, ion exchange exfoliation (IEE), and lithium-intercalated exfoliation (LIE) [18]. Detailed discussion about these methods is done ahead in this chapter.

2.2.1.1 Exfoliations

Exfoliation is a technique used in the preparation of 2D SCMs. It involves separating layers from a bulk material to obtain thin, atomically flat sheets. This process is crucial for the fabrication and study of 2D materials, allowing researchers to access their unique properties at the atomic scale [19]. An explanation of various types of exfoliation techniques is provided here.

2.2.1.1.1 Mechanical Exfoliation

One of the pioneering techniques in the top-down approach is ME, commonly known as the "scotch tape method." In this process, a sticky tape is repeatedly applied and peeled off from bulk material, such as graphite, gradually reducing thickness to the monolayer or few-layer range. This method was pivotal in isolating graphene, the first 2D material to capture widespread attention. In 2004, Andre Geim and Konstantin Novoselov obtained graphene by applying adhesive tape to a graphite crystal and then repeatedly peeled it off [8]. With each peel, thinner layers of graphene were obtained until, eventually, single-layer graphene was isolated.

The principle behind ME lies in the weak interlayer forces that hold the material together. In materials such as graphite or certain layered crystals, the layers are held together by relatively weak van der Waals forces. Applying mechanical force can overcome these forces, allowing for the separation of individual layers. ME can be performed through various methods, including tape-based methods, microcleavage, or using an ME apparatus. In tape-based methods, adhesive tapes (such as scotch tape) are used to repeatedly peel off layers from the bulk material. Microcleavage involves using a sharp object, like a razor blade, to gently scrape or cleave the material, creating thin layers. ME apparatus, such as a micro-ME tool or a micro-manipulator, can also be used for more controlled and automated exfoliation.

One of the advantages of ME is that it produces high-quality, defect-free layers, which often retain the pristine properties of the bulk material, making them suitable for a wide range of applications. Additionally, it can be performed at room temperature without complex equipment or harsh chemicals, making it a relatively simple and accessible method.

2.2.1.1.2 Liquid-Phase Exfoliation

LPE is another top-down technique that produces 2D materials by dispersing bulk-layered materials in a liquid medium and then applying various forces to separate them into individual layers or nanosheets. This method has been successfully applied to various layered materials, including transition metal dichalcogenides (TMDs), black phosphorus, and h-BN. This method has gained significant attention due to its versatility, scalability, and ability to produce large quantities of high-quality 2D materials. LPE typically involves several steps. First, the bulk-layered material, such as graphite or TMDs, is added to a suitable liquid solvent or dispersant. The choice of solvent or dispersant is crucial as it should be compatible with the layered material and facilitate its exfoliation [20].

Once the material is dispersed in the liquid medium, different forces like ultrasonication, shear mixing, or high-pressure homogenization are applied to promote the separation of the layers. Ultrasonication involves high-frequency sound waves, which create cavitation bubbles that collapse and generate intense shear forces, leading to the exfoliation of the layers. Shear mixing involves stirring or agitating the dispersion to induce shear forces between the layers, causing them to separate. High-pressure homogenization involves passing the dispersion through a narrow channel at high pressure, resulting in intense shear forces that exfoliate the layers [21].

LPE offers several advantages over other exfoliation techniques. It allows for rapidly producing large quantities of 2D materials with high yield and minimal defects, preserving their desirable properties. The choice of liquid medium is crucial in LPE, as it can influence the stability and properties of the resulting dispersion. Different solvents or dispersants, including water, organic solvents, surfactants, and polymer solutions, can be used depending on the specific layered material being exfoliated. LPE has been successfully applied to various layered materials, including graphene, MoS_2, WS_2, h-BN, and many others [22].

2.2.1.1.3 Ultrasonic Exfoliation

UE involves using high-frequency sound waves, typically in the ultrasonic range, to break down bulk materials into thin, atomically flat layers. This method is particularly effective for exfoliating materials that are difficult to cleave using ME or LPE techniques. UE begins by dispersing the bulk material in a solvent to form a suspension. The suspension is then subjected to high-intensity ultrasonic waves generated by an ultrasonic probe or bath. These pressure waves induce cavitation, the formation, growth, and implosive collapse of tiny gas-filled bubbles in the liquid. The bubbles rapidly expand and collapse during cavitation, generating localized hotspots with high temperatures and pressures. These hotspots produce intense shear forces and shock waves that cause the bulk material to fracture and exfoliate into thin flakes or nanosheets and then get separated from the suspension using centrifugation or filtration techniques [23].

Several advantages of UE over other methods are that it is a rapid and efficient process for the quick production of large quantities of 2D materials; it can be controlled by adjusting parameters such as ultrasonic power, frequency, and duration to achieve the desired thickness and quality of the exfoliated layers. Furthermore, UE can be performed at room temperature, preventing the degradation of sensitive materials' properties under high temperatures [24].

2.2.1.1.4 Ion Exchange Exfoliation

IEE is a fascinating technique to produce 2D SCMs by exploiting the ion exchange process between layered materials and counterions. It involves introducing a bulk-layered material, such as clays or layered metal oxides, into a solution containing a suitable counterion. These counterions have a higher affinity for the layered material than the original interlayer ions present within the layers. As a result, the counterions replace the interlayer ions, causing an expansion of the interlayer spacing and creating a favorable environment for the separation of layers. This method allows for the controlled separation of layers, manufacturing good-quality nanosheets with unique properties and applications.

IEE can be achieved through direct ion exchange, intercalation, and in situ ion exchange methods. In direct ion exchange, the layered material is immersed in a solution containing the desired counterions, and the exchange occurs directly at the interlayer sites. Intercalation involves the insertion of guest species, like organic molecules or polymers, into the interlayer spacing to facilitate the subsequent ion exchange process, whereas in situ ion exchange involves the introduction of a precursor solution that undergoes a chemical reaction with the layered material, leading to the formation of new interlayer ions. Once the ion exchange process is complete, the layers can be further exfoliated by applying external mechanical shearing, sonication, thermal treatment forces, or subsequent processing steps [25].

One of the significant advantages of IEE is the ability to control the thickness and properties of the resulting nanosheets by varying the counterions or intercalating species. This control allows for tailoring the materials to specific applications. Furthermore, the ion exchange process can be performed in a solution, making it amenable to large-scale production and scalable manufacturing processes [26].

2.2.1.1.5 Lithium-Intercalated Exfoliation

The LIE technique involves the introduction of lithium ions into the interlayer spacing of layered materials to facilitate the separation and exfoliation of their layers. This causes an expansion of the interlayer spacing and creates favorable conditions for the subsequent exfoliation process [27]. This expansion can be attributed to lithium ions' relatively small size and high mobility within the interlayer regions. The LIE process can be achieved through various methods, including electrochemical intercalation and CVD. This method allows precise control over the intercalation process and is commonly employed for producing lithium-ion battery electrodes. Upon completion, the layers can be further exfoliated by applying external forces or through subsequent steps like—mechanical shearing, sonication, or thermal treatment.

In the field of energy storage, these lithium-intercalated nanosheets can be used as high-performance electrode materials for lithium-ion batteries, offering improved energy storage capacity and cycling stability. Additionally, they have potential applications in supercapacitors, where their high surface area and conductivity can enhance charge storage and delivery. Beyond energy storage, lithium-intercalated nanosheets have applications in electronics, catalysis, sensors, and optoelectronics [28].

2.2.1.2 Chemical Reduction

The chemical reduction (CR) method allows for the controlled reduction of precursor compounds, forming high-quality 2D materials with tailored properties. In the CR method, a precursor compound containing the desired elements is dissolved in a solvent, and a reducing agent is added to initiate the reduction reaction. The reducing agent donates electrons, leading to the reduction of the precursor compound and the formation of the 2D SCMs. One of the most widely used CR methods is solution-phase synthesis, where the precursor compound (metal salts or metal–organic complexes) is dissolved in a solvent such as water or an organic solvent. A suitable reducing agent, such as hydrazine or sodium borohydride, is added to the solution. The reduction reaction is typically carried out under controlled temperature and reaction time conditions to ensure the formation of high-quality 2D nanosheets or nanocrystal SCMs, which can be isolated and further processed for device applications [29].

First, the CR method offers several advantages, providing a simple and scalable approach that is easily adaptable for large-scale production. Second, it allows for controlling the size, shape, and composition. Additionally, the method offers the possibility of incorporating dopants or functional groups into the material during the synthesis process, further enhancing its properties [30].

2.2.2 Bottom-up Approach

The bottom-up approach is another method for the preparation of 2D SCMs. The bottom-up approach focuses on the controlled assembly and the growth of individual building blocks to form the desired 2D structure, which is particularly valuable for materials that do not naturally occur in 2D forms or when precise control over the structure and composition is essential. One of the key advantages of the bottom-up approach is its precise control over the size, shape, and arrangement of the resulting 2D materials. The properties of the materials to meet specific requirements can be controlled by carefully selecting the precursors and adjusting the growth conditions [31].

Some popular bottom-up approaches are molecular self-assembly, epitaxial growth (EG), CVD, and so on. Researchers can create materials by carefully assembling and growing individual building blocks. The bottom-up approach holds great promise for developing novel 2D SCMs with tailored properties and precise control over their structure [32]. However, it may be challenging to scale up production compared to top-down methods. The choice between top-down and bottom-up approaches depends on the specific material, desired properties, and intended applications, with researchers often combining both strategies to achieve their goals.

2.2.2.1 Epitaxial Growth

EG is a widely used technique for the synthesis of 2D SCMs. It involves the controlled deposition of atomically thin layers of the desired material on a suitable substrate, resulting in the growth of a crystalline film with a specific crystal structure and orientation. It enables precise control over film thickness, crystal orientation, and defect density, making it crucial for developing advanced electronic and optoelectronic devices. In EG, the substrate plays a crucial role as a template for the growth of the material. The lattice structure of the substrate acts as a guide for the arrangement of atoms or molecules in the growing film. The goal is to achieve a perfect lattice match between the substrate and the growing material, ensuring minimal defects and strain in the resulting 2D structure [33].

To initiate EG, the substrate is typically prepared by cleaning and pre-treating its surface to remove contaminants or oxides and ensure a clean and well-defined surface for subsequent growth. The substrate is then exposed to a precursor gas or vapor, which contains the atoms or molecules necessary for the growth of the desired material. The choice of precursor gas or vapor depends on the synthesized material. For example, in the growth of graphene, a precursor gas such as methane (CH_4) or ethylene (C_2H_4) is commonly used. In the case of TMDs, such as molybdenum disulfide (MoS_2) or tungsten diselenide (WSe_2), metal and sulfur or selenium precursors are utilized [34, 35].

The growth process is carefully controlled by adjusting temperature, pressure, and gas composition parameters. During EG, the atoms or molecules from the precursor selectively attach themselves to the surface of the substrate, aligning with the underlying crystal lattice. This alignment is crucial for achieving epitaxy, where the crystal structure and orientation of the growing film match that of the substrate. The epitaxial relationship between the substrate and the growing material ensures structural perfection and continuity in the 2D film.

2.2.2.2 Chemical Vapor Deposition

CVD is one of the most prominent methods used to synthesize 2D SCMs. It involves the controlled deposition of atoms or molecules from a precursor gas onto a substrate, resulting in the growth of a thin film with precise control over its properties. The synthesized 2D SCMs produced using CVD have applications in various fields, including electronics, optoelectronics, energy storage, and sensing. Their unique properties, such as high carrier mobility, tunable bandgap, and excellent mechanical flexibility, make them promising candidates for next-generation devices and technologies [36].

In CVD, after cleaning and pre-treating the substrate to remove contaminants or oxides, it is placed into a reaction chamber, typically heated to a specific temperature. Inside the reaction chamber, a precursor gas is introduced containing the atoms or molecules necessary for growing the desired 2D material. The choice of precursor gas depends on the specific material being synthesized. For example, in the growth of graphene, methane (CH_4) or ethylene (C_2H_4) is commonly used. In the case of TMDs, metal and chalcogenide precursors are utilized. When the precursor gas encounters the heated substrate, it undergoes a chemical reaction, leading to the deposition of the desired material. The reaction may involve the breaking of chemical bonds in the precursor gas and the subsequent reformation of bonds on the substrate surface. This process allows the atoms or molecules from the precursor gas to attach themselves to the substrate, forming a growing film [37].

Various parameters, including temperature, pressure, precursor gas flow rate, and deposition duration, can influence CVD's growth process. These parameters are carefully controlled to achieve the desired film quality, thickness, and composition [38]. By modifying these factors, researchers can customize the 2D SCM's characteristics and enhance its performance for specific applications. CVD offers several advantages, such as large-area growth and excellent control over film thickness, composition, and crystal quality, making it suitable for industrial-scale production.

2.2.2.3 Pulsed Laser Deposition

Pulsed laser deposition (PLD) is another popular method for synthesizing 2D SCMs. It involves using laser pulses to ablate a target material, creating a plasma plume that is then directed onto a substrate, resulting in the growth of a thin film with controlled properties like tunable bandgap, high carrier mobility, and excellent mechanical flexibility [39].

The target material is placed in a vacuum chamber, and the substrate is positioned close to the target in the PLD unit. The target material is subsequently exposed to a high-energy laser pulse, which quickly heats and evaporates the material's surface. We call this procedure "laser ablation." As the target material vaporizes, it forms a plasma plume of atoms, ions, and clusters of the desired SCM. The plasma plume expands and moves toward the substrate, where the atoms and ions deposit and condense, forming a thin film. The substrate temperature is carefully controlled to facilitate the growth of the desired 2D material [40].

The growth process in PLD can be influenced by various parameters, including laser fluence, target–substrate distance, substrate temperature, oxygen partial pressure (in the case of oxide materials), and deposition time. These parameters are adjusted to optimize the film quality, thickness, and composition. Regarding advantages, PLD offers precise control over film thickness and composition and can tailor the material's properties for specific applications. PLD can also be used to grow epitaxial films.

2.2.2.4 Wet Chemical Method

Wet chemical synthesis (WCS) is a promising method for synthesizing 2D SCMs. Unlike PLD or CVD, solution-based processes are used to grow and assemble the desired 2D material. It covers template synthesis and hydro/solvothermal synthesis.

In WCS, the starting materials, usually metal salts or precursors, are dissolved to create a solution. The solvent can be water, organic solvents, or a mixture of both, depending on the specific requirements of the synthesis process. The synthesis reaction typically involves the reduction of metal ions or the reaction between different precursors to form the desired SCM. This reaction is often carried out under controlled conditions, such as temperature, pH, and reaction time, to ensure the formation of the desired 2D structure [41].

Colloidal synthesis is a popular WCS technique for preparing 2D SCMs. This method involves mixing the metal salts or precursors in a solvent along with a stabilizing and reducing agent. While the stabilizing agent keeps the particles from clumping together or expanding too much, the reducing agent reduces metal ions. The reaction mixture is then heated, promoting the nucleation and growth of the 2D SCMs. The size and shape of the resulting nanoparticles can be controlled by adjusting the reaction parameters, such as temperature, reaction time, and concentration of the precursors [42]. Post synthesis, the resulting nanoparticles can be isolated and further processed to form thin films or other desired structures. Techniques like spin-coating, drop-casting, or self-assembly can be used to deposit and arrange the nanoparticles onto a substrate.

WCS has various advantages, such as its relatively simple approach, low-cost method, and scalability for large production. Control over the size, shape, and composition of the nanoparticles makes WCS an attractive method for preparing 2D SCMs to meet some specific applications.

2.2.2.5 Microwave-Assisted Method

Microwave-assisted synthesis is an innovative and efficient method to synthesize 2D SCMs. This technique utilizes microwave radiation to facilitate the chemical reactions forming the desired material. The method can be easily adapted for large-scale production, making it suitable for industrial applications. The microwave-assisted synthesis method has gained significant attention due to its efficiency, speed, and control over the synthesis process [43]. It has been successfully employed to synthesize various 2D SCMs, including TMDs, graphene, and black phosphorus.

In microwave-assisted synthesis, the starting materials or precursors (metal salts, organic compounds, or other chemical species) are mixed in a suitable solvent or medium. The mixture is then exposed to microwave radiation, typically at a specific power and time. Microwave radiation generates heat within the reaction mixture, rapidly raising the temperature and providing localized heating to facilitate the chemical reactions. Microwave radiation's targeted heating has several benefits compared to traditional heating techniques. Because the heat is produced immediately within the reaction mixture, enabling quicker and more effective reactions, resulting in shorter reaction times and greater yields. Additionally, the controlled heating provided by microwaves can help to obtain uniform and crystalline structures of the synthesized 2D SCMs [44, 45].

This method gives precise control over reaction conditions, like temperature and pressure, that significantly influence the properties of the resulting materials. This control allows us to fine-tune the synthesis parameters and optimize the properties of the 2D SCMs for applications.

2.2.2.6 Topochemical Transformation

Topochemical transformation (TT) technique involves converting one material into another with the same chemical composition but a different crystal structure. The conditions required for TT synthesis depend on the transformed material. Factors such as temperature, pressure, and reaction time play crucial roles in controlling the transformation process and determining the properties of the resulting 2D SCMs [46].

In TT synthesis, the starting material (layered compound or a precursor) undergoes a chemical reaction under specific conditions, leading to a rearrangement of atoms within the crystal lattice. This rearrangement preserves the chemical composition of the material while creating a new 2D crystal structure that exhibits semiconducting properties. One of the critical aspects of TT synthesis is preserving the atomic arrangement and connectivity during the transformation process. This means the reaction occurs within the crystal lattice, maintaining the structural integrity and reducing material loss [47].

The layered starting material alters its atomic arrangement during TT, forming new chemical bonds or crystal structures. The advantage of TT synthesis lies in its ability to create 2D SCMs with controlled crystal structures and tailored properties, including tunable bandgap, increased carrier mobility, and enhanced electrical conductivity. This opens new possibilities for device fabrication and technological advancements.

2.3 Characterization Techniques

Characterization techniques play a pivotal role in understanding the properties, structure, and behavior of 2D SCMs. These techniques provide crucial insights for optimizing material synthesis, engineering device performance, and advancing various technological applications [48, 49]. Here's a breakdown of the role and importance of different characterization techniques for 2D SCMs.

2.3.1 X-Ray Diffraction

X-ray diffraction is employed to determine the crystalline structure and crystallographic orientation of 2D materials. It helps identify the presence of different phases, stacking arrangements, and structural defects within the materials.

2.3.2 Scanning Electron Microscopy and Transmission Electron Microscopy

Scanning electron microscopy and transmission electron microscopy allow researchers to visualize the morphology and structure of 2D materials at nanoscale resolutions. They provide information about layer thickness, defects, grain boundaries, and stacking order, which are vital for assessing material quality and understanding its electronic properties.

2.3.3 Atomic Force Microscopy

Atomic force microscopy is used to probe the surface topography and mechanical properties of 2D materials at the atomic level. It is valuable for characterizing these materials' thickness, roughness, and mechanical response, which is essential for applications like sensors and nanomechanical devices.

2.3.4 Raman Spectroscopy

Raman spectroscopy is an essential technique for investigating the vibrational and phonon modes. It provides information about the material's crystal structure, layer thickness, strain, and doping levels. Raman spectroscopy is non-destructive and widely used for assessing the quality of graphene and TMDs.

2.3.5 X-Ray Photoelectron Spectroscopy

X-ray photoelectron spectroscopy provides information about the elemental composition, chemical bonding, and electronic states of the elements present in 2D SCMs. It is particularly worth studying these materials' surface chemistry and functionalization.

2.3.6 Photoluminescence Spectroscopy

Photoluminescence (PL) spectroscopy is crucial for investigating the optical properties of 2D SCMs, including their bandgap, exciton behavior, and emission spectra. It provides insights into the materials' potential for optoelectronic applications, such as photodetectors and light-emitting devices.

2.3.7 Electrical Characterization

Electrical measurements, including current-voltage (I-V) and Hall effect measurements, are fundamental for evaluating the electronic properties and charge carrier mobility of 2D SCMs. They are essential for designing and optimizing electronic devices like transistors and sensors.

2.3.8 Optical Characterization

Techniques like UV–Vis absorption spectroscopy and ellipsometry are employed to analyze the optical properties, such as absorption and reflectance spectra, of 2D materials. These measurements help assess the materials' suitability for photonics and optical device applications.

2.3.9 Thermal Characterization

Thermal techniques like differential scanning calorimetry and thermogravimetric analysis provide information about the thermal stability and thermal conductivity of 2D SCMs, which are crucial for device design and thermal management.

In summary, characterization techniques are indispensable tools for unraveling the intricate properties of 2D SCMs, enabling researchers to design their properties for specific applications. These techniques aid in quality control during material synthesis, provide insights into fundamental material behavior, and guide the development of advanced electronic, optoelectronic, and nanoscale devices.

2.4 Perspective and Conclusion Perspective

The synthesis and characterization of 2D SCMs have rapidly gained attention in the scientific community due to their unique properties and potential applications. The ability to manipulate and control the properties of these materials at the atomic scale opens up new possibilities for advancements in electronics, optoelectronics, energy storage, and more. By continuing to explore different synthesis methods and characterization techniques, we can expect further breakthroughs in material design, device fabrication, and performance optimization of 2D SCMs.

One perspective that emerges from studying 2D SCMs is the need for scalable and reproducible synthesis methods. While numerous synthesis techniques have been developed, including ME, CVD, and MBE, there is a demand for methods to produce high-quality, large-area 2D materials with controlled thickness, composition, and crystal structure. Additionally, efforts are being made to integrate synthesis processes with device fabrication techniques to prepare functional devices out of 2D SCMs seamlessly.

Another perspective revolves around the characterization of these materials. As 2D SCMs exhibit unique properties compared to their bulk counterparts, it becomes essential to employ characterization techniques that can probe their specific features. Researchers are continuously developing and refining characterization techniques to gain a comprehensive understanding of the various properties of these materials. The advancement of these techniques will allow for more accurate and detailed characterization, leading to a better understanding of the underlying physics and improved device performance.

2.5 Conclusion

The synthesis and characterization of 2D SCMs hold immense potential for technological advancements. The development of scalable and reproducible synthesis methods will enable large-scale production of high-quality 2D SCMs, facilitating their integration into practical devices. Ongoing advancements in various characterization techniques are expanding the ideas to uncover the intricate details of these materials, such as their crystal structure, morphology, electronic properties, and optical behavior. This knowledge is crucial for designing the properties of these materials to meet device requirements. Thus, significant advancements in electronics, optoelectronics, energy storage, and sensing can be anticipated with the growing interest in research in 2D SCMs. These materials have the potential to drive technological innovation and shape the future of various industries, offering improved performance, efficiency, and versatility in a wide range of applications.

References

[1] L.I. Berger, *Semiconductor Materials* (2020) London, CRC Press.

[2] M.S. Tyagi, *Introduction to Semiconductor Materials and Devices* (2008) India, Wiley India Pvt. Limited.

[3] D. Chi, K.J. Goh, A.T. Wee, *2D Semiconductor Materials and Devices* (2019) USA, Elsevier.

[4] S. Kang, D. Lee, J. Kim, A. Capasso, H.S. Kang, J.-W. Park, C.-H. Lee, G.-H. Lee, 2D semiconducting materials for electronic and optoelectronic applications: Potential and challenge, *2D Mater.* 7 (2020) 022003.

[5] F. Wang, Z. Wang, C. Jiang, L. Yin, R. Cheng, X. Zhan, K. Xu, F. Wang, Y. Zhang, J. He, Progress on electronic and optoelectronic devices of 2D layered semiconducting materials, *Small* 13 (2017) 1604298.

[6] H. Yoo, K. Heo, M.H.R. Ansari, S. Cho, Recent advances in electrical doping of 2D semiconductor materials: Methods, analyses, and applications, *Nanomaterials*, 11 (2021) 832.

[7] Z. Lin, A. McCreary, N. Briggs, S. Subramanian, K. Zhang, Y. Sun, X. Li, N.J. Borys, H. Yuan, S.K. Fullerton-Shirey, A. Chernikov, H. Zhao, S. McDonnell, A.M. Lindenberg, K. Xiao, B.J. LeRoy, M. Drndić, J.C.M. Hwang, J. Park, M. Chhowalla, R.E. Schaak, A. Javey, M.C. Hersam, J. Robinson, M. Terrones, 2D materials advances: From large scale synthesis and controlled heterostructures to improved characterization techniques, defects and applications, *2D Mater.*, 3 (2016) 042001.

[8] K.S. Novoselov, A.K. Geim, S.V. Morozov, D. Jiang, Y. Zhang, S.V. Dubonos, I.V. Grigorieva, A.A. Firsov, Electric field in atomically thin carbon films, *Science*, 306 (2004) 666–669.

[9] K.S. Novoselov, D. Jiang, F. Schedin, T.J. Booth, V.V. Khotkevich, S.V. Morozov, A.K. Geim, Two-dimensional atomic crystals, *Proc. Natl. Acad. Sci. USA*, 102 (2005) 10451–10453.

[10] P. Avouris, T.F. Heinz, T. Low, *2D Materials* (2017) India, Cambridge University Press.

[11] A. Chaves, J.G. Azadani, H. Alsalman, D.R. da Costa, R. Frisenda, A.J. Chaves, S.H. Song, Y.D. Kim, D. He, J. Zhou, A. Castellanos-Gomez, F.M. Peeters, Z. Liu, C.L. Hinkle, S.-H. Oh, P.D. Ye, S.J. Koester, Y.H. Lee, P. Avouris, X. Wang, T. Low, Bandgap engineering of two-dimensional semiconductor materials, *npj 2D Mater. Appl.*, 4 (2020) 29.

[12] B. Liu, A. Abbas, C. Zhou, Two-dimensional semiconductors: From materials preparation to electronic applications, *Adv. Electron. Mater.*, 3 (2017) 1700045.

[13] X. Jing, Y. Illarionov, E. Yalon, P. Zhou, T. Grasser, Y. Shi, M. Lanza, Engineering field effect transistors with 2D semiconducting channels: Status and prospects, *Adv. Funct. Mater.*, 30 (2020) 1901971.

[14] C. Cong, J. Shang, X. Wu, B. Cao, N. Peimyoo, C. Qiu, L. Sun, T. Yu, Synthesis and optical properties of large-area single-crystalline 2D semiconductor WS_2 monolayer from chemical vapor deposition, *Adv. Optical Mater.*, 2 (2014) 131–136.

[15] Z. Sun, T. Liao, L. Kou, Strategies for designing metal oxide nanostructures, *Sci. China Mater.*, 60 (2017) 1–24.

[16] V. Shanmugam, R.A. Mensah, K. Babu, S. Gawusu, A. Chanda, Y. Tu, R.E. Neisiany, M. Försth, G. Sas, O. Das, A review of the synthesis, properties, and applications of 2D materials, *Part. Part. Syst. Charact.*, 39 (2022) 2200031.

[17] P. Samorì, V. Palermo, X. Feng, Chemical approaches to 2D materials, *Adv. Mater.*, 28 (2016) 6027–6029.

[18] W. Zheng, L.Y.S. Lee, Beyond sonication: Advanced exfoliation methods for scalable production of 2D materials, *Matter*, 5 (2022) 515–545.

[19] Z. Lin, Z. Wan, F. Song, B. Huang, C. Jia, Q. Qian, J.S. Kang, Y. Wu, X. Yan, L. Peng, C. Wan, J. Zhou, Z. Sofer, I. Shakir, Z. Almutairi, S. Tolbert, X. Pan, Y. Hu, Y. Huang, X. Duan, High-yield exfoliation of 2D semiconductor monolayers and reassembly of organic/inorganic artificial superlattices, *Chem*, 7 (2021) 1887–1902.

[20] Z. Li, R.J. Young, C. Backes, W. Zhao, X. Zhang, A.A. Zhukov, E. Tillotson, A.P. Conlan, F. Ding, S.J. Haigh, K.S. Novoselov, J.N. Coleman, Mechanisms of liquid-phase exfoliation for the production of graphene, *ACS Nano*, 14 (2020) 10976–10985.

[21] C. Huo, Z. Yan, X. Song, H. Zeng, 2D materials via liquid exfoliation: A review on fabrication and applications, *Sci. Bull.*, 60 (2015) 1994–2008.

[22] J. Shen, Y. He, J. Wu, C. Gao, K. Keyshar, X. Zhang, Y. Yang, M. Ye, R. Vajtai, J. Lou, P.M. Ajayan, Liquid phase exfoliation of two-dimensional materials by directly probing and matching surface tension components, *Nano Lett.*, 15 (2015) 5449–5454.

[23] K. Varoon Agrawal, X. Zhang, B. Elyassi, D.D. Brewer, M. Gettel, S. Kumar, J.A. Lee, S. Maheshwari, A. Mittal, C.-Y. Sung, M. Cococcioni, L.F. Francis, A.V. McCormick, K.A. Mkhoyan, M. Tsapatsis, Dispersible exfoliated zeolite nanosheets and their application as a selective membrane, *Science*, 334 (2011) 72–75.

[24] F.I. Alzakia, S.C. Tan, Liquid-exfoliated 2D materials for optoelectronic applications, *Adv. Sci.*, 8 (2021) 2003864.

[25] R. Ma, T. Sasaki, Two-dimensional oxide and hydroxide nanosheets: Controllable high-quality exfoliation, molecular assembly, and exploration of functionality, *Acc. Chem. Res.*, 48 (2015) 136–143.

[26] V. Nicolosi, M. Chhowalla, M.G. Kanatzidis, M.S. Strano, J.N. Coleman, Liquid exfoliation of layered materials, *Science*, 340 (2013) 1226419.

[27] J. Wang, G. Li, L. Li, P.K. Nayak, *Two-dimensional Materials: Synthesis, Characterization and Potential Applications, Synthesis Strategies about 2D Materials* (2016) Croatia, IntechOpen.

[28] J. Zheng, H. Zhang, S. Dong, Y. Liu, C. Tai Nai, H. Suk Shin, H. Young Jeong, B. Liu, K. Ping Loh, High yield exfoliation of two-dimensional chalcogenides using sodium naphthalenide, *Nat. Commun.*, 5 (2014) 2995.

[29] S.H. Choi, S.J. Yun, Y.S. Won, C.S. Oh, S.M. Kim, K.K. Kim, Y.H. Lee, Large-scale synthesis of graphene and other 2D materials towards industrialization, *Nat. Commun.*, 13 (2022) 1484.

[30] B. Mendoza-Sánchez, Y. Gogotsi, Synthesis of two-dimensional materials for capacitive energy storage, *Adv. Mater.*, 28 (2016) 6104–6135.

[31] N. Antonatos, H. Ghodrati, Z. Sofer, Elements beyond graphene: Current state and perspectives of elemental monolayer deposition by bottom-up approach, *Appl. Mater. Today*, 18 (2020) 100502.

[32] R. Dong, T. Zhang, X. Feng, Interface-assisted synthesis of 2D materials: Trend and challenges, *Chem. Rev.*, 118 (2018) 6189–6235.

[33] T.H. Choudhury, X. Zhang, Z.Y.A. Balushi, M. Chubarov, J.M. Redwing, Epitaxial growth of two-dimensional layered transition metal dichalcogenides, *Annu. Rev. Mater. Res.*, 50 (2020) 155–177.

[34] M.-Y. Li, Y. Shi, C.-C. Cheng, L.-S. Lu, Y.-C. Lin, H.-L. Tang, M.-L. Tsai, C.-W. Chu, K.-H. Wei, J.-H. He, W.-H. Chang, K. Suenaga, L.-J. Li, Epitaxial growth of a monolayer WSe$_2$-MoS$_2$ lateral p-n junction with an atomically sharp interface, *Science*, 349 (2015) 524–528.

[35] P. Yang, S. Zhang, S. Pan, B. Tang, Y. Liang, X. Zhao, Z. Zhang, J. Shi, Y. Huan, Y. Shi, S.J. Pennycook, Z. Ren, G. Zhang, Q. Chen, X. Zou, Z. Liu, Y. Zhang, Epitaxial growth of centimeter-scale single-crystal MoS$_2$ monolayer on Au(111), *ACS Nano*, 14 (2020) 5036–5045.

[36] Y. Zhang, Y. Yao, M.G. Sendeku, L. Yin, X. Zhan, F. Wang, Z. Wang, J. He, Recent progress in CVD growth of 2D transition metal dichalcogenides and related heterostructures, *Adv. Mater.,* 31 (2019) 1901694.

[37] V. Kranthi Kumar, S. Dhar, T.H. Choudhury, S.A. Shivashankar, S. Raghavan, A predictive approach to CVD of crystalline layers of TMDs: The case of MoS_2, *Nanoscale,* 7 (2015) 7802–7810.

[38] J. Zhang, F. Wang, V.B. Shenoy, M. Tang, J. Lou, Towards controlled synthesis of 2D crystals by chemical vapor deposition (CVD), *Mater. Today,* 40 (2020) 132–139.

[39] Z. Yang, Z. Wu, Y. Lyu, J. Hao, Centimeter-scale growth of two-dimensional layered high-mobility bismuth films by pulsed laser deposition, *InfoMat,* 1 (2019) 98–107.

[40] F. Godel, V. Zatko, C. Carrétéro, A. Sander, M. Galbiati, A. Vecchiola, P. Brus, O. Bezencenet, B. Servet, M.-B. Martin, B. Dlubak, P. Seneor, WS_2 2D semiconductor down to monolayers by pulsed-laser deposition for large-scale integration in electronics and spintronics circuits, *ACS Appl. Nano Mater.,* 3 (2020) 7908–7916.

[41] C. Tan, H. Zhang, Wet-chemical synthesis and applications of non-layer structured two-dimensional nanomaterials, *Nat. Commun.,* 6 (2015) 7873.

[42] A.P. Frauendorf, A. Niebur, L. Harms, S. Shree, B. Urbaszek, M. Oestreich, J. Hübner, J. Lauth, Room temperature micro-photoluminescence studies of colloidal WS_2 Nanosheets, *J. Phys. Chem. C,* 125 (2021) 18841–18848.

[43] R. Kumar, S. Sahoo, W.K. Tan, G. Kawamura, A. Matsuda, K.K. Kar, Microwave-assisted thin reduced graphene oxide-cobalt oxide nanoparticles as hybrids for electrode materials in super-capacitor, *J. Energy Storage,* 40 (2021) 102724.

[44] P.C. Kumar, S. Senapati, D. Pradhan, J. Kumar, R. Naik, A facile one-step microwave-assisted synthesis of bismuth oxytelluride nanosheets for optoelectronic and dielectric application: An experimental & computational approach, *J. Alloys Compd.,* 968 (2023) 172166.

[45] P.C. Kumar, A. Mohapatra, S. Senapati, M. Pradhan, R. Naik, A facile microwave-assisted synthesis of bismuth copper oxytelluride for optoelectronic and photodetection applications, *FlatChem,* 42 (2023) 100580.

[46] J. Zhu, L. Bai, Y. Sun, X. Zhang, Q. Li, B. Cao, W. Yan, Y. Xie, Topochemical transformation route to atomically thick Co_3O_4 nanosheets realizing enhanced lithium storage performance, *Nanoscale,* 5 (2013) 5241–5246.

[47] Y. Kuang, G. Feng, P. Li, Y. Bi, Y. Li, X. Sun, Single-crystalline ultrathin nickel nanosheets array from in situ topotactic reduction for active and stable electrocatalysis, *Angew. Chem. Int. Ed.,* 55 (2016) 693–697.

[48] Q. Li, J. Lin, T.-Y. Liu, X.-Y. Zhu, W.-H. Yao, J. Liu, Gas-mediated liquid metal printing toward large-scale 2D semiconductors and ultraviolet photodetector, *npj 2D Mater. Appl.,* 5 (2021) 36.

[49] S. Jiang, Q. Dai, J. Guo, Y. Li, In-situ/operando characterization techniques for organic semi-conductors and devices, *J. Semicond.,* 43 (2022) 041101.

3

Synthesis and Applications of Graphene Quantum Dots

Mumtaz Ali*, Zafar Arshad, and Hassan Anwer
Correspondence: mumtaz_ali_152@yahoo.com

3.1 Graphene Quantum Dots

Graphene quantum dots (GODs) are defined as graphene chunks with a size smaller than 100 nm and thickness less than ten layers [1]. GQDs are extremely small fragments of graphene sheets, specifically below 20 nm, where charge is highly confined in all dimensions. Most commonly, GQDs are synthesized via the cutting of graphene sheets using a top-down approach. In addition, the utilization of organic synthesis routes for connecting conjugated molecules via a bottom-up approach is possible. Compared to carbon quantum dots (solid spheres), the compact film formation capability, highly oxidized surface for strong dispersion, and suitable optical stability are the benefits of GQDs. In contrast, the production yield of GQDs is much smaller, and hence they are applied only in high-tech applications. During the conversion of graphene to GQDs, interesting properties are induced because of its smaller size, such as its non-zero bandgap nature. Unlike graphene being conductive, GQDs are semiconductors, offering suitable fluorescence. The bandgap of GQDs can be controlled by changing size, functional groups, and the combined effect of both. The optical and processing characteristics of GQDs are similar to molecules, unlike the colloidal characteristics of other quantum dots [1].

GQDs have been the center of various applications and have specifically maintained a great research interest in next-generation electronics. By the conversion of graphene to GQDs, the intrinsic limitation of zero bandgap can be set aside. In this regard, GQDs can potentially serve in diverse applications, such as energy storage, display, and photovoltaics. In addition, the use of GQDs in biomedical applications is emerging, because of their biocompatible nature. Therefore, these are the first choice for bioimaging applications, biosensing, and cancer treatment, thanks to their good photostability, low toxicity, and suitable photoluminescence quantum yields (PLQYs) [2]. Similarly, GQDs are applied for environmental monitoring, water/air remediation, and thermal interface materials [3, 4]. Tunable functional properties of GQDs via grafting, facile solubility, dispersion stability, chemical inertia, and optical stability are the unique features of their competence with metallic semiconductor-based quantum dots.

The GQDs offer several benefits over the quantum dots of a conventional semiconductor, including stable fluorescence qualities, low toxicity, and excellent water solubility. The fluorescence properties of the GQDs are the most significant of them. Although the size of the GQDs, surface chemical groups, and doping atoms can all be used to explain the

DOI: 10.1201/9781003439448-3

photoluminescence (PL) mechanism of GQDs, the precise PL mechanism remains a matter of debate. The surface/edge state in GQDs, the quantum confinement effect of conjugated domains, and the combined effects of these two components make up the major PL mechanism of GQDs. Their fluorescence emission wavelength can range from deep ultraviolet through blue, green, yellow, and red light, among other wavelengths. The varying diameters, surface functional groups, and excitation wavelength are considered to be the causes.

The surface of GQDs is intrinsically covered with oxygenated functional groups, which are highly electronegative. These functional groups provide dispersion stability; however, the optical properties such as PLQY are fewer for these undoped GQDs. To enhance the radiative recombination in GQDs, heteroatom doping, specifically of non-metals including nitrogen, sulfur, boron, and phosphorous, is explored. The density of electrons or holes near the Fermi-level of doped GQDs varies in their recombination probability, leading to differences in optical properties. Also, doped atoms generate intermediate energy levels between the highest-occupied molecular orbital (HOMO)–lowest-unoccupied molecular orbital (LUMO) levels, thus the energy gap is reduced. A lower HOMO–LUMO gap is associated with the absorption of a broader energy spectrum, which renders better photocatalysis. In addition, the surface reactivity and the interfacial interactions of doped GQDs are altered, as doped heteroatoms change the surface charge. Elaborately, heteroatom-doped regions can have higher or lower electronegativity, which can either attract or repel the electron cloud, rendering partial surface charge redistribution. Compared to uniform surface charge, the charge redistributed surface has a higher affinity for the reaction [2].

3.2 Synthesis of GQDs

The synthesis methods can be mainly divided into two categories: (1) top-down and (2) bottom-up approach. Both of these methods are commonly applied using larger-sized or molecular-level carbon-rich precursors for the formation of GQDs, respectively. These methods are viable for the controlled and productive synthesis of GQDs. In the next section, each synthesis route will be discussed in detail.

3.2.1 Oxidative Cleavage

It is the most common approach, also known as oxidation cutting in which sulfuric acid (H_2SO_4), or nitric acid (HNO_3) is used to break the carbon–carbon bonds between graphene, GO, or carbon nanotubes, as shown in **Figure 3.1**. Shen et al. [3] synthesized GO sheets having a size of two-dimensional micron grade, immersed in nitric acid, and then sliced into smaller pieces. The final products were reduced with hydrazine hydrate, and finally surface passivation was performed using ethylene glycol. In the process, GQDs having diameters ranging from 5 to 19 nm were eventually obtained. It has been researched that optimized 365-nm light is necessary for stimulation of the GQDs, and blue fluorescence is generated. It is intriguing that green fluorescence is achieved with a wavelength of 980 nm light having fluorescence upconversion capabilities. To describe the mechanism of fluorescence generation and upconversion, structural models and energy levels of GQDs are provided [4].

GQDs may also be produced from materials such as fullerenes C60, carbon nanotubes, and carbon fibers. Due to its well-defined dimension, Pumera et al. [5] employed fullerene as a starting material to generate very small GQDs (2–3 nm). The oxidation,

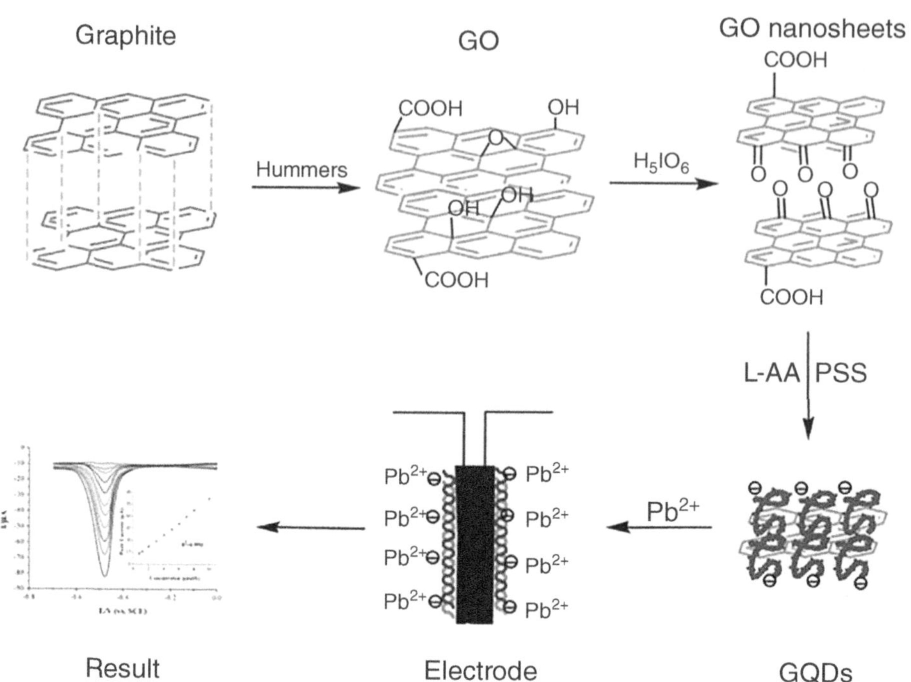

FIGURE 3.1

Mechanism synthesis of GQDs from graphene oxide using oxidative cleavage. Figure adapted with permission [7], Copyright (2023), Gruyter.

cage-opening, and fragmentation processes of fullerene were triggered by treating it with a mixture of strong acid and chemical oxidant. The products remained extensively dispersed in aqueous suspension and exhibited significant luminescence characteristics, with the greatest intensity at 460 nm under an excitation wavelength of 340 nm. Further chemical treatments with hydrazine hydrate and hydroxylamine led to GQDs and a red and blue shift in luminosity, respectively. The current study has shown that fullerene can be oxidized and its cage is opened simultaneously to produce GQDs. The ease with which this technology produces GQDs suggests that it has the potential for further refinement for incorporation into practical devices or applications such as optoelectronics and biological labeling.

In the preparation of GQDs, oxidative cleavage or oxidation cutting is easy and effective, but there are certain drawbacks, such as the strong oxidant utilized, which may cause burning or explosion, and the post-processing process, which is more complicated as shown in **Figure 3.1**. As a result, various new oxidation-cutting technologies are being developed to address the issues. Lu et al. [5], for example, used black carbon as a precursor and H_2O_2 as a facile oxidant to establish a simple and easy one-pot technique for the synthesis of GQDs without the need for a strong concentrated acid, and the complete synthetic process took only 90 minutes. Black carbon was used as a precursor in the preparation, and H_2O_2 was used as an oxidant to cut the black carbon. H_2O and GQDs were the only reaction products. As a result, the suggested synthetic technique not only avoids the use of a highly concentrated acid and the introduction of metal impurity contamination but also eliminates the requirement for any additional post-processing processes. The postulated process is that free radicals (e.g., OH and O) formed by the breakdown of H_2O_2 have high reactivity and strong oxidizing properties, allowing them to effectively oxidize and cut the graphene structure of black carbon. GQDs with diameters ranging from 3.0 to 4.5 nm show strong photostability, good salt solution resistance, and minimal toxicity [6].

3.2.2 Hydrothermal Method

This method was initially used for etching or cutting graphene sheets into extremely small (5–13 nm) chunks, labeled as GQDs [5]. First, graphene sheets were oxidized in strong sulfuric acid. The acid treatment not only functionalizes the graphene sheets with oxygenated functional groups, but the size of larger sheets is also reduced to around <50 nm. These functionalized small-sized chunks of graphene are further subjected to hydrothermal treatment at 200°C for 10 h, keeping a pH of 8. Afterward, the treated solution was subjected to dialysis for purification and separation of GQDs [5]. The same process was further enhanced using the extremely thin graphene sheets as a starting precursor and hydrothermal treatment at pH 12, which yielded the size of GQDs less than 5 nm [6]. The initial oxidation process generates linear carboxylic acid, hydroxyl, and epoxy functional groups on the carbon lattice of graphene, caused by the oxidation process. Upon final hydrothermal treatment, these functional groups are converted to carbonyl functional groups, thereby creating a cleavage in the graphene sheet. Repeated cutting of graphene due to hydrothermal treatment produces fine pieces, which are ultimately converted to GQDs [8].

3.2.3 Solvothermal Method

This synthetic route is commonly used for the bottom-up approach, where small building blocks are condensed to form a GQD skeleton. For this purpose, organic solvents are usually preferred due to their mutual compatibility with organic precursors. By varying the reaction conditions and precursors, precise control of shape can also be achieved. For instance, phloroglucinol (PG) was refluxed in ethanol (as a solvent), whereas a small amount of strong acid catalyzed to control the degree of reduction and associated size of GQDs (**Figure 3.2**) [9]. PG glucogin has three hydroxyl groups at the meta position, which condensate upon solvothermal treatment to yield triangular-shaped GQDs. The average diameter of GQDs was less than 5 nm. Due to the unique triangular shape, high color purity with a narrow full-width half maximum of emission can be achieved [9]. Other than organic solvents, weak acids such as formic acid and acetic acid are also used for solvothermal synthesis. These weak acids can reduce the carbon backbone during the synthesis process, hence determining the hydrophobicity and optical properties.

The application of organic solvents is also tested for the top-down etching of graphene via the solvothermal method. For instance, graphene sheets were dispersed in dimethyl formamide (DMF), followed by its solvothermal treatment at 200°C for 5 h [9]. After the filtration of larger particles, small GQDs were collected by the evaporation of DMF. The average thickness of GQDs was 1.2 nm and a diameter of 5 nm. The GQDs synthesized in the presence of DMF also showed a slight content of nitrogen doping. On the other hand, the optical properties of DMF-assisted GQDs were not promising. Therefore, the production yield and optical properties of GQDs were enhanced using an equal mixture of methanol and methylene chloride as a solvent.

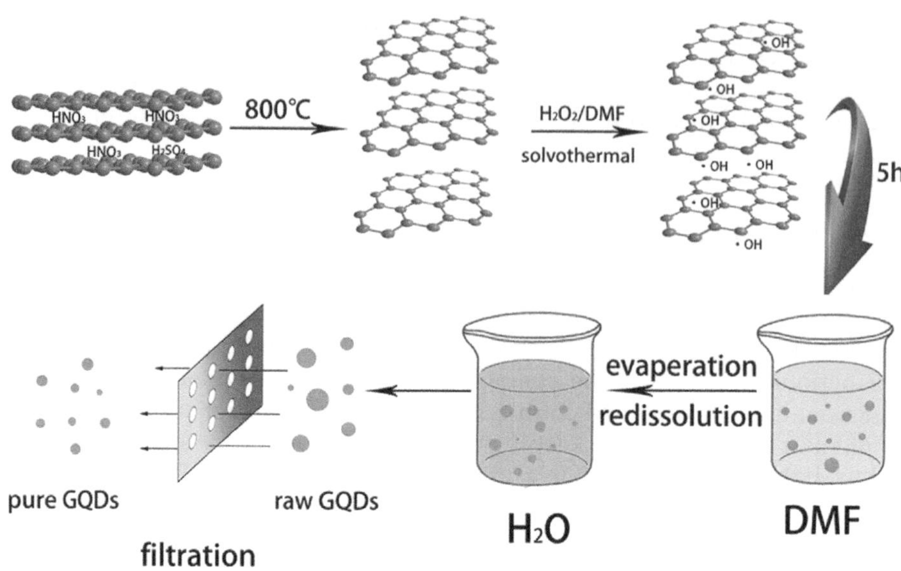

FIGURE 3.2

Solvothermal process for the synthesis of graphene quantum dots. Figure adapted with permission [10], Copyright (2023), Elsevier.

3.2.4 Ultrasound Method

Ultrasound waves produce alternate low and high pressures in liquids and are responsible for the repeated formation and destruction of vacuum bubbles. The high frequency and intense waves are responsible for impinging jets, which provide essentially the required shear force for the dispersion of graphene sheets. If this high-energy process is carried out in strongly acidic or basic conditions, the cutting of graphene sheets to GQDs can be made possible. Typically, the oxidation of graphene sheets was carried out in a mixture of sulfuric acid and nitric acid, which was then treated in an ultrasound bath for 12 hours [11]. The process resulted in uniformly dispersed GQDs, in the range of 3–5 nm. Similar to acid-oxidized graphene, the KOH-assisted activation of graphene sheets also results in the formation of similar GQDs [12]. Without KOH activation, the optical and electrochemical properties of GQDs were poor, as shown in **Figure 3.3**. It is important to mention that the energy and time of sonication are inversely related; that is, sonication time is reduced at high-energy sonication. In contrast to KOH activation, acid oxidation provides smaller-size and higher production yields [13].

In addition to the top-down cutting of graphene, the ultrasound method can be used for the ultrasonic-assisted hydrothermal synthesis of GQDs. In the bottom-up approach, ultrasound waves produce frictional heat, required for the condensation of precursors and formation of GQDs. The sheer force of ultrasound is also responsible for the uniform size distribution of synthesized GQDs.

3.2.5 Electrochemical Oxidation

Under high REDOX voltage (1.5–3 V), carbon nanotubes, graphite, or graphene are oxidatively cleaved into GQDs as the working electrode in the electrochemical oxidation

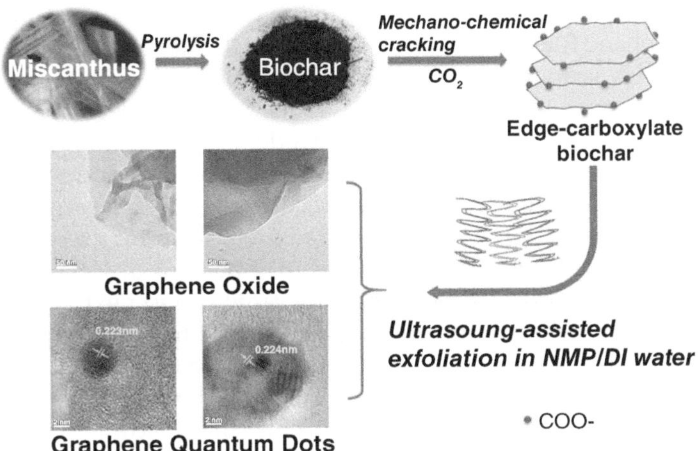

FIGURE 3.3
Ultrasonic method for the synthesis of graphene quantum dots. Figure adapted with permission [10], Copyright (2023), Elsevier.

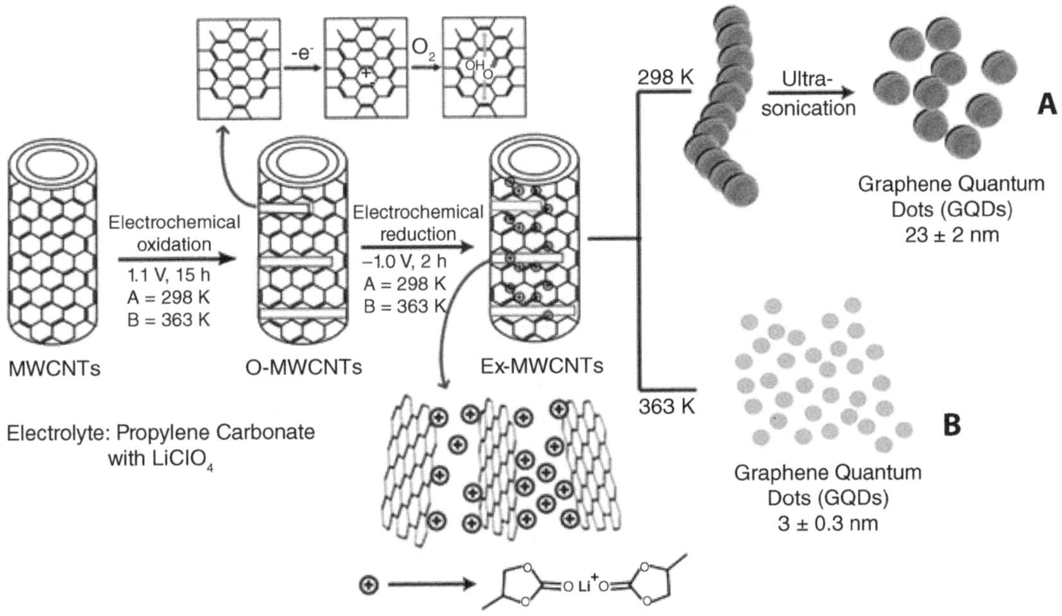

FIGURE 3.4
Electrochemical oxidation to get graphene quantum dots. Figure adapted with permission [7], Copyright (2023), De Gruyter.

process (**Figure 3.4**). Electrochemical oxidation can be accomplished in two ways. One is that carbon–carbon bonds in graphene or carbon nanotubes are instantly broken by electrochemical oxidation. The other is that oxidizing water produces hydroxyl or oxygen free radicals (OH or O), both of which can oxidatively break down GQDs. The disadvantage of the electrochemical oxidation process is that it takes a long time to purify the raw materials and convert them into GQD products [14].

In their study, Li et al. [15] produced GQDs using phosphate buffer solution (PBS) as the electrolyte and graphene filter film as the working electrode. To electrochemically prepare functional GQDs, a CV scan inside 3.0 V at a scan rate of 0.5 V s^{-1} in 0.1 m PBS was utilized. A graphene film (5 mm × 10 mm) was used as the working electrode. Pt wire and Ag/AgCl were used as counter and reference electrodes. Before making the GQDs, the graphene film was filtered and exposed to O$_2$ plasma for a brief period of time to increase its hydrophilicity. The water-soluble GQDs were recovered after filtering and dialyzed using a cellulose ester membrane bag. The as-prepared GQDs exhibit green fluorescence, have a homogeneous size distribution (3–5 nm), and may be kept in water for several months without any alteration in optical properties. The same group next synthesized nitrogen-doped GQDs (N-GQDs) with diameters ranging from 2 to 5 nm by replacing PBS with acetonitrile that included tetrabutylammonium perchlorate as the electrolyte. The newly synthesized N-GQDs, in contrast to their N-free counterparts, display electrocatalytic activity for the oxygen reduction reaction in an alkaline medium similar to a commercially available Pt/C catalyst and exhibit blue luminescence. These GQDs are very useful for cellular and molecular imaging applications due to their inherent luminous behavior, enhanced photostability, and increased fluorescence quantum yield.

3.2.6 Continuous or Batch Hydrothermal for Bottom-Up Synthesis

This method involves a controlled stepwise synthesis of GQDs from phenyl-containing molecules in an organic solvent. The GQDs are homogeneous in size and shape as well as contain a tunable and controlled number of carbon atoms. However, the preparation method comprises numerous challenging chemical steps that take a lot of time and have a low yield. For instance, Li et al. created GQDs of 168, 132, and 170 carbon atoms (denoted by 1, 2, and 3, respectively) using this technique. The 2′,4′,6′-trialkyl phenyl molecules were attached to the edges of the GQDs to prevent them from reuniting. Sulfur-doped GQDs (S-GQDs) with vivid blue emission were created using a straightforward one-pot method using 3-mercaptopropionic acid (MPA) and 1,3,6-trinitropyrene (TNP). In a nutshell, TNP and MPA were combined using ultrasound before being autoclaved for a 10-hour hydrothermal treatment at 200°C. The insoluble carbon product was then removed from the solution containing the water-soluble GQDs using a microporous barrier, and unreacted small molecules were dialyzed away. The S-GQDs were freeze-dried after being suspended in a light yellow aqueous solution following dialysis [16]. Although the optical properties of GQDs synthesized by such batch hydrothermal methods are better, long processing time is problematic in commercial applications. Therefore, the concept of continuous hydrothermal process at high temperatures and shorter times is emerging nowadays. However, this method is more effective for carbon quantum dots compared to GQDs.

3.2.7 Ambient Pyrolysis

Unlike high pressures in the hydrothermal method, a simple one-step straight-way strategy is to carbonize the carbon source at high temperatures. This causes a condensation reaction of molecules, and the final product is carbon material, which is further subjected to purification to obtain GQDs **(Figure 3.5)**. A basic and eco-friendly process known as molecular carbonization uses the proper organic molecules or polymers for dehydration and further carbonization. GQDs with polydispersity are produced because it is challenging to accurately control the size and shape in this method. Using simply deionized water and glucose as a precursor, this method, for instance, was employed to produce low-cost and high-yield green-photoluminescent single-layer GQDs.

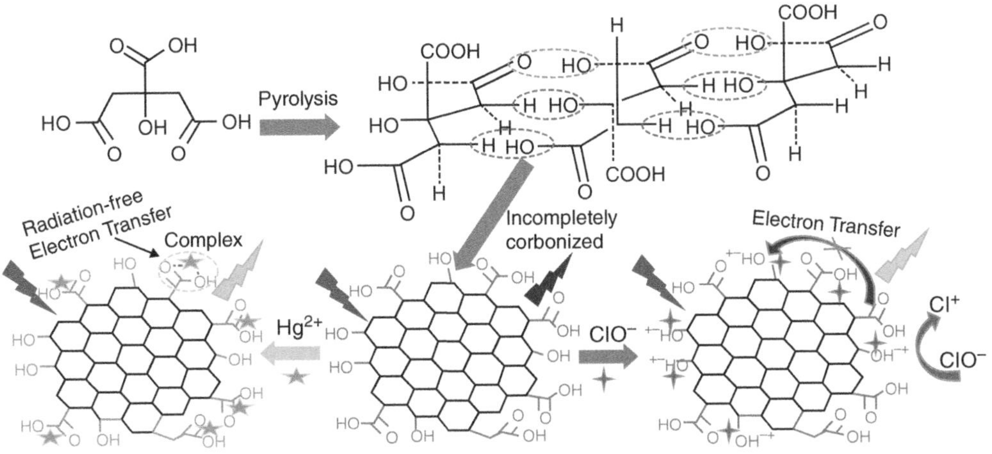

FIGURE 3.5
Graphene quantum dots achieved as a result of the condensation reaction. Figure adapted with permission [18], Copyright (2023), Elsevier.

In a typical synthesis, glucose powder was dissolved in deionized water. The mixture was heated in an oven for 8 hours at 200°C. After carbonization at high temperatures, the sample changed its color from translucent (colorless) to orange and black. Following is the mechanism: glucose molecules are dehydrated to produce C=C, the fundamental building block of graphene structure. During the production of a GQD, a glucose molecule's hydrogen atoms interact with the hydroxyl groups of a nearby glucose molecule to produce water molecules. GQDs have an average size of about 8 nm and have uniform dispersion and no discernible aggregation. The emission peaks of the sample do not alter when the GQD solution is excited at wavelengths ranging from 450 to 520 nm with a 10 nm gap, and the maximum emission wavelength stays close to 540 nm. It indicates that the product is green photoluminescent, and its size and surface condition are constant. Using just glucose powder as a precursor in deionized (DI) water, this straightforward method for making GQDs illustrates the major benefits of low-cost, high-yield, and mass manufacturing over previously described techniques [17].

Generally speaking, the top-down approach is frequently used in the building of highly pure GQDs. Cleaving carbon materials including graphene, fullerenes, and carbon nanotubes via physical or chemical procedures like laser ablation, oxidative cleavage, hydrothermal or solvothermal methods, electrochemical oxidation, ultrasonic or microwave assistance, or hydrothermal or solvothermal methods are a better option. The regulated synthesis or carbonization of the required organic compounds or polymers yields the GQDs for the bottom-up approach [19].

3.3 Applications of GQDs

3.3.1 Solar Cells

Research on solar cells is frequently conducted in the field of clean energy. The most crucial element in the use of GQDs in solar cells is their size-dependent bandgap tuning property. Due to their better dispersion, regulated band gap, excellent chemical stability, and low toxicity, GQDs have many benefits over other materials used in solar cells, such as silicon and perovskite. GQDs are perfect for solar cells since they also exhibit the quantum confinement effect and the tunable band edge alignment. Hole transport layer material, silicon/GQD heterojunction solar cells, semiconductor/GQD solar cells, and conductive polymer-doped GQD solar cells are just a few of the GQD-based solar cells that have been described thus far [20]. In the research of Li et al. [21], solution chemistry was used to create GQDs with a uniform size distribution from double-walled carbon nanotubes (CNTs). The GQDs in chlorobenzene produce a dazzling blue glow when exposed to UV light. Power conversion efficiency (PCE) is greatly increased when GQDs are added to a bulk heterojunction polymer solar cell (PSC) made of poly(3-hexylthiophene):(6,6)-phenyl-C61 butyric acid methyl ester (P3HT:PCBM). The efficiency can be raised even more by modifying the PCBM content in the active layer, leading to a maximum PCE of 5.24%. This mixed P3HT:PCBM:GQD-based ternary system offers a cutting-edge method for enhancing PSC effectiveness. With this development, PSC devices can now use GQDs for better charge separation and transport [22].

GQDs are also applied as an energy downshift layer for the Si-solar cells, converting the UV light to visible spectra. As Si-solar cells cannot efficiently utilize the UV and near-UV spectrum due to huge bandgap differences, the GQD overlayer converts these low-performance spectra to visible light, which can be effectively absorbed by solar cells. It is important to note that for such high-performance optical applications, a high quantum yield of GQDs is essential in the solid state. The conventional coating methods of GQDs induce aggregates in the GQD overlayer; hence, performance is quenched. To overcome this issue, controlled kinetic spraying is introduced, which results in a uniform thin layer of GQDs on Si-solar cells. The spectral tuning capability of GQDs is responsible for 2.7% enhanced solar cell performance [23]. As GQDs were synthesized via top-down cutting on graphene, the quantum yield and associated enhancement in performance were limited. In contrast, bottom-up synthesized GQDs increase the PCE of dye-sensitized solar cells (DSSCs) by 10%, as the PLQY was more than 70% [24]. In addition to enhanced UV light harvesting characteristics, this energy downshift is effective for suppressing the degradation of organic dyes used in the sensitization of DSSC.

GQDs are also applied as a counterelectrode of DSSC; that is, with an optimum inclusion of extremely small GQDs, a porous orientation of holey graphene was achieved. This porous architecture provides excellent electrocatalytic reduction of iodide electrolytes, related to the highly active surface of GQDs. It is important to note that the conductivity of pure GQDs is almost negligible; therefore, GQDs–holey graphene composite was subjected to a thermal reduction process before its application in the counterelectrode of DSSC. GQDs reinforce the conductivity and overall functionality of the electrode. It was observed that the size of GQDs is critical in determining the morphology of composite coating, where the smallest-sized GQDs showed strong phase separation and associated porous orientation [25]. Unlike metallic particles used for the porous orientation of graphene, GQD-induced orientation is a metal-free, harsh-chemical-free, and highly active method for enhanced electrocatalysis in electrochemical solar cells.

3.3.2 Photocatalysis

Photocatalytic degradation allows for the efficient conversion of solar energy into chemical energy. The photoabsorption characteristic has a strong influence on photocatalytic activity. Because of its low cost, long-term stability, outstanding photocatalytic efficacy, and environmental friendliness, TiO_2 can be an effective catalyst in the degradation of organic contaminants and water splitting. TiO_2 has two major limitations: (i) it can absorb only ultraviolet light due to its broad intrinsic bandgap (3.2 eV) and (ii) it has low electron mobility and a high recombination rate of electron–hole pairs, which reduces its catalytic activity. The drawbacks of TiO_2 were mitigated by building a complex system for photocatalytic application under visible-light irradiation using the visible-light absorption characteristics of GQDs. The photocatalytic behavior of the rutile TiO_2/GQD composite system in the degradation of methylene blue (Xe lamp) under visible light (>420 nm) irradiation was reported to be nine times greater than that of the anatase TiO_2 only (reference photocatalyst). The primary distinction between these two systems is that the bandgap of rutile TiO_2 is narrower than 407 nm, whereas the bandgap of anatase TiO_2 is broader than 407 nm [3]. GQDs also form a strong heterojunction for better charge separation [26].

Because of their ideal bandgap, metal sulfides are an alternate and effective possibility for visible-light photocatalysts. In particular, CdS is an intriguing photocatalyst with a bandgap of 2.4 eV and a conduction band position that is exceptionally tiny and negative in comparison to the H_2O/H_2 reduction potential. The main challenges with the photocatalysis of CdS nanoparticles are aggregation to produce larger ones, which result in a reduced surface area, an increased recombination rate of the photogenerated electron–hole pairs, and photocorrosion, which results in rapid oxidation of sulfide ion by photogenerated holes. Cocatalysts may be the ideal solution for this photocorrosion problem since they have low activation potentials and active sites for H_2 or O_2 evolution. Liang et al. presented GQDs/CdS core/shell nanospheres for photocatalysis via water splitting into hydrogen, demonstrating that they had more stable photocatalytic behavior and an excellent hydrogen evolution rate than pure CdS [27].

3.3.3 Electrochemical Energy Storage Devices

The importance of electrochemical energy storage devices—including supercapacitors and batteries—is increasing due to the exponential increase in renewable energy. Supercapacitors are electrochemical capacitors that store charge via a double layer or redox reaction for fast charging and discharging reactions. Although the discharge time of supercapacitors is faster than the battery, the high stability of supercapacitors is demanded in several energy applications. Small GQDs have highly active edge sites for charge storage reactions, providing a better interface for charge storage reactions. To further boost these electrochemical dynamics, alkali-assisted activation is a possible option [12]. Similarly, nitric acid assisted in cutting graphene sheets to GQDs and its reduction at 800°C was carried out to achieve ideal double-layer capacitor characteristics. Thermal reduction renders conductivity in GQDs as required for electrocatalytic applications, as shown in **Figure 3.6** [28]. Suitably enhanced performance of GQDs compared to graphene was related to higher surface area and suppression of stacking in ultra-small size of GQDs, as shown in Figure 3.6. GQDs are coupled with graphene, activated charcoal, and CNTs for supercapacitor electrode applications, where GQDs enhance the conductivity of the network by providing excessive linkages at a much smaller scale [29, 30]. Other than electrode applications, the potential of GQDs was also tested for the solid-state electrolytes of supercapacitors. Enhanced ionic conductivity and ion donating ability of the GQDs were realized by neutralizing the acidic functional groups using KOH [31]. Without neutralization, the GQDs are active only in solution form, whereas, after KOH treatment, the activity was enhanced in both liquid and solid states [32]. Similarly, GQD doping in TiO_2 increases the surface area of the composite, along with improving the electronic properties [33].

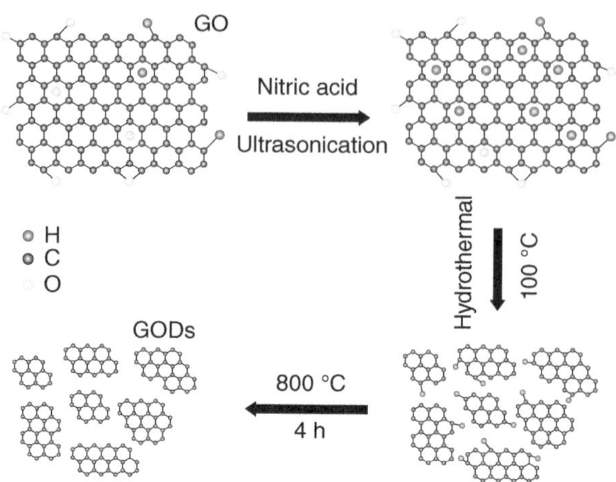

FIGURE 3.6
Uniform-sized graphene quantum dots achieved from ultrasonication. Figure adapted with permission [28], Copyright (2023), American Chemical Society.

3.3.4 Optical Applications

GQDs are used for optical applications because of their optical stability. Han et al. synthesized GQDs via the etching of graphene using vapor phase etching and tested the optical stability of GQDs for one year [34]. GQDS were dispersed in poly(acrylic acid), and its fibers and films were fabricated, which showed strong fluorescence under a UV lamp. Depending on the etching temperature, the fluorescent color of GQDs was altered from blue to yellow, showing size-dependent fluorescence.

3.3.4.1 Light-Emitting Diodes

Light-emitting diodes (LEDs) based on GQDs are important for their superb optical properties, which render low-energy solutions compared to conventional optical systems. Limited production yields and lower PLQY of GQDs (compared to carbon dots) limit their application in LEDs. Recent focus has been given to the synthesis of GQDs under controlled oxidation processes to achieve both high PLQY and production yields, without using toxic chemicals. In this regard, recent work on GQDs synthesis was carried out via etching graphene oxide in a hydrothermal process, resulting in tunable fluorescence and up to 60% production yields [35]. Reaction time and temperature alter the size and quantum confinement effect of GQDs; thus, blue-to-red emission can be achieved for white LED applications. The concentration of oxidizing agents $FeCl_3$ and H_2O_2 was adjusted in the hydrothermal process to achieve the required GQD size [35]. To further enhance the ecofriendliness of GQD-based LEDs, fullerene was cut into GQDs via ozone gas exposure, without using any harsh chemicals [36]. This oxidation process converts the fullerene into GQDs with strong orange emission, which was dispersed in the polyvinyl alcohol (PVA) matrix [36]. This fluorescent film was then coated with UV light to fabricate a white LED. Optical properties of GQDs were also tuned via crystal structure and associated edge states responsible for the high PLQY emission. Highly crystalline GQDs with precise hexagonal shapes were obtained by controlling the reaction dynamics through catalytic solution chemistry [37]. This unique control of morphology results in a better lifetime of exciton and PLQY; however, Stokes shift of the proposed GQDs was low and hence used for deep-blue LED applications.

3.3.4.2 Bioimaging

GQDs are biocompatible fluorescent materials with high optical properties, which makes their utility vital in clinical diagnosis and therapy evaluation, enabling non-invasive, exceptionally sensitive, and precise monitoring of both pathological and physiological occurrences [38]. The photoluminescence (PL) of GQDs can be influenced by a range of factors, such as their size, pH, solvent, surface groups, and heteroatom doping. To date, extensive research efforts have been dedicated to unraveling the PL mechanisms of GQDs [39, 40]. The quantum confinement effect, edge state, and surface state-induced PL are among the most widely accepted PL mechanisms. The diverse array of influencing factors and PL mechanisms associated with GQDs provides ample opportunities to devise GQD-based sensors using various strategies. These strategies can be majorly divided into five categories: ON–OFF sensors, recognition units, ratiometric sensors, reference signals, and energy donors/acceptors. In the ON–OFF strategy, the GQDs are combined with compounds that generate fluorescence quenchers upon interaction with the target analyte [41].

Extensive practical use and clinical implementation of GQDs require overcoming significant hurdles and a substantial journey ahead. Despite numerous techniques having been documented for the production of GQDs, achieving precise control over their size, shape, layering, or surface properties remains a formidable task. Additionally, most of the reported GQD synthesis methods have drawbacks like being time-consuming, limited in volume, and environmentally unfriendly. As a result, there is a pressing need for the development of scalable, efficient, environmentally friendly approaches for synthesizing GQDs with well-defined properties. Furthermore, there should be increased focus on fine-tuning and functionalizing the surfaces of GQDs. Specifically, high Stokes shift emissive GQDs are essential for deep tissue imaging; therefore, near-infrared emissive GQDs are highly desired in the future.

3.4 Summary

As described in this chapter, various GQD preparation techniques have been developed. The fundamental concept and conventional process for making GQDs is the repetitive oxidation and reduction of carbonaceous materials. Additional techniques include controlled synthesis and carbonization. Graphene oxide is the common starting material in a hydrothermal etching process to synthesize GQDs; however, the approach that can make GQDs in large quantities is the strong acidic treatment of graphite. For the purpose of furthering the development of GQDs, we must identify practical, readily available, and cost-effective fabrication techniques. GQDs can help devices and applications because of their distinct optical and electrical properties. Promising solutions for current use and upcoming research have been given by extensive designs of microstructures and device structures. However, research on GQDs is still in its early stages compared to graphene. Although there is still a long way to go until there are widespread practical applications, there is still a ton of potential for research by academics.

Before any further work can start, there are a number of issues that must be addressed. For instance, establishing scalable yet straightforward synthetic methods to produce high-quality GQDs remains challenging despite substantial research on fabrication techniques. Additionally, there is still no conclusive evidence or compelling justification for the PL mechanism of GQDs. Further, it is not apparent how optical properties are influenced by factors like size, doping, crystallinity, and surface functionalization. Furthermore, detailed theoretical computations and experimental verification are greatly desired. For instance, a catalytic model should provide a conceptual framework for comprehending the interactions between GQDs and active components. The best locations or defect types to modify the interactions and their impact on the response process must also be determined. Moreover, all of these fascinating optical properties are drastically quenched in the solid state, thus limiting their use in a variety of applications. Strategies to retain the performance of GQDs in the solid state are another concern to be addressed.

References

[1] S. C. Dhanabalan, B. Dhanabalan, J. S. Ponraj, Q. Bao, and H. Zhang, "2D–materials-based quantum dots: Gateway towards next-generation optical devices," *Advanced Optical Materials*, vol. 5, p. 1700257, 2017.

[2] A. Manikandan, Y.-Z. Chen, C.-C. Shen, C.-W. Sher, H.-C. Kuo, and Y.-L. Chueh, "A critical review on two-dimensional quantum dots (2D QDs): From synthesis toward applications in energy and optoelectronics," *Progress in Quantum Electronics*, vol. 68, p. 100226, 2019.

[3] X. Wang, G. Sun, N. Li, and P. Chen, "Quantum dots derived from two-dimensional materials and their applications for catalysis and energy," *Chemical Society Reviews*, vol. 45, pp. 2239–2262, 2016.

[4] B. Shao, Z. Liu, G. Zeng, H. Wang, Q. Liang, Q. He, *et al.*, "Two-dimensional transition metal carbide and nitride (MXene) derived quantum dots (QDs): Synthesis, properties, applications and prospects," *Journal of Materials Chemistry A*, vol. 8, pp. 7508–7535, 2020.

[5] S. J. Phang and L.-L. Tan, "Recent advances in carbon quantum dot (CQD)-based two dimensional materials for photocatalytic applications," *Catalysis Science & Technology*, vol. 9, pp. 5882–5905, 2019.

[6] W. Tao, X. Ji, X. Xu, M. A. Islam, Z. Li, S. Chen, *et al.*, "Antimonene quantum dots: Synthesis and application as near-infrared photothermal agents for effective cancer therapy," *Angewandte Chemie*, vol. 129, pp. 12058–12062, 2017.

[7] W. Chen, G. Lv, W. Hu, D. Li, S. Chen, and Z. Dai, "Synthesis and applications of graphene quantum dots: A review," *Nanotechnology Reviews*, vol. 7, pp. 157–185, 2018.

[8] P. Atkin, T. Daeneke, Y. Wang, B. Carey, K. Berean, R. Clark, *et al.*, "2D WS 2/carbon dot hybrids with enhanced photocatalytic activity," *Journal of Materials Chemistry A*, vol. 4, pp. 13563–13571, 2016.

[9] F. Yuan, T. Yuan, L. Sui, Z. Wang, Z. Xi, Y. Li, *et al.*, "Engineering triangular carbon quantum dots with unprecedented narrow bandwidth emission for multicolored LEDs," *Nature Communications*, vol. 9, p. 2249, 2018/06/08.

[10] R. Tian, S. Zhong, J. Wu, W. Jiang, Y. Shen, and T. Wang, "Solvothermal method to prepare graphene quantum dots by hydrogen peroxide," *Optical Materials*, vol. 60, pp. 204–208, 2016.

[11] H. Li, X. He, Y. Liu, H. Huang, S. Lian, S.-T. Lee, *et al.*, "One-step ultrasonic synthesis of water-soluble carbon nanoparticles with excellent photoluminescent properties," *Carbon*, vol. 49, pp. 605–609, 2011.

[12] M. Hassan, E. Haque, K. R. Reddy, A. I. Minett, J. Chen, and V. G. Gomes, "Edge-enriched graphene quantum dots for enhanced photo-luminescence and supercapacitance," *Nanoscale*, vol. 6, pp. 11988–11994, 2014.

[13] J. Kabel, S. Sharma, A. Acharya, D. Zhang, and Y. K. Yap, "Molybdenum disulfide quantum dots: Properties, synthesis, and applications," *C*, vol. 7, p. 45, 2021.

[14] M. Bat-Erdene, A. S. Bati, J. Qin, H. Zhao, Y. L. Zhong, J. G. Shapter, *et al.*, "Elemental 2D Materials: Solution-Processed Synthesis and Applications in Electrochemical Ammonia Production," *Advanced Functional Materials*, vol. 32, p. 2107280, 2022.

[15] N. S. Arul and V. Nithya, "Molybdenum disulfide quantum dots: Synthesis and applications," *RSC Advances*, vol. 6, pp. 65670–65682, 2016.

[16] K. P. Musselman, K. H. Ibrahim, and M. Yavuz, "Research update: Beyond graphene—Synthesis of functionalized quantum dots of 2D materials and their applications," *APL Materials*, vol. 6, 2018.

[17] M. Shaker, R. Riahifar, and Y. Li, "A review on the superb contribution of carbon and graphene quantum dots to electrochemical capacitors' performance: Synthesis and application," *FlatChem*, vol. 22, p. 100171, 2020.

[18] H. Zhang, Z. Yuan, M. Wang, L. Zhu, X. Cheng, D. Cao, *et al.*, "Application of graphene quantum dots in the detection of Hg^{2+} and ClO^- and analysis of detection mechanism," *Diamond and Related Materials*, vol. 117, p. 108454, 2021.

[19] Z. Sun, H. Xie, S. Tang, X. F. Yu, Z. Guo, J. Shao, *et al.*, "Ultrasmall black phosphorus quantum dots: Synthesis and use as photothermal agents," *Angewandte Chemie International Edition*, vol. 54, pp. 11526–11530, 2015.

[20] Q. A. Alsulami, Z. Arshad, M. Ali, and S. Wageh, "Efficient tuning of the opto-electronic properties of sol–gel-synthesized Al-doped titania nanoparticles for perovskite solar cells and functional textiles," *Gels*, vol. 9, p. 101, 2023.

[21] M. Batmunkh, M. Bat-Erdene, and J. G. Shapter, "Black phosphorus: Synthesis and application for solar cells," *Advanced Energy Materials*, vol. 8, p. 1701832, 2018.

[22] W. Chen, K. Li, Y. Wang, X. Feng, Z. Liao, Q. Su, *et al.*, "Black phosphorus quantum dots for hole extraction of typical planar hybrid perovskite solar cells," *The Journal of Physical Chemistry Letters*, vol. 8, pp. 591–598, 2017.

[23] K. D. Lee, M. J. Park, D.-Y. Kim, S. M. Kim, B. Kang, S. Kim, *et al.*, "Graphene quantum dot layers with energy-down-shift effect on crystalline-silicon solar cells," *ACS Applied Materials & Interfaces*, vol. 7, pp. 19043–19049, 2015/09/02.

[24] R. Riaz, M. Ali, T. Maiyalagan, A. S. Anjum, S. Lee, M. J. Ko, *et al.*, "Dye-sensitized solar cell (DSSC) coated with energy down shift layer of nitrogen-doped carbon quantum dots (N-CQDs) for enhanced current density and stability," *Applied Surface Science*, vol. 483, pp. 425–431, 2019/07/31.

[25] M. Ali, R. Riaz, A. S. Anjum, K. C. Sun, H. Li, S. H. Jeong, *et al.*, "Graphene quantum dots induced porous orientation of holey graphene nanosheets for improved electrocatalytic activity," *Carbon*, vol. 171, pp. 493–506, 2021/01/01.

[26] K. H. Lee, U. Arfa, Z. Arshad, E.-J. Lee, M. Alshareef, M. M. Alsowayigh, *et al.*, "The comparison of metal doped TiO2 Photocatalytic active fabrics under sunlight for waste water treatment applications," *Catalysts*, vol. 13, p. 1293, 2023.

[27] M. Humayun, C. Wang, and W. Luo, "Recent progress in the synthesis and applications of composite photocatalysts: A critical review," *Small Methods*, vol. 6, p. 2101395, 2022.

[28] S. Zhang, L. Sui, H. Dong, W. He, L. Dong, and L. Yu, "High-performance supercapacitor of graphene quantum dots with uniform sizes," *ACS Applied Materials & Interfaces*, vol. 10, pp. 12983–12991, 2018.

[29] Y. Qing, Y. Jiang, H. Lin, L. Wang, A. Liu, Y. Cao, *et al.*, "Boosting the supercapacitor performance of activated carbon by constructing overall conductive networks using graphene quantum dots," *Journal of Materials Chemistry A*, vol. 7, pp. 6021–6027, 2019.

[30] Y. Hu, Y. Zhao, G. Lu, N. Chen, Z. Zhang, H. Li, *et al.*, "Graphene quantum dots–carbon nanotube hybrid arrays for supercapacitors," *Nanotechnology*, vol. 24, p. 195401, 2013/04/12.

[31] S. Zhang, Y. Li, H. Song, X. Chen, J. Zhou, S. Hong, *et al.*, "Graphene quantum dots as the electrolyte for solid state supercapacitors," *Scientific Reports*, vol. 6, p. 19292, 2016/01/14.

[32] J. Liu, H. Xia, D. Xue, and L. Lu, "Double-shelled nanocapsules of V2O5-based composites as high-performance anode and cathode materials for Li ion batteries," *Journal of the American Chemical Society*, vol. 131, pp. 12086–12087, 2009/09/02.

[33] W. Zhang, T. Xu, Z. Liu, N.-L. Wu, and M. Wei, "Hierarchical TiO2–x imbedded with graphene quantum dots for high-performance lithium storage," *Chemical Communications*, vol. 54, pp. 1413–1416, 2018.

[34] H. Park, S. Hyun Noh, J. Hye Lee, W. Jun Lee, J. Yun Jaung, S. Geol Lee, *et al.*, "Large Scale Synthesis and Light Emitting Fibers of Tailor-Made Graphene Quantum Dots," *Scientific Reports*, vol. 5, p. 14163, 2015/09/18.

[35] B. Lyu, H.-J. Li, F. Xue, L. Sai, B. Gui, D. Qian, *et al.*, "Facile, gram-scale and eco-friendly synthesis of multi-color graphene quantum dots by thermal-driven advanced oxidation process," *Chemical Engineering Journal*, vol. 388, p. 124285, 2020/05/15.

[36] Y.-X. Chen, D. Lu, G.-G. Wang, J. Huangfu, Q.-B. Wu, X.-F. Wang, *et al.*, "Highly efficient orange emissive graphene quantum dots prepared by acid-free method for white LEDs," *ACS Sustainable Chemistry & Engineering*, vol. 8, pp. 6657–6666, 2020/05/04.

[37] S. H. Lee, D. Y. Kim, J. Lee, S. B. Lee, H. Han, Y. Y. Kim, *et al.*, "Synthesis of single-crystalline hexagonal graphene quantum dots from solution chemistry," *Nano Letters*, vol. 19, pp. 5437–5442, 2019/08/14.

[38] J. Shaya, P. R. Corridon, B. Al-Omari, A. Aoudi, A. Shunnar, M. I. H. Mohideen, *et al.*, "Design, photophysical properties, and applications of fluorene-based fluorophores in two-photon fluorescence bioimaging: A review," *Journal of Photochemistry and Photobiology C: Photochemistry Reviews*, vol. 52, p. 100529, 2022/09/01.

[39] R. Rabeya, S. Mahalingam, A. Manap, M. Satgunam, M. Akhtaruzzaman, and C. H. Chia, "Structural defects in graphene quantum dots: A review," *International Journal of Quantum Chemistry*, vol. 122, p. e26900, 2022.

[40] P. R. Kharangarh, N. M. Ravindra, G. Singh, and S. Umapathy, "Synthesis of luminescent graphene quantum dots from biomass waste materials for energy-related applications—An overview," *Energy Storage*, vol. 5, p. e390, 2023.

[41] X. Xie, Y. Lian, L. Xiao, and L. Wei, "Facile and label-free fluorescence sensing of β-galactosidase activity by graphene quantum dots," *Spectrochimica Acta Part A: Molecular and Biomolecular Spectroscopy*, vol. 240, p. 118594, 2020/10/15.

4

Wide Bandgap 2D Semiconductors and Their Applications

Ghulam Dastgeer*, Sobia Nisar, Laraib Sajjad, Eom Jonghwa, and Deok-kee Kim
*Correspondence: gdastgeer@sejong.ac.kr

4.1 Introduction

In the realm of materials science and electronics, two-dimensional (2D) materials have emerged as a revolutionary class of substances that promise to reshape the technological landscape. These materials, consisting of single or few layers of atoms arranged in a planar structure, have garnered considerable attention due to their remarkable electronic and optoelectronic properties [1–4]. Among the pioneers of this material class, graphene stands tall as a trailblazer and an archetype of the 2D world. Graphene, composed of a single layer of carbon atoms arranged in a hexagonal lattice, took the scientific community by storm upon its discovery in 2004. Its extraordinary electrical conductivity, mechanical strength, and thermal properties quickly earned it the reputation of a "wonder material." Graphene's zero bandgap, a fundamental characteristic resulting from its linear energy dispersion, paved the way for groundbreaking applications in fields such as electronics, sensors, and materials science [5–7]. However, as promising as graphene is, its lack of an intrinsic bandgap poses significant challenges for certain applications, limiting its full potential. The merits of graphene are undeniable. Its exceptional electron mobility, thermal conductivity, and mechanical strength have spurred numerous innovations, including ultrafast transistors, flexible electronic devices, and advanced composite materials. Yet, the absence of a bandgap—essential for controlling the flow of electrons—has hindered its application in transistors, leading to substantial leakage currents and low ON/OFF ratios [8–12].

4.1.1 Wide Bandgap 2D Materials: A New Horizon

2D semiconductor materials, composed of atomically thin layers arranged in a planar structure, have become the focus of intense research and innovation in recent years. Among the intriguing members of this material class are those with wide bandgaps, a defining characteristic that sets them apart from their narrow bandgap counterparts. In this brief overview, we will delve into the fundamental aspects and significance of wide bandgap 2D semiconductor materials.

Wide bandgap 2D materials, often endowed with bandgaps ranging from a few electron volts to beyond ultraviolet energies, encompass a diverse family of compounds [7–8, 13]. Transition metal dichalcogenides (TMDCs), hexagonal boron nitride (h-BN), and black phosphorus (BP) are just a few examples of this burgeoning class. These materials exhibit

DOI: 10.1201/9781003439448-4

a rich palette of electronic behaviors, enabling tailored control over charge transport and optical properties. Their atomically precise interfaces, similar to graphene, make them appealing for various applications, particularly in the field of electronic and optoelectronic devices. Wide bandgap 2D semiconductors are characterized by their substantial energy bandgaps, typically exceeding 1 electron volt (eV). This wide energy gap sets them apart from narrow bandgap 2D semiconductors like graphene, which lack this intrinsic property. The bandgap, an energy range in which electrons are not allowed to exist, plays a pivotal role in controlling the electrical and optical properties of semiconductors. Enter the wide bandgap 2D materials—a class of 2D substances characterized by substantial energy bandgaps. These materials offer an elegant solution to the bandgap limitation, opening doors to a broader spectrum of applications in electronics and beyond. The significance of wide bandgap 2D materials lies in their unique ability to combine the advantages of 2D materials, such as atomically thin structures and high carrier mobility, with the control over electronic properties that bandgaps provide.

4.1.2 Applications of Wide Bandgap 2D Materials

Wide bandgap 2D materials are poised to revolutionize several technological domains. They find applications in high-frequency electronics, where their intrinsic bandgap facilitates the design of transistors with superior ON/OFF ratios, enabling faster-switching speeds. In optoelectronics, these materials exhibit exceptional photoluminescence (PL) and light–matter interactions, making them ideal candidates for light-emitting diodes (LEDs), photodetectors, and lasers. Furthermore, their energy bandgap enables their integration into power electronics, promising enhanced energy efficiency and reduced power consumption. Additionally, wide bandgap 2D materials hold great potential for quantum technologies, sensors, and advanced materials.

4.1.3 Importance and Significance

The wide bandgap in these materials is of paramount importance, as it endows them with unique electronic, photonic, and thermal properties. This characteristic allows for precise control over charge transport, enabling the development of high-performance electronic devices with excellent ON/OFF ratios and reduced leakage currents. Moreover, the wide bandgap facilitates efficient light emission and absorption, making these materials ideal candidates for optoelectronic applications, including LEDs, photodetectors, and lasers.

4.1.4 Properties of Wide Bandgap 2D Semiconductor Materials

Among the wide bandgap 2D materials, MoS_2, a 2D TMDC semiconductor, has gained substantial attention for its exceptional properties. It stands out as one of the most stable 2D TMDC materials, boasting high electron mobility and two distinct crystal phases: the 2H phase (**Figure 4.1a**), known for its stability and n-type semiconductor behavior, and the octahedral 1T phase (**Figure 4.1b**) with metallic properties. MoS_2 possesses a non-centro-symmetric structure and exhibits a direct energy bandgap of 1.8 eV in its monolayer form, while the indirect bandgap increases to 1.2 eV in multilayer MoS_2. Its crystalline structure features a trigonal prismatic configuration of molybdenum (Mo) atoms within hexagonal planes of sulfur (S) atoms. The Raman spectra of MoS_2 flakes (**Figure 4.1c**) reveal in-plane and out-of-plane vibrational modes, with distinctive shifts in wavenumbers as the number of layers varies. These attributes, including the intriguing vibrational characteristics,

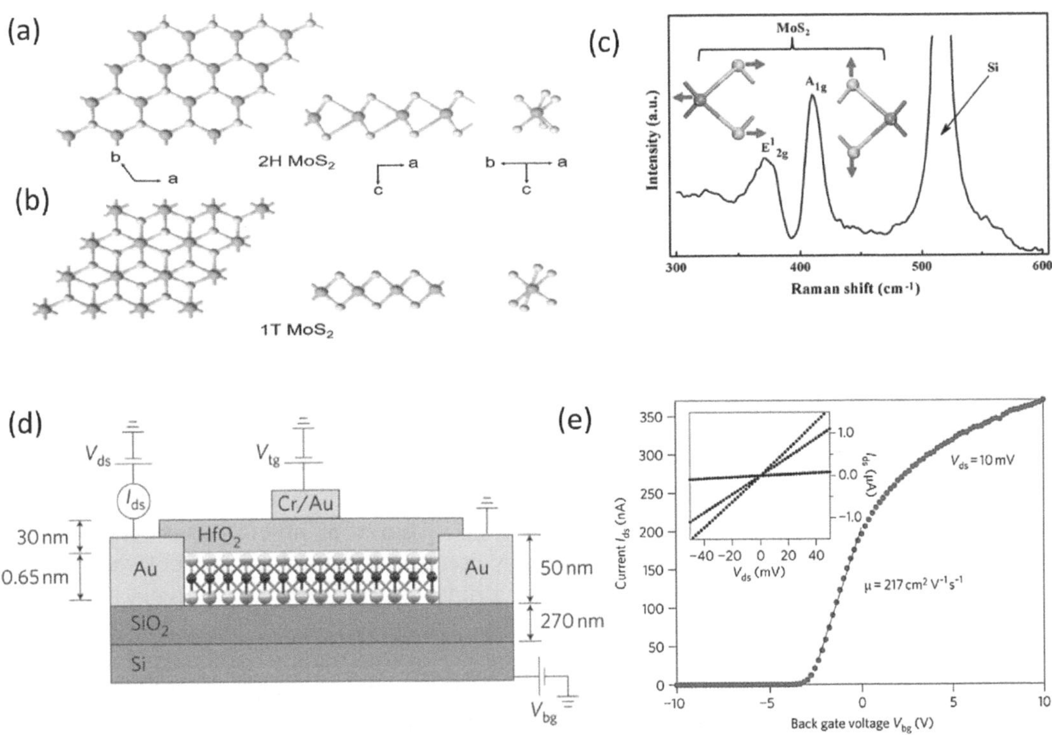

FIGURE 4.1
(a) and (b) Schematic illustrations depict the crystal structures of MoS_2 in the 2H and 1T phases, where Mo atoms are cyan and S atoms are yellow [14]. Copyright 2020, *Inorganic Chemistry Communications*. (c) The Raman peaks of MoS_2 show the in-plane and out-of-plane modes of vibrations, over the Si/SiO_2 substrate [15]. Copyright 2015, *Journal of Alloys and Compounds*. (d) A schematic image portrays the HfO_2-covered MoS_2 device featuring gold metal electrodes. (e) The gate-dependent electrical characterization of the monolayer MoS_2 with I_{ds}–V_{ds} curve in the inset figure. Adapted with permission [16]. Copyright 2011, Nature Publishing Group.

underscore MoS_2's significance in various applications and research endeavors. The gate-assisted field-effect transistor (FET) based on MoS_2 and its electrical characterization are presented in **Figure 4.1d and e.**

4.1.5 Optoelectronic Applications

2D semiconducting materials have become the keystones of innovation in the field of materials science, opening incredible possibilities that close the gap between theoretical promise and practical reality. Due to graphene's exceptional performance in several domains, these materials have earned a lot of interest in the scientific community [17]. While graphene clearly leads the pack in terms of 2D materials, a pantheon of alternate 2D semiconducting materials such as TMDCs, hexagonal boron nitride (h-BN) [18], and BP [18–20] have also emerged, broadening the scope of scientific inquiry. These 2D crystals have great potential for groundbreaking applications in the fields of electronics and optoelectronics [21]. This is because of their outstanding electron mobility [22], suitable bandgap [23], effective light absorption, and impressive photoresponsivity properties, all of which individually raise the possibility of new and significant technological improvements.

Traditional photodetectors work on the photoconductivity principle. However, there has recently been a rise in the development of photodetectors based on 2D materials, demonstrating amazing advances in photodetection capabilities [24]. Graphene is a promising 2D material which has enormous potential for photodetectors since it absorbs a wide spectrum of light wavelengths, from ultraviolet to terahertz. However, using pure graphene to make photodetectors has flaws such as short-lived photoexcited carriers and a low absorption efficiency of incoming light. Furthermore, employing pristine graphene with no bandgap would result in a strong dark current, which would make it unsuitable for sensitive photodetectors [25]. Aside from graphene, semiconducting TMDCs such as WS_2, WSe_2, MoS_2, and $MoTe_2$. demonstrated outstanding optical and electrical properties. These materials are regarded as building blocks for highly sensitive photodetectors since they have a range of bandgap values varying from less than 1 eV to well above 2.5 eV [26, 27]. However, they are not well-suited for detecting infrared light due to their very high bandgaps.

BP with a direct bandgap of ≈1.5 eV for the monolayer form and 0.3 eV for bulk form, like other 2D materials, in the field of optoelectronics is a strong contender for optoelectronic applications which cover near and mid-infrared band [28]. But photodetectors based on BP have a weak response to visible light due to bandgap. Therefore, an effort has been made by researchers to achieve a visible light response by creating a heterojunction between BP and MoS_2 [29]. However, there is a dilemma. Only monolayer and bilayer MoS_2 (very thin layers of molybdenum disulfide) have a direct bandgap. This property is critical for the efficient interaction of light and matter, since it makes it easier for the material to absorb and respond to light. However, creating a heterojunction between BP and mono- or bilayer MoS_2 that consistently results in high photo response efficiency is challenging [30]. Therefore, it is critical that we focus our attention on a material that excels in this domain: selenium indium.

2D indium selenide (InSe), due to its high carrier mobility, small effective electron mass, and broad optical absorption, has recently attracted interest in nanoelectronics and optoelectronic applications. Photodetectors based on InSe have exceptional performance, with a broad photoresponse (400–1,000 nm), strong photoresponsivity (up to 105 A W1 at 633 nm), and fast response time (down to 2 s). These distinguishing characteristics imply that InSe is a good contender for broadband photodetectors with fast response.

Researchers recently demonstrated dominating lasers based on single-layer and few-layer TMDCs using various approaches and configurations. For instance, utilizing a photonic-crystal nanocavity, ultra-low-threshold and continuous-wave lasing was achieved in single-layer WSe_2 at temperatures below 160 K [31]. Moreover, at 10 K, lasing in single-layer WS_2 was achieved with a microdisk resonator and femtosecond pulsed laser stimulation [18]. Also, using a linked microdisk-microsphere cavity, multiple continuous-wave lasing peaks were recorded from few-layer MoS_2 at ambient temperatures [32]. Researchers proposed utilizing 2D semiconductors as the active medium for vertical-cavity surface-emitting lasers based on the intense and stable PL of single-layer WS_2 [33]. Furthermore, low-threshold lasing of few-layer $MoTe_2$ on a silicon photonic-crystal nanocavity at an emission wavelength of 1305 nm has been observed, making it acceptable for traditional fiber-optic communication [34]. These advancements in 2D semiconductor lasers hold great promise for developing a variety of technological applications.

In a recent study, authors fabricated a p–n junction based on wide bandgap n-type WS_2 and p-type BP. The schematic diagram and the optical images are presented in **Figure 4.2a** and **b**. In this study, the BP/WS_2 p–n diode is subjected to illumination with a light of λ = 600 nm wavelength and 1.2 µW irradiation power. At Vg = 0, the device displays noteworthy characteristics, including a short-circuit current (I_{sc}) of 0.6 µA and an open-circuit voltage (V_{oc}) of 0.35 V at room temperature, as illustrated in **Figure 4.2c**. Remarkably, the

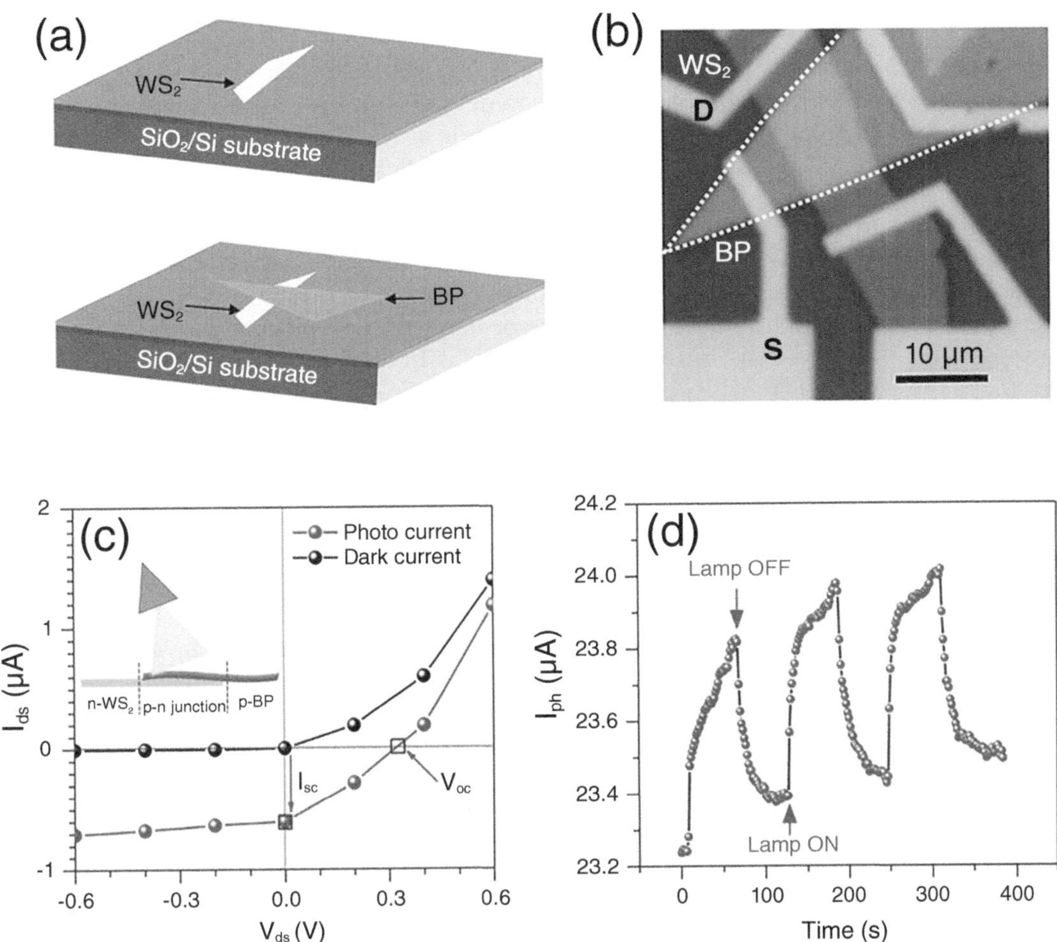

FIGURE 4.2
(a) and (b) The schematic and optical image of the van der Waals heterostructure based on n-type WS_2 and p-type black phosphorus. (c) The photovoltaic behavior of the BP/WS_2 is illustrated under a light of $\lambda = 600$ nm and 1.2 μW irradiation power. (d) The device response is recorded as a function of time with the sequential ON and OFF state of the light source.

photoresponsivity of the BP/WS_2 van der Waals p–n diode reaches 500 mA/W, surpassing prior heterojunctions of BP with other TMDs by a factor of 100. The external quantum efficiency (EQE) is determined for this BP/WS_2 heterojunction, which reaches the highest EQE (%) value of 103% among all reported BP heterojunctions, indicative of its exceptional photodetection capabilities. The observed increase in reverse- and forward-bias currents (**Figure 4.2c**) is attributed to the generation of electron–hole pairs upon light illumination, with charge separation occurring at the interface. This leads to the formation of dipoles within the junction area, further enhancing electron–hole pair generation. The sequential light ON and OFF experiments (**Figure 4.2d**) reveal photocurrent switching behavior, suggesting potential applications as a photo-switching device. Overall, the remarkable photoresponsivity and high EQE of our BP/WS_2 heterojunctions highlight their substantial promise for both electronic and optoelectronic device applications.

4.1.6 2D Wide Bandgap Semiconductors for Biosensing and Switching Applications

Elemental 2D materials' distinct and remarkable properties are poised to transform advances in electronics and sensing for silicon devices. Many of these elemental 2D materials have beneficial properties such as high electron and hole mobilities, outstanding ON/ OFF ratios, topological insulating behavior, material flexibility, and increased sensitivity to adsorbed particles, and they have the potential to improve on present state-of-the-art systems. To date, active FETs have been synthesized using elemental 2D materials such as $MoTe_2$, $MoSe_2$, SnS_2, and WSe_2 [4,10,35–37]. These 2D FET-based devices have a wide range of applications, including usage as biosensors. For example, in a study conducted in 2023, researchers developed an FET-based biosensor using $MoSe_2$ material for the detection of streptavidin as depicted in **Figure 4.3(a)** and **(b)** [35]. In this research, the biosensor, which employs the $MoSe_2$ technology, is intended to detect the streptavidin protein as the target analyte. This is accomplished while maintaining steady biasing and gate voltage. A specific receptor molecule is used to preserve the sensitivity and appropriate alignment of the device during the functionalization of $MoSe_2$-based FETs. Using stacking, this receptor stably attaches to the $MoSe_2$ surface, allowing for the detection of the specified analyte without interference.

The pyrene–lysine–biotin (PLB) supporting molecule is added by solid-phase peptide synthesis to demonstrate the practical implementation of the $MoSe_2$-based FET biosensor. Surprisingly, this newly designed structure can precisely interact with streptavidin at picomolar levels in less than a minute. The results demonstrate that this approach enables precise and rapid identification of the target. The $MoSe_2$-based FET is functionalized with the supporter molecules of PLB and then the streptavidin is captured due to non-covalent interaction as represented by transfer and output curves presented in **Figure 4.3© and (d)**. The device response against the PLB and target analyte is recorded in the form of an electric signal (I_{ds}).

Moreover, recently the van der Waals stacking of the wide bandgap 2D semiconductor materials has been introduced to fabricate a bipolar junction transistor (BJT) composed of n-type $MoTe_2$ and p-type GeSe. In this study, the authors explore the performance of a van der Waals (vdW) BJT device in a common-base configuration, with the base grounded. This configuration allows for a thorough investigation of the device's characteristics and behavior. The study utilizes a measurement setup depicted in **Figure 4.4a** to conduct a series of experiments.

In **Figure 4.4b**, the input characteristics are presented, showing how the emitter current (I_e) varies with the base-emitter voltage (V_{be}) at different collector-base voltages (V_{cb}). Notably, as both V_{be} and V_{cb} increase, I_e exhibits a gradual rise. This behavior is attributed to the reduction in the barrier height between the base and emitter junction due to the electric field, leading to increased charge carrier diffusion from the emitter to the base region.

Moving on to the output characteristics, **Figure 4.4c** illustrates the relationship between collector current (I_c) and V_{cb} at a fixed V_{be}, demonstrating the significant influence of V_{be} on both the collector–base junction and I_c. As V_{be} increases, a substantial number of carriers are driven from the emitter–base junction toward the collector, resulting in exponential growth of both I_c and I_e. The chapter introduces the concept of current gain (α) in the common-base configuration, representing the ratio between I_c and I_e. **Figure 4.4d** shows the extraction of α as a function of V_{be} at various V_{cb} values. Interestingly, α decreases with increasing V_{be} but increases with higher V_{cb} values. Notably, the highest α value of 0.95 is achieved at V_{be} = 0.5 V and V_{cb} = 2 V, a remarkable improvement compared to previously reported vdW 2D BJT devices. This superior performance is attributed to the clean, residue-free interface and high carrier density of the emitter.

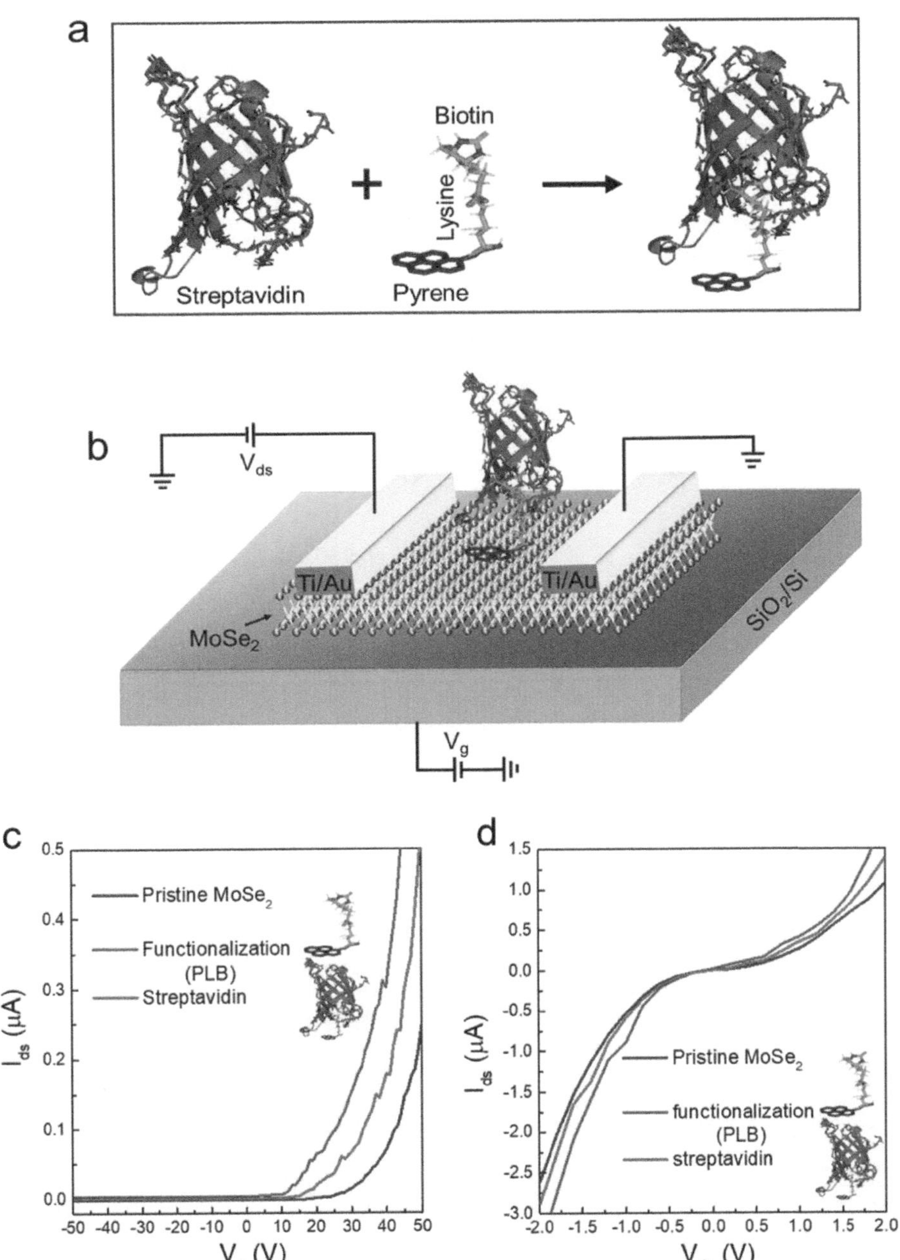

FIGURE 4.3

(a) Schematic representation of streptavidin and PLB supporter constructs, showing non-covalent conjugation. (b) A MoSe$_2$ field-effect transistor (FET) loaded with a PLB supporter molecule for streptavidin detection. (c) MoSe$_2$ FET transfer characteristics: Comparison before and after drop-casting PLB molecules (green) and streptavidin (red). (d) Current–voltage (I$_{ds}$–V$_{ds}$) characeristics of the pristine MoSe$_2$ FET, MoSe$_2$ FET with introduced PLB construct (Green), and streptavidin (red). Adapted with permission [38], Copyright 2023, Elsevier.

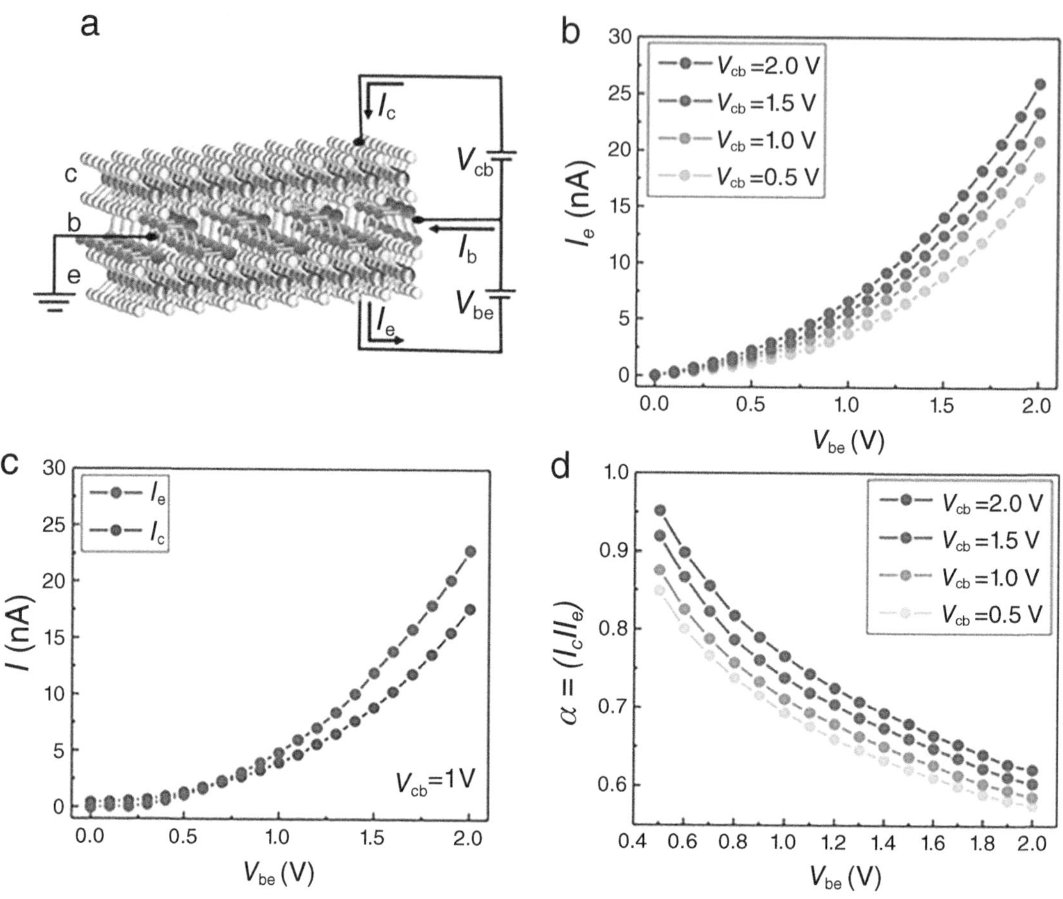

FIGURE 4.4
(a) Measurement setup of the vdW BJT device in a common-base configuration, showing the forward-biased emitter-base region and reverse-biased collector-base region. (b) Input characteristics depicting the relationship between emitter current (I_e) and base-emitter voltage (V_{be}) at various collector-base voltages (V_{cb}). (c) Output characteristics illustrating the correlation between collector current (I_c) and collector-base voltage (V_{cb}) at a fixed base-emitter voltage (V_{be}). (d) Current gain (α) extraction as a function of V_{be} at different V_{cb} values, showcasing the remarkable α value of 0.95 achieved at specific V_{be} and V_{cb} conditions due to the high-quality emitter interface. Adapted with permission [39], Copyright 2023, John Wiley & Sons.

In summary, the chapter provides a comprehensive examination of the vdW BJT device in a common-base configuration, offering valuable insights into its input and output characteristics. The observed behaviors and performance metrics, particularly the impressive α value, underscore the promising potential of this device for various electronic applications.

4.1.7 Energy and Power Electronics

Owing to their remarkable properties like higher breakdown voltage, less leakage current, higher carrier mobility, better thermal stability, and the ability to operate at higher frequencies, various wide bandgap 2D materials like TMDCs (WSe_2, MoS_2, etc.), phosphorene (single-layer BP), GaN, and AlN [40] have gathered tremendous attention in the field of electronics and semiconductors. TMDs, for instance, are employed in a variety

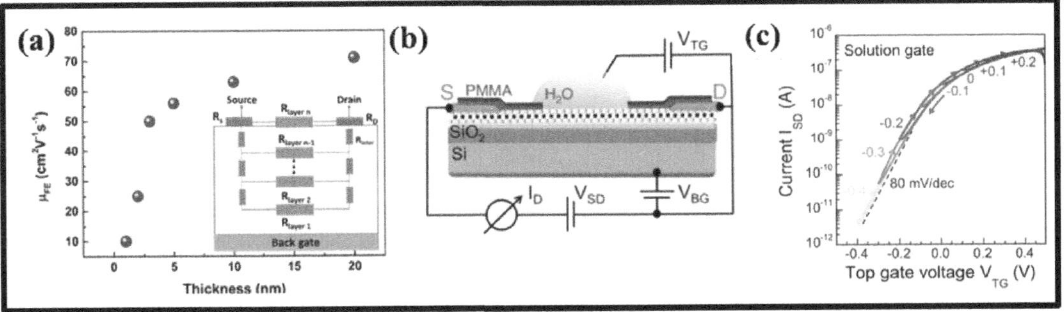

FIGURE 4.5

(a) Thickness dependence of the field-effect mobility calculated from the transfer curve. The inset shows the resistor network model of InSe FETs. Reproduced with permission [43] (b) Schematic diagram of the geometry of SnS$_2$ FET devices with SiO$_2$/Si back gate and H$_2$O solution top gate. (c) Transfer characteristics (I$_{SD}$ vs V$_{TG}$) of a deionized (DI) water top-gated SnS$_2$ device, measured with dual top gate voltage sweeps showing minimal hysteresis.

Reproduced with permission [44].

of optoelectronic devices, such as solar cells, photodetectors, LEDs, and phototransistors, since their tunable bandgap is accompanied by a strong PL and a substantial exciton binding energy [26]. For instance, unique properties of MoS$_2$ include direct bandgap ($\sim$1.8 eV), good mobility ($\sim$700 cm^{2}V^{-1}s^{-1}), high current ON/OFF ratio of $\sim$10^7–10^8, and large optical absorption ($\sim$10^7 m^{-1} in the visible region) [41].

Due to their dangling-bond-free surface and potential for further device downsizing into the sub-nanometer range, 2D semiconductors are seen to be particularly attractive candidates for channel materials in high-efficiency power electronics. While traditional 3D semiconductors like Si or Ge suffer from significant carrier transport scattering effects due to surface flaws, 2D materials, such as Si or Ge, are appropriate for electronic applications that aim to extend Moore's law because of their atomically smooth surface. Materials like few-layer InSe, which are atomically thin, allow for accurate electrostatic gating control [42]. The mobility varies as the thickness is changed as shown in **Figure 4.5a**. Few-layer InSe exhibited extraordinary mobility, reaching 10^4 cm^2 V^{-1} s^{-1} and having an energy gap (E$_g$) of 1.2 eV. Several variables, including a light electron effective mass and weak electron-phonon scattering, contribute to this amazing mobility. Additionally, monolayer InSe exhibits exceptional electrical characteristics and possesses an energy gap of 2.2 eV. In a work by Yang et al., InSe films with thicknesses ranging from 1 to 70 nm were made using pulsed laser deposition, and the layer-dependent electronic characteristics were assessed. SiO$_2$ was used as the dielectric material in bottom-gate monolayer InSe FET devices to achieve a mobility of 10 cm^2 V^{-1} s^{-1}. It's important to notice that increasing mobility by 50 cm^2 V^{-1} s^{-1} by decreasing the thickness to 5 nm [43]. For instance, utilizing a liquid top gate with deionized (DI) water ($\varepsilon\sim$80ε_0) as the top dielectric, Huang et al. created ultrathin SnS$_2$ FET devices (schematic and transfer curves are presented in **Figure 4.6(b) and (c)**) with exceptional mobility, reaching 230 cm^2 V^{-1} s^{-1}. The nearly perfect sub-threshold swing displayed by these devices' transfer curves, which measured roughly 80 mV/decade, suggested the potential for exceeding a meaningful switching ratio of 10^6. Notably, bottom-gate ultrathin SnS$_2$ FET devices maintained an average mobility of 5 cm^2 V^{-1} s^{-1} [44].

2D semiconductors provide a number of benefits over traditional semiconductors as superior luminescent materials. The efficiency of LED devices would be impacted by the competing radiative and non-radiative processes that the injected electrons and holes would undergo. Due to ambient dielectric screening, 2D semiconductors exhibit higher exciton binding energies than their 3D bulk equivalents. This is because stronger electron–hole interactions result, suppressing non-radiative recombination. The layered structure can be easily incorporated into well-liked device structures and has excellent compatibility with modern semiconductor fabrication techniques [45]. A potential material for light-emitting devices that can generate light in the orange-to-violet spectrum is 2D perovskite. Its remarkable PL quantum yield and ability to adjust its broad bandgap are the main causes of this. Phenylbutylammonium bromide (PBABr) and $CsPbX_3$ (X = Cl, Br) were used to create a quasi-2D perovskite film in a work conducted under the direction of Wang et al. When they used an optical ratio of 72% PBABr, they were able to attain an astounding maximum PL quantum yield of 35%. The passage of charge carriers was shown to be hampered by a larger concentration of long-chain organic groups. The emission color might be controlled by altering the PBABr ratio by altering the number of layers. For instance, when 30% Cl was added, the exciton peak changed from 496 to 472 nm, moving toward the blue. This demonstrates that changing the number of layers does, in fact, change the color of the emission. They made it simple to adjust the peak location of the emitting light by carefully regulating the Cl:Br ratio and perfecting the device architecture. This led to the electroluminescence displayed by 2D perovskite-based LEDs revealing certain emission wavelengths: 490 nm with 30% Cl, 481 nm with 40% Cl, and 473 nm with 50% Cl. The changes in the material's number of layers can be blamed for the fluctuation in emission color. For instance, when the Cl concentration reached 30%, the exciton peak moved toward the blue, moving from 496 to 472 nm. This indicates that it is possible to precisely regulate the peak location of the produced light by adjusting the Cl:Br ratio and improving the device architecture. In conclusion, electroluminescence was seen in 2D perovskite-based LEDs at 490 nm with 30% Cl, 481 nm with 40% Cl, and 473 nm with 50% Cl [46].

2D TMDCs are gaining a lot of attention in the field of energy conversion and storage as electrode materials for both supercapacitors and Li-ion batteries. Their atomically layered architectures, huge surface area, and remarkable electrochemical characteristics are likely the cause of this increased attention. These materials' multilayer construction provides more ion storage sites while retaining structural stability throughout charge and discharge cycles, improving device performance. They are a particularly good option for energy storage electrodes due to the combination of their large surface area, surface functionality, and electrical conductivity [47]. Because of its stacked-sheet-like structure and capacity to display significant electrical double-layer capacitance due to numerous oxidation states of molybdenum, which range from +2 to +6, MoS_2 stands out in particular. Tour et al. employed a novel strategy by fabricating edge-oriented and vertically aligned MoS_2 nanosheets. By offering areas for electrolyte ions to engage with reactive dangling bonds, this design created more van der Waals gaps, which eventually boosted capacitance. These vertically aligned, flexible MoS_2 electrodes with a sponge-like appearance showed an impressive areal capacitance of up to 12.5 mF cm^{-2} [48]. Graphite anodes and lithium cobalt oxide ($LiCoO_2$) cathodes are generally used in conventional commercial-grade Li-ion batteries, which have restricted theoretical specific capacities of 372 mAh g^{-1} and observed capacity of 150 mAh g^{-1}. These capacities are insufficient to satisfy the present energy storage requirements. Due to their distinctive layered architectures, 2D materials like MoS_2 and WeS_2 have become attractive options for Li-ion battery anode materials. Li^+ ions can intercalate into the structure without significantly changing its volume, preventing the disintegration of active components during the charging and discharging processes. MoS_2 has a high theoretical capacity of approximately 670 mAh g^{-1} [49].

4.2 Emerging Applications

Nowadays scientists are working on employing wide bandgap 2D semiconductors in various emerging applications like quantum technology, photodetectors, gas sensors, and piezo electronics. The sub-nanometer dimension in 2D semiconductors gives rise to various charming quantum effects that are unseen in bulk counterparts, such as enhanced excitonic effects, quantum Hall effect, and valleytronics, which is beneficial for progress in quantum technologies. In one study, Weng et al. prepared a $ZnIn_2S_4$ 2D semiconductor with a direct bandgap of about 2 eV, a high ON/OFF ratio of (>10^4), together with an ultralow dark current (less than 10^{-9} A) at room temperature. ZnIn2S4 acts as a promising material for research into novel quantum phenomena and applications in the semiconductor sector due to its outstanding air stability and good optoelectronic performance [50].

In a different research, Zhang and colleagues used first-principles calculations to compare the characteristics of completely hydrogenated and semi-hydrogenated AlN nanosheets. According to the investigation, the thinner bilayer AlN nanotube is a ferromagnetic metal, whereas the completely hydrogenated single bilayer AlN nanotube is an indirect bandgap semiconductor. These results suggested that the properties of 2D AlN (both electronic and magnetic) can be controlled through chemical functionalization and thus they can be employed in electronics and spintronics.

4.3 Challenges and Future Prospects

As we explore the exciting realm of wide bandgap 2D semiconductor materials and their applications, it is crucial to acknowledge the challenges that accompany their utilization and to contemplate the promising future prospects that lie ahead.

4.3.1 Fabrication and Scalability Challenges

One of the foremost challenges in the integration of wide bandgap 2D materials into practical devices is the fabrication process. The precise control required to produce defect-free, large-scale 2D material structures can be technically demanding. Achieving uniformity across large areas while maintaining the intrinsic properties of these materials remains a formidable task. Researchers are actively working on scalable fabrication methods to overcome this challenge, with promising results in the development of growth techniques and transfer processes.

4.3.2 Environmental Sensitivity

Wide bandgap 2D materials, like their narrow bandgap counterparts, are susceptible to environmental factors such as humidity and oxygen exposure. These materials can undergo degradation or oxidation when exposed to ambient conditions, which can compromise their performance and stability. Encapsulation and passivation techniques are being explored to mitigate these issues, ensuring that the materials retain their properties under real-world conditions.

4.3.3 Heterostructure Integration

Wide bandgap 2D materials often find applications in heterostructures, where precise alignment with other materials is critical. Achieving atomically sharp interfaces between different materials poses a challenge but is essential for optimizing device performance. Researchers are devising innovative approaches to integrate wide bandgap 2D materials seamlessly with other materials, enabling novel device architectures with enhanced functionalities.

4.3.4 Defects and Quality Control

Defects and structural imperfections can significantly impact the electronic and optoelectronic properties of wide bandgap 2D materials. Identifying, characterizing, and minimizing defects are ongoing challenges in the field. Advanced characterization techniques and defect engineering methods are being developed to enhance the quality control of these materials.

4.3.5 Novel Applications and Multidisciplinary Collaboration

Future prospects for wide bandgap 2D materials are highly promising, with opportunities spanning various fields, including electronics, photonics, energy, and quantum technologies. Realizing these prospects requires multidisciplinary collaboration and innovative thinking. Researchers, engineers, and scientists from diverse backgrounds must work together to explore and exploit the full potential of these materials in emerging applications.

4.3.6 Sustainable Growth and Commercialization

For wide bandgap 2D materials to make a meaningful impact on technology, sustainable and cost-effective production methods must be established. Commercialization efforts are underway to bridge the gap between lab-scale research and scalable manufacturing. Sustainable sourcing of raw materials and eco-friendly synthesis techniques are integral components of this endeavor.

In summary, while wide bandgap 2D semiconductor materials hold immense promise, they also present significant challenges that require concerted efforts from the scientific community. Overcoming fabrication hurdles, ensuring environmental stability, and optimizing heterostructure integration are key steps in realizing the full potential of these materials. With ongoing research and innovation, we anticipate breakthroughs in addressing these challenges and unlocking the transformative possibilities that wide bandgap 2D materials offer. The future of electronics, optoelectronics, energy, and quantum technologies is poised to be shaped by these remarkable materials, and their journey holds great promise for technological advancement and innovation.

4.4 Conclusion

Wide bandgap 2D semiconductors have created new opportunities in a variety of fields of science and technology. We have explored the synthesis, characterization, and applications of wide bandgap 2D semiconductor materials throughout this book chapter in depth. The distinctive structural, optical, and electrical characteristics that underlie their significance have been covered in detail, with a focus on the wide bandgap and the significant effects it has on the optoelectronic behavior of these materials. Wide bandgap 2D semiconductors have a variety of applications, making them important participants in current research and innovation. In this regard, we've looked at their potential applications in a wide range of industries, from photovoltaics, where they promise to redefine the effectiveness and sustainability of energy conversion, to biosensors, where their incredibly sensitive properties enable advancements in medical diagnosis. Additionally, their use in electrical switching devices provides a preview of high-performance electronics with lower power consumption and improved functionality in the future.

Finally, we have endeavored to provide a comprehensive overview of the current state-of-the-art research on wide bandgap 2D semiconductors and their applications. We have illuminated the opportunities and challenges that lie ahead, emphasizing that our journey has only just begun in this exciting and transformative field. As we navigate this path together, we do so with a shared vision of unlocking the boundless potential of wide bandgap 2D semiconductors for the betterment of science, technology, and society.

Acknowledgment

This research was supported by the National Research Foundation of Korea funded by the Ministry of Science and ICT (2022R1G1A1009887).

References

[1] K. S. Novoselov, A. Mishchenko, A. Carvalho, A. H. Castro Neto, *Science* **2016**, 353, aac9439.
[2] G. Dastgeer, A. M. Afzal, G. Nazir, N. Sarwar, *Advanced Materials Interfaces* **2021**, 8, 2100705.
[3] B. Anasori, M. R. Lukatskaya, Y. Gogotsi, *Nature Reviews Materials* **2017**, 2, 16098.
[4] G. Dastgeer, S. Nisar, Z. M. Shahzad, A. Rasheed, D. k. Kim, S. H. A. Jaffery, L. Wang, M. Usman, J. Eom, *Advanced Science* **2023**, 10, 2204779.
[5] A. K. Geim, K. S. Novoselov, *Nature Materials* **2007**, 6, 183.
[6] G. Dastgeer, A. M. Afzal, J. Aziz, S. Hussain, S. H. Jaffery, D.-k. Kim, M. Imran, M. A. Assiri, *Materials* **2021**, 14.
[7] G. Dastgeer, A. M. Afzal, S. H. A. Jaffery, M. Imran, M. A. Assiri, S. Nisar, *Journal of Alloys and Compounds* **2022**, 919, 165815.
[8] G. Dastgeer, M. F. Khan, J. Cha, A. M. Afzal, K. H. Min, B. M. Ko, H. Liu, S. Hong, J. Eom, *ACS Applied Materials & Interfaces* **2019**, 11, 10959.
[9] G. Dastgeer, M. F. Khan, G. Nazir, A. M. Afzal, S. Aftab, B. A. Naqvi, J. Cha, K.-A. Min, Y. Jamil, J. Jung, S. Hong, J. Eom, *ACS Applied Materials & Interfaces* **2018**, 10, 13150.
[10] M. S. Zafar, G. Dastgeer, A. Kalam, A. G. Al-Sehemi, M. Imran, Y. H. Kim, H. Chae, *Nanomaterials* **2022**, 12, 1305.
[11] S. Nisar, B. Basha, G. Dastgeer, Z. M. Shahzad, H. Kim, I. Rabani, A. Rasheed, M. S. Al-Buriahi, A. Irfan, J. Eom, D.-k. Kim, *Advanced Science* **2023**, n/a, 2303654.
[12] S. Nisar, M. Shahzadi, Z. M Shahzad, D.-k. Kim, G. Dastgeer, A. Irfan, *ACS Applied Electronic Materials* **2023**, 5, 5714.
[13] G. Dastgeer, A. M. Afzal, G. Nazir, N. Sarwar, *Advanced Materials Interfaces* **2021**, n/a, 2100705.
[14] D. Gupta, V. Chauhan, R. Kumar, *Inorganic Chemistry Communications* **2020**, 121, 108200.

[15] Y. Liu, L. Hao, W. Gao, Q. Xue, W. Guo, Z. Wu, Y. Lin, H. Zeng, J. Zhu, W. Zhang, *Journal of Alloys and Compounds* **2015**, 631, 105.

[16] B. Radisavljevic, A. Radenovic, J. Brivio, V. Giacometti, A. Kis, *Nature Nanotechnology* **2011**, 6, 147.

[17] V. Georgakilas, J. N. Tiwari, K. C. Kemp, J. A. Perman, A. B. Bourlinos, K. S. Kim, R. Zboril, *Chemical Reviews* **2016**, 116, 5464.

[18] Y. Ye, Z. J. Wong, X. Lu, X. Ni, H. Zhu, X. Chen, Y. Wang, X. Zhang, *Nature Photonics* **2015**, 9, 733.

[19] M. Zhang, Q. Wu, F. Zhang, L. Chen, X. Jin, Y. Hu, Z. Zheng, H. Zhang, *Advanced Optical Materials* **2019**, 7, 1800224.

[20] J. Pang, A. Bachmatiuk, Y. Yin, B. Trzebicka, L. Zhao, L. Fu, R. G. Mendes, T. Gemming, Z. Liu, M. H. Rummeli, *Advanced Energy Materials* **2018**, 8, 1702093.

[21] S. Lu, L. Miao, Z. Guo, X. Qi, C. Zhao, H. Zhang, S. Wen, D. Tang, D. Fan, *Optics Express* **2015**, 23, 11183.

[22] S. Bai, C. Sun, H. Yan, X. Sun, H. Zhang, L. Luo, X. Lei, P. Wan, X. Chen, *Small* **2015**, 11, 5807.

[23] X. Peng, Q. Wei, A. Copple, *Physical Review B* **2014**, 90, 085402.

[24] S. H. Jo, D. H. Kang, J. Shim, J. Jeon, M. H. Jeon, G. Yoo, J. Kim, J. Lee, G. Y. Yeom, S. Lee, *Advanced Materials* **2016**, 28, 4824.

[25] L. Jiao, X. Wang, G. Diankov, H. Wang, H. Dai, *Nature Nanotechnology* **2010**, 5, 321.

[26] W. Choi, N. Choudhary, G. H. Han, J. Park, D. Akinwande, Y. H. Lee, *Materials Today* **2017**, 20, 116.

[27] C. Mai, A. Barrette, Y. Yu, Y. G. Semenov, K. W. Kim, L. Cao, K. Gundogdu, *Nano Letters* **2014**, 14, 202.

[28] J. Ma, S. Lu, Z. Guo, X. Xu, H. Zhang, D. Tang, D. Fan, *Optics Express* **2015**, 23, 22643.

[29] T. Hong, B. Chamlagain, T. Wang, H.-J. Chuang, Z. Zhou, Y.-Q. Xu, *Nanoscale* **2015**, 7, 18537.

[30] R. Cao, H. D. Wang, Z. N. Guo, D. K. Sang, L. Y. Zhang, Q. L. Xiao, Y. P. Zhang, D. Y. Fan, J. Q. Li, H. Zhang, *Advanced Optical Materials* **2019**, 7, 1900020.

[31] S. Wu, S. Buckley, J. R. Schaibley, L. Feng, J. Yan, D. G. Mandrus, F. Hatami, W. Yao, J. Vuckovic, A. Majumdar, *arXiv preprint arXiv:1502.01973* **2015**.

[32] J. Shang, C. Cong, L. Wu, W. Huang, T. Yu, *Small Methods* **2018**, 2, 1800019.

[33] O. Salehzadeh, M. Djavid, N. H. Tran, I. Shih, Z. Mi, *Nano Letters* **2015**, 15, 5302.

[34] C. Chang, W. Chen, Y. Chen, Y. Chen, Y. Chen, F. Ding, C. Fan, H. J. Fan, Z. Fan, C. Gong, *Acta Phys. Chim. Sin* **2021**, 37, 2108017.

[35] S. Nisar, G. Dastgeer, M. Shahzadi, Z. M. Shahzad, E. Elahi, A. Irfan, J. Eom, H. Kim, D.-k. Kim, *Materials Today Nano* **2023**, 100405.

[36] M. Shahzadi, S. Nisar, D.-K. Kim, N. Sarwar, A. Rasheed, W. Ahmad, A. M. Afzal, M. Imran, M. A. Assiri, Z. M. Shahzad, *Chemosensors* **2023**, 11, 83.

[37] E. Elahi, G. Dastgeer, G. Nazir, S. Nisar, M. Bashir, H. A. Qureshi, D.-k. Kim, J. Aziz, M. Aslam, K. Hussain, *Computational Materials Science* **2022**, 213, 111670.

[38] S. Nisar, G. Dastgeer, M. Shahzadi, Z. M. Shahzad, E. Elahi, A. Irfan, J. Eom, H. Kim, D.-k. Kim, *Materials Today Nano* **2023**, 24, 100405.

[39] G. Dastgeer, Z. M. Shahzad, H. Chae, Y. H. Kim, B. M. Ko, J. Eom, *Advanced Functional Materials* **2022**, 32, 2204781.

[40] Z. Wang, G. Wang, X. Liu, S. Wang, T. Wang, S. Zhang, J. Yu, G. Zhao, L. Zhang, *Journal of Materials Chemistry C* **2021**, 9, 17201.

[41] M. S. Fuhrer, J. Hone, *Nature Nanotechnology* **2013**, 8, 146.

[42] M.-Y. Li, S.-K. Su, H.-S. P. Wong, L.-J. Li, *Nature* **2019**, 567, 169.

[43] Z. Yang, W. Jie, C.-H. Mak, S. Lin, H. Lin, X. Yang, F. Yan, S. P. Lau, J. Hao, *ACS Nano* **2017**, 11, 4225.

[44] Y. Huang, E. Sutter, J. T. Sadowski, M. Cotlet, O. L. Monti, D. A. Racke, M. R. Neupane, D. Wickramaratne, R. K. Lake, B. A. Parkinson, *ACS Nano* **2014**, 8, 10743.

[45] Y. Lu, J. H. Warner, *ACS Applied Electronic Materials* **2020**, 2, 1777.

[46] K.-H. Wang, Y. Peng, J. Ge, S. Jiang, B.-S. Zhu, J. Yao, Y.-C. Yin, J.-N. Yang, Q. Zhang, H.-B. Yao, *ACS Photonics* **2018**, 6, 667.

[47] S. Ratha, C. S. Rout, *ACS Applied Materials & Interfaces* **2013**, 5, 11427.

[48] Y. Yang, H. Fei, G. Ruan, C. Xiang, J. M. Tour, *Advanced Materials* **2014**, 26, 8163.

[49] J. Zhou, J. Qin, X. Zhang, C. Shi, E. Liu, J. Li, N. Zhao, C. He, *ACS Nano* **2015**, 9, 3837.

[50] S. Weng, W. Zhen, Y. Li, X. Yan, H. Han, H. Huang, L. Pi, W. Zhu, H. Li, C. Zhang, *physica status solidi (RRL)–Rapid Research Letters* **2020**, 14, 2000085.

5

Interfacial Properties and Geometry of 2D Semiconducting Materials

Shiv Kumar Pal and Neeraj Mehta*
**Correspondence: dr_neeraj_mehta@yahoo.co.in*

5.1 Introduction

The physical and chemical features of the border or interface between two different substances or phases are referred to as interfacial properties. These interfaces can exist between a variety of substances, including solids, liquids, and gases, as well as between various phases of matter, such as the interface between a liquid and a gas [1]. Chemistry, physics, materials science, biology, and many other scientific and technical disciplines depend heavily on interfacial qualities. Due to their atomically thin nature and the existence of surfaces or interfaces that can dramatically affect their electronic, optical, and chemical characteristics, 2D semiconducting materials (SCMs) are a group of materials that display special interfacial features [2–5]. Surface states, surface roughness, substrate interaction, dielectric interface, heterostructures, chemical functionalization, quantum confinement, interlayer interactions, surface charge transfer, and edge effects are some important interfacial characteristics and factors for 2D semiconductors [6].

Figure 5.1 summarizes the important interfacial characteristics and related fields of 2D semiconductors. Surface states or surface traps, which can have an impact on charge carrier mobility, recombination dynamics, and electronic band structure, can be introduced into the surface of 2D semiconductors. To improve device performance, these surface states can be passivated or altered. The quality of the interface and surface roughness of 2D semiconductor materials can have a substantial impact on the electrical and optical properties. Reducing surface roughness is critical for boosting the performance of devices like field-effect transistors (FETs) and photodetectors. In FET devices, 2D semiconductors frequently need a dielectric layer as a gate insulator (e.g., silicon dioxide or hafnium oxide). The dielectric interface affects the vital FET operating parameters of gate capacitance, threshold voltage, and carrier transport [6, 7]. Typically, 2D semiconductors are grown on substrates, and the properties of the 2D-SCMs may be impacted by interactions between them and the substrates. The substrate-induced strain or lattice mismatch, for instance, can alter the electrical band structure.

Designer materials with specialized interfacial properties can be produced by stacking several 2D materials in heterostructures [7]. In photodetectors and light-emitting devices, heterostructures of 2D semiconductors can display distinctive electronic features, such as type-II band alignment. The electrical characteristics and reactivity of 2D-SCMs can be changed by functionalizing the surface with molecules or atoms. This can be utilized to

DOI: 10.1201/9781003439448-5

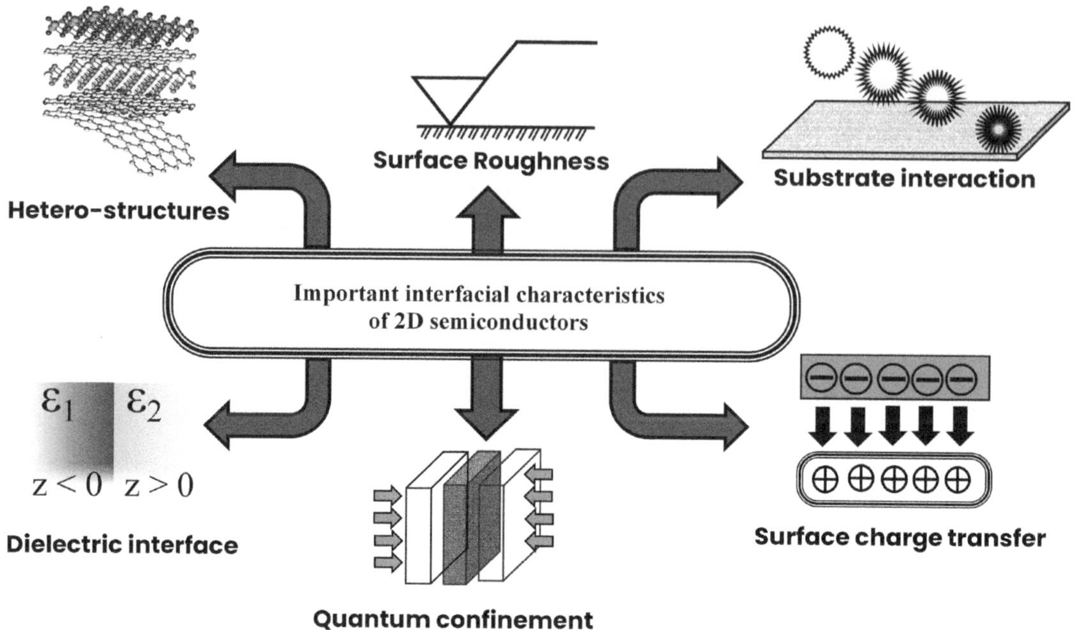

FIGURE 5.1
Illustration of significant interfacial properties of 2D semiconductors.

develop hybrid materials with new functionalities, for chemical sensing or passivation. Due to the material's extreme thinness, quantum confinement effects become relevant in 2D semiconductors [8]. Due to those consequences, 2D semiconductors have distinct energy levels and size-dependent electrical characteristics, which makes them desirable for quantum dots and other nanostructured devices. The interlayer contacts are critical in influencing the electrical characteristics and band alignment between the layers in van der Waals heterostructures (vdWHs) made of 2D-SCMs. To build new gadget features, these interactions can be changed.

The phenomenon of charges moving from a material's surface to its surroundings is known as surface charge transfer. When a 2D semiconducting material comes into contact with another material or is exposed to certain circumstances, such as temperature fluctuations or the presence of specific chemicals, this charge transfer may take place. Further, the charge transfer between 2D semiconductors and adsorbates or neighboring materials at the interface can result in doping or manipulation of the material's electronic characteristics, making it valuable for sensors and adjustable devices. The special characteristics and actions that take place at the edges or boundaries of 2D semiconducting materials are referred to as the "edge effect" on their surface. The edges of 2D semiconductors can possess different properties from the central regions; that is, these edge effects have significant effects on the material's electrical, optical, and chemical characteristics and are unique from the bulk properties of the substance. Edge states and edge reactivity can be harnessed for applications like nanoribbon-based transistors or catalysis.

The interfacial properties of different kinds of 2D semiconductors can lead to various consequences that significantly impact their electronic, optical, and chemical behavior. The interfacial properties of 2D-SCMs can alter their electrical band structures. For instance, band alignment and bandgap engineering can be accomplished when vdWHs

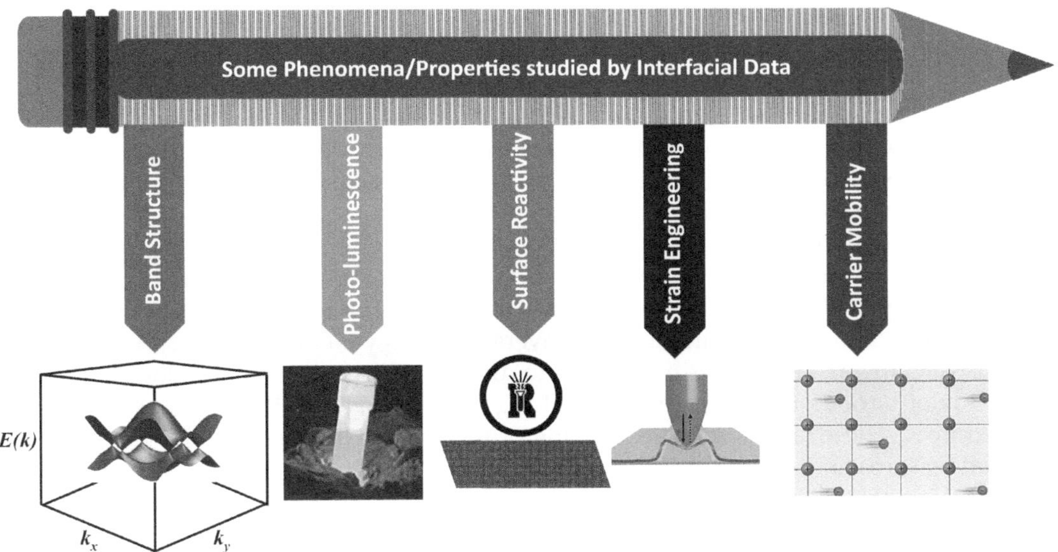

FIGURE 5.2
Flow chart showing some significant phenomena/properties investigated by using interfacial data as a tool.

are formed with other 2D materials or substrates. This makes it possible to design devices with unique electronic characteristics, such as tunable bandgaps and innovative electronic states. Photoluminescence qualities can be improved by specific interfacial states or imperfections. As a result, 2D-SCMs may emit light that is brighter and more effective, making them appropriate for lasers and other optoelectronic devices.

FETs and other high-speed electronic devices like them find 2D semiconductors appealing because properly designed interfaces can lower scattering and increase charge carrier mobility [7]. The interfacial features of 2D-SCMs can make them particularly sensitive to changes in their surroundings, such as the adsorption of gas molecules. This sensitivity is exploited in gas sensors and chemical sensors based on 2D materials. It is possible to alter the electronic and mechanical characteristics of 2D-SCMs by introducing interfacial strain, which is brought on by the interaction with substrates or nearby materials. This is useful for developing strain-engineered gadgets with specialized features. The interfacial characteristics of 2D semiconductors can affect how chemically reactive they are [9]. To adjust the surface chemistry for use in chemical sensing and catalysis, functional groups at the interface can be employed. Numerous properties, for instance, band structure, photoluminescence, carrier mobility, sensing, strain engineering, and surface reactivity, can be tailored by studying the interfacial properties in different types of 2D semiconductors as shown in the flow chart of Figure 5.2.

The two-dimensional atomic structure of 2D-SCMs, which differs significantly from the three-dimensional (3D) crystal lattice of conventional bulk semiconductors, is referred to as their geometry [10]. Understanding the geometry of 2D semiconductors is crucial because it has a fundamental contribution in establishing their electronic, optical, and chemical properties and applications [11]. Atomic monolayers, the type of lattice, the unit cell, layer stacking, stacking sequences, edge terminations, curvatures, defects, and grain boundaries are some of the essential features of the geometry of 2D semiconductors [12–14].

A single atomic layer of atoms or molecules organized in a 2D lattice often makes up a 2D-SCM. They have distinct qualities that differentiate them from their bulk counterparts, such as quantum confinement effects, thanks to their monolayer structure. The most prevalent lattice forms for the atomic arrangement in 2D semiconductors are hexagonal and rectangular. While transition metal dichalcogenides (TMDCs) like MoS_2 frequently have a trigonal prismatic lattice, graphene, for example, has a hexagonal lattice. The smallest repeating unit that captures the structure of a 2D semiconductor lattice is the unit cell. The unit cell normally comprises two carbon atoms in hexagonal lattices like graphene, while four silicon atoms are typically present in rectangular lattices like silicene.

The stacking configuration between these layers might affect the properties of some 2D semiconductors that can create multiple layers (e.g., bilayers and trilayers). For instance, various stacking arrangements might lead to distinct electronic band topologies. Multiple 2D layers of various materials are stacked together in vdWHs, thanks to weak van der Waals forces. These layers' relative stacking order can produce heterojunctions with specialized features and special electrical properties.

The edges of 2D semiconductors are fundamental in their geometry. Edge atoms or vacancies can introduce edge states or modify the local chemical reactivity. Edge sites can be functionalized or passivated to alter the material's characteristics. The 2D lattice can have flaws like vacancies, substitutions, or dislocations. Grain boundaries, where multiple orientations of the lattice meet, are also crucial in the geometry of polycrystalline 2D materials. These faults and boundaries can influence charge transport and device performance. The geometry of 2D-SCMs that are bent or folded undergoes changes that affect their electrical band structures and cause strain effects [15]. The characteristics of 2D semiconductors in nanoscale devices can be modified via curvature. Figure 5.3 illustrates the key elements that are related to the geometry of 2D-SCMs.

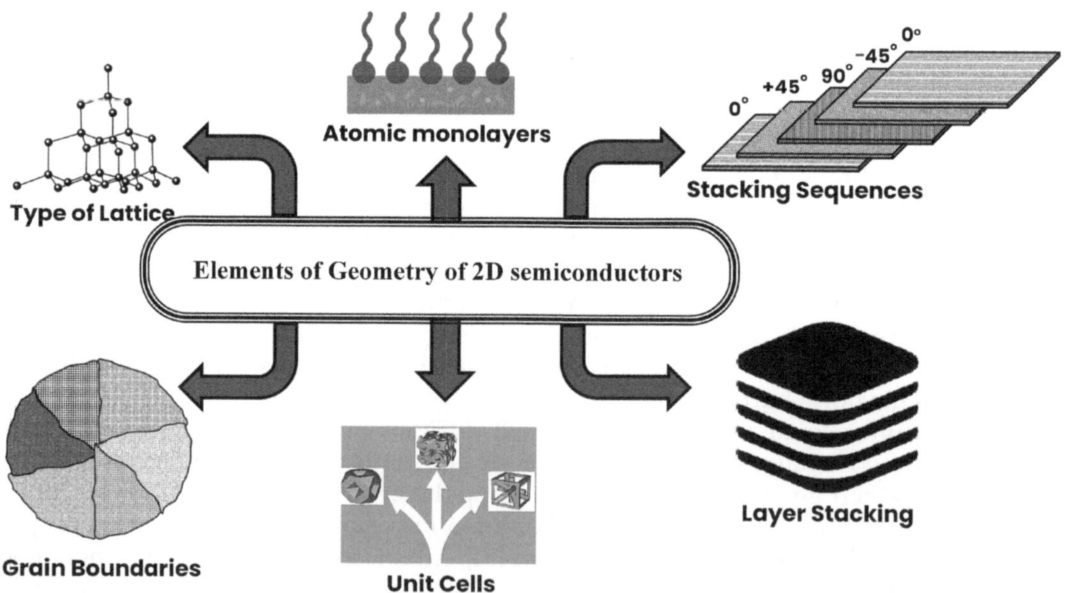

FIGURE 5.3
Some key elements of geometry of 2D semiconductors.

5.2 Consequences of Interfacial Properties and Geometry of Different Kinds of 2D Semiconductors

The characteristics of the interface between multilayer MoS_2 crystals that received UV/O_3 treatment and atomic layer-deposited (ALD) Al_2O_3 thin films were described by Park et al. in their study [16]. They observed that the surface energy enhanced monotonically by increasing the time of UV/O_3 treatment due to the development of either Mo–O bonds (at high UV/O_3 power) or S–O bonds (at low UV/O_3 power) on the external layer of MoS_2. The capacitance–voltage measurement reveals the presence of an electron density of order 10^{17} cm^{-3} and the lowest interface trap density of order 10^{11} cm^{-2} eV^{-1} in n-type MoS_2 [16]. It is well known that the stacking of 2D-SCMs is so-called vdWHs. The growth of multilayered GaSe on graphene/SiC heterostructure has been demonstrated by Aziza et al. via the vdW epitaxy method. They investigated GaSe to bring a reduction in the number density of the electrons in the underlayer of graphene [17]. The interfacial properties of HfN_2 monolayer and $MoTe_2$ have been studied by Mohanta et al. [18]. They have reported their prospective applications in several fields like photovoltaics, thermo-electricity, piezo, and digital/analog electronics. They developed an excitonic solar cell in $HfN_2/MoTe_2$ via a type-II vdWH with a monolayer of 1H-$MoTe_2$ [18].

The charge transport characteristics of FETs made of 2D-SCMs of TMDCs were reviewed by Li et al [19]. They studied several properties like interfacial impurities, the role of thickness on its carrier mobility, and electronic performance [19]. Peng et al. reported the photocatalytic performance of GaS/arsenic and GaSe/arsenic vdWHs to generate hydrogen by water splitting [20]. The study of different interfacing 2D materials and their utilization in several fields such as alkali ion batteries, energy storage, and biosensing applications has been performed by Ashraf et al. [21]. Due to less contact resistance and high gate tunability of the semiconducting 2H phase (as a channel) and the metallic 1T or semi-metallic 1T′ phase (as an electrode) in 2D TMDCs, FETs have gained increasing visibility from the scientific communities as suggested by Liu et al. [22]. They investigated the gate-tunable interfacial properties with the aid of "ab initio" simulations of quantum transport [22]. Monolayer (ML) $MoSe_2$'s interfacial characteristics were investigated by Pan et al. They utilized numerous metallic interfaces (e.g., palladium, silver, nickel, aluminum, chromium, and titanium) in their simulations to calculate their Schottky barrier height (SBH). In aluminum, silver, and palladium metallic interfaces with ML $MoSe_2$, weak or medium adsorption was found. However, they observed strong adsorption in titanium, nickel, and chromium metallic interfaces with ML $MoSe_2$ [23].

Yang et al. proposed a convenient and efficient approach, examining the interfacial characteristics of 2D-SCMs and high-k dielectrics [24] by using a selective hydrogenated process. By using density functional theory, Yang et al. [25] studied the interfacial characteristics of $b_{12\text{-phase}}$ borophene contacts with widely used 2D materials such as TMDCs, group IV-enes, and group V-enes. With the exception of the 12 borophene/graphene contact, they observed the zero-tunneling barrier in all investigated $b_{12\text{-phase}}$ borophene contacts [25]. Li et al. examined the progress made in the controlled synthesis of hexagonal boron nitride (h-BN) and graphene heterostructures that are layered vertically and in planes and highlighted their structure of interfaces, electronic properties, and novel device constructions [26]. A comparative study was performed by Zhong et al. in different metal contacts (such as Sc, Ti, Ag, Pt, Ni, and Au) and monolayer (ML) as well as bilayer (BL) MoS_2 [27]. They found that ML MoS_2–metal contacts have an enhanced SBH in comparison to BL MoS_2–metal contacts as a result of the interlayer coupling [27].

Stormer et al. investigated the 2D electron gas at a semiconductor–semiconductor (GaAs–AlGaAs) contact first time and observed that the 2D carrier concentration regularly varies from (1.1 to 1.6) $\times$ 10^{12} cm^{-1} because of a novel, high mobility, and photoconductive effects [28]. Nagashio studied the gate stack properties in FETs by comparing the interfacial properties of the 2D/SiO$_2$ interface with the SiO$_2$/Si interface. He provided evidence of recent developments in the analysis of the gate stack properties for bilayer graphene and MoS$_2$ FETs [29]. Guo et al. proposed that high performance in ML GeSe FETs with graphene electrodes can be achieved by the addition of graphene between ML GeSe and Cu electrodes [30]. Qi et al. demonstrated the interfacial characteristics of 2D semiconductor PtSe$_2$ and metallic interfaces of aluminum, silver, copper, gold, platinum, palladium, scandium, and titanium. For the contact of silver, gold, platinum, and palladium, the formation of a vertical Schottky barrier is found. However, they observed the development of a lateral Schottky barrier when they used the contacts of aluminum, copper, scandium, and titanium. In both cases, the barrier comprises a strong Fermi level pinning (FLP) effect which strictly restricts the practical performance of 2D electronics [31]. By using an example of the WSe$_2$/MoSe$_2$ and WSe$_2$/MoS$_2$ interface, Ponomarev et al. [32] reported how interfacial characteristics can be quantitively defined by the properties of the constituent 2D-SCMs. They observed the direct and indirect band gaps for the former and the later interfaces [32]. Using first-principles calculation, Liu et al. examined the interfacial characteristics of the SrRuO$_3$ and MoS$_2$ contacts. They discovered two different kinds of terminated interfaces: SrO- and Ru-terminated interfaces. As reported by them, the SBH surprisingly dropped to zero for the SrO-terminated contact [33].

The InGaAs passivation by ALD of aluminum oxide, gadolinium oxide, and scandium oxide has been investigated by Ameen et al. [34]. In these contacts, they have documented a number of interfacial characteristics, including current leakage, frequency dispersion, interface trap density, and field breakdown [34]. To study the channel resistance and interface defects, a metal-oxide-semiconductor (MOS) capacitor model for ultra-thin 2D-SCMs has been suggested by Gaur et al. [35]. This model successfully captures the static and dynamic trap density behavior and the length dependency of MOS capacitors [35]. The effect of interfacial InAs layers on the characteristics of the MOS interface between gallium nitride and aluminum oxide has been investigated by Yokoyama et al. They found that the interface properties in both the valence/conduction band sides have been improved by the InAs layer [36]. Wu et al. examined the interfacial characteristics of a GaN MOS instrument by coating of gate dielectric film of aluminum oxide on gallium nitride substrates with polar c-plane and non-polar m-plane surfaces. They observed better interface properties in the devices on the non-polar surface rather than those on the polar surface. By insertion of the AlN passivation layer, the interfacial properties of both devices are intensely improved [37]. The impact of process temperature on interfacial properties of GaSb MOS device made up of an ex-situ ALD process has been studied by Yokoyama et al. [38]. They have reported in the synthesis of GaSb MOS interfaces that it is preferable to use a low-temperature technique to prevent the degradations induced at high temperatures in the ex-situ ALD process [38].

Liu et al. reviewed the methods for widely characterizing and exactly controlling surfaces and interface phenomena up to the atomic scale, in addition to approaches that allow modification and optimization of interfacial interactions in vdWHs [39]. Jiang et al. developed thermally induced self-assembly of 2D organic crystalline films unveiling well-defined molecular layer numbers over a millimeter-sized area via molecule–substrate vdW interactions [40]. They observed the fabricated organic FETs with bilayer films display outstanding electrical performance [40]. The weak FLP effect between monolayer

In$_2$Ge$_2$Te$_6$ (or Janus In$_2$Ge$_2$Te$_3$Se$_3$) 2D-SCMs and 2D metal interfaces has been demonstrated by Li et al. [41]. They discovered that the potential build between the semiconducting and metallic sides controls this weak FLP [41].

5.3 Applications of Multiple 2D Semiconducting Layers

For 2D semiconductors to fully realize their promise in a diversity of electrical and optoelectronic applications, such as transistors, photodetectors, sensors, and quantum devices, it is essential to comprehend and manage these interfacial phenomena. The performance of 2D semiconductor-based devices can be enhanced by manipulating and optimizing these features to enable new functionalities. This section highlights the significant applications that become possible because of interfacial characteristics and multiple 2D layer geometry of semiconducting materials.

It is known that 2D-SCMs (like graphene, TMDCs, and black phosphorus) are typically single or few-layered materials possessing remarkable electronic, optical, and mechanical properties due to their atomically thin nature, quantum confinement effects, large surface-to-volume ratio, and the presence of van der Waals gaps between layers. The quantum confinement in 2D semiconductors can be utilized to create quantum dots and other quantum devices for applications in quantum computing and quantum communication. They have drawn a lot of attention as a result of the distinctive interfacial features that appear at the interfaces between 2D-SCMs and other materials or environments. Several applications leverage the interfacial properties of 2D semiconductors in various fields of science and technology such as:

5.3.1 Field-Effect Transistors

The interface and geometry of 2D semiconducting layers are critically important in the design and functionality of FETs, a fundamental building block of modern electronics. They provide atomically thin channels for the operations of FETs. As a result, the FET's current flow channel is incredibly thin, providing superb electrostatic control over the channel. Better gate control and more effective switching are made possible as a result, which helps to reduce power consumption and speed up operation. The mechanical flexibility of 2D semiconductor materials facilitates the miniaturization and scalability to fabricate smaller, highly energy-efficient, and flexible FETs.

Due to their outstanding charge carrier mobility of TMDCs, graphene, and black phosphorus, 2D-SCMs make great options for creating exceedingly thin channels. The interface between the gate dielectric and the 2D semiconductor channel permits efficient electrical control of the conductivity of the channel resulting in faster and more energy-efficient transistors. Their atomically flat interfaces and high carrier mobility make them capable candidates for next-generation, high-performance, and low-power electronics and integrated circuits [16, 22, 30, 33].

5.3.2 Photodetectors and Photovoltaics

The unique electronic band structures of 2D semiconductors lead to strong light–matter interactions. The interfacing between 2D semiconductors and other materials can boost light absorption due to the atomically thin nature of 2D semiconductors, facilitate charge separation and collection, and improve the efficiency of photodetectors and solar cells. Their robust light absorption and effective charge transport properties are beneficial for creating lightweight and flexible solar panels. Thus, a very efficient and thin-film solar cell can be designed due to the tunable bandgap of 2D semiconductors. It is also utilized in the fabrication of sensitive photodetectors due to strong light–matter interaction in 2D materials where their electronic properties change in response to incident photons, enabling high-speed and sensitive light detection for various wavelength ranges, from visible to infrared [18].

5.3.3 Optoelectronics and Light Emission

The stacking of different 2D semiconductors in vdWHs produces type-II heterojunctions with staggered band arrangements. Due to efficient charge separation and emission of light in the visible and infrared range, these structures promising for light-emitting diodes (LEDs) and lasers. The 2D semiconductors can be utilized for modulating the light in integrated optical circuits, enabling applications in data communication and optical computing. 2D semiconductors can emit light due to the recombination of charge carriers by applying an electric field to it. Due to this unique property, it is exploited in the development of LEDs for display and lighting applications to create efficient light sources with tunable emission properties. TMDs, such as MoS_2 and WS_2, exhibit strong light–matter interactions because of their direct bandgap and atomically thin nature [17, 24].

5.3.4 Sensing Applications

2D-SCMs are susceptible to surface interactions due to their "high surface-to-volume ratio" characteristics. This high sensitivity and selectivity are beneficial in different fields of sensing devices like gas sensors to detect toxic gases or environmental pollutants, biosensors, and chemical sensors. The conductivity and other properties that can be used to sense analytes have been changed by interaction at the interface. In biosensors, the interaction between biomolecules and the surface of 2D materials can be utilized for biosensing applications, including medical diagnostics. 2D semiconductors can be incorporated into nano-electro-mechanical systems devices, enabling ultrasensitive sensors and resonators to detect tiny forces or mass changes [21, 28].

5.3.5 Catalysis

To enhance the chemical reactions, the interfacial properties between 2D semiconductors and catalytic materials are very beneficial. 2D semiconductors can not only act as catalysts themselves but also modify the activity of other catalysts, which is very valuable in applications such as water splitting and carbon dioxide reduction. To act as a catalyst in various electrochemical reactions, the exposed edges and active sites of 2D semiconductors can be utilized [20].

5.3.6 Flexible and Transparent Electronics

The flexibility due to their atomic thickness and mechanical strength of 2D semiconductors make them suitable to integrate into flexible, bendable, and transparent electronic devices, which creates the possibilities for wearable technology and conformable electronics.

Graphene is an example of such a 2D semiconductor, that exhibits excellent electrical conductivity and transparency, making it suitable for applications in transparent conductive films for displays, touchscreens, and solar cells. To design wearable electronics and bendable devices, the thin layers of 2D semiconductor materials can be integrated into flexible substrates [24, 26, 31].

5.3.7 Memory Devices

The charge trapping, storage capabilities, and high on/off ratio of 2D semiconductors at interfaces with dielectrics are useful for non-volatile memory devices and the switching behavior of charge in these structures can be exploited for memory applications [10]. 2D semiconducting layers can be combined with other substances, including ferroelectric substances, to produce hybrid memory devices with improved functionality and performance. The geometry of thin layers can be stacked vertically, enabling 3D memory architectures that can store more data in a smaller space. 2D materials can be utilized to design memristors (resistors with memory), with variable resistance resulting in memory storage, artificial intelligence, mimicking certain aspects of brain function, and neuromorphic computing. Due to the exceptional interfacial characteristics of 2D semiconducting materials like high carrier mobility, and excellent electrical properties, they offer the designs of memory devices having low power consumption during write and read operations, and reduced access times for fast switching between different memory states [10].

5.3.8 Spintronics

The combination of inherent spin properties of electrons with their interfacial properties in 2D semiconductors enables the creation of spintronic devices. The electron spin and charge are utilized for data storage, processing, and manipulation in these devices, potentially leading to more energy-efficient electronics. In certain 2D semiconductors, the intrinsic spin-polarized electronic states can be used for spintronic applications like spin-based transistors and memory devices. The interface between different 2D semiconducting materials or between a 2D material and a magnetic material can be engineered to control and manipulate the spin of electrons effectively. Spin-polarized carriers can be injected, detected, and manipulated at these interfaces, enabling the creation of spintronic devices like spin valves and spin transistors. The isolation of individual atomic layers in 2D materials makes them attractive candidates for quantum bits (Qubits) in quantum computing systems [33]. Due to the exceptional interfacial characteristics of 2D semiconducting materials like TMDCs, they have high electron mobility and low spin-orbit coupling, making them excellent candidates for efficient spin transport. The 2D geometry allows for long spin diffusion lengths, which are crucial for spintronics devices.

5.3.9 Thermal Interface Materials

The interface and geometry of these materials play a crucial role in enhancing their thermal conductivity and overall effectiveness in dissipating heat. One of the most significant advantages of 2D semiconducting layers is their exceptionally high thermal conductivity. The 2D nature of these materials means they have a large surface area compared to their volume. This large surface area provides more contact points for efficient heat transfer between the two surfaces they are placed between. The thermal resistance at the interface between two materials can be a significant bottleneck in heat dissipation. 2D semiconducting layers help reduce this resistance due to their high thermal conductivity and excellent surface contact. The good electrical conductivity of 2D materials coupled with their low thermal conductivity enables the creation of thermoelectric devices that convert waste heat into electricity. Thus, the interfacing of 2D semiconductors with other materials is being explored for their potential applications in thermal management such as heat sinks. In electronic and optoelectronics devices, they provide efficient pathways for heat transport by enhancing heat dissipation [38].

5.3.10 Energy Storage

In batteries and supercapacitors, the interfaces between 2D semiconductors and electrode materials can enhance energy storage capabilities by improving charge transfer kinetics and higher surface area. In energy storage devices, 2D materials can be utilized as electrode material offering high surface area and excellent charge transport properties. Their large surface area is beneficial to boost the charge storage capacity and to improve overall device performance [21, 23].

The versatility and tunability of the interfacial properties of 2D-SCMs have a noteworthy role in all these applications to determine and improve performance, and efficiency, and offer a wide range of opportunities for innovation across numerous devices [3]. The unique interfacing between the surfaces of different kinds of 2D semiconducting layers and their geometrical features (e.g., high surface area per unit volume and atomically thin nature) offer several advantages in energy storage applications, including lower self-discharge rates, improved ion diffusion, adaptable properties, flexible designs, and enhanced thermal management. This enables the design of heterostructures with synergistic properties, such as improved charge separation and electron transport, which can be advantageous for the growth of flexible and wearable energy storage devices, expanding their potential applications [10].

5.4 Characterization Techniques and Approaches for Dealing with Interfacial Properties and Geometry of 2D Semiconducting Layers

Researchers can employ various techniques and approaches to deal with the interface and geometry of 2D semiconducting layers to control and optimize their properties for specific applications. In this section, we have discussed some common methods and strategies that are used to deal with the interfacial properties and geometry of 2D semiconducting layers while engaged in some specific research problems and materials synthesis.

Chemical vapor deposition is a widely used technique to grow 2D semiconducting layers with precise control over layer thickness and crystal orientation. Another method for creating tiny layers of semiconducting materials atom by atom is molecular beam epitaxy. It enables the fabrication of heterostructures using various 2D materials and fine control over layer thickness. Researchers manipulate growth conditions in both methods to control the geometry and stacking of such layers. Exfoliation can also be manipulated to create the desired layer geometries and interfaces. While dealing with the interfaces between the two 2D semiconducting layers, we can impart mechanical strain to 2D layers by bending or stretching them. This strain can change the electrical band structure, leading to desired features. Controlled strain is produced using methods like microfabrication and transfer printing.

Doping is also an effective approach that involves introducing foreign atoms or molecules into their lattices to modify their properties. Surface functionalization with chemical groups can also be used to tailor the chemical reactivity of these materials. Furthermore, heterostructures can be made by stacking various 2D semiconducting layers having complementary characteristics. To obtain particular electrical and optical features, the materials chosen and the order in which they are stacked can be modified.

We can predict the behavior of 2D semiconducting materials having diversified geometries and interfaces, thanks to available computational simulations and modeling software. This helps to clarify the fundamental physics and direct experimental efforts to manipulate the interface and geometry of 2D semiconducting layers so that we can customize their properties, optimize performance, and develop innovative materials and devices for an extensive range of applications.

5.5 Summary

To customize the electrical, optical, and mechanical properties of 2D semiconducting layers, their interfacial and geometrical properties are crucial. This expands a wide range of applications in the fields of electronics, photonics, sensors, and energy-related technologies. To fully realize the promise of 2D semiconductors in numerous sectors, researchers are still investigating and experimenting with these features. Table 5.1 summarizes the consequences and/or application of different kinds of 2D-SEMs discussed in Sections 5.2 and 5.3 and their interfacial properties and geometrical aspects.

The interface and geometry of 2D semiconducting layers, such as graphene and TMDCs, play a vital role in determining their properties and, subsequently, their applications in numerous industries. TMDCs have a thickness-dependent bandgap, similar to MoS_2 and WS_2. The bandgap grows when the number of layers is decreased. This variable bandgap enables the construction of semiconducting materials with precise electronic properties, making them appropriate for different electronic and optoelectronic applications, including transistors and photodetectors.

The reduced dimensionality of 2D materials leads to quantum confinement effects. Electrons in these materials are confined in two dimensions, which results in discrete energy levels and enhanced electron mobility. This property is advantageous for high-speed electronic devices and quantum computing applications.

Due to their atomic-scale thickness, 2D semiconducting materials have a very high surface-to-volume ratio. With a wide surface area needed for interactions with other molecules or materials in applications like sensors and catalysis, this feature is advantageous. The exceptional mechanical properties, including high tensile strength and flexibility, make 2D semiconducting materials useful for applications in flexible electronics and wearable devices that require lightweight yet strong components. 2D semiconductors provide ideal candidates for sensors and detectors due to their high surface sensitivity. They are excellent materials for gas sensors, biosensors, and environmental monitoring systems because changes in the environment or the adsorption of molecules on their surfaces can result in

TABLE 5.1

Importance of Different kinds of 2D-SEMs

S. No.	Interfacing Components	Importance	References
1	Al_2O_3/MoS_2	Providing their integration into field-effect transistors	[16]
2	GaSe/graphene	GaSe films on graphene substrate used for nano-electronic and optoelectronic applications with high photo-response	[17]
3	$HfN_2/MoTe_2$	Utilized in fabricating a solar cell having a high-power conversion efficiency of 21.44%	[18]
4	GaX/As (X = S, Se)	In a photocatalytic performance for water splitting to produce hydrogen	[20]
5	ML 1T'/1T-2H MX_2	Interfacial gate-tunable characteristics in FET	[22]
6	ML $MoSe_2$-metals (Al, Ag, Pt, Cr, Ni, and Ti) interface	Offers choice of suitable metal electrodes	[23]
7	High-k dielectrics/2D semiconductors	Promote toward high-performance nanoelectronics	[24]
8	b_{12} phase borophene/TMDs, group IV-enes, and group V-enes	Guidance for the design of borophene-based 2D materials for future devices	[25]
9	Graphene/hexagonal boron nitride (h-BN)	Applications in nanoelectronics	[26]
10	ML and BL MoS_2-metals (Sc, Ti, Ag, Pt, Ni, and Au) contacts	Reveals the standing of a more advanced conceptual strategy used in the interface study beyond the energy band computation	[27]
11	GaAs–AlGaAs interface	Offers a promising substitute for studying various fundamental physical properties of a 2D electron gas	[28]
12	BL graphene/MoS_2 FET	Provides conceptual and experimental approaches for controlling the 2D hetero-interface properties	[29]
13	ML GeSe/(Cu, Ag, Ti, Au, Pd, Pt, graphene, and graphene–Cu) electrodes	More appropriate as a channel material for high-performance transistors of the future	[30]
14	Metals/$PtSe_2$ contacts	Beneficial for creating cutting-edge 2D semiconductor-based circuits	[31]
15	$WSe_2/MoSe_2$ and WSe_2/MoS_2	Explain how the properties of the underlying 2D-SCMs can be used to quantitatively determine interfacial characteristics	[32]
16	$SrRuO_3/MoS_2$ heterojunction	Applications in spintronics devices and favor of scaling of FETs	[33]
17	$Al_2O_3/GaSb$ MOS interface	The interfacial properties have been improved by the InAs layer	[36]
18	ML $In_2Ge_2Te_6$ (or Janus $In_2Ge_2Te_3Se_3$)/2D metal	Explains how to reduce the FLP to get the adjustable SBH	[41]

observable electrical or optical responses. To build heterostructures with certain features, 2D semiconductors can be stacked in many different ways. They can be miniaturized and are suited for next-generation nanoelectronics because of their atomic-scale thickness. They are capable of being included in nanoscale devices, increasing device density and enhancing performance. This makes it possible to create novel electronics with distinctive properties, including memristors and tunnel diodes.

References

[1] C. Rosslee, N. L. Abbott, Active control of interfacial properties, *Cur. Opin. Colloid Interface Sci.*, 5 (2000) 81–87. https://doi.org/10.1016/S1359-0294(00)00035-2

[2] B. G. Sumpter, L. Liang, A. Nicolai, V. Meunier, Interfacial properties and design of functional energy materials, *Acc. Chem. Res.*, 47 (2014) 3395–3405. https://doi.org/10.1021/ar500180h

[3] X. Huang, C. Liu, P. Zhou, 2D semiconductors for specific electronic applications: From device to system, *npj 2D Mater. Appl.*, 6 (2022) 51. https://doi.org/10.1038/s41699-022-00327-3

[4] Z. Yang, J. Hao, Recent progress in 2D Layered III–VI semiconductors and their heterostructures for optoelectronic device applications, *Adv. Mater. Tech.*, 4 (2019) 1900108. https://doi.org/10.1002/admt.201900108

[5] Y. Zhao, S. Bertolazzi, P. Samorì, A universal approach toward light-responsive two-dimensional electronics: Chemically tailored hybrid van der Waals heterostructures, *ACS Nano*, 13 (2019) 4814–4825. https://doi.org/10.1021/acsnano.9b01716

[6] B. Jiang, Z. Yang, X. Liu, Y. Liu, L. Liao, Interface engineering for two-dimensional semiconductor transistors, *Nano Today*, 25 (2019) 122–134. https://doi.org/10.1016/j.nantod.2019.02.011

[7] X. Jing, Y. Illarionov, E. Yalon, P. Zhou, T. Grasser, Y. Shi, M. Lanza, Engineering field effect transistors with 2D semiconducting channels: Status and prospects, *Adv. Funct. Mater.*, 30 (2020) 1901971. https://doi.org/10.1002/adfm.201901971

[8] M. G. Stanford, P. D. Rack, D. Jariwala, Emerging nanofabrication and quantum confinement techniques for 2D materials beyond graphene, *npj 2D Mater. Appl.*, 2 (2018) 20. https://doi.org/10.1038/s41699-018-0065-3

[9] J. Yang, X. Liu, Q. Dong, Ya. Shen, Y. Pan, Z. Wang, K. Tang, X. Dai, R. Wu, Y. Jin, W. Zhou, S. Liu, J. Sun, Oxidations of two-dimensional semiconductors: Fundamentals and applications, *Chinese Chem. Lett.*, 33 (2022) 177–185. https://doi.org/10.1016/j.cclet.2021.06.078

[10] L. Gao, Flexible device applications of 2D semiconductors, *Small*, 13 (2017) 1603994. https://doi.org/10.1002/smll.201603994

[11] S. Kang, D. Lee, J. Kim, A. Capasso, H. Kang, J.-W. Park, C.-H. Lee, G.-H. Lee, 2D semiconducting materials for electronic and optoelectronic applications: Potential and challenge, *2D Mater.*, 7 (2020) 022003. https://doi.org/10.1088/2053-1583/ab6267

[12] K. -C. Chiu, X.-Q. Zhang, X. Liu, V. Menon, Y. Chen, J. M. Wu, Y. H. Lee, Synthesis and application of monolayer semiconductors, *IEEE J. Quantum Electron.*, 51 (2015) 0600110. https://doi.org/10.1109/JQE.2015.2476360.

[13] T. Schram, S. Sutar, I. Radu, I. Asselberghs, Challenges of wafer-scale integration of 2D semiconductors for high-performance transistor circuits, *Adv. Mater.*, 34 (2022) 2109796. https://doi.org/10.1002/adma.202109796

[14] S.-J. Yang, M.-Y. Choi, C.-J. Kim, Engineering grain boundaries in two-dimensional electronic materials, *Adv. Mater.*, 35 (2023) 2203425. https://doi.org/10.1002/adma.202203425

[15] Z. Peng, X. Chen, Y. Fan, D. J. Srolovitz, D. Lei, Strain engineering of 2D semiconductors and graphene: From strain fields to band-structure tuning and photonic applications, *Light Sci. Appl.*, 9 (2020) 190. https://doi.org/10.1038/s41377-020-00421-5

[16] S. Park, S. Y. Kim, Y. Choi, M. Kim, H. Shin, J. Kim, W. Choi, Interface properties of atomic-layer-deposited Al_2O_3 thin films on ultraviolet/ozone-treated multilayer MoS_2 crystals, *ACS Appl. Mater. Inter.*, 8 (2016) 11189–11193. https://doi.org/10.1021/acsami.6b01568

[17] Z. B. Aziza, H. Henck, D. Pierucci, M. G. Silly, E. Lhuillier, G. Patriarche, F. Sirotti, M. Eddrief, A. Ouerghi, Vander Waals epitaxy of GaSe/Graphene heterostructure: Electronic and interfacial properties, *ACS Nano*, 10 (2016) 9679–9686. https://doi.org/10.1021/acsnano.6b05521

[18] M. K. Mohanta, A. Rawat, A. D. Sarkar, Atomistic manipulation of interfacial properties in HfN_2/$MoTe_2$ van der Waals heterostructure via strain and electric field for next generation multifunctional nanodevice and energy conversion, *Appl. Surf. Sci.*, 568 (2021) 150928. https://doi.org/10.1016/j.apsusc.2021.150928

[19] S. -L. Li, K. Tsukagoshi, E. Orgiu, P. Samori, Charge transport and mobility engineering in two-dimensional transition metal chalcogenide semiconductors, *Chem. Soc. Rev.*, 45 (2016) 118–151. https://doi.org/10.1039/C5CS00517E

[20] Q. Peng, Z. Guo, B. Sa, J. Zhou, Z. Sun, New gallium chalcogenides/arsenic van der Waals heterostructures promising for photocatalytic water splitting, *Inter. J. Hydrog. Energy*, 43 (2018) 15995–16004. https://doi.org/10.1016/j.ijhydene.2018.07.008

[21] N. Ashraf, M. I. Khan, A. Majid, M. Rafique, M. B. Tahir, A review of the interfacial properties of 2-D materials for energy storage and sensor applications, *Chin. J. Phys.*, 66 (2020) 246–257. https://doi.org/10.1016/j.cjph.2020.03.035

[22] S. Liu, J. Li, B. Shi, X. Zhang, Y. Pan, M. Ye, R. Quhe, Y. Wang, H. Zhang, J. Yan, L. Xu, Y. Guo, F. Pan, J. Lu, Gate-tunable interfacial properties of in-plane ML MX_2 1T'-2H heterojunctions, *J. Mater. Chem. C*, 6 (2018) 5651–5661. https://doi.org/10.1039/C8TC01106K

[23] Y. Pan, S. Li, M. Ye, R. Quhe, Z. Song, Y. Wang, J. Zheng, F. Pan, W. Guo, J. Yang, J. Lu, Interfacial properties of monolayer $MoSe_2$-metal contacts, *J. Phys. Chem. C*, 120 (2016) 13063–13070. https://doi.org/10.1021/acs.jpcc.6b02696

[24] Y. Yang, T. Yang, T. Song, J. Zhou, J. Chai, L. M. Wong, H. Zhang, W. Zhu, S. Wang, M. Yang, Selective hydrogenation improves interface properties of high-k dielectrics on 2D semiconductors, *Nano Res.*, 15 (2022) 4646–4652. https://doi.org/10.1007/s12274-021-4025-4

[25] J. Yang, R. Quhe, S. Feng, Q. Zhang, M. Lei, J. Lu, Interfacial properties of borophene contacts with two-dimensional semiconductors, *Phys. Chem. Chem. Phys.*, 19 (2017) 23982. https://doi.org/10.1039/C7CP04570K

[26] Q. Li, M. Liu, Y. Zhang, Z. Liu, Hexagonal boron nitride–graphene heterostructures: Synthesis and interfacial properties, *Small*, 12 (2016) 32–50. https://doi.org/10.1002/smll.201501766

[27] H. Zhong, R. Quhe, Y. Wang, Z. Ni, M. Ye, Z. Song, Y. Pan, J. Yang, L. Yang, M. Lei, J. Shi, J. Lu, Interfacial properties of monolayer and bilayer MoS_2 contacts with metals: Beyond the energy band calculations, *Sci. Rep.*, 6 (2016) 21786. https://doi.org/10.1038/srep21786

[28] H. L. Stormer, R. Dingle, A. C. Gossard, W. Wiegmann, M. D. Sturge, Two-dimensional electron gas at a semiconductor-semiconductor interface, *Solid State Comm.*, 29 (1979) 705–709. https://doi.org/10.1016/0038-1098(79)91010-X

[29] K. Nagashio, Understanding interface properties in 2D heterostructure FETs, *Semicond. Sci. Technol.*, 35 (2020) 103003. https://doi.org/10.1088/1361-6641/aba287

[30] Y. Guo, F. Pan, Y. Ren, Y. Wang, B. Yao, G. Zhao1, J. Lu, Anisotropic interfacial properties of monolayer GeSe-metal contacts, *Semicond. Sci. Technol.*, 34 (2019) 095021. https://doi.org/10.1088/1361-6641/ab37cc

[31] L. Qi, M. Che, M. Liu, B. Wang, N. Zhang, Y. Zou, X. Sun, Z. Shi, D. Li, S. Li, Mechanistic understanding of the interfacial properties of metal–$PtSe_2$ contacts, *Nanoscale*, 15 (2023) 13252. https://doi.org/10.1039/D3NR02466K

[32] E. Ponomarev, N. Ubrig, I. Gutierrez-Lezama, H. Berger, A. F. Morpurgo, Semiconducting van der Waals interfaces as artificial semiconductors, *Nano Lett.*, 18 (2018) 5146–5152. https://doi.org/10.1021/acs.nanolett.8b02066

[33] B. Liu, L.-J. Wu, Y.-Q. Zhao, L.-Z. Wang, M.-Q. Cai, The interfacial properties of $SrRuO_3$/MoS_2 heterojunction: A first-principles study, *Eur. Phys. J. B*, 89 (2016) 80. https://doi.org/10.1140/epjb/e2016-60584-x

[34] M. Ameen, L. Nyns, S. Sioncke, D. Lin, T. Ivanov, T. Conard, J. Meersschaut, M. Y. Feteha, S. V. Elshocht, A. Delabie, Al_2O_3/InGaAs Metal-oxide-semiconductor interface properties: Impact of Gd_2O_3 and Sc_2O_3 interfacial layers by atomic layer deposition, *ECS J. Solid State Sci. Tech.*, 3 (2014) 133–141. https://doi.org/10.1149/2.0021411jss

[35] A. Gaur, T. Agarwal, I. Asselberghs, I. Radu, M. Heyns, A MOS capacitor model for ultra-thin 2D semiconductors: The impact of interface defects and channel resistance, *2D Mater.*, 7 (2020) 035018. https://doi.org/10.1088/2053-1583/ab7cac

[36] M. Yokoyama, H, Yokoyama, M. Takenaka, S. Takagi, Impact of interfacial InAs layers on Al_2O_3/GaSb metal-oxide-semiconductor interface properties, *Appl. Phys. Lett.*, 106 (2015) 122902. https://doi.org/10.1063/1.4914453

[37] X. Wu, R. Liang, L. Guo, L. Liu, L. Xiao, S. Shen, J. Xu, J. Wang, Improved interface properties of GaN metal-oxide-semiconductor device with non-polar plane and AlN passivation layer, *Appl. Phys. Lett.*, 109 (2016) 232101. https://doi.org/10.1063/1.4971352

[38] M. Yokoyama, Y. Asakura, H. Yokoyama, M. Takenaka, S. Takagi, Impact of process temperature on GaSb metal-oxide-semiconductor interface properties fabricated by ex-situ process, *Appl. Phys. Lett.*, 104 (2014) 262901. https://doi.org/10.1063/1.4884950

[39] X. Liu, M. C. Hersam, Interface characterization and control of 2D Materials and heterostructures, *Adv. Mater.*, 30 (2018) 1801586. https://doi.org/10.1002/adma.201801586

[40] S. Jiang, J. Qian, Y. Duan, H. Wang, J. Guo, Y. Guo, X. Liu, Q. Wang, Y. Shi, Y. Li, Millimeter-sized two-dimensional molecular crystalline semiconductors with precisely defined molecular layers via interfacial-interaction-modulated self-assembly, *J. Phys. Chem. Lett.*, 9 (2018) 6755–6760. https://doi.org/10.1021/acs.jpclett.8b03108

[41] J. Li, W. Zhou, L. Xu, J. Yang, H. Qu, T. Guo, B. Xu, S. Zhang, H. Zeng, Revealing the weak Fermi level pinning effect of 2D semiconductor/2D metal contact: A case of monolayer $In_2Ge_2Te_6$ and its Janus structure $In_2Ge_2Te_3Se_3$, *Mater. Today Phys.*, 26 (2022) 100749. https://doi.org/10.1016/j.mtphys.2022.100749

6

Molecular Orbital Delocalization and Stacking Effect on 2D Semiconducting Materials

Mostafa Moslempoor and Esmaeil Sheibani*

**Correspondence: e.sheibani@sci.ui.ac.ir*

6.1 Introduction

The emergence of graphene atomic crystals and the study of two-dimensional (2D) materials has garnered significant attention in the field of 2D semiconductor science. The properties of compounds, in addition to the materials from which they are made, depend on the size and dimensions of the materials, which determine their behavior. These materials can be classified according to **Figure 6.1(a)** in terms of nano. A restraint on one of the dimensions produces a 2D material with sheet structure; restraining two dimensions in terms of their size, one-dimensional material is obtained, and when all the dimensions are restrained, zero-dimensional materials are produced. Among these materials, 2D materials can be integrated with existing semiconductor technologies and provide new physical and chemical properties. Electrically, they can behave as insulators, semiconductors, metals, or even superconductors, according to **Figure 6.1(b)**. 2D semiconductors have emerged as promising candidates for next-generation electronic and optoelectronic devices. The properties of 2D semiconductors are significantly influenced by their electronic structure, which, in turn, is heavily impacted by two key phenomena: molecular orbital (MO) delocalization and stacking effects. MO delocalization refers to the extended spatial distribution of electrons within a molecule or a material, facilitated by the overlap of atomic orbitals. In the context of 2D semiconductors, MO delocalization plays a crucial role in determining their electronic band structure and, consequently, their electronic transport properties. The presence of delocalized orbitals can give rise to interesting phenomena, such as bandgap engineering, enhanced carrier mobility, and the emergence of novel electronic states. Moreover, the stacking order can give rise to diverse electronic structures, band alignments, and interlayer coupling in 2D materials. The stacking effect also plays an important role in the formation of interlayer excitons, which are combined electronic excitations involving electrons and holes in adjacent layers. Understanding and controlling these two key phenomena are of paramount importance in tailoring the electronic properties of 2D semiconductors for various applications in electronics and beyond [1–5].

DOI: 10.1201/9781003439448-6

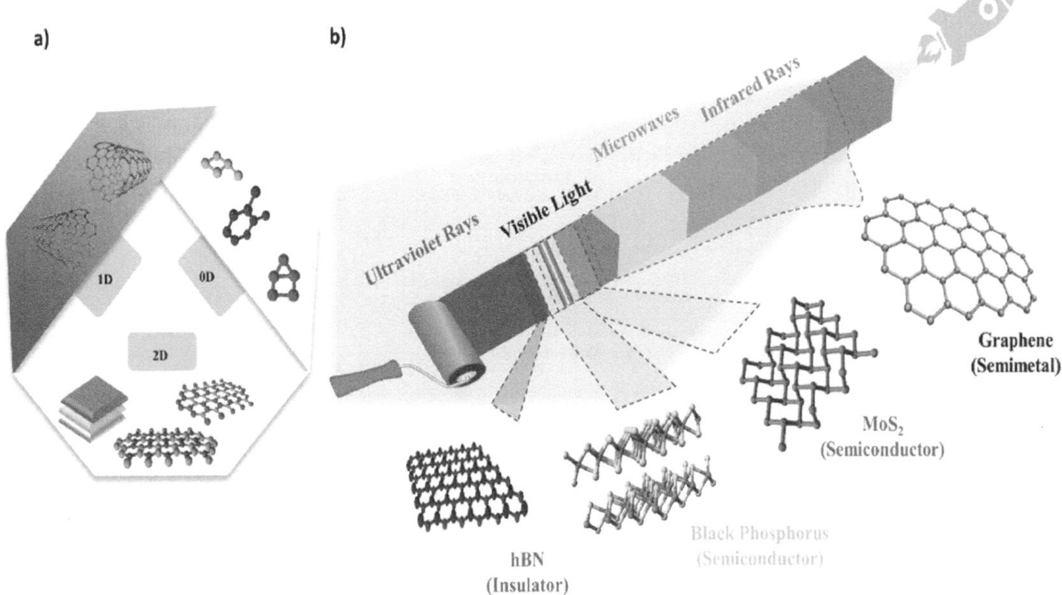

FIGURE 6.1
(a) Classification of the size and dimensions of materials in terms of nanoscale. (b) Physical properties and energy spectrum of two-dimensional semiconductors and their atomic crystal structures.

6.2 Stacking in Two-Dimensional Semiconductor Materials

The "stacking effect" in 2D semiconductors refers to the unique features and characteristics that emerge when multiple layers of 2D semiconductor materials are stacked on top of each other in a specific arrangement. Graphene is the most well-known example of a 2D semiconductor, comprising a single layer of carbon atoms arranged in a honeycomb pattern. Despite its popularity, single-layer graphene (SLG) has limited practical use as a semiconductor due to its zero bandgap (**Figure 6.2**), which is a key determinant of its electrical properties. To overcome this limitation, various methods can be employed. Here are some approaches that have been explored. (1) Substrate-induced bandgap opening: by placing graphene on specific substrates, such as hexagonal boron nitride (hBN) or silicon carbide (SiC) [6], a bandgap can be induced in graphene. (2) Chemical substitution doping: introducing dopant atoms or molecules onto the graphene lattice can modify its electronic properties and create a bandgap. For example, nitrogen doping or hydrogenation of graphene can induce a bandgap. (3) Quantum confinement: patterning graphene into narrow ribbons or nanoribbons can create a quantum confinement effect, leading to the opening of a bandgap. It is worth mentioning that the width and edge structure of the nanoribbons determines the size of the bandgap. (4) Electric field tuning: applying a strong vertical electric field to a graphene bilayer can open up a bandgap, and this approach offers the advantage of tunability. (5) Functionalization: modifying the graphene surface with functional groups or molecules can alter its electronic properties and introduce a

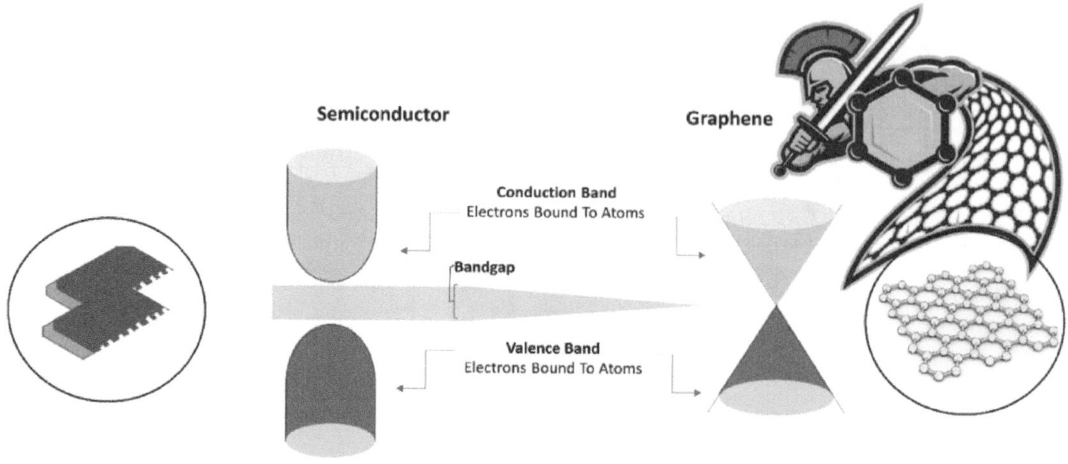

FIGURE 6.2
Contrasting bandgap characteristics of graphene and semiconductors. Semiconductor has a bandgap between the valence and conduction bands. However, graphene lacks a discernible bandgap.

bandgap. As a result, functionalization can be achieved through chemical reactions or surface adsorption [7–9]. Further, researchers have investigated the use of other 2D materials that exhibit a bandgap, such as hBN and transition metal dichalcogenides (TMDCs). When these 2D materials are layered on top of each other with particular orientations and alignments, a phenomenon known as van der Waals (vdW) stacking occurs. vdW forces are weak intermolecular forces that originate from temporary fluctuations in electron distribution within atoms and molecules. These forces are weaker than covalent or ionic bonds, but they can still hold the 2D layers together, helping them to form a stable structure. As a result, as previously mentioned, stacking configurations can significantly affect the electronic, optical, and mechanical properties of multilayer structures, and they can be classified into different types based on relative orientation and layer alignment. Each stacking configuration creates different electronic band structures and properties, and the most common ones will be discussed next [4, 5].

6.2.1 AA and AB Stacking Effect

AA and AB stacking refers to two different ways of arranging atomic layers in 2D materials like graphene and other similar materials. AA stacking involves aligning the layers in a manner where each atom in one layer directly overlays an atom in the adjacent layer. This alignment results in a repeating pattern, often referred to as Bernal stacking. The most iconic example of AA stacking is found in graphene. In graphene's AA stacking, each carbon atom lines up directly above or below another carbon atom in the neighboring layer. AB stacking, on the other hand, involves a slight offset between the layers. In this arrangement, atoms in one layer are positioned directly above the empty spaces or centers of hexagons in the adjacent layer. AB stacking is responsible for creating different atomic arrangements in comparison to AA stacking. Notably, this type of stacking leads to unique patterns known as moiré patterns, which can have significant effects on the material's electronic properties. An example of AB stacking can be observed in hBN [4, 10], a material often used to encapsulate other 2D materials due to its insulating nature and weak interaction with other layers. Consequently, choosing between AA and AB stacking holds

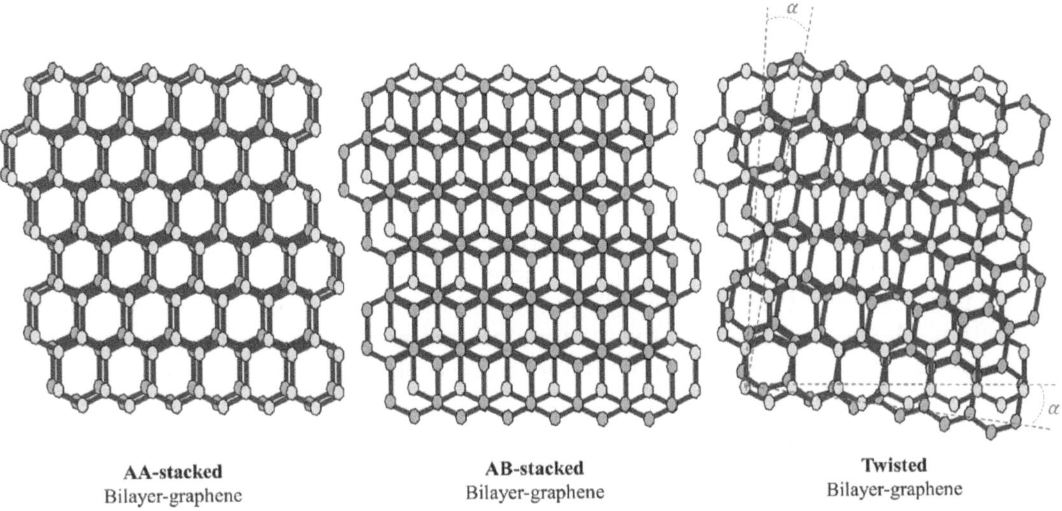

FIGURE 6.3
The scheme illustrates the structure of AA-stacked, AB-stacked, and twisted BLG.

Permission line: We designed this shape, and it is not copied from anywhere.

considerable value for the behaviors and functionalities of 2D materials. For instance, AB stacking in bilayer graphene (BLG) leads to tunable bandgaps, making it possible to manipulate its electronic properties for diverse applications. Furthermore, the interaction between these stacked layers and the angle at which they are aligned—referred to as the "twist angle"—can give rise to fascinating effects such as moiré patterns. In the following, to better understand, we specifically discuss the effects of stacking on graphene [11].

In another definition, based on the difference in the number of graphene layers, graphene can be categorized into different types: SLG, BLG, and few-layer graphene. The bandgap of SLG is zero, which greatly limits its application to electronic devices [12]. In contrast to SLG, BLG not only inherits the majority of SLG's benefits but also exhibits its own distinctive features. When a sufficient electric field is applied, the bandgap of BLG can be well opened, which is of great importance for practical uses. After that, the magic angle twisted BLG has been revealed in recent years as a superconductor. BLG can be generally described as a 2D material in which two SLGs are stacked in parallel with each other. According to the relative positions and angles of the axis of the two layers of graphene, BLG can be mainly divided into three categories: AA-stacked BLG, AB-stacked BLG, and twisted BLG. **Figure 6.3** shows the corresponding structure diagram of three types of BLG. AA stacking is the simplest form of BLG [12–14] in which the layer spacing is ~3.6 Å. In AB-stacked BLG, the layers are covered in a different way from AA-stacked BLG. Only half of the atoms on the top sheet are positioned directly above the atoms, while the other half is situated above the center of the hexagons in the lattice of atoms. AB-stacked BLG has a layer spacing of 3.4 Å, which closely resembles that of graphite. Consequently, AB-stacked BLG exhibits superior thermodynamic stability compared to AA-stacked BLG. In twisted BLG, one layer has a certain angle related to the other that in turn has a crucial impact on specific properties. Among them, AB-stacked BLG is easier to prepare because of its higher thermodynamic stability. However, AA-stacked BLG and twisted BLG have specific properties [13, 14]

Ohta et al. first discovered that BLG has an adjustable bandgap through potassium doping and found that the bandgap of doped BLG can be adjusted by controlling the density of carriers [15]. Subsequently, a different research team recognized the potential for direct adjustment of the bandgap in BLG, obviating the need for doping. Through their experimental endeavors, they crafted field-effect transistors utilizing BLG and subjected them to analysis via infrared micro-spectroscopy. Their observations revealed the existence of a modifiable bandgap, demonstrating the capability to vary its magnitude by up to 250 mV [16]. Based on the outcomes of these two experiments, it can be seen that the critical reason BLG has an adjustable bandgap is that the symmetry between graphene layers is broken. Morozov et al. interpreted this property as an electric field effect and also showed that this phenomenon existed in asymmetric biased BLG such as AB-stacked BLG. Meyer and colleagues conducted an intricate investigation into the surface roughness of graphene membranes using advanced techniques such as transmission electron microscopy and electron diffraction. The results show that the microscopic curvature of SLG is the strongest and is stronger than that of BLG. On the other hand, twisted BLG has some differences in properties due to the relative angles between layers. Some studies regard twisted BLG as a vdW heterostructure; vdW heterostructures are vertically stacked binary building units that allow for more engineering manipulation based on the rich functionality of 2D materials. Cao et al. and Mele et al. found that twisted BLG produced two completely new electronic states [17].

6.2.2 Vertical and Lateral Stacking Effect

In 2D semiconductors, vertical stacking can occur in two primary configurations: heterostructures and homostructures. These stacking arrangements involve layering different or identical 2D semiconductor materials on top of each other to create novel materials with unique electronic properties. Heterostructures involve stacking different types of 2D semiconductor materials on top of each other. These materials can have distinct electronic properties and bandgaps. When combined in a heterostructure, they can create unique band alignment, enabling functionalities not possible in individual layers. In addition, the vdW forces between layers help maintain the integrity of the structure, because these forces are weak, it becomes possible to assemble different materials without strong chemical bonds. The weak interaction allows for relative flexibility between the layers, resulting in the formation of sharp interfaces and preserving the individual properties of each material. The weak vdW forces between layers also facilitate the alignment of the crystal lattices of the stacked materials. Proper lattice alignment is crucial in achieving desirable electronic band alignments that lead to efficient charge transfer and separation. One of the most well-known examples of a 2D heterostructure is the "MoS_2-WSe_2" heterostructure. Here, two different TMDCs, molybdenum disulfide (MoS_2) and tungsten diselenide (WSe_2), are vertically stacked [18, 19]. The band alignment in this heterostructure leads to efficient charge separation and potential applications in photodetectors, solar cells, and light-emitting diodes (LEDs). Heterostructures can be tailored by selecting specific 2D semiconductor materials, controlling the number of layers, and adjusting the relative stacking orientations, resulting in precise control over their electronic and optoelectronic properties. These properties make heterostructures promising candidates for various electronic and optoelectronic devices. Homostructures, on the other hand, involve stacking multiple layers of the same 2D semiconductor material on top of each other. This results in a repeating pattern of identical layers and the vdW forces again play a critical role in stabilizing the vertical stacking. The weak interlayer interactions allow for the formation of regular repeating patterns, with precise alignment (AB stacking) or a specific rotational angle (AA stacking) to create different electronic properties, which we mentioned in the previous section. Moreover, lateral stacking in 2D semiconductors also refers to the arrangement of individual semiconductor layers side by side on a substrate, forming a lateral stack. The properties of lateral stacks of 2D semiconductors can be influenced by the interlayer interactions and stacking order, which may influence their electronic, optical, and transport properties. Lateral stacking, like vertical stacking, can be controlled to create heterogeneous structures with specific properties by combining different 2D semiconductor materials. In the following, to better understand, we specifically discuss the effects of stacking on TMDCs [20, 21].

TMDCs crystallize as layered compounds with metal atoms sandwiched between chalcogen atoms with a stoichiometry of 1:2, respectively. TMDCs with the formula MX_2 (M: transition metal; X: chalcogen) have emerged as aces of the post-graphene era. About 40 distinct compounds of TMDCs can be identified, featuring transition metals belonging to Groups 4, 5, 6, 7, 9, and 10 on the periodic table (as shown in **Figure 6.4(a)**), in addition to

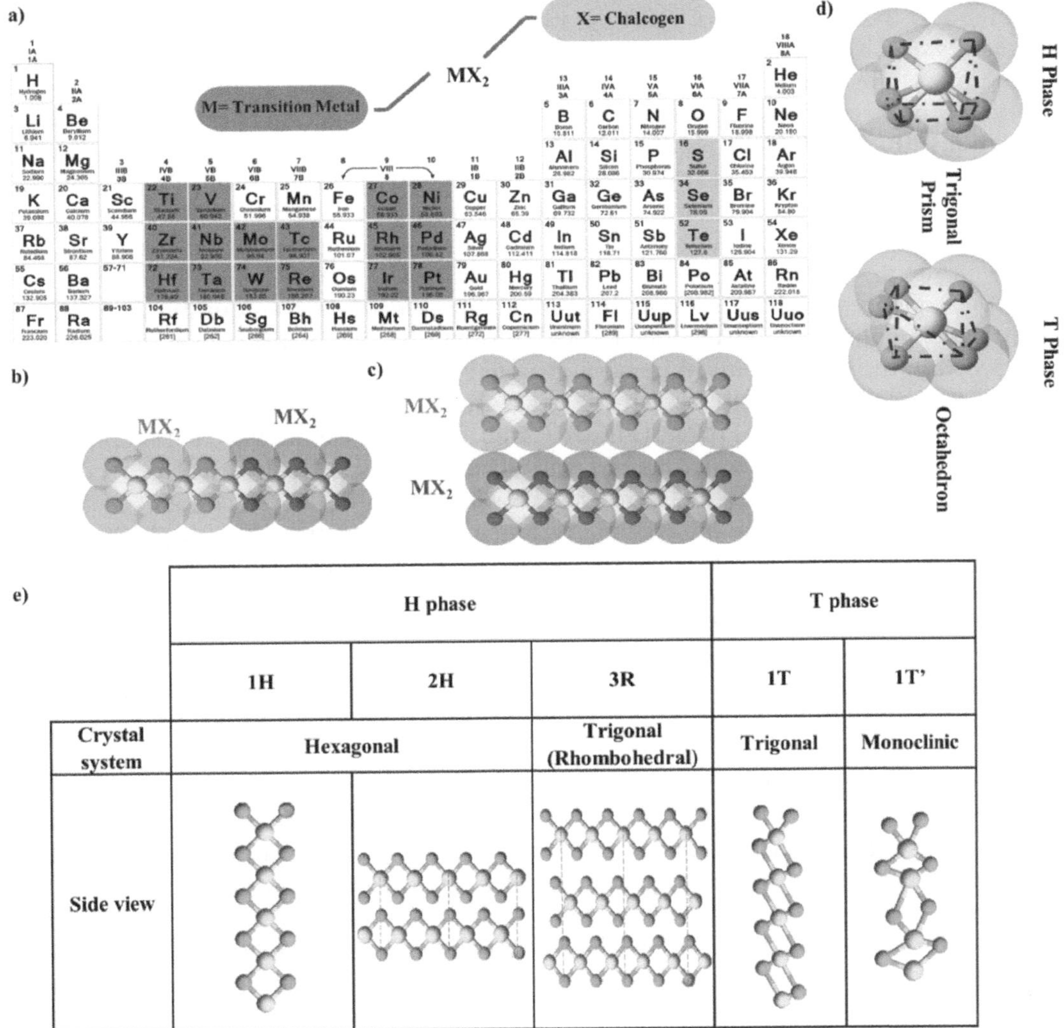

	H phase			T phase	
	1H	2H	3R	1T	1T'
Crystal system	Hexagonal		Trigonal (Rhombohedral)	Trigonal	Monoclinic
Side view					

FIGURE 6.4

(a) Within the periodic table, you will find 16 transition metals and 3 chalcogen elements. These elements collectively serve as the foundational components for constructing the diverse family of 40 distinct transition metal dichalcogenides, all of which exhibit crystalline structures in a two-dimensional layered arrangement. (b) Lateral heterostructures. (c) Vertical heterostructures. (d) Primary phases for metal: chalcogen TMDC: H phase and T phase. (e) crystal structures of common TMDC phases in side view.

sulfur (S), selenium (Se), and tellurium (Te) [22]. In the realm of chemistry, the metal and chalcogens are covalently bonded within a layer and the layers are held together by vdW bonds. This structural characteristic enables various types of layered TMDCs to be likened to atomically thin building blocks similar to Legos, which can be vertically stacked or laterally stitched to construct engineered structures. This Lego-inspired concept has served as the driving force for the synthesis of lateral (**Figure 6.4(b)**) and vertical (**Figure 6.4(c)**) heterostructures with unprecedented properties. If the two layers are connected in-plane, they will form lateral heterostructures. If one layer is stacked onto another layer, a vertical heterostructure will be constructed. The initial experimental instances involved vertically stacking dissimilar 2D materials, achieved through the controlled transfer of exfoliated films or by directly growing distinct TMDCs on pre-existing layers. Nonetheless, the achievement of lateral heterostructures featuring in-plane variations faced impediments attributed to the complexity of refining advanced growth methodologies. Nevertheless, progress was ultimately made through the application of the edge epitaxy technique. Furthermore, the bonding arrangement of metal and chalcogen atoms within a TMDC can be comprehended as comprising two tetrahedrons arranged in opposing orientations (**Figure 6.4(d)**). The distinct phases, labeled as the H and T phases, are defined by the configuration of these tetrahedral. In the H phase, the upper and lower tetrahedrons are symmetrically organized, yielding a trigonal prismatic structure. Conversely, the T phase entails a 180-degree rotation of the upper tetrahedron, leading to an octahedral structure that typically exhibits distortion. The arrangement order of atomic layers, known as the stacking sequence or polytype, exerts influence on the electronic configuration, vibrational spectra, and light-interaction characteristics. The identification of different atomic arrangements within the same chemical phase is denoted by a numerical value in the nomenclature. The single-layer structure of the H phase demonstrates a hexagonal crystal lattice, thus being labeled as the 1H phase (**Figure 6.4(e)**). The H phase can have varied polytypes due to different stacking sequences of layers. The 2H phase, which is the most common one, adds a screw rotation in the symmetry resulting from the AB stacking of the second layer while maintaining the hexagonal crystal system. In contrast, the 3R phase reduces the rotational symmetry from sixfold to threefold, yielding a rhombohedral lattice. On the other hand, the T phase of both monolayer and multilayer have the same symmetry owing to the AA stacking, and therefore they can all be called 1T [21–23].

According to the aforementioned explanations, the characteristics of some TMDCs and the effect of the type of stacking in them have been discussed. Recently, an innovative approach involved the manipulation of the out-of-plane structural symmetry in monolayers of $MoSe_2$ and MoS_2, leading to an asymmetric Janus MosSe monolayer structure that has been synthesized and characterized. Song et al. conducted a study revealing the influence of dipole moments tied to distinct stacking orientations within the MoSSe bilayer on interlayer interactions, thereby providing a mechanism for finely adjusting carrier lifetimes. Within their experimentation, they successfully synthesized a range of vertical and lateral heterostructures, which encompassed lateral MoSSe/WSSe, vertical MoSSe/WSSe, and the classical Janus/MoSSe/MoS_2 arrangement. These achievements indicate a significant advancement toward the fabrication of heterogeneous structures through the utilization of 2D Janus layers. Through vdW interactions, vertical and lateral Janus heterostructures with type-II band alignment spontaneously separate free electrons and holes to further improve high-efficiency photocatalytic activity.

Despite early milestones in this domain, a fundamental understanding of these fascinating Janus heterostructures, especially for the underlying mechanisms of excited carrier relaxation and recombination, remains unknown. In recent years, based on

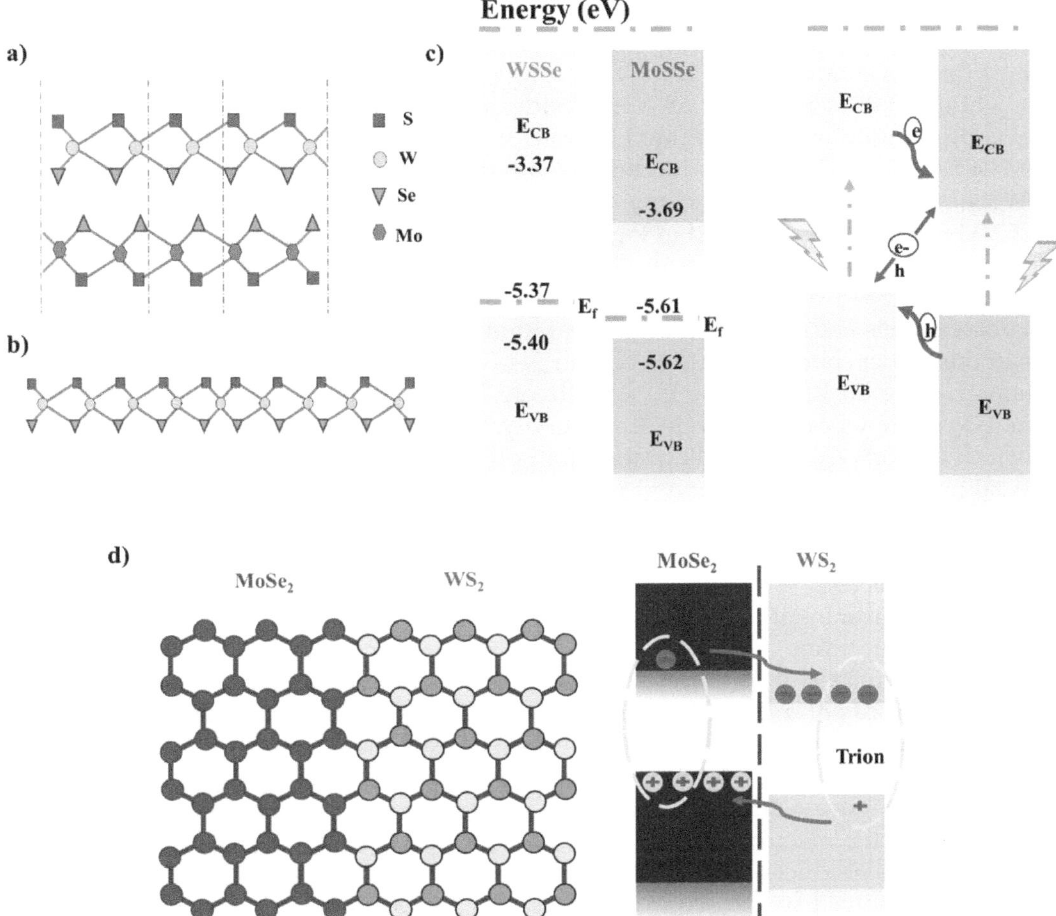

FIGURE 6.5
Side views of the MoSSe/WSSe (a) vertical and (b) lateral heterostructure (pink, yellow, blue, and orange balls represent S, W, Se, and Mo atoms, respectively). (c) Computed optical absorption spectra for monolayers of MoSSe and WSSe, as well as vertical and lateral heterostructures of MoSSe/WSSe. (d) An arrangement of $MoSe_2$/WS_2 heterostructure showcasing predicted band alignment and tightly bound trion in transition metal dichalcogenide heterostructure.

time-dependent nonadiabatic molecular dynamics (NAMD), it has become possible to gain valuable insights into photocarrier dynamics in many structures, including 2D monolayer semiconductors, halide perovskites, and heterostructure interfaces. Inspired by this, TMDC heterostructures, including vertical and lateral heterostructures formed by MoSSe and WSSe monolayers (**Figure 6.5(a) and (b)**), were investigated as templates.

To investigate some features, several theoretical calculations, including the composite approach of Heyd, Scuseria, and Ernzerhof (HSE06), have been employed to investigate the properties of heterostructures. The HSE06 calculated valance band maximum (VBM) and conduction band minimum (CBM) of MoSSe and WSSe monolayers exhibiting type-II band alignment and thus can effectively separate the electron–hole pairs. The work function difference between two semiconductors can cause charge redistribution, thereby facilitating photogenerated charge transfer at the interface. Specifically, the work function of the WSSe monolayer (5.37 eV) is lower than that of the MoSSe monolayer (5.61 eV), while the VBM and CBM of WSSe reside above the corresponding levels of MoSSe. This configuration implies that hot electrons from the CBM of WSSe can migrate to the CBM of MoSSe, and, likewise, hot holes can move from the VBM of MoSSe to the VBM of WSSe; and electron–hole recombination takes place between the VBM of WSSe and the CBM of MoSSe, as depicted in **Figure 6.5(c)**. Further, through the integration of time-dependent density functional theory with NAMD, lateral and vertical heterostructures of MoSSe and WSSe exhibiting type-II band alignments showcase impressive optical trapping capabilities and efficient charge separation properties. The electron and hole transfers in the vertical heterostructure occur in 544 fs and 2 ps, respectively, and in the lateral heterostructure, in 103 and 181 fs, respectively. The driving force behind this transfer is an out-of-plane vibrational mode. Importantly, the recombination of electron–hole pairs demands a relatively extended duration for both MoSSe and WSSe heterostructures. Due to weak nonadiabatic coupling and rapid decoherence between the CBM and VBM, the recombination time of the vertical heterostructure (11.42 ns) is threefold longer than that of the lateral heterostructure (4.43 ns), making them excellent candidates for photocatalyst applications [24].

One of the prominent features of TMDC monolayers is the strong Coulomb interaction between electrons and holes, which results in reduced dielectric screening due to the small dimensions. It produces strongly bound excitons with binding energies in the range of hundreds of meV. In these materials, the robust Coulomb interactions can also give rise to the emergence of charged exciton species referred to as trions. The schematic representation of a compact trion formation within the $MoSe_2/WS_2$ heterostructure is illustrated in **Figure 6.5(d)**. Since TMDCs are semiconductors, photoexcitation should produce excitons, electrostatically electron–hole pair bound. Recombination of electron–hole pairs produces a photon, which is observed as photoluminescence radiation. Momentum-resolved spectroscopy of single-layer TMDCs shows that exciton dipoles are confined only in the in-plane direction, with lifetime in the range of 100s before recombination. Early investigations into charged excitons or trions in MoS_2 have been quite clear in demonstrating the strong Coulomb interaction. Nanoscale excitonic spectra have been observed in single-layer TMDC semiconductors and heterogeneous structures. These excitons are very important in 2D TMDCs, as the reduced dimensions lead to stronger Coulomb interactions due to spatial confinement and less dielectric screening than in bulk crystals. Consequently, the behavior of excitons unequivocally governs the optical phenomena observed in TMDCs [22].

In general, the obtained heterogeneous structures offer new properties and applications beyond their 2D atomic crystals, and exciting experimental results have been reported in the past few years (Table 6.1).

6.3 Delocalization in Two-Dimensional Semiconductor Materials

Delocalization in the context of 2D materials refers to the spreading out or distribution of electronic charge, particularly electrons, across the entire material, rather than being confined to specific atomic sites.

TABLE 6.1

Future Research Prospects and Stacking Types for 2D Heterogeneous TMDC Structures Based on Different Stackings Are Presented

Heterostructure	Type	Year	Reference
WS_2/graphene	Lateral	2023	[25]
MoS_2/WS_2	Lateral	2022	[26]
NbS_2/MoS_2	Lateral	2023	[27]
WS_2/WSe_2	Lateral	2021	[28]
WSe_2/$MoSe_2$	Lateral	2022	[29]
SnS_2/MoS_2	Vertical	2023	[30]
MoO_2/MoS_2	Vertical	2022	[31]
$NiTe_2$/MoS_2	Vertical	2020	[32]
$MoSe_2$/WSe_2	Vertical	2023	[33]
SnS/SnS_2	Vertical and lateral	2020	[34]
MoSSe/WSSe	Vertical and lateral	2023	[24]
WS_2/MoS_2	Vertical and lateral	2021	[35]
Pd_2Se_3/MoS_2	Vertical and lateral	2022	[36]

6.3.1 Effect of π-Bonding in 2D Semiconducting Materials

In graphene, each carbon atom forms three σ-bonds with its neighboring carbon atoms and an additional π-bond that arises from the overlapping of the p orbitals of the carbon atoms. This π-bonding is responsible for the unique electronic properties of graphene. The π-bonds in graphene enable the formation of a delocalized cloud of π-electrons that can move relatively freely throughout the lattice. This delocalization (secondary electrical bonds) creates excellent electronic conductivity and also controls the structural stability of graphene [37, 38]. In addition, the π-bonding is responsible for the formation of the antibonding π* state through electronic excitation. As mentioned, each carbon atom in graphene has one π electron, which has a high probability of being found in a region just above or below the plane formed by the carbon atoms in a graphene layer. Overall, the π-bonding in graphene is crucial for its remarkable properties, including its high electrical conductivity, mechanical strength, and unique electronic band structure [39].

6.3.2 Effect of Low Dimensionality in 2D Semiconducting Materials

The dimensional reduction of 2D materials has significant implications for electron bonding and behavior, which can be referred to as limiting the number of atomic sites available for electrons. In bulk materials, electrons are subject to scattering and collisions with impurities, defects, and lattice vibrations, which can hinder their motion and lead to lower conductivity. However, in 2D materials, the confinement in one or two dimensions reduces the extent of scattering, allowing electrons to move more freely and exhibit higher mobility [40, 41]. It is worth noting that the lack of surface groups or dangling bonds in 2D materials reduces charge carrier scattering, which is common in bulk materials. Moreover, the reduction in dimensions affects the bonding energy and cohesion of 2D materials, and the high electrical conductivity of 2D materials, driven by their small dimensions, makes them suitable for applications in electronics and photonics. These materials offer unprecedented design freedom for novel p-n junction device topologies, surpassing conventional bulk semiconductors [40, 42].

6.3.3 Quantum Effects in 2D Semiconducting Materials

2D materials have one or two layers of atoms, which gives them extraordinary properties compared to their bulk counterparts. These materials show quantum confinement at least in one dimension, which leads to the quantization of electronic energy levels and changes in the behavior of electrons in them. So, the small thickness of these materials allows for a greater influence of quantum effects. In this phenomenon, particles, such as electrons, can "tunnel" through energy barriers that would be insurmountable according to classical physics [43]. This occurrence arises due to the wave-like nature of particles at the quantum level. When a particle encounters an energy barrier, its wavefunction extends into the classically forbidden region, and there is a finite probability that the particle can appear on the other side of the barrier without actually having to traverse it in the classical sense. The energy barriers that electrons encounter can be comparable in size to the characteristic energy levels of these confined systems [44]. Eventually, electrons can tunnel through these barriers and exhibit behaviors that deviate significantly from classical physics.

This delocalization has implications for electrical conductivity, energy transport, and other electronic properties. Additionally, quantum tunneling contributes to effects such as quantum capacitance, where the charge density and voltage relationship becomes quantized due to the discrete energy levels available to the electrons in the confined dimensions. For example, the delocalization of electrons due to quantum effects, as well as the unique electronic band structure of some 2D materials, can lead to the formation of so-called Dirac cones or other special energy dispersion relationships. In materials with such properties, the behavior of charge carriers (electrons and holes) is more similar to relativistic particles than traditional electrons in bulk materials. This can result in linear energy–momentum relationships and extremely high carrier mobility, contributing to the high conductivity of 2D materials [41, 45].

Due to the significant progress of the perovskite material in the photovoltaic industry, In addition to the aforementioned 2D structures [46–49], there are other 2D semiconductors such as high-membered quasi-two-dimensional (quasi-2D) tin halide perovskites (**Figure 6.6(a)**) and structural order that have improved crystal lattice organization. For the high-member quasi-2D film, the formamidinium iodide precursor is replaced with mixed phenylethylammonium bromide and phenethylammonium thiocyanate. This substitution resulted in a layered perovskite with a chemical formula represented as $PEA_2FA_{n-1}Sn_nX_{3n+1}$, where X can be either a halide or pseudohalide anion (such as Br^-, I^-, or SCN^-), and n denotes the number of perovskite layers encapsulated by the ligands. Quasi-2D tin halide perovskites are a subset of perovskite materials (**Figure 6.6(b)**) known for their promising optoelectronic properties, particularly in photovoltaic and LED applications. Understanding their electronic structure and exciton dynamics is crucial for optimizing their performance; consequently, the presence of exciton displacement is prominent in these materials. This means that the electron and hole forming the exciton are spread out over multiple atomic sites rather than being confined to a single site (**Figure 6.6(c)**). Therefore, the observed exciton delocalization suggests that high-member quasi-2D tin halide perovskites have enhanced optoelectronic properties, making them potentially more efficient for applications such as solar cells (over fivefold increase in exciton lifetime and much improved solar cell efficiency in devices) and LEDs. These results

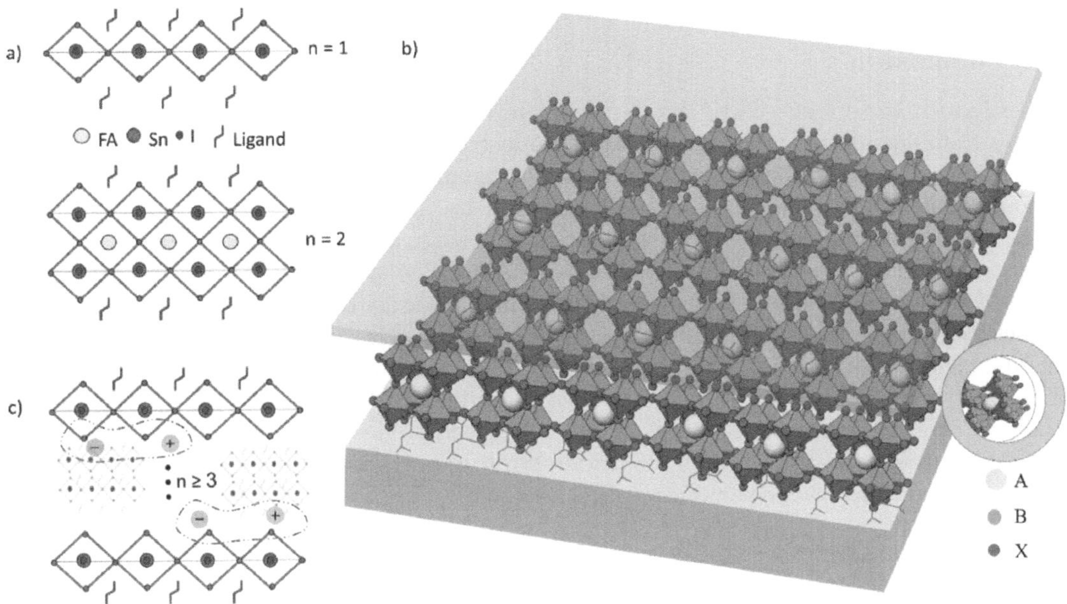

FIGURE 6.6
(a) In cases, n = 1 and 2 denote pure 2D phases, while n ≥ 3 represents quasi-two-dimensional (quasi-2D) phases. (b) Schematic illustration of perovskite total view. (c) Schematic illustrations of photogenerated excitons in the formed high-member quasi-2D perovskite film.

provide insights into the structure–property relationship of high-performance quasi-2D tin perovskite optoelectronic devices [50].

In a recent study, researchers used electronic delocalization effects to demonstrate that topological factors are effective for the nanoscale engineering of bandgaps and thermodynamic stability in partially oxidized graphene. Since conductivity arises from the delocalized π-electrons, chemical intuition suggests that selective saturation of some sp^2 carbons will allow strategic control over the bandgap. However, the logical cognition of different 2D π-delocalization topologies is complex. Using partially oxidized graphene with its facile and reversible epoxides, it was shown that delocalization crucially influences the nature of the frontier bands. Therefore, two distinct substructures, the >C=C< unit and the aromatic sextet, open the largest possible bandgap in the partially oxidized graphene. Progressive extension of conjugation from these fundamental units will result in a proportional reduction in the bandgap and a subsequent increase in their thermodynamic stability. Isolated molecular fragments of sp^2 carbons from sp^3 carbons of the epoxide groups have the bandgap mostly direct with its magnitude proportional to the HOMO–LUMO gap of corresponding conjugated hydrocarbon with a minor increase in the hyperconjugative effects. The bandgaps of one-dimensionally extended conjugated systems mostly have indirect bandgaps, whose magnitude is proportional to the degree of asymmetry in their π-bond order. Network topologies that allow fixed single or double bonds help open the bandgap in that direction but are detrimental to their stability. Higher thermodynamic stability requires aggregating sp^2 regions together as it enhances delocalization. Totally, the systematic inquiry into the isomeric nanosheets modeled in this work can be generalized to comprehensively understand the nature of π-delocalization and other electronic effects on the stabilities and bandgaps of partially oxidized graphene, irrespective of the degree of saturation [3].

6.4 Conclusion

Investigating the effects of MO displacement and stacking in 2D semiconducting materials shows their significant potential for various electronic and optical applications; for example, AB-stacked BLG has a layer spacing of 3.4 Å, which closely resembles that of graphite. Consequently, AB-stacked BLG exhibits superior thermodynamic stability to AA-stacked BLG. On the other side, in MoSSe and WSSe heterostructures, due to the weak nonadiabatic coupling and fast misalignment between the CBM and the VBM, the recombination time of the vertical heterostructure is three times longer than that of the lateral heterostructure. This feature makes them excellent candidates for photocatalyst applications or in MO delocalization of high-membered quasi-2D tin halide perovskites that enhanced optoelectronic properties by the observed exciton localization effect, making them potentially more efficient for applications such as solar cells. These results provide insights into the structure–property relationship, understanding and controlling electronic properties of 2D-semiconductors for various applications in electronics and beyond.

References

[1] Z. Peng, X. Chen, Y. Fan, D.J. Srolovitz, D. Lei, Strain engineering of 2D semiconductors and graphene: From strain fields to band-structure tuning and photonic applications, *Light: Science & Applications*, 9 (2020) 190.

[2] J.Y. Lee, J.H. Shin, G.H. Lee, C.H. Lee, Two-dimensional semiconductor optoelectronics based on van der Waals heterostructures, *Nanomaterials (Basel)*, 6 (2016).

[3] G. Jhaa, P.D. Pancharatna, M.M. Balakrishnarajan, Topological impact of delocalization on the stability and band gap of partially oxidized graphene, *ACS Omega*, 8 (2023) 5124–5135.

[4] H.W. Guo, Z. Hu, Z.B. Liu, J.G. Tian, Stacking of 2D materials, *Advanced Functional Materials*, 31 (2020).

[5] A. Chaves, J.G. Azadani, H. Alsalman, D.R. da Costa, R. Frisenda, A.J. Chaves, S.H. Song, Y.D. Kim, D. He, J. Zhou, A. Castellanos-Gomez, F.M. Peeters, Z. Liu, C.L. Hinkle, S.-H. Oh, P.D. Ye, S.J. Koester, Y.H. Lee, P. Avouris, X. Wang, T. Low, Bandgap engineering of two-dimensional semiconductor materials, *npj 2D Materials and Applications*, 4 (2020).

[6] S.Y. Zhou, G.-H. Gweon, A. Fedorov, P. First, de, W. De Heer, D.-H. Lee, F. Guinea, A. Castro Neto, A. Lanzara, Substrate-induced bandgap opening in epitaxial graphene, *Nature Materials*, 6 (2007) 770–775.

[7] Q. Wan, Z. Xiao, A. Kursumovic, J.L. MacManus-Driscoll, C. Durkan, Ferrotronics for the creation of band gaps in graphene, *Physical Review B*, 107 (2023) 045428.

[8] K. Panda, E. Inami, Y. Sugimoto, K. Sankaran, I.-N. Lin, Straight imaging and mechanism behind grain boundary electron emission in Pt-doped ultrananocrystalline diamond films, *Carbon*, 111 (2017) 8–17.

[9] D. Jariwala, A. Srivastava, P.M. Ajayan, Graphene synthesis and band gap opening, *Journal of Nanoscience and Nanotechnology*, 11 (2011) 6621–6641.

[10] C. Dai, D. Popple, C. Su, J.-H. Park, K. Watanabe, T. Taniguchi, J. Kong, A. Zettl, Evolution of nanopores in hexagonal boron nitride, *Communications Chemistry*, 6 (2023) 108.

[11] Y. Huang, X. Li, H. Cui, Z. Zhou, Bi-layer graphene: Structure, properties, preparation and prospects, *Current Graphene Science*, 2 (2019) 97–105.

[12] S. Debroy, V.P. Kumar, K.V. Sekhar, S.G. Acharyya, A. Acharyya, Synergistic effect of temperature and point defect on the mechanical properties of single layer and bi-layer graphene, *Superlattices and Microstructures*, 110 (2017) 205–214.

[13] M. Ould Ne, M. Boujnah, A. Benyoussef, A.E. Kenz, Electronic and electrical conductivity of AB and AA-stacked bilayer graphene with tunable layer separation, *Journal of Superconductivity and Novel Magnetism*, 30 (2017) 1263–1267.

[14] M.L. Ould Ne, M. Boujnah, A. Benyoussef, A.E. Kenz, Electronic and electrical conductivity of AB and AA-stacked bilayer graphene with tunable layer separation, *Journal of Superconductivity and Novel Magnetism*, 30 (2016) 1263–1267.

[15] T. Ohta, A. Bostwick, T. Seyller, K. Horn, E. Rotenberg, Controlling the electronic structure of bilayer graphene, *Science*, 313 (2006) 951–954.

[16] Y. Zhang, T.-T. Tang, C. Girit, Z. Hao, M.C. Martin, A. Zettl, M.F. Crommie, Y.R. Shen, F. Wang, Direct observation of a widely tunable bandgap in bilayer graphene, *Nature*, 459 (2009) 820–823.

[17] J.C. Meyer, A. Geim, M. Katsnelson, K. Novoselov, D. Obergfell, S. Roth, C. Girit, A. Zettl, On the roughness of single-and bi-layer graphene membranes, *Solid State Communications*, 143 (2007) 101–109.

[18] B. Peng, G. Yu, X. Liu, B. Liu, X. Liang, L. Bi, L. Deng, T.C. Sum, K.P. Loh, Ultrafast charge transfer in MoS2/WSe2 p–n heterojunction, *2D Materials*, 3 (2016) 025020.

[19] Z. Chen, M. Zhu, T. Ren, J. He, K.P. Loh, Q.-H. Xu, Transient reflection spectroscopy on ultrafast interlayer charge transfer processes in a MoS2/WSe2 van der Waals heterojunction, *The Journal of Physical Chemistry C*, 125 (2021) 26575–26582.

[20] K. Khan, A.K. Tareen, M. Aslam, R. Wang, Y. Zhang, A. Mahmood, Z. Ouyang, H. Zhang, Z. Guo, Recent developments in emerging two-dimensional materials and their applications, *Journal of Materials Chemistry C*, 8 (2020) 387–440.

[21] H. Taghinejad, A.A. Eftekhar, A. Adibi, Lateral and vertical heterostructures in two-dimensional transition-metal dichalcogenides [Invited], *Optical Materials Express*, 9 (2019).

[22] S. Joseph, J. Mohan, S. Lakshmy, S. Thomas, B. Chakraborty, S. Thomas, N. Kalarikkal, A review of the synthesis, properties, and applications of 2D transition metal dichalcogenides and their heterostructures, *Materials Chemistry and Physics*, 297 (2023).

[23] W. Yu, K. Gong, Y. Li, B. Ding, L. Li, L. Xu, R. Wang, L. Li, G. Zhang, S. Lin, Flexible 2D materials beyond graphene: Synthesis, properties, and applications, *Small*, 18 (2022) e2105383.

[24] T. Bao, X. Yu, X. Wang, J. Zhao, Y. Su, Photoinduced carrier transfer dynamics in MoSSe/WSSe vertical and lateral heterostructures, *The Journal of Physical Chemistry C*, 127 (2023) 2078–2087.

[25] Z. Razaghi, S.A. Hosseini, A. Simchi, Photoresponsivity of ultrathin 2D WS2/graphene heterostructures, *Physica E: Low-dimensional Systems and Nanostructures*, 147 (2023).

[26] X. Wan, S. Xu, M. Gao, T. Huang, Y. Duan, R. Zhan, K. Chen, X. Gu, W. Xie, J. Xu, Gate-tunable junctions within monolayer MoS2–WS2 lateral heterostructures, *ACS Applied Nano Materials*, 5 (2022) 15775–15784.

[27] Z. Wang, M. Tripathi, Z. Golsanamlou, P. Kumari, G. Lovarelli, F. Mazziotti, D. Logoteta, G. Fiori, L. Sementa, G.M. Marega, Substitutional p-type doping in NbS2–MoS2 Lateral heterostructures grown by MOCVD, *Advanced Materials*, 35 (2023) 2209371.

[28] C. Herbig, C. Zhang, F. Mujid, S. Xie, Z. Pedramrazi, J. Park, M.F. Crommie, Local electronic properties of coherent single-layer WS(2)/WSe(2) lateral heterostructures, *Nano Letters*, 21 (2021) 2363–2369.

[29] M. Shimasaki, T. Nishihara, K. Matsuda, T. Endo, Y. Takaguchi, Z. Liu, Y. Miyata, Y. Miyauchi, Directional exciton-energy transport in a lateral heteromonolayer of WSe(2)-MoSe(2), *ACS Nano*, 16 (2022) 8205–8212.

[30] L. Chen, W. Yao, T. Su, R. Xu, S. Wan, J. Yang, X. Peng, D. Li, H. Yuan, Y. Fu, X. Wang, Layer-dependent valley depolarization and Raman phonon softening of SnS2/MoS2 vertical van der Waals heterostructures, *ACS Applied Electronic Materials*, 5 (2023) 3489–3498.

[31] Q. Wu, Y. Luo, R. Xie, H. Nong, Z. Cai, L. Tang, J. Tan, S. Feng, S. Zhao, Q. Yu, Space-confined one-step growth of 2D MoO2/MoS2 vertical heterostructures for superior hydrogen evolution in alkaline electrolytes, *Small*, 18 (2022) 2201051.

[32] X. Zhai, X. Xu, J. Peng, F. Jing, Q. Zhang, H. Liu, Z. Hu, Enhanced optoelectronic performance of CVD-grown metal-semiconductor NiTe(2)/MoS(2) heterostructures, *ACS Applied Materials & Interfaces*, 12 (2020) 24093–24101.

[33] Z. Li, F. Tabataba-Vakili, S. Zhao, A. Rupp, I. Bilgin, Z. Herdegen, B. Marz, K. Watanabe, T. Taniguchi, G.R. Schleder, A.S. Baimuratov, E. Kaxiras, K. Muller-Caspary, A. Hogele, Lattice reconstruction in MoSe(2)-WSe(2) heterobilayers synthesized by chemical vapor deposition, *Nano Letters*, 23 (2023) 4160–4166.

[34] Y. Cheng, P. Tang, P. Liang, X. Liu, D. Cao, X. Chen, H. Shu, Sulfur-driven transition from vertical to lateral growth of 2D SnS–SnS2 heterostructures and their band alignments, *The Journal of Physical Chemistry C*, 124 (2020) 27820–27828.

[35] B. Pielic, D. Novko, I.S. Rakic, J. Cai, M. Petrovic, R. Ohmann, N. Vujicic, M. Basletic, C. Busse, M. Kralj, Electronic structure of quasi-freestanding WS(2)/MoS(2) heterostructures, *ACS Applied Materials & Interfaces*, 13 (2021) 50552–50563.

[36] H. Park, G.S. Jung, K.M. Ibrahim, Y. Lu, K.L. Tai, M. Coupin, J.H. Warner, Atomic-scale insights into the lateral and vertical epitaxial growth in two-dimensional Pd(2)Se(3)-MoS(2) heterostructures, *ACS Nano*, 16 (2022) 10260–10272.

[37] W. Yu, L. Sisi, Y. Haiyan, L. Jie, Progress in the functional modification of graphene/graphene oxide: A review, *RSC Advances*, 10 (2020) 15328–15345.

[38] S.K. Tiwari, S. Sahoo, N. Wang, A. Huczko, Graphene research and their outputs: Status and prospect, *Journal of Science: Advanced Materials and Devices*, 5 (2020) 10–29.

[39] A.R. Urade, I. Lahiri, K. Suresh, Graphene properties, synthesis and applications: A review, *JOM*, 75 (2023) 614–630.

[40] P.V. Pham, S.C. Bodepudi, K. Shehzad, Y. Liu, Y. Xu, B. Yu, X. Duan, 2D heterostructures for ubiquitous electronics and optoelectronics: Principles, opportunities, and challenges, *Chemical Reviews*, 122 (2022) 6514–6613.

[41] T. Dutta, N. Yadav, Y. Wu, G.J. Cheng, X. Liang, S. Ramakrishna, A. Sbai, R. Gupta, A. Mondal, Z. Hongyu, Electronic properties of 2D materials and their junctions, *Nano Materials Science*, 6 (2024) 1–23.

[42] R. Gutzler, D.F. Perepichka, π-Electron conjugation in two dimensions, *Journal of the American Chemical Society*, 135 (2013) 16585–16594.

[43] F. Trixler, Quantum tunnelling to the origin and evolution of life, *Current Organic Chemistry*, 17 (2013) 1758–1770.

[44] X. Zou, Y. Xu, W. Duan, 2D materials: Rising star for future applications, *Innovation (Camb)*, 2 (2021) 100115.

[45] S.H. Mir, V.K. Yadav, J.K. Singh, Recent advances in the carrier mobility of two-dimensional materials: A theoretical perspective, *ACS Omega*, 5 (2020) 14203–14211.

[46] E. Sheibani, L. Yang, J. Zhang, Recent advances in organic hole transporting materials for perovskite solar cells, *Solar RRL*, 4 (2020).

[47] X. Li, M. Haghshenas, L. Wang, J. Huang, E. Sheibani, S. Yuan, X. Luo, X. Chen, C. Wei, H. Xiang, G. Baryshnikov, L. Sun, H. Zeng, B. Xu, A multifunctional small-molecule hole-transporting material enables perovskite QLEDs with EQE exceeding 20%, *ACS Energy Letters*, 8 (2023) 1445–1454.

[48] E. Sheibani, M. Moslempoor, F. Arami Ghahfarokhi, Investigation of hole transporting materials based on p-type polymers in invert perovskite solar cells (in press), *Iranian Journal of Polymer Science and Technology - IPPI (Persian)*, 1 (2023).

[49] J. Deng, H. Ahangar, Y. Xiao, Y. Luo, X. Cai, Y. Li, D. Wu, L. Yang, E. Sheibani, J. Zhang, Side-group-mediated small molecular interlayer to achieve superior passivation strength and enhanced carrier dynamics for efficient and stable perovskite solar cells, *Advanced Functional Materials*, (2023) 2309484.

[50] Z. Xing, Z. Zang, H. Li, Z. Ning, K.S. Wong, P.C.Y. Chow, Improved structural order and exciton delocalization in high-member quasi-two-dimensional tin halide perovskite revealed by electroabsorption spectroscopy, *The Journal of Physical Chemistry Letters*, 14 (2023) 4349–4356.

7

Properties of 2D Semiconducting Materials

Surinder Pal Kaur and T. J. Dhilip Kumar*
Correspondence: dhilip@iitrpr.ac.in

7.1 Introduction

The discovery of graphene in 2004 led to the exploration of numerous two-dimensional materials and activated a new era in materials sciences [1]. Depending on their chemical composition and structural arrangements, these nanomaterials have metallic, semi-metallic, semiconducting, and insulating properties [2]. Among all the nanomaterials, two-dimensional semiconducting materials (2D-SCMs) are of great importance because of their use in various sectors such as optoelectronics, electronics, photonics, and sensors [2,3].

In general, 2D-SCMs are a group of materials that are arranged in layered structures with a few thicknesses of individual monolayers, bonded to each other via long-range interactions. The family of 2D-SCMs possesses a unique range of electronic, optical, thermal, and mechanical properties, making them promising candidates for semiconducting devices, photocatalysis, sensing devices, and so on [2].

In recent years, 2D-SCMs attained the attention of researchers in materials sciences to utilize these materials for different applications. For example, silicene, phosphorene, borophene, and two-dimensional dichalcogenides (2D-TMDs) have attained research focus. The different types of two-dimensional materials and energy diagrams of selected 2D-SCMs are shown in Figure 7.1.

Because of their high surface-to-volume ratio and semiconducting properties, such materials show potential to be used in ultra-small and low-power transistors. Moreover, the tuning of electronic band structures via layer modulation, alloying, and doping accelerates the interest in 2D-SCMs. The ease of exfoliation, optical properties, mechanical properties, thermal properties, and electrical properties of 2D-SCMs have been the topic of extensive studies, both theoretically and experimentally. An in-depth understanding of these properties is crucial to maximize their potential in various applications.

In this chapter, the unique optical, electrical, thermal, and mechanical properties of these nanomaterials are discussed explicitly. The effect of modulation of bandgaps with doping and external stimuli opens new avenues for tailored applications, which are discussed in this chapter. The progress and applications of 2D-SCMs in various sectors are also discussed. This chapter aims to provide an overview of the properties of 2D-SCMs.

DOI: 10.1201/9781003439448-7

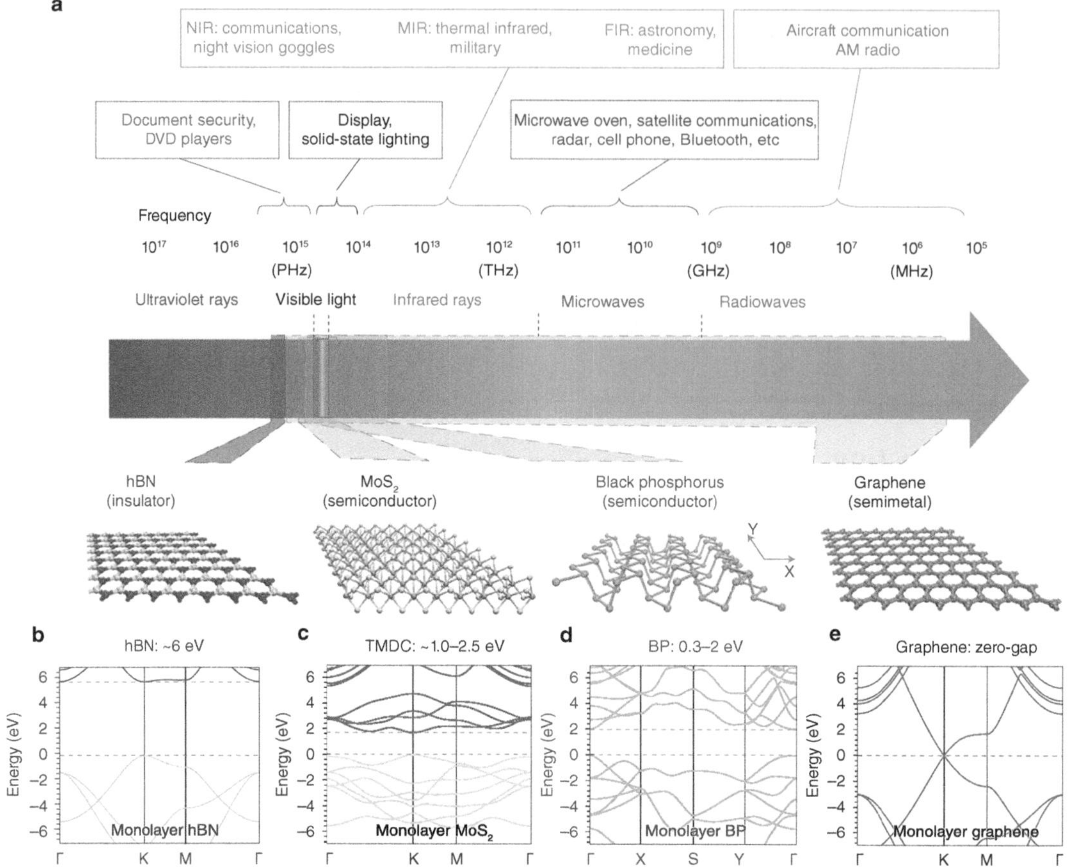

FIGURE 7.1

(a) The spectral range of 2D materials and their applications; (b) the structural geometries and band structures of (b) hBN, (c) MoS_2, (d) black phosphorus, and (e) graphene [2(b)]. Adapted with permission [2(b)]. Copyright (2014), Springer Nature Publications.

7.2 Optical Properties

The notable optical behavior of graphene is responsible for the investigation of the optical properties of 2D-SCMs [4,5]. The optical properties of 2D-SCMs are of significant importance because of their utilization in optoelectronic and photonic applications [6,7]. Moreover, the optical properties of these nanomaterials are dependent on bandgaps and electron transitions [8].

7.2.1 Linear Optical Properties

The optical properties of nanomaterials were investigated through the interaction of light with thin-layered nanomaterials and helped to determine the feasibility of materials for optical products [9]. Spectroscopic ellipsometry has been used as a competent method to characterize the 2D-SCMs and to understand their bandgaps. The optical constants and

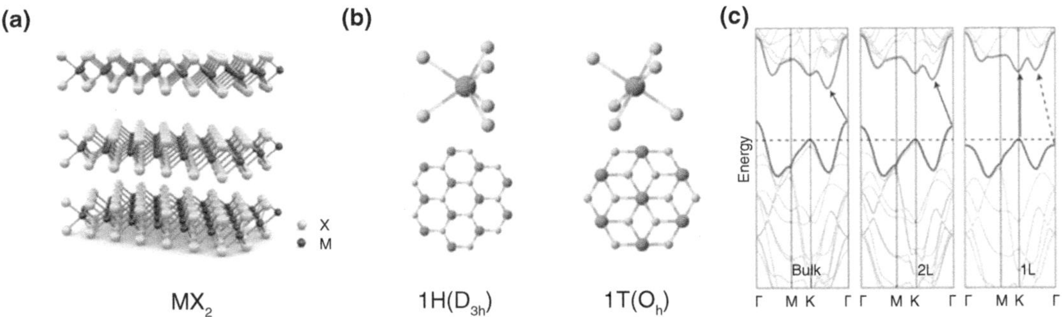

FIGURE 7.2
(a) The structure of MX$_2$ nanomaterial, where M = transition metal atom and X = chalcogenide [12 (a, b)], (b) the hexagonal and tetragonal polymorphs of MoS$_2$ [12(c)], and (c) band structure of bulk, bilayered (2L), and monolayer (1L) MoS$_2$ [12(d)]. Adapted with permissions from (a) Copyright (2023) Springer Nature [12(b)], (b) Copyright (2023) Springer Nature [12(c)], and (c) Copyright (2010) American Chemical Society [12(d)].

thickness can be determined with the help of the evaluation of change in the polarization state of the reflected film from the surface of the film [10,11].

The bandgap of 2D-SCMs determines its tendency to absorb and emit light [12]. The bandgap modulation is viable by strain engineering, doping, and forming heterostructures. For example, semiconducting TMDs show unique and notable optical properties compared to bulk semiconductors [13]. TMDs show the shift from indirect to direct bandgaps when lowered down from bulk phase to monolayers [14]. The correlation between band states and the number of layers is shown in Figure 7.2. Because of the alteration in the band structures by tuning a number of layers, a change in the photoluminescence can also be observed [15]. For example, the photoluminescence of TMDs can be improved considerably compared to the multilayer [16]. To improve the quantum efficiency, the MoS$_2$ monolayer can be dipped into an organic-based superacid solution [17]. These outcomes showed the utilization of TMDs as optoelectronic properties.

The 2D-SCMs have strong light–matter interactions, which lead to the absorption of light to a significant extent in the visible range. Studies have shown that 2D-SCMs such as MoS$_2$, MoSe$_2$, WS$_2$, and WSe$_2$ tend to absorb ~5–10% of visible sunlight [18–20]. Moreover, due to the quantum confinement effect, the 2D-SCMs show extremely large exciton energy compared to inorganic semiconductors. Theoretical studies have shown that exciton binding energies of TMDs vary between 0.3 and 1.0 eV [14].

The optical emission changes with a change in the thickness of 2D-SCMs due to the drastic change in the band structures from bulk to monolayer as mentioned earlier [12]. The tuning of photoluminescence properties of 2D-SCMs such as MoS$_2$ with oxygen and water have also been studied in the past [21]. The study showed that the adsorption of p-type dopants helps to enhance the photoluminescence intensity of the MoS$_2$ monolayer.

The change in band structures with applied strain for 2D-TMDs was studied by Lau et al [22]. The report suggested the possibility of tuning the bandgap of the material with 100 meV with 1% external pressure on its surface.

In the case of the WS$_2$, which was synthesized by chemical vapor deposition (CVD), photoluminescence was observed to be inversely proportional to the number of layers [23]. Along with that, the edges of WS_2 were observed to emit 25% higher photoluminescence compared to the central part. The studies also showed that defects enhance photoluminescence.

7.2.2 Nonlinear Optical Properties

The optical absorption of nanomaterials changes with the intensity of the incident light. This phenomenon is called nonlinear optical absorption, and it is typically observed in strong nonlinear optical materials. These absorption effects are of immense importance for photonics and the design of optical limiters, optical switchers, modulators, and so on. TMDs show two-photon absorption properties indicating their application potential in the near future. Both WS_2 and MoS_2 monolayers showed that the nonlinear absorption varies with the variation in the number of layers [24]. For the WS_2 monolayer, saturation in two-photon absorption was observed after three layers.

7.2.3 Photodetectors

Photodetectors are a prominent technology in today's optoelectronic devices. The internal photoelectric effect is a process in which an incoming photon of sufficient energy interacts with an electron and promotes it to the conduction band [25]. The quantum efficiency of a photodetector is determined by the ratio of the number of electrons emitted to the total number of incident photons. The number of electron–hole pairs or carriers shows a direct correlation with absorption energy. With time, the number of carriers continues to increase with the increase in the absorbed energy, and it leads to the formation of the linear relationship between absorption energy and time. The electron–hole combination has a finite lifetime in the excited state and eventually recombines. The carrier pairs are separated by the finite bandgap of the material [25].

Among 2D-SCMs, TMDs have attained research attention for acting as promising candidates for optoelectronic devices. As mentioned earlier, TMDs tend to absorb and emit photons at fundamental bandgaps with promising photoluminescence and electroluminescence effects [26]. Additionally, during absorption, the existence of intense interband transition peaks can be explained by the heavy effective mass of d-electrons and by van Hove peaks in the band plots [27]. It has been suggested that a 300 nm thick film of TMD tends to absorb ~95% of the incident rays due to its optical absorption values being higher than 10^7 m^{-1} [28]. The double-layered TMDs showed low density of states which leads to enhanced carrier mobility [29]. The optoelectronic properties of 2D-SCMs can further be enhanced by forming heterostructures [25]. Thus, the effective absorption and emission of light make 2D-SCMs efficient photodetectors. The strong bound carriers enable better charge separation and improve the sensitivity of the photodetectors. The 2D-SCMs have the potential for novel and innovative designs which could improve performance and applications.

7.2.4 Laser

Materials possessing nonlinear optical properties are of immense importance in laser devices. A suitable nonlinear nanomaterial should have a quick response time, high linearity, wide wavelength limit, little optical loss, and low energy consumption and be cheaper [30, 31]. Previously, single-walled carbon nanotubes were considered a suitable material for this application as they fulfill all the aforementioned requirements [25]. However, its difficulty in obtaining high modulation depths leads to the exploration of better alternatives such as 2D-SCMs.

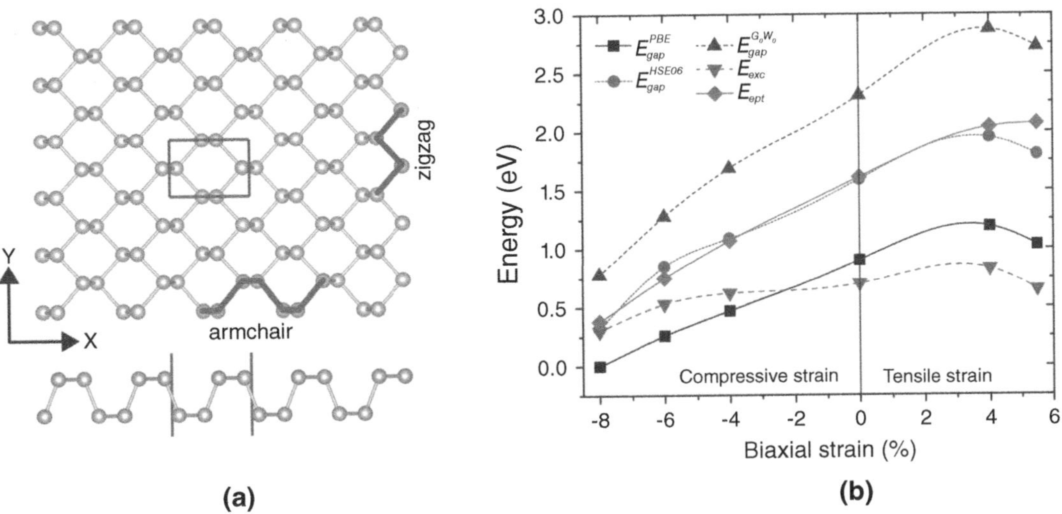

FIGURE 7.3

(a) The top and side views of black phosphorus monolayer. (b) The electronic bandgap (E_{gap}), optical gap (E_{opt}), and exciton binding energy (E_{exc}) as a function of biaxial strain. Adapted with permission [32]. Copyright (2023), American Physical Society.

7.2.5 Effect of Strain

The tuning of bandgaps of 2D-SCMs with homogenous strain and their implication on different physical properties have been studied in the past using theoretical methods. For example, for black phosphorous, the optical properties and exciton binding energies are directly proportional to tensile strain as shown in Figure 7.3 [32]. A comparable trend is also observed for uniaxial strain, that is, the transition from direct to indirect gaps. Studies have shown that for both silicene and germanene, band structures of both can be tuned by homogenous stress. The Dirac point gaps can be tuned with the help of strain. It increases with compression and reduces with tensile stress.

The optoelectronic behaviors of 2D-SCMs can also be tuned by imposing non-uniform or inhomogenous strains, which are of utmost importance for the excellent functioning of these materials in optoelectronic devices. A non-uniform external strain results in continuous spatial deviancy in the band structures of 2D-SCMs, which has been suggested to design a solar energy funnel to collect the energy of a broad range.

7.3 Electrical Properties of 2D-SCMs

A deeper understanding of the electrical properties of 2D-SCMs is important for utilizing them in electronic machines. The high charge carrier mobility in 2D-SCMs compared to their bulk counterparts makes them suitable for high-speed electronic devices [33]. The ease with which charge carriers move through 2D-SCMs determines their field-effect mobility. This in turn is important for the performance of field-effect transistors (FETs) which use an electrical field to control the current. The electrical properties of bulk TMDs were investigated for decades and were used for FETs for ~10 years [34]. The role of TMD monolayers in FETs was investigated only after the discovery of graphene. At first, field-effect mobility for MoS_2 monolayer was reported to be quite smaller compared to graphene FETs [35]. Later, the field-effect mobility of top-gated SL-MoS_2 was observed for ambient conditions [36]. The results showed that the top dielectric helps to enhance the carrier mobility in the material by screening out the effects of the surrounding environment.

The carrier mobility in ultrathin 2D-SCMs such as TMDs is equally sensitive to the interface on each side of the semiconducting layer, which may introduce defects and impurities that can scatter the charge carriers [33]. Moreover, the increase in charge mobilities is observed by increasing the thickness of TMDs.

The comparative analysis for field-effect mobilities and on/off ratios for different semiconducting materials was made in the past. 2D-SCMs have followed similar trends to organics, oxides, and nanotubes. This suggests that 2D-SCMs can be used for FETs and can perform as efficiently as FETs built from other semiconducting materials. The 2D-SCMs can also be used in flexible FETs and can withstand mechanical bindings and, hence, are essential for wearable electronics and flexible displays.

7.4 Thermal Properties of 2D-SCMs

The 2D-SCMs have attained research interest because of their exceptional thermal properties. The TMDs have a different lattice structure than that of graphene and, hence, have different phonon transport behavior and lower conductivity [37]. Unlike graphene, the thermal conductivities of TMDs showed no dependence on dimensions and coarseness. This mainly occurs due to the short phonon mean free path (MFP) [38]. Among all the TMD nanomaterials, the thermal properties of TMDs formed with Mo and W transition metal atoms have been extensively studied in the past. The theoretical studies have shown that the thermal conductivities are not dependent on the transition metal atoms.

The change in thermal conductivity with the change in the number of layers was also studied previously by different research groups. Although theoretical studies showed a correlation between the layers and thermal conductivities, a much weaker dependency of these two factors was reported by experimental studies. Research has shown that isotopic defects could affect the phonon transport. For example, in the case of isotopic defected MoS_2, thermal conductivity was found to be ~50% higher than that of pristine monolayer [39].

The black phosphorene nanomaterial showed orientation-dependent phonon transport properties as it has honeycomb-like arrangements, whereas, in the blue phosphorene, the phonon transport is independent of the orientation because of its zigzag structure [37]. Similarly, the thermal properties of more 2D-SCMs have also been studied in the past. For instance, Bi_2Te_3 showed bulk-like thermal conductivity with a thickness of ~5 nm [40]. It has also been observed that the thermal conductivity of the material showed nonmonotonic dependence on layer thickness. However, later experimental studies have shown a rise in the thermal conductivity as the thickness increases.

The thermal stability behavior for N-graphdiyne was computed for various temperatures using density functional theory (DFT) methods [41]. The ab initio molecular dynamics results showed that the material remained undistorted up to 2000 K. The understanding of the thermal conductivity of such materials is also important to understanding the usage of these materials in nanodevices. High thermal conductivities are helpful

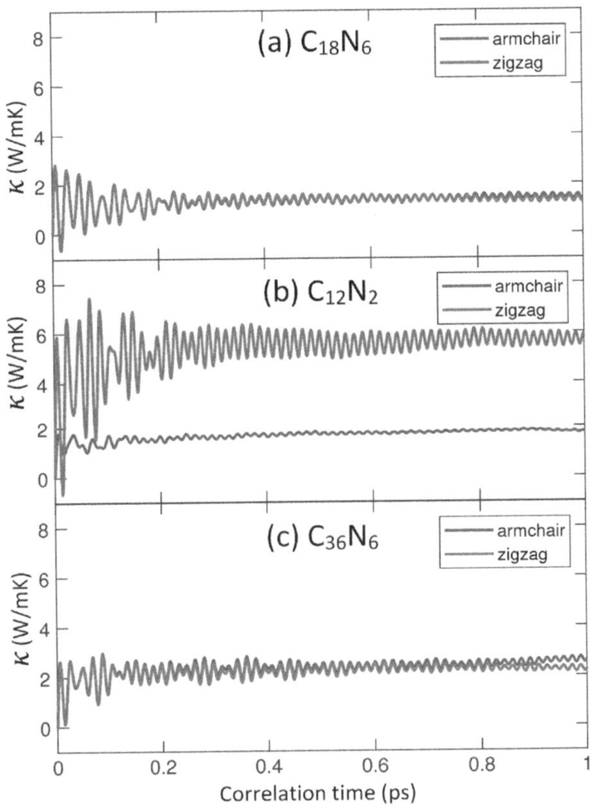

FIGURE 7.4
The thermal conductivity with the function of correlation time of $C_{18}N_6$, $C_{12}N_2$, and $C_{36}N_6$ monolayers. Adapted with permission [41]. Copyright (2023) Elsevier.

to prevent overheating issues, whereas low thermal conductivities of 2D-SCMs are desirable in thermoelectric materials. The authors have computed the thermal conductivity of C- and N-derived nanomaterial utilizing the equilibrium molecular dynamics (EMD) method [42]. The change in thermal conductivities with period is provided in Figure 7.4. The figure indicates that thermal conductivity values are approximately three times lower than that of pristine graphene.

Along with that, the heat transfer is dependent on the orientation for $C_{18}N_6$ and $C_{36}N_6$ whereas it is anisotropic for $C_{12}N_2$. The EMD results also suggested that the N-graphdiyne nanomaterials are not sensitive to temperature change. The insensitivity of these nanomaterials to the varied temperatures due to the contributions of the phonon–phonon scatterings is quite negligible.

Young's modulus and thermal conductivities were computed for BeN_4, MgN_4, and PtN_4 monolayers earlier [43]. The material is anisotropic and has higher values in the armchair direction compared to the zigzag path. Along with that, the dependence of anisotropy of Young's modulus on temperature was also computed. The study reported that, for BeN_4 and MgN_4 monolayers, anisotropy of Young's modulus is independent of temperature, whereas, in the case of PtN_4 monolayer, the decrease in anisotropy of Young's modulus with the increase in temperature was observed [43].

7.5 Mechanical Properties of 2D-SCMs

The mechanical characteristics of these materials are essential in understanding the role of flexible electronics and energy harvesting. The mechanical parameters of 2D-SCMs can be adjusted by altering the structural configuration of 2D-SCMs. For instance, in the case of MXenes, the report suggested that the structural and electronic properties can be modified via surface modulations with fluoride and oxygen atoms [44]. In general, MXenes are metallic and tend to become semiconductors via surface passivation. For example, Sc_2C possesses zero bandgap, whereas functionalized Sc_2CT_2 where T = OH, F, and O become semiconducting with bandgap ranges between 0.45 and 1.80 eV. This suggests that Sc_2C can become semiconducting via layer modulation. Thermal conductivity values of functionalized Sc_2C nanomaterial are higher than MoS_2 and phosphorene. Unlike fluoride and oxygen-functionalized Sc_2C, nitride-containing MXenes modulated with oxygen atoms were observed to be dynamically unstable. The electronic and mechanical characteristics of $2D-M_2X$, where M = Sc/Ti/V/Mn/Nb/Mo/Hf and X = C/N systems were studied using DFT methods. The studies showed that the mechanical stability of the monolayer could be enhanced via surface modulation. The reports have suggested that the uneven modulation creates varied chemical environments on both sides of the monolayer, which might tune the elastic behaviors.

TMDs have attracted research attention because of their tunable bandgaps. The bandgaps can be converted from indirect to direct with the change in the structure from bulk phase to few-layered phase [45]. The mechanical properties of TMDs are of immense importance to study the utilization of these materials in flexible electronics. The initial studies were conducted to analyze the mechanical behavior of TMDs, particularly, impure MoS_2 flakes.

Researchers have used atomic force microscopy (AFM) to determine the mechanical characteristics of mono-/bilayer MoS_2 [46]. For the monolayer, Young's modulus was 180 ± 60 N/m, which is lower than that of graphene monolayer, whereas the 2D modulus for the MoS_2 bilayer is 260 ± 70 N/m.

The CVD method is considered a cost-effective way to grow large-area TMDs. Unlike single crystals, CVD samples are observed to be imperfect. To date, CVD methods have been improved to produce larger surface areas of TMDs such as MoS_2. Liu and his coworkers used the polydimethylsiloxane (PDMS) process for converting the CVD-grown TMDs into holey surfaces as they did in the case of the CVD of graphene [46]. The purpose of doing this is to avoid the ripples that may impact the mechanical properties. The AFM was employed for analyzing the modulus for both MoS_2 and WS_2, which was reported to be 171 ± 11 and 117 ± 12 N/m, respectively [47]. The similar values in the case of both TMDs are due to their similar lattice constants and binding energy values.

The 2D modulus of both of these materials was also computed using DFT methods in the past. For MoS_2 and WS_2 monolayers, the values were computed to be 123 and 137 N/m, respectively, and these computed values are lower than the experimental reports. The difference between the experimental and computational values is because DFT-generalized gradient approximation (GGA) computations underestimate the bulk modulus of semiconducting nanomaterials [48]. Like graphene, the modulus of CVD MoS_2 was reported to be ~5% lower than its exfoliated counterpart, implying that its mechanical properties did not show drastic change with the introduction of point deficiencies.

The stacking of two different 2D-SCM monolayers via long-range interactions forms two-dimensional heterostructures. The heterostructures are considered a promising advantage of the utilization of 2D-SCMs in optoelectronic devices. However, the impact of the interlayer coupling between both monolayers on the mechanical properties is still not investigated properly. In the past, different ways were used to investigate interlayer coupling from the mechanical point of aspect. For homostructures, such as multilayer graphene, the interlayer shear Raman modes were calculated. The study showed the shift in the Raman peaks as moved from the bilayer graphene system to bulk graphite, indicating that the peaks are correlated to the interlayer interactions. This indirect method was used to calculate the shear modulus of graphite, suggesting that this method could also be used for 2D heterostructures. Koren and his coworkers used AFM to calculate the force required to slide a graphene layer by applying the sideways force to the graphene. Using this, interface adhesion energy was calculated to be 0.227 ± 0.005 J/m^2, which was in accordance with the theoretical results [49]. Although this method can precisely quantify the interlayer interactions in the case of 2D homostructures, it is unsuitable for 2D heterostructures because of the difficulty in the formation of two-dimensional bulky heterostructures.

To calculate the mechanical behaviors of heterostructures, Liu and his coworkers used the elastic modulus (E^{2D}) using the nanoindentation process. The results suggested that E^{2D} values for bilayer structures are lower than the addition of E^{2D} of each monolayer, inferring the sliding of monolayers which is subjected to interactions between layers as shown in Figure 7.5. In the experimental setup, the authors suspended a bilayered material on the spherical hole. The underside monolayer was firmly clamped to the substrate, while the upper side monolayer was permitted to slide. The authors measured the 2D modulus of bilayer, which measures its stiffness. They found that the interaction coefficient (α), which measures the strength of interactions between monolayers, was 0.80 for MoS$_2$–WS$_2$ heterostructure, 0.75 in MoS$_2$–MoS$_2$ homostructure, and 0.69 in MoS$_2$–graphene heterostructure (Figure 7.5). The reported approach may not be an explicit way for the calculation of such interactions; however, it helps to provide a simpler way to determine the interlayer interactions.

Being ultrathin and atomically flat, 2D-SCMs are considered suitable for in-plane mechanical nanodevices such as resonators. The resonators are electromechanical machines which stimulate the exterior force and can be used as a sensor for mass, force, and so on. Bunch and his coworkers developed a prototype of the graphene resonator. Graphene nanomaterial is considered a suitable candidate for resonators because of its high Young's modulus,

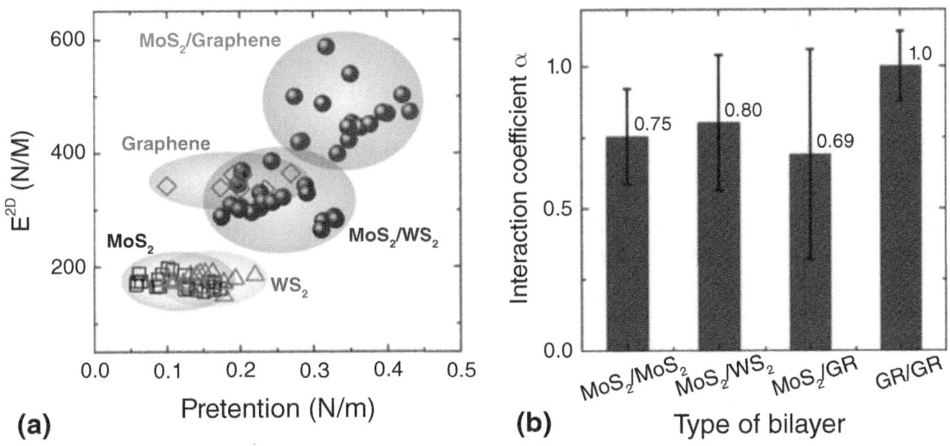

FIGURE 7.5
(a) The elastic properties (E^{2D}) of 2D-SCM structures. (b) Interaction coefficients for different heterostructures. Adapted with permission [47]. Copyright (2014) American Chemical Society (ACS) publication.

very little mass, and huge surface area [50]. Similarly, other 2D-SCMs also have the potential to be adopted as mechanical resonators. Based on the studied mechanical properties, it has been reported that MoS_2 can be used as a resonator.

7.6 Conclusion

The discovery of 2D-SCMs has led to the development of a new era in the field of materials sciences. The 2D-SCMs acquire exceptional optical, electrical, thermal, and mechanical characteristics that make these materials useful for numerous applications such as optoelectronics, photonics, and flexible electronics. Their distinctive features, for instance, strong light–matter interactions and high charge carrier mobility, make them favorable candidates for photodetectors, lasers, and so on. Researchers have also explored their nonlinear optical properties and the potential for tuning their bandgaps through strain engineering, doping, and heterostructures. Furthermore, the thermal properties of 2D-SCMs have been a subject of study, with implications for thermoelectric materials and heat management. The 2D-SCMs with their outstanding charge carrier mobilities compared to the respective bulk phases tend to modernize next-generation electronic devices by facilitating fast performance. The mechanical properties of 2D-SCMs propose their potential in flexible electronics, resonators, energy harvesting, and sensing applications.

However, further research needs to be done to resolve the stability concerns, which will help to fully utilize the capability of these nanomaterials. Along with that, in heterostructures, the correlation between defects and mechanical properties needs further evaluation as the existence of defects in these materials may improve or weaken the interlayer interactions and, hence, will impact the properties. Thus, further improvement in 2D-SCMs needs to be done to utilize these materials in varied spaces of flexible and stretchable electronics.

References

1. K. S. Novoselov, A. K. Geim, S. V. Morozov, D. Jiang, Y. Zhang, S. V. Dubonos, I. V. Grigorieva, A. A. Firsov, Electric Field in Atomically Thin Carbon Films. *Science* 306 (2004) 666–669.
2. (a) X. Song, J. Hu, H. Zeng, Two-Dimensional Semiconductors: Recent Progress and Future Perspectives. *J. Mater. Chem. C* 1 (2013), 2952–2969.; (b) F. Xia, H. Wang, Di Xiao, M. Dubey, A. Ramasubramaniam, A. Two-Dimensional Material Nanophotonics. *Nat. Photonics* 8 (2014) 899–907.
3. K. S. Novoselov, D. Jiang, F. Schedin, T. J. Booth, V. V. Khotkevich, S. V. Morozov, A. K. Geim, Two-Dimensional Atomic Crystals. *Proc. Natl. Acad. Sci. USA.* 102 (2005) 10451–10453.
4. H. S. S. Ramakrishna Matte, A. Gomathi, A. K. Manna, D. J. Late, R. Datta, S. K. Pati, C. N. R Rao, MoS_2 and WS_2 Analogues of Graphene. *Angew. Chem. Int. Ed.* 49 (2010) 4059–4062.
5. K. Wang, J. Wang, J. Fan, M. Lotya, A. O'Neill, D. Fox, Y. Feng, X. Zhang, B. Jiang, Q. Zhao, H. Zhang, J. Coleman, L. Zhang, W. J. Blau, Ultrafast Saturable Absorption of Two-Dimensional MoS_2 Nanosheets. *ACS Nano.* 7 (2013) 9260–9267.
6. A. Luo, M. Liu, X. Wang, Q. Ning, W. Xu, Z. Luo, Few-Layer MoS_2-Deposited Microfiber as Highly Nonlinear Photonic Device for Pulse Shaping in a Fiber Laser. *Photon. Res.* 3 (2015) A69–A78.
7. J. Wang, Two-Dimensional Semiconductors for Ultrafast Photonic Applications. SPIE, 9359 (2015) 935902.

8. L. Zhang, A. Zunger, Evolution of Electronic Structure as a Function of Layer Thickness in Group-VIB Transition Metal Dichalcogenides: Emergence of Localization Prototypes. *Nano Lett.* 15 (2015) 949–957.

9. H. Shi, H. Pan, Y. W. Zhang, B. I. Yakobson, Quasiparticle Band Structures and Optical Properties of Strained Monolayer MoS_2 and WS_2. *Phys. Rev. B.* 87 (2013) 155304.

10. S. Liu, J. Ye, Y. Cao, Q. Shen, Z. Liu, L. Qi, X. Guo, Tunable Hybrid Photodetectors with Superhigh Responsivity. *Small* 5 (2009) 2371–2376.

11. W. Tian, C. Zhang, T. Zhai, S. Li, X. Wang, X. Jie, D. Liu, M. Liao, Y. Koide, D. Golberg, Y. Bando, Flexible Ultraviolet Photodetectors with Broad Photoresponse Based on Branched ZnS-ZnO Heterostructure Nanofilms. *Adv. Mater.* 26 (2014) 3088–3093.

12. (a) J. Y. Lee, J. H. Shin, G. H. Lee, C. H. Lee, Two-Dimensional Semiconductor Optoelectronics Based on van Der Waals Heterostructures. *Nanomater.* 6 (2016) 193. (b) B. Radisavljevic, A. Radenovic, J. Brivio, I. V. Giacometti, A. Kis, Single-Layer MoS_2 Transistors. *Nat. Nanotechnol.* 6 (2011) 147–150. (c) R. Kappera, D. Voiry, S. E. Yalcin, B. Branch, G. Gupta, A. D. Mohite, M. Chhowalla, Phase-Engineered Low-Resistance Contacts for Ultrathin MoS2 Transistors. *Nat. Mater.* 13 (2014) 1128–1134. (d) A. Splendiani, L. Sun, Y. Zhang, T. Li, J. Kim, C.-Y. Chim, G. Galli, F. Wang, Emerging Photoluminescence in Monolayer MoS_2. *Nano Lett.* 10 (2010) 1271–1275.

13. K. F. Mak, C. Lee, J. Hone, J. Shan, T. F. Heinz, Atomically Thin MoS2: A New Direct-Gap Semiconductor. *Phys. Rev. Lett.* 105 (2010) 136805.

14. T. Cheiwchanchamnangij, W. R. L. Lambrecht, Quasiparticle Band Structure Calculation of Monolayer, Bilayer, and Bulk MoS_2. *Phys. Rev. B.* 85 (2012) 205302.

15. A. Splendiani, L. Sun, Y. Zhang, T. Li, J. Kim, C. Y. Chim, G. Galli, F. Wang, Emerging Photoluminescence in Monolayer MoS_2. *Nano Lett.* 10 (2010) 1271–1275.

16. M. Amani, D. H. Lien, D. Kiriya, J. Xiao, A. Azcatl, J. Noh, S. R. Madhvapathy, R. Addou, K. C. Santosh, M. Dubey, K. Cho, R. M. Wallace, S. Lee, J. He, J. W. Ager, X. Zhang, E. Yablonovitch, A. Javey, Near-Unity Photoluminescence Quantum Yield in MoS2. *Science* 350 (2015) 1065–1068.

17. M. Amani, R. A. Burke, X. Ji, P. Zhao, D. H. Lien, P. Taheri, G. H. Ahn, D. Kirya, J. W. Ager, E. Yablonovitch, J. Kong, M. Dubey, A. Javey, High Luminescence Efficiency in MoS_2 Grown by Chemical Vapor Deposition. *ACS Nano* 10 (2016) 6535–6541.

18. A. Carvalho, R. M. Ribeiro, A. H. Castro Neto, Band Nesting and the Optical Response of Two-Dimensional Semiconducting Transition Metal Dichalcogenides. *Phys. Rev. B.* 88 (2013) 115205.

19. M. Bernardi, M. Palummo, J. C. Grossman, Extraordinary Sunlight Absorption and One Nanometer Thick Photovoltaics Using Two-Dimensional Monolayer Materials. *Nano Lett.* 13 (2013) 3664–3670.

20. A. Arora, M. Koperski, K. Nogajewski, J. Marcus, C. Faugeras, M. Potemski, Excitonic Resonances in Thin Films of WSe_2: From Monolayer to Bulk Material. *Nanoscale* 7 (2015) 10421–10429.

21. S. Mouri, Y. Miyauchi, K. Matsuda, Tunable Photoluminescence of Monolayer MoS_2 via Chemical Doping. *Nano Lett.* 13 (2013) 5944–5948.

22. Y. Y. Hui, X. Liu, W. Jie, N. Y. Chan, J. Hao, Y-Te. Hsu, L. J. Li, W. Guo, S. P. Lau, Exceptional Tunability of Band Energy in a Compressively Strained Trilayer MoS_2 Sheet. *ACS Nano.* 7 (2013) 7126–7131.

23. H. R. Gutiérrez, N. Perea-López, A. L. Elías, A. Berkdemir, B. Wang, R. Lv, F. López-Urías, V. H. Crespi, H. Terrones, M. Terrones, Extraordinary Room-Temperature Photoluminescence in Triangular WS_2 Monolayers. *Nano Lett.* 13 (2013) 3447–3454.

24. S. Zhang, N. Dong, N. McEvoy, M. O'Brien, S. Winters, N. C. Berner, C. Yim, Y. Li, X. Zhang, Z. Chen, L. Zhang, G. S. Duesberg, J. Wang, Direct Observation of Degenerate Two Photon Absorption and Its Saturation in WS_2 and MoS_2 Monolayer and Few-Layer Films. *ACS Nano* 9 (2015) 7142–7150.

25. J. S. Ponraj, Z. Q. Xu, S. C. Dhanabalan, H. Mu, Y. Wang, J. Yuan, P. Li, S. Thakur, M. Ashrafi, K. McCoubrey, Photonics and Optoelectronics of Two-Dimensional Materials Beyond Graphene. *Nanotech.* 27 (2016) 462001.

26. G. Eda, S. A. Maier, Two-Dimensional Crystals: Managing Light for Optoelectronics. *ACS Nano.* 7 (2013) 5660–5665.

27. L. F. Mattheiss, Band Structures of Transition-Metal-Dichalcogenide Layer Compounds. *Phys. Rev. B.* 8 (1973) 3719.

28. S. R. Cohen, L. Rapoport, E. A. Ponomarev, H. Cohen, T. Tsirlina, R. Tenne, C. Lévy-Clément, The Tribological Behavior of Type II Textured MX_2 (M=Mo, W; X=S, Se) Films. *Thin Solid Films* 324 (1998) 190–197.

29. V. Podzorov, M. E. Gershenson, C. Kloc, R. Zeis, E. Bucher, High-Mobility Field-Effect Transistors Based on Transition Metal Dichalcogenides. *Appl. Phys. Lett.* 84 (2004) 3301–3303.

30. F. Bonaccorso, Z. Sun, T. Hasan, A. C. Ferrari, Graphene Photonics and Optoelectronics. *Nat. Photonics* 4 (2010) 611–622.

31. R. Roldán, A. Castellanos-Gomez, E. Cappelluti, F. Guinea, Strain Engineering in Semiconducting Two-Dimensional Crystals. *J. Phys.: Cond. Matter.* 27 (2015) 313201.

32. D. ÇakIr, H. Sahin, F. M. Petters, Tuning of the Electronic and Optical Properties of Single-Layer Black Phosphorus by Strain. *Phys. Rev. B.* 90 (2014) 205421.

33. D. Jariwala, V. K. Sangwan, L. J. Lauhaon, T. J. Marks, M. C. Hersam, Emerging Device Applications for Semiconducting Two-Dimensional Transition Metal Dichalcogenides. *ACS Nano* 8 (2014) 1102–1120.

34. V. Podzorov, M. E. Gershenson, C. Kloc, R. Zeis, E. Bucher, High-Mobility Field-Effect Transistors Based on Transition Metal Dichalcogenides. *Appl. Phys. Lett.* 84 (2004) 3301–3303.

35. K. S. Novoselov, D. Jiang, F. Schedin, T. J. Booth, V. V. Khotkevich, S. V. Morozov, A. K. Geim, Two-Dimensional Atomic Crystals. *Proc. Natl. Acad. Sci.* 102 (2005) 10451–10453.

36. B. Radisavljevic, A. Radenovic, J. Brivio, V. Giacometti, A. Kis, Single-Layer MoS_2 Transistors. *Nature Nanotech.* 6 (2011) 147–150.

37. Y. Zhao, Y. Cai, L. Zhang, B. Li, G. Zhang, J. T. L. Thong, Thermal Transport in 2D Semiconductors—Considerations for Device Applications. *Adv. Funct. Mat.* 30 (2020) 1903929.

38. X. Wei, Y. Wang, Y. Shen, G. Xie, H. Xiao, J. Zhong, G. Zhang, Phonon Thermal Conductivity of Monolayer MoS_2: A Comparison with Single Layer Graphene. *Appl. Phys. Lett.* 105 (2014) 103902.

39. X. Li, J. Zhang, A. A. Puretzky, A. Yoshimura, X. Sang, Q. Cui, Y. Li, L. Liang, A. W. Ghosh, H. Zhao, R. R. Unocic, V. Meunier, C. M. Rouleau, B. G. Sumpter, D. B. Geohegan, K. Xiao, Isotope-Engineering the Thermal Conductivity of Two-Dimensional MoS_2. *ACS Nano* 13 (2019) 2481–2489.

40. B. Qiu, X. Raun, Thermal Conductivity Prediction and Analysis of Few-Quintuple Bi_2Te_3 Thin Films: A Molecular Dynamics Study. *Appl. Phys. Lett.* 97 (2010) 183107.

41. B. Mortazavi, M. Makaremi, M. Shahrokhi, Z. Fan, T. Rabczuk, N-Graphdiyne Two-Dimensional Nanomaterials: Semiconductors with Low Thermal Conductivity and High Stretchability. *Carbon* 137 (2018) 57–67.

42. Z. Fan, L. F. C. Pereira, H. Q. Wang, J. C. Zheng, D. Donadio, A. Harju, Force and Heat Current Formulas for Many-Body Potentials in Molecular Dynamics Simulations with Applications to Thermal Conductivity Calculations. *Phys. Rev. B.* 92 (2015) 094301.

43. K. Ghorbani, P. Mirchi, S. Arabha, S. Volz, Lattice Thermal Conductivity and Young's Modulus of XN_4 (X= Be, Mg and Pt) 2D Materials Using Machine Learning Interatomic Potentials. *Phys. Chem. Chem. Phys.* 25 (2023) 12923–12933.

44. W. Jin, S. Wu, Z. Wang, Structural, Electronic and Mechanical Properties of Two-Dimensional Janus Transition Metal Carbides and Nitrides. *Physica E Low Dimens. Syst. Nanostruct.* 103 (2018) 307–313.

45. K. Liu, J. Wu, Mechanical Properties of Two-Dimensional Materials and Heterostructures. *J. Mater. Res.* 31 (2016) 832–844.

46. K. Liu, Q. Yan, M. Chen, W. Fan, Y. Sun, J. Suh, D. Fu, S. Lee, J. Zhou, S. Tongay, J. Ji, J. B. Neaton, J. Wu, Elastic Properties of Chemical-Vapor-Deposited Monolayer MoS_2, WS_2, and Their Bilayer Heterostructures. *Nano Lett.* 14 (2014) 5097–5103.

47. S. Bertolazzi, J. Brivio, A. Kis, Stretching and Breaking of Ultrathin MoS_2. *ACS Nano* 5 (2011) 9703–9709.

48. C. Filippi, D. J. Singh, C. J. Umrigar, All-Electron Local-Density and Generalized-Gradient Calculations of the Structural Properties of Semiconductors. *Phys. Rev. B* 50 (1994) 14947.

49. E. Koren, E. Lörtscher, C. Rawlings, A. W. Knoll, U. Duerig, Surface Science: Adhesion and Friction in Mesoscopic Graphite Contacts. *Science* 348 (2015) 679–683.

50. J. S. Bunch, A. M. Van Der Zande, S. S. Verbridge, I. W. Frank, D. M. Tanenbaum, J. M. Parpia, H. G. Craighead, P. L. McEuen, Electromechanical Resonators from Graphene Sheets. *Science* 315 (2007) 490–493.

Optical, Electrical, Thermal, and Mechanical Properties of 2D Semiconducting Materials

Yang Li, Cheng-Yan Xu*, Liang Zhen, and Jing-Kai Qin
*Correspondence: cy_xu@hit.edu.cn

8.1 Optical Property of Two-dimensional Semiconductors

Excitons, hydrogen-like bound states, form via Coulomb attraction between a negatively charged electron and a positively charged hole [1, 2]. Typically, excitons arise from photo-excitation in semiconductors, exhibiting a narrow spectral linewidth [3]. The large oscillator strength of excitons and their enhanced light–matter interaction enable efficient recombination and light emission [4]. Although excitons offer a rich foundation for both physics and device applications, their practical utilization has traditionally posed challenges [5]. In conventional bulk semiconductor crystals like silicon, the combination of substantial dielectric screening and small quasiparticle effective mass leads to minimal exciton binding energies, typically in the range of ~1–10 meV [6]. Consequently, the bound behavior of an exciton becomes negligible when compared to thermal fluctuations unless the system is cooled to low temperatures.

Conversely, bound exciton states frequently dictate the optical properties of 2D materials, exemplified by monolayer transition metal dichalcogenides (TMDCs) [7]. The pronounced Coulomb interaction in reduced dimensions, along with reduced dielectric screening in comparison to bulk crystals, inherently leads to exciton binding even at room temperature, with binding energies in the hundreds of meV range [8]. Furthermore, the electronic valleys within TMDCs host excitons and establish selection rules for the excitation and emission of light [9]. The 2D nature of the material allows for easy tuning of excitons through various external stimuli, facilitating the development of numerous 2D photonic devices, including light-emitting diodes and lasers [10]. In this section, we present recent progress in understanding the physical properties of excitons in 2D semiconductors.

8.1.1 Excitons and Trions

TMDCs, including molybdenum and tungsten-based disulfide and diselenides (MoS_2, WSe_2), are characterized by their compositions as layered van der Waals (vdWs) crystals [10]. Within each layer, chemical bonding occurs, while the layers are held together by weak vdWs interlayer interactions. These materials are known as semiconductors in bulk form, characterized by an indirect band gap. However, a significant discovery was made when these crystals were mechanically exfoliated into monolayers, leading to an indirect-to-direct band gap transition due to the absence of interlayer interactions in the TMDC monolayer.

DOI: 10.1201/9781003439448-8

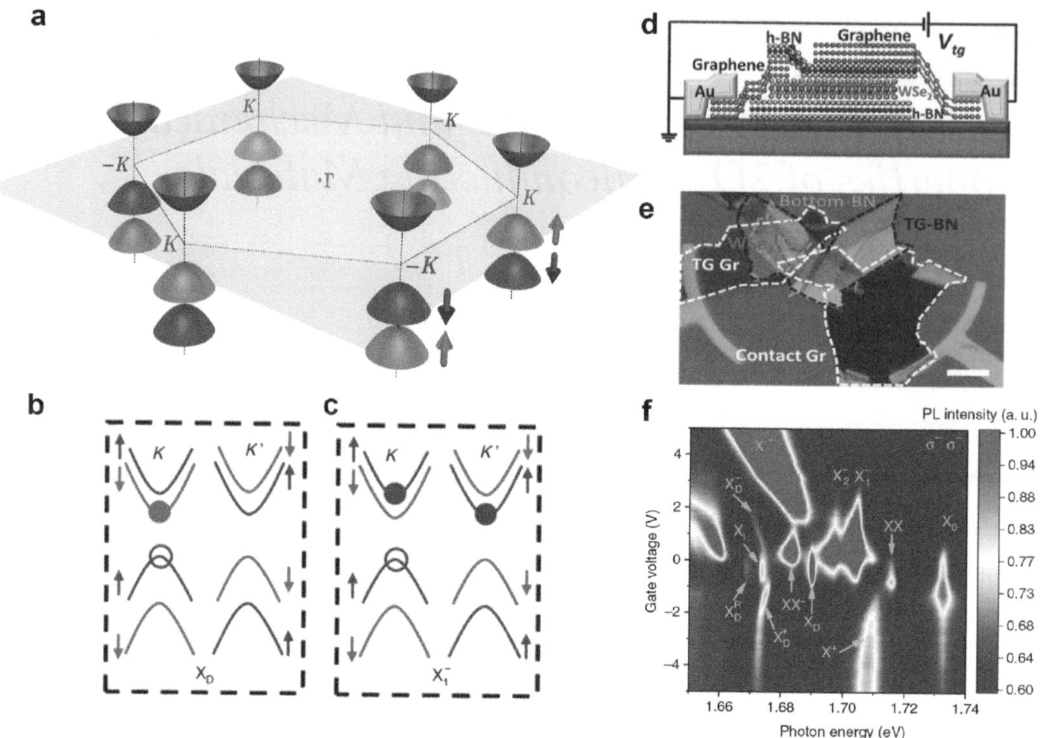

FIGURE 8.1

Excitonic states in 2D tungsten-based TMDC monolayer. (a) Valley-dependent optical selection rules for interband transitions in 2D monolayer tungsten-based TMDCs. Adapted with permission [6], Copyright (2012), American Physical Society. (b) Configurations of the bright exciton (X_0). (c) Configurations of the dark exciton (X_D). (d) The hexagonal boron nitride-sandwiched monolayer WSe$_2$ device with a top gate. (e) A representative optical image of a top-gated WSe$_2$ device. (f) Electrostatic doping–dependent color plot of the PL. (Figure 1b–f) Adapted with permission [12], Copyright (2018), Springer Nature.

This reduction in interlayer interactions results in a lower valence band energy, thereby forming a direct band gap. This direct band gap exhibits greater absorption and robust photoluminescence (PL) emission compared to multilayered crystals. Importantly, due to the unique electronic structure shown in Figure 8.1(a), the PL exhibits valley-dependent behavior [11]. Given that TMDCs are semiconductors, it is expected that photo-excitation will generate excitons, comprising pairs of bound electrons and holes that exhibit behavior akin to hydrogen atoms, as shown in Figure 8.1(b) [12, 13]. The recombination of these electron–hole pairs emits photons, which are observed as PL. A close examination within a monolayer, using momentum-resolved spectroscopy, reveals that these exciton dipoles are confined exclusively to the in-plane direction, with a lifetime in the range of 100 ps before recombination [9]. Consequently, it becomes evident that excitons play a dominant role in observable optical phenomena.

The strong Coulomb interaction became strikingly evident during the initial work aimed at demonstrating charged excitons, also known as trions, in MoS$_2$ [14], as shown in Figure 8.1(c). While trions have been studied in traditional quantum well systems, their observability was limited to low temperatures due to the small binding energy. When an electric field is applied via the transistor gate, the neutral state can be dynamically shifted toward either positivity or negativity. In the case of MoS_2, shifting the Fermi level toward the conduction band leads to an increase in electron population. Consequently, neutral excitons begin binding with these additional electrons, resulting in the formation of charged excitons [15]. The recombination of these charged excitons also leads to PL, albeit at lower energy compared to neutral excitons. The determination of trion binding energy, representing the energy required to dissociate the extra charge from the neutral exciton, involves analyzing the energy difference between trion luminescence and neutral exciton energy. In MoS_2, this binding energy is approximately 20 meV, significantly higher than observed in quantum wells, resulting in notable spectral weights even at room temperature. The significant Coulomb interaction, indicated by such large trion binding energy, implies an even more substantial exciton binding energy. This energy represents the force required to separate the electron and hole (the energy difference between the exciton level and the conduction band), and its magnitude is further substantiated by the absorption spectrum, which lacks an absorption step indicative of a transition to the conduction band. Moreover, theoretical estimates place the true quasi-particle band gap over 500 meV above the exciton emission energy, nearly two orders of magnitude greater than traditional exciton binding energies. Consequently, considerable efforts have been dedicated to unraveling the true exciton binding energy and quasiparticle band gap.

8.1.2 Biexciton and Trion–Exciton Complexes

The robust Coulomb interaction observed in monolayer TMDCs results in strong exciton–exciton interactions, making the emergence of biexcitons likely [16]. Although biexcitons were experimentally confirmed early in 2015, disparities between theoretical calculations and experimentally determined biexciton binding energies necessitate a comprehensive and clear understanding of biexcitons in TMDCs [17]. Recent research reports the identification of both neutral biexcitons and trion–exciton complexes (equivalent to negatively charged biexcitons) in an h-BN-encapsulated monolayer WSe_2 device shown in Figure 8.1(d) and (e) [12]. As shown in Figure 8.1(f), the researchers demonstrate that two new excitonic states, XX and XX^-, exhibit a superlinear fluent dependence (almost a quadratic power law), while the bright exciton displays nearly linear behavior. XX is positioned between the exciton and the trions, with a binding energy of approximately 17 meV, closely matching the theoretically predicted biexciton binding energy. To further elucidate the nature of XX and XX^-, researchers employ efficient electrostatic gating, transitioning from hole-doping to electron-doping [12]. Consequently, the biexciton configuration is deduced to consist of a bright exciton and a dark exciton, considering valley polarization and the inverted PL intensity behavior. These advancements in biexciton understanding lay the foundation for nonlinear quantum optoelectronics based on TMDCs.

8.1.3 Interlayer Exciton and Moiré Exciton in vdWs Heterostructures

Comprising stacked atomically thin 2D materials, the properties of vdWs heterostructures are not solely determined by their constituent monolayers but also by layer interactions [18]. Notably, these vdWs heterostructures exhibit unique excited-state dynamics, including the formation of interlayer excitons, rapid charge transfer between layers, the persistence of long-lived spin and valley polarization in resident carriers, and the emergence of moiré-trapped valley excitons within moiré superlattices. Among 2D vdWs heterostructures, semiconducting ones composed of stacked TMDC layers are extensively studied

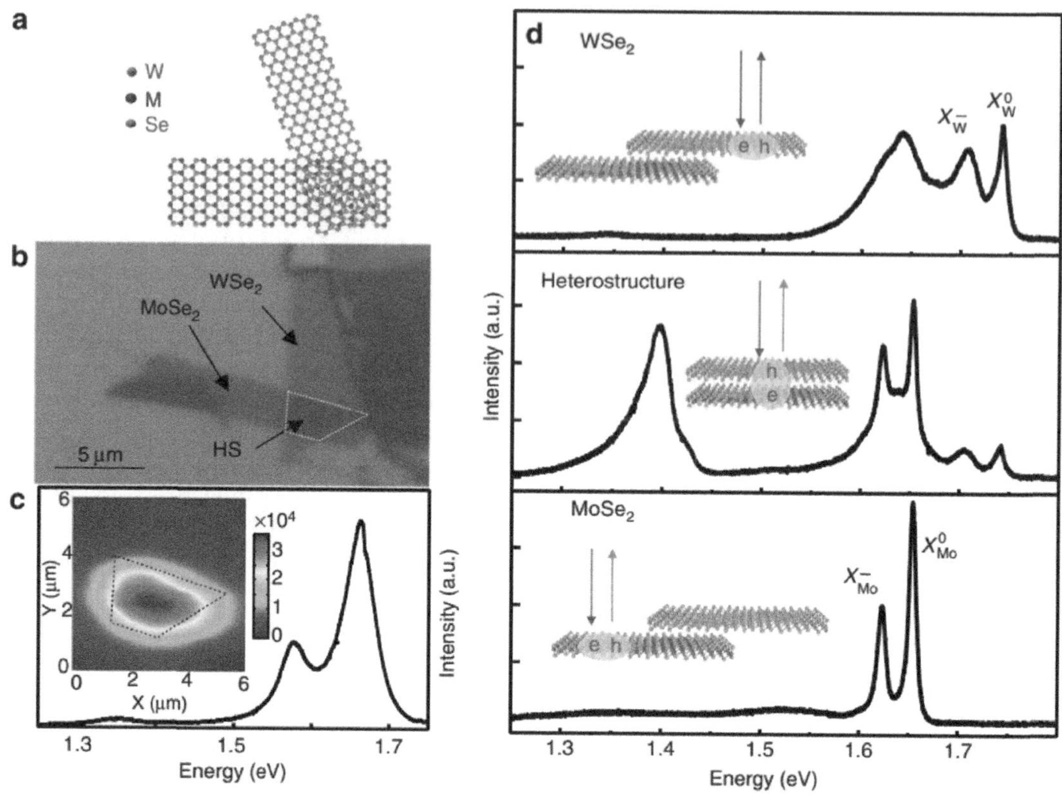

FIGURE 8.2

Intralayer and interlayer excitons of a monolayer MoSe₂–WSe₂ vertical vdWs heterostructure. (a) Cartoon depiction of a MoSe₂–WSe₂ heterostructure. (b) Optical image of a MoSe₂–WSe₂. (c) Room-temperature photoluminescence of the heterostructure under 20 mW laser excitation at 2.33 eV. (d) Photoluminescence of individual monolayers and the heterostructures (HS) at 20 K under 20 mW excitation at 1.88 eV. Adapted with permission [16]. Copyright (2015), Springer Nature.

due to their distinctive exciton states and accessibility to the valley degree of freedom. Intriguingly, the introduction of moiré superlattices, periodic patterns created by stacking two monolayer 2D materials with a lattice mismatch or rotational misalignment, offers the means to manipulate the electronic band structure and optical properties of vdWs heterostructures.

The exploration of exciton dynamics in vdWs heterostructures gained momentum after observing interlayer excitons in PL spectra [19]. The discovery of interlayer excitons in 2D materials dates back to 2015, with the demonstration of long-lived interlayer excitons in monolayer MoSe₂/WSe₂ heterostructures, featuring a distinct additional resonance below the intralayer excitons [20], as illustrated in Figure 8.2(a) and (b). In contrast to intralayer excitons under weak excitation, this low-energy peak exhibits significantly enhanced PL intensity, attributed to highly populated interlayer excitons, as illustrated in Figure 8.2(c). Moreover, the direct measurement of the binding energy of interlayer excitons is demonstrated in WSe₂/WS₂ hetero-bilayers, where a novel 1s–2p resonance is detected using phase-locked mid-infrared pulses, as illustrated in Figure 8.2(d). Regarding other aspects of excited-state dynamics, such as ultrafast kinetics, prolonged lifetimes, and moiré excitons, some research suggests their connection to interlayer excitons [16]. While charge transfer between vertically stacked layers of vdWs heterostructures is empirically expected to be slow, transient absorption measurements, implemented by resonantly injecting excitons with ultrafast laser pulses, reveal sub-picosecond charge separation in vdWs heterostructures.

Recent research delves into the impact of moiré potentials on light emission and absorption in TMDC hetero-bilayers, offering fresh insights compared to studies involving graphene–boron nitride hetero-bilayers [21]. Notably, the variation in band energies between different atomic registries is more pronounced. Light emission and absorption in TMDC structures are governed by excitons, bound states formed by electrons and holes. These excitons interact with light of specific polarization, dictated by the local symmetry of atoms arranged periodically in the hetero-bilayer. Emission lines from excitons serve as fingerprints of the magnetic moment and valley characteristics of the electron–hole pairs that constitute them. These investigations provide an initial glimpse into the possibilities unlocked by combining two materials and altering their optical properties simply by adjusting the twist angle.

8.2 Electrical Properties of Two-Dimensional Semiconductors and Their Tunability

8.2.1 Electrical Performance of TMDC Field-Effect Transistors

While the electrical properties of bulk semiconducting TMDCs have been a subject of investigation for decades, their utilization in field-effect transistors began only around a decade ago, following the groundbreaking work on graphene by the Manchester group [22]. Initially, monolayer MoS_2 exhibited field-effect mobility of at least three orders of magnitude lower than graphene. However, interest in monolayer semiconducting TMDC FETs was rekindled in 2011 when top-gated monolayer-MoS_2 FETs with room temperature moderate mobilities (60–70 cm^2 V^{-1} s^{-1}), large on/off ratios (~10^8), and low subthreshold swings (74 mV/dec) were demonstrated, as illustrated in Figure 8.3(a) and (b) [23]. The two-dimensional channel, combined with a substantial bandgap (>1 eV) and ultrathin top-gated dielectric, facilitated excellent gate control, resulting in small off-currents and substantial switching ratios.

The carrier mobility in ultrathin TMDCs is significantly influenced by the interfaces on both sides of the semiconductor film, particularly evident in vacuum versus ambient measurements of unencapsulated monolayer MoS_2 FETs [24]. Vacuum conditions induce a substantial negative shift in threshold voltage and increased on-currents, indicative of p-type doping and charge scattering due to atmospheric adsorbates. The ambipolar behavior is attributed to the ultrahigh capacitance of the electrical double layer, allowing Fermi level tuning across the band gap for carrier injection into both the valence and conduction bands. Mobility improvements have also been noted with increasing TMDC thickness, attributed to substrate impurity screening and access to the third dimension for charge transport.

The sensitivity of carrier mobilities to the local dielectric environment has prompted more comprehensive investigations into their role in charge transport. Initially, temperature-dependent transport in ultrathin MoS_2 revealed variable range hopping or thermally activated transport as opposed to the band-like transport observed in bulk

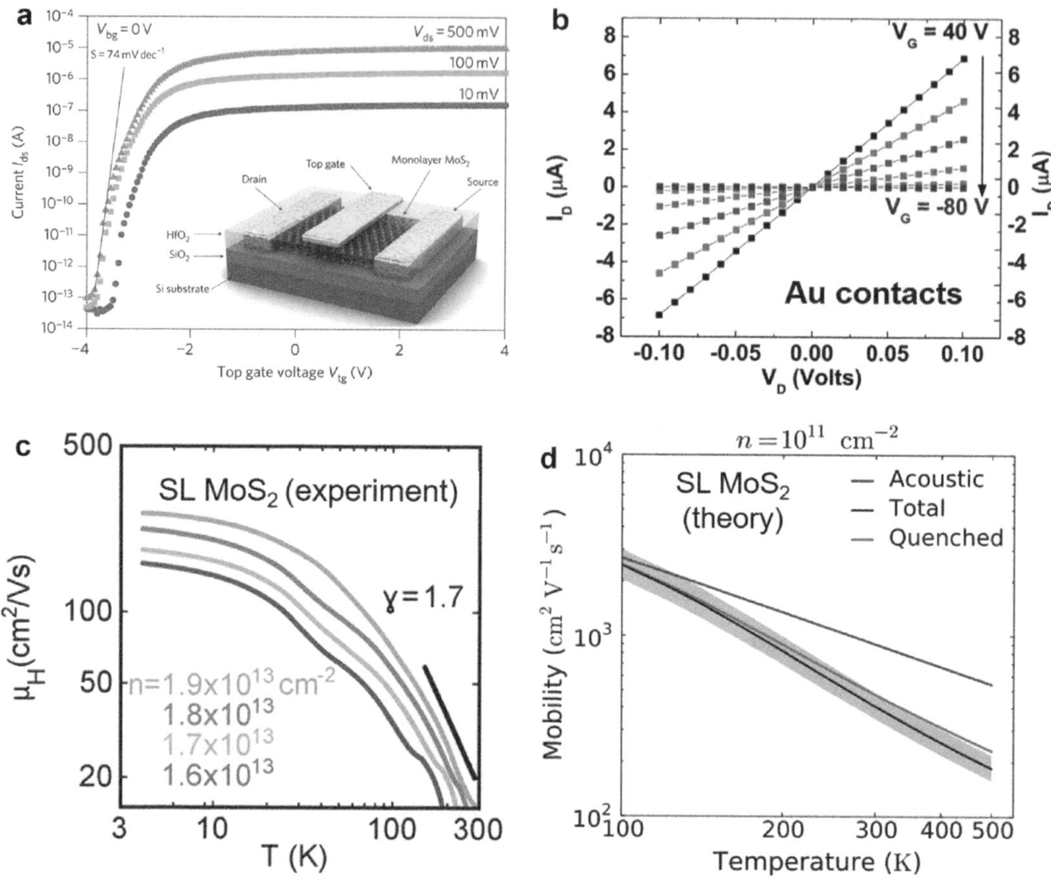

FIGURE 8.3

(a) Semi-log transfer characteristics of a top-gated monolayer-MoS$_2$ FET. Adapted with permission [23], Copyright (2011), Springer Nature. (b) Linear I_d–V_{ds} output curves. Adapted with permission [26], Copyright (2013), American Institute of Physics. (c) Hall mobility values in single layer MoS$_2$. Adapted with permission [27], Copyright (2013), American Chemical Society. (d) Calculated carrier mobility as a function of temperature for single-layer MoS$_2$ with a slope of 1.69. Adapted with permission [25], Copyright (2012), American Physical Society.

single crystals, as illustrated in Figure 8.3(d) [25]. With advancements in sample quality and device processing, band-like transport was observed in monolayer-MoS$_2$, with phonon scattering dominating for T > 100 K and charged impurity scattering for T < 100 K, as illustrated in Figure 8.3(c). A transition from variable range hopping to band-like transport with increasing carrier density was observed in monolayer MoS$_2$. The mobility (µ) follows an inverse power law relation with temperature (µ= T$^\gamma$) for T > 100 K, where the exponent γ decreases from 1.4 to 0.7 after encapsulation with the HfO$_2$ dielectric layer. However, the experimentally observed reduction in γ is much larger than predicted, suggesting the involvement of additional scattering mechanisms at higher temperatures. Recent theoretical work has taken into account the interfacial phonons from the oxide dielectric, leading to better alignment with experimental observations [25]. Particularly, the mobility displays a weak dependence on carrier density at all temperatures except T < 10 K, where the dependence becomes pronounced. This temperature dependence can be attributed to charged impurity-dominated scattering at low temperatures, where higher carrier densities more effectively screen charges, resulting in higher mobility.

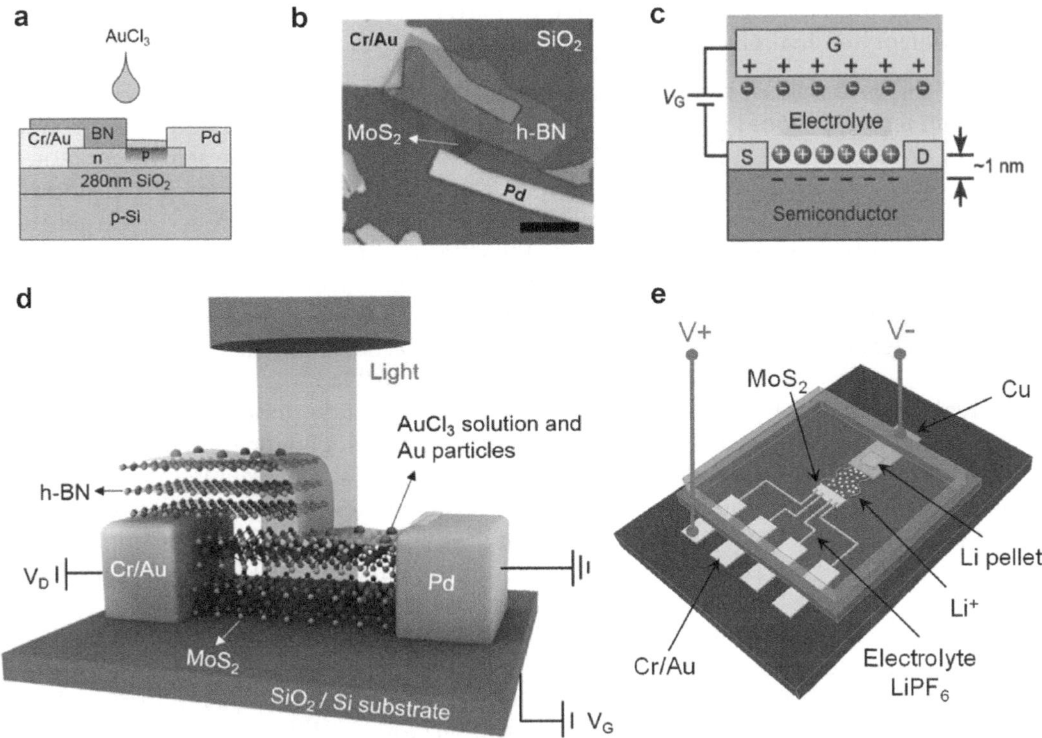

FIGURE 8.4
(a)–(c) Fabrication of the lateral MoS$_2$ p–n diode. Adapted with permission from Reference [29], Copyright (2014), American Chemical Society. (d) Schematics showing the comparison with electric double layer transistors. Adapted with permission [30], Copyright (2017), Wiley-VCH. (e) Experimental setup for electrochemical tuning of Li concentration in MoS$_2$. Adapted with permission [32], Copyright (2015), American Chemical Society.

8.2.2 Tuning the Carrier Density of TMDCs via Molecular Doping

Controlled doping of few-layer MoS$_2$ has been extensively explored within the research community [28]. Investigating doping in 2D layered materials, particularly exemplified by MoS$_2$, is a rapidly expanding and captivating area of interest, with the potential to facilitate the creation of cutting-edge nanodevices relying on variously doped 2D materials. Dopants can either donate electrons to the host semiconductor (n-type dopant or donor) or accept additional electrons from the host semiconductor, forming covalent bonds while creating holes within the host semiconductor (p-type dopant or acceptor). Broadly, doping in 2D semiconductors falls into two categories: surface charge transfer doping and substitutional doping. In surface charge transfer doping, doping occurs through electron exchange between materials adsorbed on its surface (dopants) and MoS$_2$. Generally, this process preserves the MoS$_2$ structure, and doping can be reversed by introducing different dopant materials on the MoS$_2$ surface, as shown in Figure 8.4(a)–(c) [29]. Conversely, substitutional doping involves the replacement of Mo or S atoms in the MoS$_2$ honeycomb lattice with atoms carrying different valence electrons. Such substitutions can disrupt the sp^2 honeycomb structure. Consequently, MoS$_2$ can be doped as either n-type or p-type depending on the specific doping strategy employed. Doping with p-type dopants elevates the Dirac point of MoS$_2$ above the Fermi level, while doping with n-type dopants shifts the Dirac point below the Fermi level.

8.2.3 Tuning the Electrical Performance of TMDCs via Electrochemical Gating and Intercalation

The working principle of electrolyte gating can be illustrated as shown in Figure 8.4(d) [30]. When a conventional n-type semiconductor is interfaced with a liquid electrolyte, it can form an electric double-layer (EDL) capacitor at the surface of the semiconductor. On application of a positive gate bias, the anions in the electrolyte migrate toward the interface of the gate electrode and electrolyte to compensate for the charge build-up in the gate electrode. On the other hand, simultaneously, the cations can move to the electrolyte–semiconductor interface to accumulate and form a Helmholtz layer. The ions are accumulated because they are blocked by the solid semiconductor. To balance the formation of the EDL at the interface, electrons inside the semiconductors accumulate at the solid part of the interface.

Intercalation methods in bulk layered materials have undergone extensive study over recent decades, encompassing solvent, gaseous, and electrochemical intercalations, with a particular focus on materials like graphene and TMDCs [31]. Detailed descriptions of these intercalation techniques are available in recent reviews. Electrochemical intercalation stands out due to its superior controllability and reversibility. In this method, ions, atoms, or molecules are precisely introduced into the vdWs gaps of host materials via liquid, gel-like, or solid-state electrolytes driven by electrochemical potential. This approach offers a means to investigate the structural and electronic evolution of intercalated compounds using state-of-the-art techniques.

Figure 8.4(e) illustrates typical electrochemical platforms in 2D materials [32]. Taking MoS_2 as an example, few-layer MoS_2 and Li metal serve as working and counter electrodes, respectively, enclosed by a liquid electrolyte, creating an open-circuit voltage (V_{OC}) akin to lithium-ion battery setups [32]. Intercalation (de-intercalation) occurs when the voltage between MoS_2 and Li metal is smaller (larger) than V_{OC}. This method has been employed to monitor the reversible (de-intercalation) process of MoS_2, graphite, and InSe, despite the relatively slow intercalation kinetics. Recognizing that liquid electrolytes might not be suitable for low-temperature transport measurements and can introduce complexities at the liquid (solid) interface, ion-gel-like electrolytes have been introduced as gate-controlled intercalation media. With this approach, devices are covered with a gel-like electrolyte containing Li or Na ions.

8.2.4 Structural Evolution and Phase Transition Induced by Intercalation

2D materials, ranging from metals to insulators, display a wealth of intriguing physics, including phenomena such as charge density waves (CDWs), superconductivity, ferromagnetism, and ferroelectricity [30]. These behaviors often arise from electron–electron or electron–phonon interactions within these materials. Intercalation processes can profoundly impact the carrier densities of these host materials by facilitating charge transfer between ions and hosts, ultimately inducing structural changes and, in some cases, phase transitions. In this section, we will focus on the phase transitions of atomically thin layered materials triggered by the intercalation of ions, atoms, and molecules.

Electrochemical intercalation, known for its precise controllability and reversibility, plays a pivotal role in tuning the superconducting properties of 2D materials [33]. Utilizing the solid-ion-conductor gating technique, researchers have effectively introduced Li ions between ultrathin $SnSe_2$ layers, forming a single reservoir layer that provides electrons. This results in the emergence of superconductivity in Li-intercalated $SnSe_2$ with a critical temperature (T_c) of 4.8 K. CDWs represent standing waves of electron density associated with local lattice distortions, primarily driven by electron–phonon interactions. These CDW phenomena, observed in various TMDCs like TaS_2, $TaSe_2$, and $TiSe_2$, can be significantly modulated by foreign-ion intercalation.

From the 1980s onward, researchers have extensively explored the impact of metal ions or atoms (e.g., Na, Rb, K, Cu, Cr, Mn, Fe) intercalation in bulk TMDCs such as 1T-TaS_2, $TaSe_2$, VSe_2, and $TiSe_2$ on CDW performance [34, 35]. These studies employed low-temperature transport measurements, low-energy electron diffraction, photoelectron spectroscopy, and scanning tunneling microscopy. However, the lack of exact controllability and reversibility in solvent and gaseous intercalation for bulk TMDCs has limited the ability to gain in-depth insights into the underlying structural changes down to the atomic layer limit. Electrochemical intercalation in atomically thin layered materials overcomes these limitations, providing an ideal platform for investigating the relationship between structural evolution and transport properties.

8.3 Thermal and Mechanical Properties of Two-Dimensional Semiconductors

8.3.1 Approaches to Measure the Thermal Property

TMDCs exhibit exceptional structural, mechanical, and physical properties, making them highly attractive for both fundamental scientific research and practical engineering applications [36, 37]. Their remarkable mechanical properties, in particular, make them promising candidates for use in ubiquitous electronics, flexible displays, smart health diagnostics, wearable computing, and integrated circuits (ICs). As ICs continue to advance with faster switching speeds, greater transistor counts, and higher integration densities, managing power consumption becomes increasingly critical. Elevated local temperatures within ICs can lead to performance degradation. Therefore, comprehending the thermal properties of 2D materials is of paramount importance.

Generally, there are several kinds of approaches to evaluate the thermal properties of two-dimensional materials, including Raman spectrometer, time-domain thermoreflectance (TDTR), and scanning thermal microscopy (SThM). As shown in Figure 8.5(a), leveraging the sensitive response of the phonon frequency of TMDCs to local heating induced by a laser, the thermal conductivity of these 2D materials can be estimated using a non-contact method [38]. While Raman-active optical phonons themselves may not significantly conduct heat along the flake, their vibrational frequency is sensibly responsive to local temperature variations caused by external factors.

As shown in Figure 8.5(b), the SThM employs a specialized scanning probe microscopy mode that employs a thermal probe. This probe is a vital component of a biased Wheatstone bridge configuration, comprising two known resistors (R_1 and R_2), an adjustable resistor (R_3), and the thermal probe resistor (R_{probe}) situated within the SThM tip [39]. The apex of the SThM probe serves as a local heat source for two-dimensional materials. Upon bringing the probe into proximity with the flake's surface, the probe's temperature decreases due to heat transfer to the sample, leading to a reduction in resistance R_{probe}. This alteration unbalances the bridge, resulting in a change in the output voltage V_{out}.

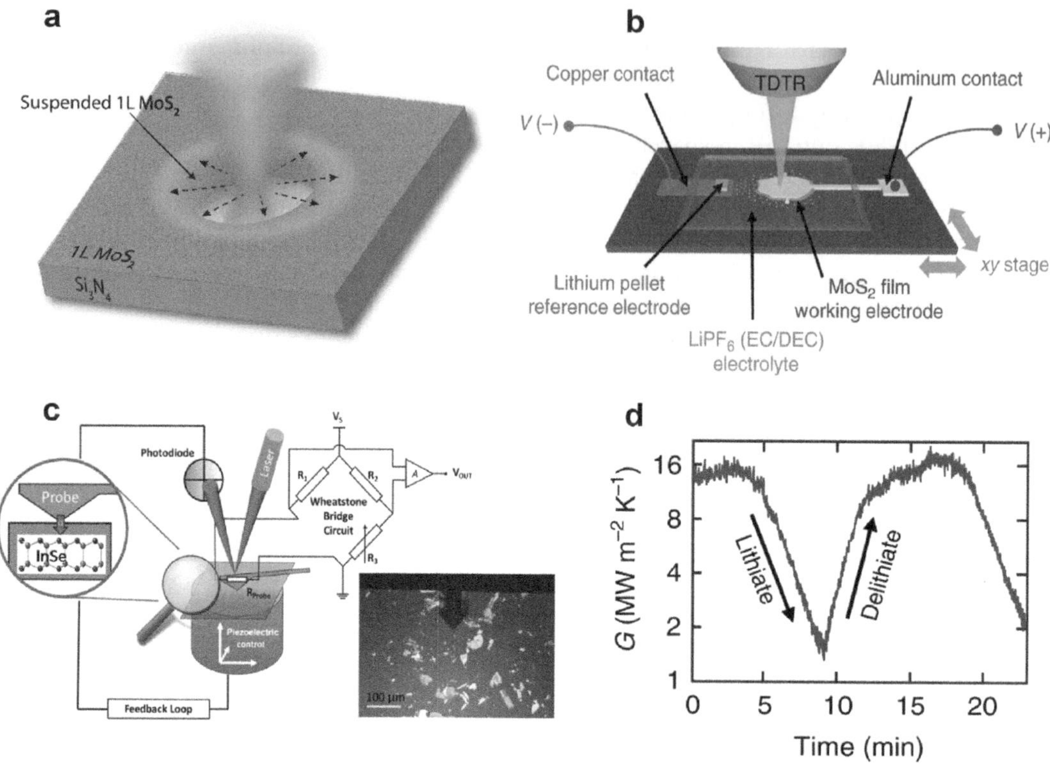

FIGURE 8.5

(a) Schematic of suspended monolayer MoS_2 over a perforated Si_3N_4. Adapted with permission [43], Copyright (2014), American Chemical Society. (b) Schematic of the electrochemical cell used for operando TDTR experiments. Adapted with permission [38], Copyright (2018), Springer Nature. (c) SThM setup showing the probe as a resistor in a Wheatstone bridge. Adapted with permission [39], Copyright (2021), Wiley-VCH. (d) Cross-plane thermal conductance measured during the electrochemical cycle. Adapted with permission [38], Copyright (2018), Springer Nature.

As the probe resistance R_{probe} decreases due to tip cooling, the output voltage V_{out} undergoes variation [39]. The corresponding change in electrical power dissipated by the probe resistor, denoted as $\Delta P = R_{probe} \Delta V_{out}$, is directly proportional to the heat flux directed toward the sample and its thermal conductivity. However, it is also subject to influences from factors such as heat loss from the cantilever, air conduction/convection, tip-sample interfacial thermal resistance, and heat propagation radius. In contrast to far-field optical techniques like Raman spectroscopy, SThM isn't constrained by optical diffraction, allowing it to capture the thermal response of a sample with high spatial resolution, reaching levels as low as 50 nm. This attribute proves indispensable for thermal investigations of 2D materials with diverse sizes and/or edges, as well as systems where the layers are supported by substrates with spatially varying topographies, as elucidated in the following sections.

TDTR is an effective approach to measure thermal conductivity. As shown in Figure 8.5(c), in a standard TDTR measurement, a femtosecond pulsed laser serves as the light source and is split into two beams: one acting as the pump beam to heat the sample and the other as the probe beam for thermoreflectance measurement [38, 40]. Typically, the pump beam undergoes modulation at a frequency ranging from 0.1 to 20 MHz through the use of an electro-optic modulator [41]. This modulation facilitates phase-sensitive detection via a lock-in amplifier.

8.3.2 Electrochemical Thermal Transistor

Given that the foreign species can be intercalated into the vdWs gap, the researchers demonstrated the capability to reversibly modulate thermal conductance by Li-ion intercalation, achieving a nearly tenfold change, as shown in Figure 8.5(d) [38]. The core of this innovation lies in nanoscale MoS_2 films that can be dynamically controlled through the reversible electrochemical intercalation of Li ions [38]. The "on" state corresponds to pristine MoS_2, while the "off" state corresponds to Li-intercalated MoS_2. The approach involves conducting thermal conductance measurements in real time using TDTR while the device functions akin to a MoS_2 nanobattery, undergoing cyclic and reversible electrochemical processes, as shown in Figure 8.5(d). Employing operando TDTR microscopy allows us to investigate the spatial distribution of Li ions within MoS_2 at various stages of intercalation within an electrochemical cycle.

8.3.3 Ultrahigh Thermal Isolation across vdWs Heterostructures

The integration of nanomaterials has been transformative for advanced electronics and photonics applications, yet applying similar advancements to thermal applications has posed significant challenges [42]. This is partly due to the shorter wavelengths of heat carriers, or phonons, compared to electrons and photons. Pop et al. showcase a remarkable achievement of exceptionally high thermal isolation in ultrathin heterostructures [42]. By layering atomically thin 2D materials such as monolayer graphene, MoS_2, and WSe_2, the researchers have created artificial stacks with thermal resistance surpassing that of materials over 100 times thicker than SiO_2 and an effective thermal conductivity lower than that of air at room temperature. The approach, which employs Raman thermometry, allows us to simultaneously determine the thermal resistance between any pair of 2D monolayers within the stack. This ultrahigh thermal isolation is the result of disparities in mass density and phonon density of states between these 2D layers. These innovative thermal metamaterials represent a significant development in the emerging field of photonics and offer potential applications in scenarios where extremely thin thermal insulation is required, such as in thermal energy harvesting or for heat management in ultra-compact designs.

8.3.4 Mechanical Properties of TMDCs and vdWs Heterostructures

Graphene, despite being the strongest material ever measured, has faced limitations in semiconductor applications due to its inherent lack of a band gap [44]. This has led to a surge of interest in TMDCs like MoS_2, WS_2, and WSe_2. These TMDCs exhibit the advantage of having a band gap, transitioning from an indirect band gap in bulk to a direct band gap in monolayers, expanding their potential applications in electronics and optoelectronics. In addition to these electronic properties, the mechanical characteristics of 2D TMDCs are crucial, especially in flexible electronics. Initial investigations into the mechanical properties of TMDC layers focused on less-defective, exfoliated flakes of MoS_2. For instance, Bertolazzi et al. [44] used atomic force microscope (AFM) nanoindentation (shown in Figure 8.6(a)) to determine that monolayer MoS_2 possesses

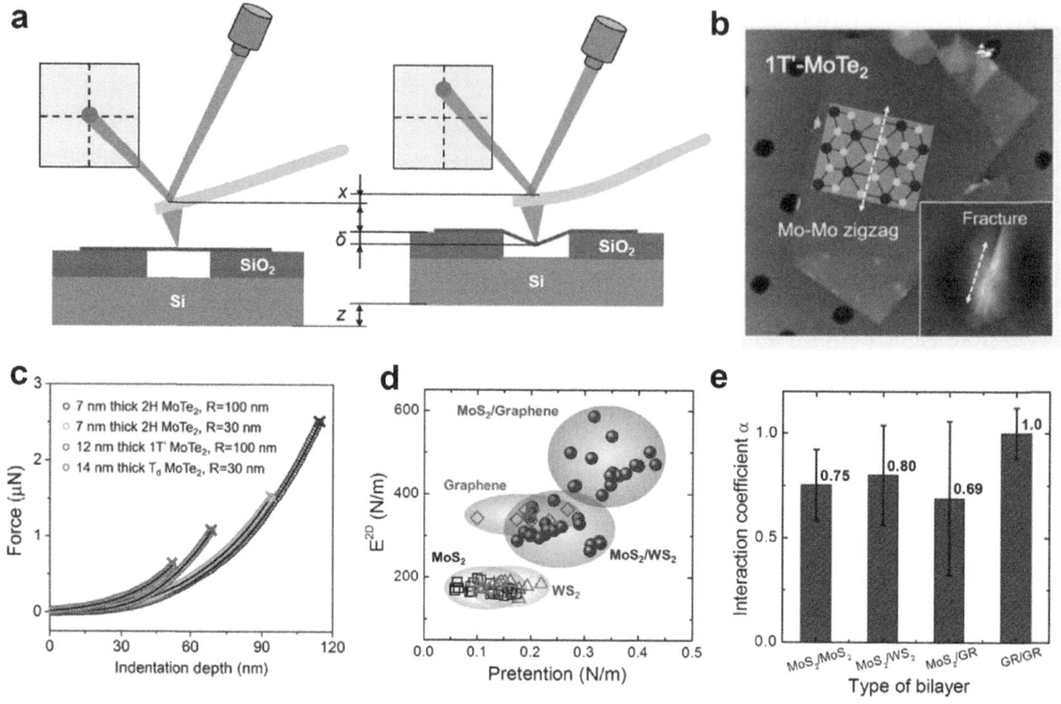

FIGURE 8.6

(a) Schematic diagram of the nanoindentation process. Adapted with permission [44], Copyright (2019), American Chemical Society. (b) Dependence of indentation force on indentation depth for the three MoTe₂ phases. Adapted with permission [44], Copyright (2019), American Chemical Society. (c) Experimental data of 2D modulus and pretension for various 2D layers and heterostructures. Adapted with permission [48], Copyright (2014), American Chemical Society. (d) Interaction coefficients for different types of bilayers. Adapted with permission [48], Copyright (2014), American Chemical Society. (e) Comparison of Young's modulus of 2D monolayers with multilayers and bulk in the literature. Adapted with permission [48], Copyright (2014), American Chemical Society.

an in-plane 2D modulus of 180 ± 60 N/m (equivalent to 270 ± 100 GPa) and average breaking strength of 15 ± 3 N/m (23 GPa), values several times lower than monolayer graphene but still substantially stronger than steel [45]. Bilayer MoS_2 exhibited a 2D modulus of 260 ± 70 N/m, corresponding to a lower 3D modulus of 200 ± 60 GPa, likely due to defects or interlayer sliding. They determined a mean Young's modulus of 330 ± 70 GPa and a pretension of 0.13 ± 0.10 N/m. Notably, both groups' modulus values exceed the bulk MoS_2 modulus (approximately 240 GPa) [46].

The biaxial deformation of suspended membranes is a common technique used in nanoindentation to explore the elastic properties of structurally isotropic 2D materials [47]. However, when it comes to the biaxial deformation of anisotropic 2D materials, their elastic properties and fracture behaviors have remained largely unexplored. Liu et al. investigated the elastic properties and fracture behaviors of 2H-, 1T'-, and T_d-MoTe₂ using temperature-variant biaxially deformed nanoindentation in combination with first-principles calculations, as illustrated in Figure 8.6(b) and (c) [44]. We found that the effective E_{2D} (2D elastic modulus) of these three phases was quite similar, with minor deviations of less than 15%. However, their breaking strengths exhibited significant variations. Density functional theory calculations supported consistent E_{2D} values for all three phases, and we attributed the lower strengths in the 1T′ and T_d phases to their distorted structures, leading to an uneven distribution of bonding strengths. These findings underscore the influence of structural symmetry on both elastic properties and fracture behaviors in these three phases, suggesting that biaxial deformation via nanoindentation is a suitable method for probing the properties of both isotropic and anisotropic 2D materials.

Liu et al. assessed the elastic modulus of chemical vapor deposition (CVD)-grown monolayer MoS_2 and WS_2, shedding light on the interlayer interaction within their heterostructures, as illustrated in Figure 8.6(d) and (e) [48]. These CVD-grown MoS_2 and WS_2 monolayers exhibited high 2D elastic moduli, approximately 170 N/m, which closely resemble the values observed in exfoliated MoS_2 and are roughly half of the graphene's modulus. Theoretical simulations confirmed that MoS_2 and WS_2 possess nearly identical lattice constants and elastic properties. While the 2D moduli of heterostructures were slightly lower than the sum of the moduli of individual layers, they were comparable to the corresponding bilayer homostructures. This implies that the interactions between heteromonolayers are similar to those between homomonolayers.

References

1. X.L. Liu, M.C. Hersam, 2D materials for quantum information science, *Nat. Rev. Mater.* 4 (2019) 669–684.
2. A. Splendiani, L. Sun, Y.B. Zhang, T.S. Li, J. Kim, C.Y. Chim, G. Galli, F. Wang, Emerging photoluminescence in monolayer MoS_2, *Nano Lett.* 10 (2010) 1271–1275.
3. T. Cao, G. Wang, W.P. Han, H.Q. Ye, C.R. Zhu, J.R. Shi, Q. Niu, P.H. Tan, E. Wang, B.L. Liu, J. Feng, Valley-selective circular dichroism of monolayer molybdenum disulphide, *Nat. Commun.* 3 (2012) 5.
4. X. Chen, T.F. Yan, B.R. Zhu, S.Y. Yang, X.D. Cui, Optical control of spin polarization in monolayer transition metal dichalcogenides, *ACS Nano* 11 (2017) 1581–1587.
5. T. Lamountain, J. Nelson, E.J. Lenferink, S.H. Amsterdam, A.A. Murthy, H.F. Zeng, T.J. Marks, V.P. Dravid, M.C. Hersam, N.P. Stern, Valley-selective optical stark effect of exciton-polaritons in a monolayer semiconductor, *Nat. Commun.* 12 (2021) 7.
6. D. Xiao, G.B. Liu, W.X. Feng, X.D. Xu, W. Yao, Coupled spin and valley physics in monolayers of MoS_2 and other group-VI dichalcogenides, *Phys. Rev. Lett.* 108 (2012) 5.
7. G.B. Liu, D. Xiao, Y.G. Yao, X.D. Xu, W. Yao, Electronic structures and theoretical modelling of two-dimensional group-VIB transition metal dichalcogenides, *Chem. Soc. Rev.* 44 (2015) 2643–2663.
8. G.H. Su, X. Wu, W.Q. Tong, C.G. Duan, Two-dimensional layered materials-based spintronics, *Spin* 5 (2015) 9.
9. H.Y. Shi, R.S. Yan, S. Bertolazzi, J. Brivio, B. Gao, A. Kis, D. Jena, H.G. Xing, L.B. Huang, Exciton dynamics in suspended mono layer and few-layer MoS_2 2D crystals, *ACS Nano* 7 (2013) 1072–1080.
10. J. Xiao, M. Zhao, Y. Wang, X. Zhang, Excitons in atomically thin 2D semiconductors and their applications, *Nanophotonics* 6 (2017) 1309–1328.
11. K.F. Mak, C. Lee, J. Hone, J. Shan, T.F. Heinz, Atomically thin MoS_2: A new direct-gap semiconductor, *Phys. Rev. Lett.* 105 (2010) 4.
12. Z.P. Li, T.M. Wang, Z.G. Lu, C.H. Jin, Y.W. Chen, Y.Z. Meng, Z. Lian, T. Taniguchi, K. Watanabe, S.B. Zhang, D. Smirnov, S.F. Shi, Revealing the biexciton and trion-exciton complexes in bn encapsulated WSe_2, *Nat. Commun.* 9 (2018) 7.
13. K.F. Mak, K.L. He, J. Shan, T.F. Heinz, Control of valley polarization in monolayer MoS_2 by optical helicity, *Nat. Nanotechnol.* 7 (2012) 494–498.
14. F. Ceballos, M.Z. Bellus, H.Y. Chiu, H. Zhao, Ultrafast charge separation and indirect exciton formation in a MoS_2-$MoSe_2$ van der Waals heterostructure, *ACS Nano* 8 (2014) 12717–12724.

15. K.F. Mak, K.L. He, C. Lee, G.H. Lee, J. Hone, T.F. Heinz, J. Shan, Tightly bound trions in mono-layer MoS_2, *Nat. Mater.* 12 (2013) 207–211.

16. P. Rivera, J.R. Schaibley, A.M. Jones, J.S. Ross, S.F. Wu, G. Aivazian, P. Klement, K. Seyler, G. Clark, N.J. Ghimire, J.Q. Yan, D.G. Mandrus, W. Yao, X.D. Xu, Observation of long-lived inter-layer excitons in monolayer $MoSe_2$-WSe_2 heterostructures, *Nat. Commun.* 6 (2015) 6.

17. A.A. Tedstone, D.J. Lewis, P. O'brien, Synthesis, properties, and applications of transition metal-doped layered transition metal dichalcogenides, *Chem. Mater.* 28 (2016) 1965–1974.

18. X.Z. Xiaoyang Zheng, Excitons in two-dimensional materials, *Mater. Sci. Eng. R* 6 (2019) 51–64.

19. Z.Y. Cai, B.L. Liu, X.L. Zou, H.M. Cheng, Chemical vapor deposition growth and applications of two-dimensional materials and their heterostructures, *Chem. Rev.* 118 (2018) 6091–6133.

20. B. Urbaszek, A. Srivastava, Materials in flatland twist and shine, *Nature* 567 (2019) 39–40.

21. Y. Pan, S.T. Li, M. Rahaman, I. Milelthin, D.R.T. Zahn, Signature of lattice dynamics in twisted 2D homo/hetero-bilayers, *2D Mater.* 9 (2022) 10.

22. D. Jariwala, V.K. Sangwan, L.J. Lauhon, T.J. Marks, M.C. Hersam, Emerging device applications for semiconducting two-dimensional transition metal dichalcogenides, *ACS Nano* 8 (2014) 1102–1120.

23. B. Radisavljevic, A. Radenovic, J. Brivio, V. Giacometti, A. Kis, Single-layer MoS_2 transistors, *Nat. Nanotechnol.* 6 (2011) 147–150.

24. B.L. Li, C.B. Gong, W. Shen, J.D. Peng, H.L. Zou, H.Q. Luo, N.B. Li, Engineering metallic MoS_2 monolayers with responsive hydrogen evolution electrocatalytic activities for enzymatic reac-tion monitoring, *J. Mater. Chem. A* 9 (2021) 11056–11063.

25. K. Kaasbjerg, K.S. Thygesen, K.W. Jacobsen, Phonon-limited mobility in n-type single-layer MoS_2 from first principles, *Phys. Rev. B* 85 (2012) 16.

26. D. Jariwala, V.K. Sangwan, D.J. Late, J.E. Johns, V.P. Dravid, T.J. Marks, L.J. Lauhon, M.C. Hersam, Band-like transport in high mobility unencapsulated single-layer MoS_2 transistors, *Appl. Phys. Lett.* 102 (2013) 4.

27. B.W.H. Baugher, H.O.H. Churchill, Y.F. Yang, P. Jarillo-Herrero, Intrinsic electronic transport properties of high-quality monolayer and bilayer MoS_2, *Nano Lett.* 13 (2013) 4212–4216.

28. Y. Li, H. Yan, B. Xu, L. Zhen, C.Y. Xu, Electrochemical intercalation in atomically thin van der Waals materials for structural phase transition and device applications, *Adv. Mater.* 33 (2021) 18.

29. M.S. Choi, D. Qu, D. Lee, X. Liu, K. Watanabe, T. Taniguchi, W.J. Yoo, Lateral MoS_2 p-n junction formed by chemical doping for use in high-performance optoelectronics, *ACS Nano* 8 (2014) 9332–9340.

30. S.Z. Bisri, S. Shimizu, M. Nakano, Y. Iwasa, Endeavor of iontronics: From fundamentals to applications of ion-controlled electronics, *Adv. Mater.* 29 (2017) 48.

31. V.P. Pham, G.Y. Yeom, Recent advances in doping of molybdenum disulfide: Industrial applica-tions and future prospects, *Adv. Mater.* 28 (2016) 9024–9059.

32. F. Xiong, H.T. Wang, X.G. Liu, J. Sun, M. Brongersma, E. Pop, Y. Cui, Li intercalation in MoS_2: In situ observation of its dynamics and tuning optical and electrical properties, *Nano Lett.* 15 (2015) 6777–6784.

33. Y.P. Song, X.W. Liang, J.G. Guo, J. Deng, G.Y. Gao, X.L. Chen, Superconductivity in li-interca-lated 1T-$SnSe_2$ driven by electric field gating, *Phys. Rev. Mater.* 3 (2019) 7.

34. W. Yu, J. Li, T.S. Herng, Z.S. Wang, X.X. Zhao, X. Chi, W. Fu, I. Abdelwahab, J. Zhou, J.D. Dan, Z.X. Chen, Z. Chen, Z.J. Li, J. Lu, S.J. Pennycook, Y.P. Feng, J. Ding, K.P. Loh, Chemically exfoli-ated VSe_2 monolayers with room-temperature ferromagnetism, *Adv. Mater.* 31 (2019) 8.

35. H. Li, G. Lu, Y.L. Wang, Z.Y. Yin, C.X. Cong, Q.Y. He, L. Wang, F. Ding, T. Yu, H. Zhang, Mechanical exfoliation and characterization of single- and few-layer nanosheets of WSe_2, TaS_2, and $TaSe_2$, *Small* 9 (2013) 1974–1981.

36. G. Zhang, Y.W. Zhang, Thermal properties of two-dimensional materials, *Chin. Phys. B* 26 (2017) 12.

37. Z.W. Zhang, P. Chen, X.D. Duan, K.T. Zang, J. Luo, X.F. Duan, Robust epitaxial growth of two-dimensional heterostructures, multiheterostructures, and superlattices, *Science* 357 (2017) 788.

38. A. Sood, F. Xiong, S.D. Chen, H.T. Wang, D. Selli, J.S. Zhang, C.J. Mcclellan, J. Sun, D. Donadio, Y. Cui, E. Pop, K.E. Goodson, An electrochemical thermal transistor (vol 9, 4510, 2018), *Nat. Commun.* 10 (2019) 1.

39. D. Buckley, Z.R. Kudrynskyi, N. Balakrishnan, T. Vincent, D. Mazumder, E. Castanon, Z.D. Kovalyuk, O. Kolosov, O. Kazakova, A. Tzalenchuk, A. Patane, Anomalous low thermal conductivity of atomically thin InSe probed by scanning thermal microscopy, *Adv. Funct. Mater.* 31 (2021) 10.
40. P.Q. Jiang, B. Huang, Y.K. Koh, Accurate measurements of cross-plane thermal conductivity of thin films by dual-frequency time-domain thermoreflectance (TDTR), *Rev. Sci. Instrum.* 87 (2016) 8.
41. S. Warkander, J.Q. Wu, Transducerless time domain reflectance measurement of semiconductor thermal properties, *J. Appl. Phys.* 131 (2022) 12.
42. S. Vaziri, E. Yalon, M.M. Rojo, S.V. Suryavanshi, H.R. Zhang, C.J. Mcclellan, C.S. Bailey, K.K.H. Smithe, A.J. Gabourie, V. Chen, S. Deshmukh, L. Bendersky, A.V. Davydov, E. Pop, Ultrahigh thermal isolation across heterogeneously layered two-dimensional materials, *Sci. Adv.* 5 (2019) 7.
43. R.S. Yan, J.R. Simpson, S. Bertolazzi, J. Brivio, M. Watson, X.F. Wu, A. Kis, T.F. Luo, A.R.H. Walker, H.G. Xing, Thermal conductivity of monolayer molybdenum disulfide obtained from temperature-dependent Raman spectroscopy, *ACS Nano* 8 (2014) 986–993.
44. Y.F. Sun, J.B. Pan, Z.T. Zhang, K.N. Zhang, J. Liang, W.J. Wang, Z.Q. Yuan, Y.K. Hao, B.L. Wang, J.W. Wang, Y. Wu, J.Y. Zheng, L.Y. Jiao, S.Y. Zhou, K.H. Liu, C. Cheng, W.H. Duan, Y. Xu, Q.M. Yan, K. Liu, Elastic properties and fracture behaviors of biaxially deformed, polymorphic $MoTe_2$, *Nano Lett.* 19 (2019) 761–769.
45. S. Bertolazzi, J. Brivio, A. Kis, Stretching and breaking of ultrathin MoS_2, *ACS Nano* 5 (2011) 9703–9709.
46. A. Castellanos-Gomez, M. Poot, G.A. Steele, H.S.J. Van Der Zant, N. Agrait, G. Rubio-Bollinger, Elastic properties of freely suspended MoS_2 nanosheets, *Adv. Mater.* 24 (2012) 772–775.
47. Z.H. Dai, L.Q. Liu, Z. Zhang, Strain engineering of 2D materials: Issues and opportunities at the interface, *Adv. Mater.* 31 (2019) 11.
48. K. Liu, Q.M. Yan, M. Chen, W. Fan, Y.H. Sun, J. Suh, D.Y. Fu, S. Lee, J. Zhou, S. Tongay, J. Ji, J.B. Neaton, J.Q. Wu, Elastic properties of chemical-vapor-deposited monolayer MoS_2, WS_2, and their bilayer heterostructures, *Nano Lett.* 14 (2014) 5097–5103.

Metal-Oxide-Semiconductor Devices

Sudipta Ray and Dilip K. Maiti*
**Correspondence: dkmchem@caluniv.ac.in*

9.1 Introduction

The number of technical applications utilizing metal-oxide-semiconductor (MOS)-based devices has grown quickly during the past 20 years. These applications cover a range of fast-moving consumer goods and display technologies. These technological solutions' underlying principles have not been comprehensively examined. This chapter seeks to give a clear picture of these changes starting from the very beginning and summarizes where they are now. Although thin-film transistors for flat-panel displays (FPDs) frequently use MOSs, their uses don't seem to be restricted to more traditional ones. Emerging applications utilizing MOS thin-film devices are listed and briefly illustrated in this chapter. The MOS device has several very important benefits. Primarily, the synthesis of a MOS has been discussed with an emphasis on the vapor-phase method and liquid-phase method. Then the different fields of application are collectively listed and firstly the developments of metal-oxide-based thin-film transistors (TFTs) are described. Next, power devices utilizing thin-film MOS devices are discussed since they can withstand high voltage and have high carrier mobility. This includes important factors in material selection, processes for material deposition, and various structural designs created for functioning devices. Sensing devices are then discussed because thin-film devices can be made in huge quantities at a reasonable cost and because their electrical characteristics can be sensitive to the surrounding environment. Also, there are representations of numerous fundamental advantages available in photocatalysis and photo voltaic applications as well as computing applications, covering many different types of linked applications. The MOS thin-film devices are potentially important parts of neuromorphic artificial intelligence systems, which are actually fundamental ideas in smart societies.

9.1.1 What Is a MOS?

MOSs represent a class of unique materials due to their electronic charge transport properties when compared to conventional covalent semiconductors such as silicon (Si). MOSs are valence compounds with a high degree of ionic bonding. Their conduction band minimum (CBM) and valence band maximum (VBM) mainly consist of the metal (M) ns and oxygen (O) 2p orbital, respectively. Nowadays, microelectronic architectures made of metal-oxide (SiO_2) and semiconductor (Si) are the most prevalent. The two terminals of a MOS capacitor, which make up the majority of MOS devices' architecture, are their most basic form. Because of this, it's crucial to comprehend MOS's workings, traits, and

DOI: 10.1201/9781003439448-9

application procedure. Two diagrams can be used to see the mechanisms under static biasing situations.

1. Energy band diagram
2. Block-charge diagram

   C–V (capacitance versus voltage) curves are a visual representation of the features of MOS.

The principles of creating a MOS structure are comparable to those of metal-semiconductor (MS) contact structures, but a thin layer of silicon oxides is sandwiched between the metal and semiconductor (Si) layers in a MOS structure. An ideal MOS device is schematically depicted in Figure 9.1. The following characteristics ought to be present in a perfect MOS construction.

1. Under both AC and DC biasing conditions, the thickness of the metallic gate should be sufficient to form an equipotential zone; that is, every point in the region should be in the same potential.
2. Under all static bias, the middle layer of oxides should be a perfect insulator; it should not allow any current to flow through it.
3. The oxide–semiconductor contact shouldn't have any charge centers.
5. The semiconductor (Si) needs to be thick enough so that charges can pass through a bulk of silicon that is free of fields before they reach the back contact.
6. The backside of the MOS device should have ohmic contacts.
7. As shown in Figure 9.1, MOS is a one-dimensional structure with variation in its layers only along a single direction (say along the x-direction).
8. The difference between the conduction band (E_C) and Fermi energy (E_F) in the forward flat band along with the electron affinity (χ) should be equal to the metal work function (Φ_M), and this is expressed as, $\Phi_M = \Phi_S = \chi + (E_C - E_F)_{FB}$. Although this feature is optional, including it makes it simpler to grasp the static behavior of MOS devices at first.

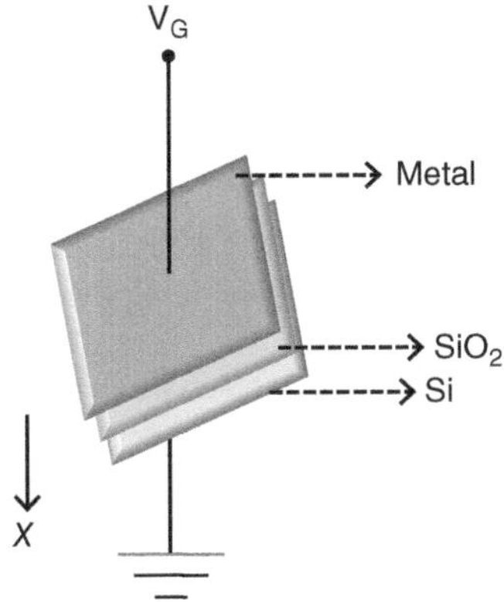

FIGURE 9.1
The schematic representation of an ideal MOS device.

9.1.2 Diagrams of the Energy Band and Block Charge

The band diagram displays the materials' band energies and how, as a function of process and spatial dimensions, their energy levels fluctuate. In the cross-sectional image of the MOS devices, the X-axis stands for the distance (x). The energy is shown on the Y-axis. There are only relative values for Y, not arbitrary ones. The associated energy is higher as the Y-axis grows. The MOS flat band energy band diagram with n-type semiconductor is shown in Figure 9.2. The lowest energy an electron needs to achieve to break free from a substance is known as the vacuum level (E_{vc}). The difference between Fermi level (E_F) and E_{vc} is represented by the terms Φ_M and Φ_S, which represent the work required to remove an electron. The energy barrier of the insulator (SiO_2 oxides) is determined by the electron affinity (χ), which determines the height of the energy barrier in semiconductors. The flat band diagram refers to the zero-bias band diagram in Figure 9.2(a). Though theoretically similar to the equilibrium MS contact, this band diagram's creation is separated by a distance equal to the oxide layer's thickness, x_0. Due to its specification in the aforementioned property 8, the Fermi level was aligned between the metal and semiconductor. The flat band MOS device has no charge or electric field; therefore, the inserted insulator can only minimally lower the energy barrier. There is also no block-charge diagram for the flat band like that shown in Figure 9.2(b). The charge density distribution inside the MOS structure is displayed in the block-charge diagram. The cross-section view's X-axis, like the energy band diagram, displays the distance (x). The precise charge distribution Q in the MOS devices, however, is represented by the Y-axis. The metal–oxide interface and the oxide–semiconductor interface are each represented by a horizontal Y-axis. The positive charge Q is formed above the X-axis by the concentration of holes, while the negative charge Q is created below the X-axis by the concentration of electrons. When compared to traditional covalent semiconductors like silicon (Si), MOSs represent a class of distinct materials due to their electronic charge transport capabilities. The valence chemicals that make up MOSs have strong ionic bonds. Their metal (M) ns and oxygen (O) 2p orbitals, respectively, make up most of their CBM and VBM. The interaction of the metal and oxide orbitals causes a significant difference in the transport of charge carriers. Since the O 2p orbital is confined and the M ns orbitals are often strongly dispersive, electrons have a lower effective mass than holes. The smaller electron effective mass in a metal oxide suggests greater electron

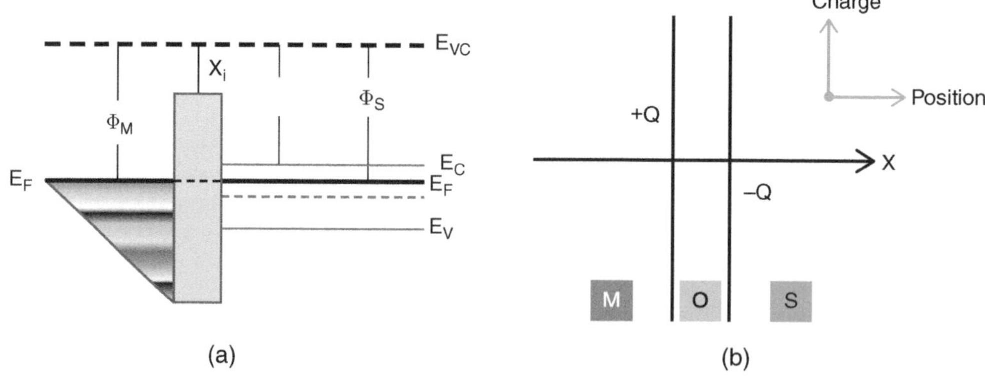

(a)   (b)

FIGURE 9.2

(a) The flat band diagram of MOS in equilibrium with n-type semiconductor and (b) the block-charge diagrams of flat band MOS.

transport compared to hole transport because the carrier mobility is inversely related to the electron effective mass, m*, via $\mu = e\tau/m^*$, where τ is the free carrier scattering time. Most common MOSs have n-type conductivity, including In_2O_3, ZnO, and SnO_2, as well as their solid solutions. On the other hand, the hole-effective mass needs to be considerably lower in metal oxides to obtain p-type conductivity. A more dispersed VBM is necessary for this. Designing materials with metal cations that introduce occupied d or s states close to the VBM is essential for realizing dispersive VBM in metal oxide. As a result, there is an increase in the dispersion of VBM and a decrease in the hole-effective mass due to the p–d and p–s interaction between the metal cation and oxygen orbitals. So far, promising p-type semiconductors have been discovered in various families of metal oxides. Since 1993, nickel oxide (NiO) has been the first known p-type transparent conductive oxide (TCO) [1]. Cu_2O and $CuMO_2$ (M = Al, Ga, or In) are two Cu(I)-based oxides that have received a lot of interest as p-type semiconductors [2]. The Cu 3d state is near the VBM in these metal oxides, which lowers the hole-effective mass. Additionally, SnO, whose valence band (VB) is made up of the hybridization of Sn 5s and O 2p states, has also been the subject of intense research in recent years for its p-type semiconductor characteristics [3].

9.2 Synthesis of MOSs

MOSs can be made using either a top-down or bottom-up approach.

The top-down method involves vapor-phase deposition techniques. The bottom-up approach is composed of mainly solution-based processes such as the sol–gel process. Different processes of synthesis of MOSs are listed in Figure 9.3.

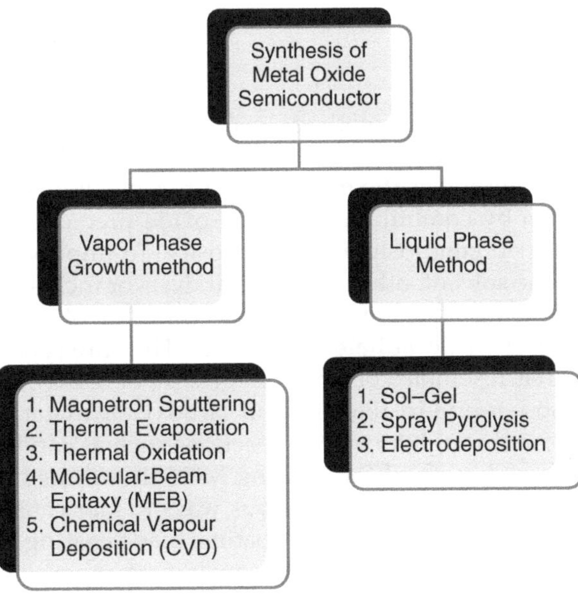

FIGURE 9.3
Different methods of synthesis of metal-oxide semiconductor.

9.2.1 Vapor-Phase Growth Methods

Numerous methods are used in the vapor-phase process that allow for the nucleation and growth of p-type MOS films with precise control over their crystal structure and thickness. Depending on the type of reaction, a material source (solid, powder, or gaseous) can be evaporated and deposited on a substrate, where nucleation and growth processes may take place. Vapor-phase growth includes both physical vapor deposition (PVD) and chemical vapor deposition (CVD) [4]. PVD entails the condensation of the desired material's vaporized gaseous molecules onto a substrate. PVD can precisely regulate the thickness, crystal structure, and microstructure of thin films despite its high cost. It uses several techniques, although magnetron sputtering, thermal evaporation, and molecular-beam epitaxy are the most extensively studied. Another PVD method is thermal oxidation, which creates thin films by using oxidation as opposed to condensation. With CVD, a chemical interaction with the gaseous molecules in the vapor takes place on the substrate's surface. The term "CVD" refers to a variety of processes, including atomic layer deposition (ALD), atmospheric-pressure CVD, low-pressure CVD, and plasma-assisted CVD. Large-scale production is made possible by CVD, which offers considerable control over morphology and defects and encourages the formation of heterojunctions. However, the emergence of hazardous byproducts hurts CVD. The vacuum-based process includes sputtering, thermal evaporation, and ALD methods [5]. Particles of the material are separated from a target surface using plasma in the sputtering process, and the substrate is then coated with the substance. Excellent adherence to the substrate and homogeneity of the deposited film are two benefits of sputter-fabricated films. Materials evaporated using a resistive heater and an electron beam can be used in thermal evaporation and electron beam evaporation processes. The evaporation process can be used with most materials and has the benefit of minimal consumption. By repeatedly repeating precursor and reactant supply and purge cycles, the ALD process creates an atomic layer unit thin film. Pinholes and step coverage problems in thin films made with the ALD technique are quite unusual.

9.2.2 Liquid-Phase Method

As previously mentioned, various methods can be used to produce MOS thin films. However, in terms of cost-efficiency, large-area electronics, and mass production, the liquid-phase method has emerged as the most competitive one [6]. The solutions are created in their most basic form by combining the metal-oxide precursors with an appropriate solvent to a predetermined composition and viscosity. To improve control over how the precursor dissolves in the solvent, other kinds of catalysts or inclusions may be introduced to the solution. Simple metal-oxide powders are just one type of precursor. Others include combinations of acetate, chloride, sulfate, and nitrate. They are typically employed to create liquid precursors that resemble colloidal sol–gels [7] or MOS complexes encapsulated by organics, such as those found in alkoxide and carboxylate materials. This makes it easier for the deposit to form through formation reactions such as condensation or hydrolysis processes. A solution-based method for creating MOS thin films has recently received a lot of attention. Additionally, there are numerous ways to fabricate MOSs using the solution process, including spin coating [8], spray coating, blade coating, flexographic printing, and inkjet printing. Due to the low annealing temperature, most solution-based manufacturing methods are advantageous for producing flexible and printable devices. Solution-based In–Ga–Zn–O (IGZO) films were produced by Moreira et al. using an annealing procedure at 300°C [9].

9.3 MOS Device

Due to their favorable electrical characteristics, including high electron mobility, chemical resistance in liquids, transparency, and ease of fabrication for display and sensor technologies, MOSs have been drawing interest. Although TFTs based on oxide semiconductors have already been taken into consideration for the backplanes of liquid-crystal displays and active-matrix organic light-emitting displays, these have yet to be taken into consideration for other applications, such as memory device applications and biosensors.

9.3.1 MOS as the Thin-Film Transistor Materials

TFTs [10] for FPDs, including liquid-crystal displays [11] and organic light-emitting diode displays, are frequently made of MOSs. IGZO is the standard material, followed by ZnO, and many additional materials are being researched [12]. MOS thin-film devices have several advantages. The MOSs are multi-element compounds. Therefore, it might be able to optimize the electrical characteristics by modifying the element compositions, crystal structures, and so on. By increasing the carrier mobility, controlling the carrier density, and optimizing the density of states, it is possible to raise the Seebeck coefficient. Peripheral environment sensitivity can also be attained, for instance, by including an optional revision.

- The bandgap energy of MOSs is high. As a result, thin-film devices may have low leakage current, exceptional high voltage tolerance, and transparency for visible light.
- Even in an atmospheric environment, MOSs are comparatively chemically stable. As a result, the fabrication process can be made simple. For instance, low-temperature fabrication techniques like thermal annealing at not-too-high temperatures and sputtering deposition without high-temperature heating are sufficient. Additionally, printing manufacture is possible because the MOSs have already undergone oxidation, meaning that even if they are deposited using precursors that have been dissolved in liquid solvents, their electrical properties will not be compromised. As a result, it is inexpensive to construct MOS thin-film devices on large, flexible substrates, and it may even be possible to build them in three dimensions by using various printing techniques. Numerous suggestions for new applications are introduced by numerous firms based on the key benefits. Power devices are suggested, for instance, due to their great mobility and high voltage tolerance.

9.3.2 MOS as a Power Device

The use of MOS thin-film devices for power devices is suggested since they can withstand high voltages and have high carrier mobilities. For on-chip high voltage I/Os in typical complementary metal-oxide-semiconductor (CMOS) large-scale integration (LSI), a back-end-of-line (BEOL) transistor with an n-type IGZO film and a p-type SnO film is proposed [13]. As an interface bridge for signal conversion between CMOS core-logics with low voltages and peripheral devices powered with high voltages, the capability of high voltage operation with a tiny feature size is anticipated. As an addition to n-type IGZO TFTs, a p-type amorphous SnO TFT is created. With a gate-to-drain offset arrangement and a temperature of 104°C, oxide–semiconductor processes with a high V_d capability of >40 V are exhibited. There are additional reports of other BEOL transistors [14]. Additionally, IGZO thin films are used to validate Schottky functionalities [15].

9.3.3 MOS as a Sensing Device

As the electrical characteristic can be sensitive to the surrounding environment by adding some optional revisions, where it is advantageous to adjust the carrier density to detect the conductivity change, sensing devices employing MOS thin-film devices are proposed. Additionally, flexible substrates are convenient to use wherever they are placed, thin-film devices may be produced on huge scales at low cost, and layered structures enable stacked structures with control circuit systems.

9.3.3.1 MOS as a Biosensor

Recently, biosensors have evolved into sophisticated sensors that support new diagnostic approaches, real-time monitoring, and early illness identification. For various cancer, virus, degenerative brain conditions, diabetes, and depression detections, they have used invasive or non-invasive methods. Selectivity for a particular target, sensitivity to a precise level or trace quantity, stable signal detection, and reproducibility in varied conditions are the four properties that can be used to evaluate biosensors. Biosensors are essential elements in the healthcare, environmental, chemical, agricultural, and energy-efficient systems industries. Biosensors have been advocated as an effective instrument for quick measurement and analysis due to the requirement of using continuous onsite monitoring with adaptable and reliable properties. The materials for biosensors need to be adapted for use in a variety of applications (quality control, screening techniques, safety gear, and environmental evaluation), which is a significant area of study with many challenging issues to resolve. MOS materials constitute a special instance. Based on their morphologic versatility [16], chemical stability [17], physicochemical interfacial properties [18], and ability to combine in composite structures [19], these have a high potential to become highly competitive materials in the biosensors market. The electrochemical sensitivity and energy band alignment of several materials, including TiO_2 [20], WO_3 [21], SnO_2 [22], and ZnO [23], have drawn significant attention. These materials are suitable for enzyme-based biosensors. MOS-based enzyme-based biosensors have multiple benefits, including the flexibility to adapt to different working situations, great energy efficiency, good sensitivity, and chemical stability in a variety of environments. In the field of biosensor application, mainly mono-component MOSs are used; however, there are several examples of multi-component semiconductors or connected semiconductors (composite, tandem, heterostructures, etc.) are available. These multi-component semiconductor materials have been doped with other metal ions or coupled with metal nanoparticles to improve certain properties. Here, in this chapter, we consider only four metal oxides (TiO_2, SnO_2, ZnO, and WO_3) as typical for biosensor applications.

9.3.3.1.1 TiO_2-Based Biosensors

Due to its great chemical stability, biocompatibility, and plasticity in terms of morphology, TiO_2 is an n-type semiconductor that is widely used in a variety of applications, including photocatalysis, biosensors, photovoltaics, and energy storage. Based on black TiO_2 deposited on an indium tin oxide substrate, the microRNA sensor is enhanced with Au nanoparticles. Dip-coating technique was employed to obtain TiO2 films serving as sensors for microRNA [20] or glucose [24].

9.3.3.1.2 Biosensors Based on SnO$_2$

SnO$_2$ was employed in various applications, including light energy conversion, biosensors, smart windows, and electrochemistry, because of qualities including high surface area, strong biocompatibility, nontoxicity, great chemical stability, and catalytic activity. An H$_2$O$_2$ sensor was developed using a SnO$_2$ nanowires-based MOSs synthesized by the thermal evaporation method [25].

9.3.3.1.3 Biosensors Based on ZnO

ZnO is a straight wide-band-gap semiconductor that displays n-type conductivity when exposed to UV light. Forms a hexagonal wurtzite structure during crystallization (see Figure 9.1), and because of its no centrosymmetric crystal structure, it exhibits unique piezoelectric capabilities. Future medical uses of biosensors will require better biological binding capabilities than tin oxide, which ZnO possesses. ZnO can be used as a permanent human sensor in chronic conditions like diabetes because it is harmless and compatible with human skin. Chemical bath deposition has been used [23] to obtain ZnO nanostars for detecting microRNA-21 in cancer cells. Another technique that has been extensively used for the synthesis of ZnO (both nanorods and nanoparticles) with biosensing applications for glucose detection is the hydrothermal procedure.

9.3.3.1.4 Biosensors Based on WO$_3$

Depending on the synthesis temperature, the crystalline structure of WO$_3$ can range from cubic to octahedral. Physical and chemical methods can be used to create WO$_3$-based materials with well-controlled dimensions, sizes, and crystal structures for sensor research. The hydrothermal method has been used to produce WO$_3$ sensors with different morphologies: flower-like foraflatoxin B1, nanorods for bisphenol A, and nanosheets for cardiac biomarker TroponinI.

9.3.4 Gas Sensor Using MOS

The importance of MOS in gas sensors can be attributed to their distinctive features, including high sensitivity, stability, low cost, ease of synthesis, low power consumption, resilience to high temperatures, and catalytic activity. Toxic gases like H$_2$S, NO$_2$, and CO, ambient gases like O$_2$, NH$_3$, CO$_2$, and O$_3$, and combustible gases like LPG, H$_2$, and CH$_4$ may all be detected using MOS chemiresistors. These sensors rely on the band theory, and their resistances change as a result of diverse surface reactions and kinetics and as a result of exposure to the adsorption of target gases. A conductometric MOS sensor consists of a receptor and a transducer that are controlled by the microstructure of the oxide and gas–solid interactions, respectively. Metal oxides have a reputation for being potentially sensitive resources. They were introduced to the market by Taguchi [26], who founded the semiconducting metal oxides (SMOX) sensors' foundational company. There are two types of metal oxides: transitional metal oxides (Fe$_2$O$_3$, NiO, and Cr$_2$O$_3$), and non-transitional metal oxides (ZnO, SnO$_2$), which contain pre-transitional metal oxides (Al$_2$O$_3$) and post-transitional metal oxides (SnO$_2$). Metal oxides in the pre-transition state (MgO) are likely to be very inactive due to their large energy band gaps and challenging electron–hole formation. They are rarely proposed as gas sensor materials because they have complicated electrical conductivity tests. The minor energy difference between a cation (dn) configuration and the configuration of dn^{-1} or dn^{+1} causes transition-metal oxides to behave differently. As a result, they are more susceptible to the environment than metal oxides. However, structural instability and non-optimality of additional elements make it necessary to restrict the use of conductometric gas sensors. Only transition-metal oxides with

d^{10} and d^0 electrical configurations find their true use as gas sensors. Dual transition-metal oxides use the d^0 arrangement [27]. SMOX gas sensors work by adsorbing oxygen onto the surfaces of metal oxides at temperatures between 150°C and 400°C. By trapping electrons, which are the most numerous charge carriers, the sensor's resistance will either increase (for n-type materials) or decrease (for p-type materials) [28]. Since oxygen on the semiconductor surface interacts with gases in the atmosphere to affect the sensor resistance, the resistance change will be seen as a signal (sensor signs). This signal's strength is related to the gas concentration. The chemical interaction between the gas and the material's surface as well as the conversion of this reaction to corresponding changes in the sensor's electric resistance must therefore be researched to produce a highly sensitive sensor [29].

9.3.5 MOS as a Photocatalyst

The primary mechanism for converting solar energy into the chemical energy required for the breakdown of dyes or organic contaminants is photocatalysis. The surfaces of semiconductors are often where the photocatalytic processes take place. Taking the situation into account, MOS photocatalysts are used as activators to help catalyze the intricate radical chain reaction involved in the photocatalytic oxidation processes. Due to benefits like (1) low or no toxins, (2) affordability, (3) tunable physiochemical properties by modifying the nanoparticle size and doping concentration, and (4) good photocatalytic lifetime without experiencing significant loss over time, this technology is primarily used in photocatalytic dye degradation [30]. Due to its high degradation efficiency, low toxicity, and favorable physical and chemical features, the use of MOSs in the advanced oxidation process for the treatment of dye wastewater has recently attracted considerable interest.

As seen in Figure 9.4, photocatalytic reactions are essentially a multi-step process comprising oxidation and reduction processes. Three main reaction pathways make up the photocatalytic processes: (1) Upon illumination from the light source, photocatalysts absorb photons.

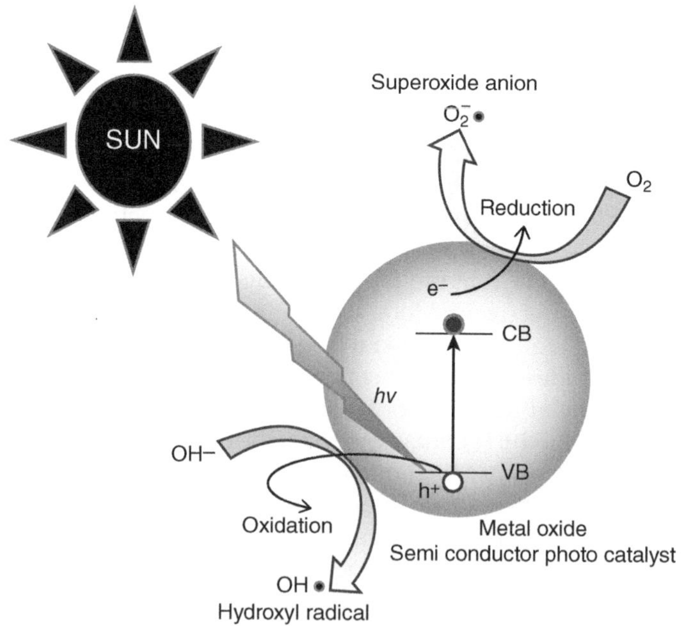

FIGURE 9.4
Schematic diagram for explaining the process of photocatalysis.

The electrons from the VB are stimulated to the conduction band (CB), generating an electron–hole pair, when the photons have higher energy than the band gap of the photocatalyst material. (2) The photo-generated carriers (holes and electrons) have a propensity to recombine on the semiconductor's surface or bulk. On the other hand, in a surface space charge region, the electron–hole pairs might potentially separate. Charge carriers would eventually make it to the surface to start chemical reactions via charge transfer from the photocatalyst to an adsorbate if the diffusion of the electrons and holes is not impeded by any trap states or defect states. (3) The reduction reaction then takes place when the photo-generated electrons meet the molecules that have been absorbed on the semiconductor (photocatalyst) surface. The CB minimum of the photocatalyst must be greater than the reduction potential of the adsorbate to allow the transport of electrons from the photocatalyst to the adsorbate. Similar to this, the photo-generated holes could interact with the adsorbed molecules on the surface to produce potent oxidizing agents like OH radicals. For effective hole transfer in this case, the VB maximum of the photocatalyst must be lower than the oxidation potential of the adsorbate. Photocatalysts need certain materials and electrical characteristics. The ability to prevent back-electron transfer or the recombination of electron–hole pairs is the only requirement for the creation of effective photocatalysts with high chemical conversion efficiencies. So long as they have the following characteristics, the electron–hole pair produced can be effectively used for photocatalysis: (1) the band gap or energy separation is sufficient or greater than the energy needed for the desired reaction; (2) the electron and hole's respective redox potentials for their valence and conduction bands are suitable for inducing redox processes; (3) The electron–hole pair recombination rate is slower than the reaction rates of the redox processes. Titanium dioxide (TiO_2) and zinc oxide (ZnO) are the two most common MOS photocatalysts, and they have attracted the most attention in the field of photocatalysis due to fascinating characteristics like chemical stability, nontoxicity, suitable band-edge alignment to the redox potential of water, and multiphasic structures [31, 32].

9.3.5.1 *Photocatalysts Using Titanium Dioxide*

TiO_2 is frequently used as a photocatalyst in the treatment of dye effluent because of its capacity to produce a highly oxidizing electron–hole pair. Additionally, it is non-toxic, long-lasting photo-stable, and has strong chemical stability [33]. Only high-energy light in the UV area with wavelengths of about 387 nm may initiate the electron–hole separation process, in TiO_2 as it has a wide band gap (e.g., 3.2 eV), which limits its potential. Therefore, one of the key problems in this sector is creating a photocatalyst that can effectively capture the energy from natural sunshine, namely from the visible region. To achieve the following goals, numerous changes have been made to TiO_2's structure: (1) lower the bandgap energy to capture photons from the visible region; (2) boost electron–hole production efficiency; and (3) increase the absorbency of organic pollutants onto TiO_2 by suitable surface modifications [34]. One method for obtaining the qualities is doping. As dopants, metal ions of noble (Pt, Pd, Ag, and Au) and transition (Cr, Cu, Mn, Zn, Co, Fe, and Ni) metals are employed [35]. For this aim, even non-metals like C, N, S, and P are employed [36].To cut costs, transition metals are employed instead of noble metals. It has been discovered that Fe-doped TiO_2 has a 90% dye degradation efficiency [37].

9.3.5.2 *Zinc oxide as Photocatalyst*

Due to its unusual electrical and optoelectronic properties and large band gap (3.2 eV), ZnO is a promising candidate for use as a photocatalyst. According to studies, ZnO performs significantly better than TiO_2 under visible light [38]. ZnO can be employed as a visible light photocatalyst with the appropriate physiochemical changes or by doping, while being very effective when exposed to UV radiation. In addition, it has been discovered that using visible light with a higher intensity (500 W) increases the photocatalytic activity of ZnO nanoparticles [39]. For the degradation of dye under UV and visible light, ZnO photocatalyst is also discovered to be superior to SnO_2, CdS, and ZnS [40].

9.3.6 MOSs as a Photovoltaic Material

MOS nanostructures are used for effective charge extraction and transportation between the electrodes and organic molecules in recent breakthroughs in solar energy conversion technologies based on organic semiconductors (OSC) as the light-harvesting layer, such as dye-sensitized solar cells and organic solar cells (OSCs). OSCs using MOSs. Depending on where the CB and VB are located, the MOSs for OSCs can be either p-type or n-type materials. The need for an n-type material is electron transport from the acceptor's LUMO to the MOS's CB. The VB of the MOS must match the HOMO of the polymer for a p-type contact material. The interface materials' broadband gaps operate as a barrier for other types of carriers, enhancing the contacts' carrier selectivity [41]. Interface materials' primary functions are as follows: (1) to align and adjust the energetic barrier height between the photoactive layer and the surrounding electrodes; (2) to create a contact that is specific for one type of carrier (either holes or electrons); (3) modify the device's polarity (to create a normal or inverted device structure); (4) to stop the polymer and electrode from reacting physically or chemically; (5) to act as a spacer in optical systems. Excitonic solar cells (ESCs) are the aggregate name for first- and second-generation solar cells. The exciton dissociates into mobile carriers (or free carriers) at the material system interface when an exciton is a quasi-particle that has a hole in the VB (or highest occupied molecular orbital, HOMO) and an electron in the CB (or lowest unoccupied molecular orbital, LUMO). A semiconductor (molecule, crystal, or cluster) is anchored with another material whose CB (LUMO) sits at lower energy as shown in Figure 9.5.

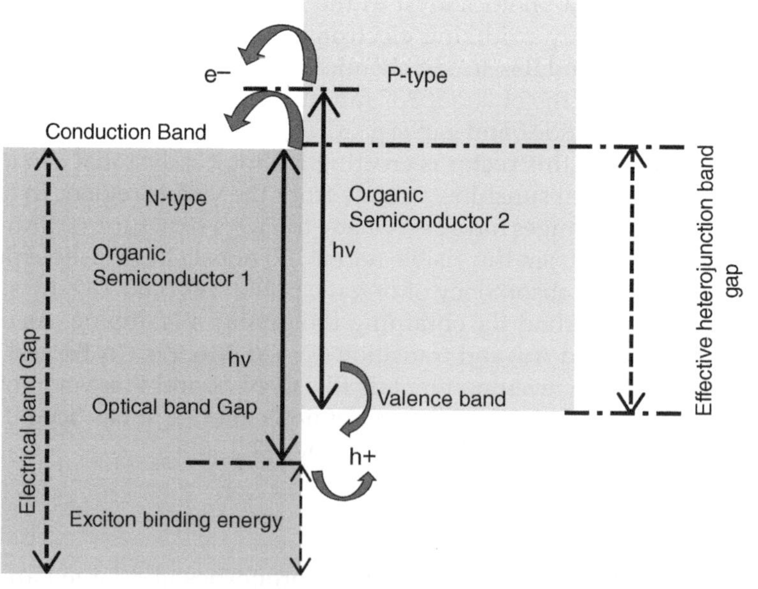

FIGURE 9.5
Schematic representation of an excitonic solar cell.

The foundation of ESCs is this procedure. OSCs, dye-sensitized solar cells (DSSCs) [42], and quantum dot solar cells are a few examples of this kind of ESCs. The preferred materials for OSCs are conjugated polymers and/or organic compounds, such as PCBM and P3HT. A wide-band-gap MOS, such as TiO_2, is tethered to a dye in DSSCs. O'Regan and Graetzel reported on the cathode interface DSSCs with an efficiency of 7.9% [43]. For the creation of DSSCs, wide-band-gap MOSs (e.g., 3 eV) with an appropriate band position in relation to dye (or photosensitizer) have been employed. The MOSs used for the manufacturing of DSSCs have absorption in the UV range due to their wide band gap. Therefore, the absorption of light in the visible and near-infrared area is caused by photosensitizer/dye. Additionally, the greater surface area of nanoporous MOS leads to an increase in dye loading, which improves light absorption and boosts DSSC performance. In addition to the aforementioned physical properties, another significant benefit of MOS for application in DSSCs is the inexpensive cost, abundant natural supply, and simple synthesis methods of MOS. TiO_2 is the most adaptable of the many wide-band-gap oxide semiconductors (TiO_2, ZnO, and SnO_2) that have been investigated as potential electron acceptors for DSSCs. It offers the highest levels of efficiency, is chemically stable, non-toxic, and widely accessible [44].

9.3.7 MOS as a Photodetector

MOS as photodetector [45], for example, NiO, ZnO, and IGZO photodiodes, is used for ultraviolet photodetectors, where the lack of sensitivity to visible light and the reduction of the detection error make the huge bandgap energies of these materials advantageous. To further increase photosensitivity, a phototransistor with a hybrid film of an IGZO film and graphene quantum dots, which serve as a light-absorbing material, is also proposed, where the material affinities of MOSs are used. Since MOSs can have a range of sensitivity by changing the constituent material, element compositions, crystal structure, device architecture, fabrication process, electrical characteristics, and so on, there is also some opportunity to use them as a photodetector.

9.4 Conclusion

MOS devices have a versatile and promising technological field with a wide range of applications, such as gas sensors, biosensors, photocatalysts, photovoltaics, and TFTs. One of the key advantages of MOS devices is their synthesis flexibility, using vapor and liquid-phase methods. This flexibility allows for tailoring materials to specific applications, making them cost-effective and adaptable to various requirements. One crucial attribute of MOSs is their band gap, a property that renders them effective in these various roles. The chapter elucidates how the band gap characterizes these materials as semiconductors and underpins their functionalities from electronic devices to catalysis. This chapter also explores how the band gap facilitates electron–hole pair generation upon exposure to light, initiating catalytic reactions, and advancing the field of photocatalysis. Their versatility, ease of fabrication, and a diverse range of applications make them a crucial technology for addressing current and future challenges across various industries. As research and development in this field continue to advance, we can anticipate even more innovative and transformative applications in the years to come.

References

1. D. A. Wruck, M. Rubin. Structure and electronic properties of electrochromic NiO films, *J. Electrochem. Soc.*, 140, [**1993**], 1097.

2. P. Raksa, S. Nilphai, A. Gardchareon, S. Choopun. Copper oxide thin film and nanowire as a barrier in ZnO dye-sensitized solar cells, *Thin Solid Films*, 517, [**2009**], 17, 4741.

3. W. Zhang, R. Hong, W. Qin, Y. Lv, J. Ma, L. Liao, K. Li, C. Jiang, Enhanced performance of p-type SnOx thin film transistors through defect compensation. *J. Phys.: Condens. Matter*, 34, [**2022**], 404003.

4. A. Moumen, G. C. W. Kumarage, E. Comini. P-type metal oxide semiconductor thin films: Synthesis and chemical sensor applications, *Sensors*, 22, [**2022**], 1359.

5. G. Mahendra, R. Malathi, S.P. Kedhareswara, A. Lakshmi-Narayana, M. Dhananjaya, N. Guruprakash, O.M. Hussain, A. Mauger, C.M. Julien, Sputter-Deposited nanostructured CuO films for micro-supercapacitors, *Appl. Nano*, 2, [**2021**], 46.

6. A. Liu, H. Zhu, Z. Guo, Y. Meng, G. Liu, E. Fortunato, R. Martins, F. Shan, Solution combustion synthesis: Low-temperature processing for p-type Cu:NiO thin films for transparent electronics. *Adv. Mater*, 29, [**2017**], 1701599.

7. A.E. Danks, S.R. Hall, Z. Schnepp, The evolution of "sol-gel" chemistry as a technique for materials synthesis. *Mater. Horizons*, 3, [**2016**], 91.

8. R.R. Prabhu, A.C. Saritha, M.R. Shijeesh, M.K. Jayaraj, Fabrication of p-CuO/n-ZnO heterojunction diode via sol-gel spin coating technique. *Mater. Sci. Eng. B Solid State Mater. Adv. Technol*, 220, [**2017**], 82.

9. M. Moreira, E. Carlos, C. Dias, J. Deuermeier, M. Pereira, P. Barquinha, R. Branquinho, R. Martins, E. Fortunato, Tailoring IGZO composition for enhanced fully solution-based thin film transistors, *Nanomaterials*, 9, [**2019**], 1273.

10. (a) K. Nomura, H. Ohta, K. Ueda, T. Kamiya, M. Hirano, H. Hosono, Thin-film transistor fabricated in single-crystalline transparent oxide semiconductor, *Science* 300, [**2003**], 1269. (b) K. Nomura, H. Ohta, A. Takagi, T. Kamiya, M. Hirano, H. Hosono. Room-temperature fabrication of transparent flexible thin-film transistors using amorphous oxide semiconductors, *Nature*, 432, [**2004**], 488.

11. J.-H. Lee, World's largest (15-inch) XGA AMLCD panel using IGZO oxide TFT, *J. Soc. Inf. Disp*, 39, [**2008**], 625.

12. E. Fortunato, P. Barquinha, R. Martins, Oxide semiconductor thin-film transistors: A review of recent advances. *Adv. Mater.*, 24, [**2012**], 2945.

13. H. Sunamura, K. Kaneko, N. Furutake, S. Saito, M. Narihiro, N. Ikarashi, M. Hane, Y. Hayashi, High on/off-ratio P-type oxide-based transistors integrated onto Cu-interconnects for on-chip high/low voltage-bridging BEOL-CMOS I/Os, 2012 *International Electron Devices Meeting, San Francisco, CA, USA*, [**2012**], 18.8.1–18.8.3, doi: 10.1109/IEDM.2012.6479070.

14. H.J. Kim, B.S. Song, W.-J. Cho, J.T. Park, Reliability of amorphous InGaZnO TFTs with ITO local conducting buried layer for BEOL power transistors, *Microelectron. Reliab.*, 333, [**2017**], 76.

15. J. Zhang, Y. Li, B. Zhang, H. Wang, Q. Xin, A. Song, Flexible indium–gallium–zinc–oxide Schottky diode operating beyond 2.45 GHz, *Nat. Commun.*, 6, [**2015**], 7561.

16. H. Song, Y. Zhang, S. Wang, K. Huang, W, Xu, Label-free polygonal plate fluorescent-hydrogel biosensor for ultrasensitive microRNA detection. *Sensor. Actuat. B-Chem.*, 306, [**2019**], 27554.

17. G. Hernández-Cancel, D. Suazo-Dávila, J. Medina-Guzmán, M. Rosado-González, K. Griebenow, Chemically glycosylation improves the stability of an amperometric horseradish peroxidase biosensor. *Anal. Chim. Acta*, 854, [**2015**], 129.

18. V. Scognamiglio, A. Antonacci, F. Arduini, D. Moscone, G. Palleschi, An eco-designed paper-based algal biosensor for nano formulated herbicide optical detection. *J. Hazard. Mater.*, 373, [**2019**], 483.

19. H. Zheng, M. Liu, Z. Yan, J. Chen, Highly selective and stableglucose biosensor based on incorporation of platinum nanoparticles into polyaniline-montmorillonite hybrid composites. *Microchem. J.*, 152, [**2020**], 104266.

20. M. Wang, H. Yin, Y. Zhou, C. Sui, Y. Wang, X. Meng, Photoelectrochemical biosensor for microRNA detection based on a MoS2/g-C3N4/black TiO2 heterojunction with Histostar @ AuNPs for signal amplification. *Biosens. Bioelectron.*, 128, [**2019**], 137.

21. H. Liu, C. Duan, C. Yang, X. Chen, W. Shen, Z. Zhu, A novel nitritebiosensor based on the direct electron transfer hemoglobin immobilized in theWO3 nanowires with high length–diameter ratio, *Mater. Sci. Eng. C* 53, [**2015**], 43.

22. Y. Dong, J.A. Zheng, Nonenzymatic L-cysteine sensor based on SnO2-MWCNTs nanocomposites. *J. Mol. Liq.*, 196, [**2014**], 280.
23. X. Zhang, W. Li, Y. Zhou, Y. Chai, R. Yuan, An ultrasensitive electro chemiluminescence biosensor for microRNA detection basedonluminol-functionalized Au NPs@ZnO nanomaterials as signal probe and dissolved O2 as coreactant. *Biosens. Bioelectron.*, 135, [**2019**], 8.
24. S. Rajendran, D. Manoj, K. Raju, D. D. Dionysiou, M. Naushad, F. Gracia, Influence of mesoporous defect induced mixed-valent NiO(Ni2+/Ni3+)-TiO2 nanocomposite for non-enzymatic glucose biosensors. *Sens. Actuat. B*, 264, [**2018**], 27.
25. L. Li, J. Huang, Y. Wang, H. Zhang, Y. Liu, J. Li, An excellentenzyme biosensor based on Sb-doped SnO2 nanowires. *Biosens. Bioelectron.*, 25, [**2010**], 2436.
26. N. Taguchi, *U.S. Patent No. 3,631,436*. Washington, DC: U.S. Patent and Trademark Office, [**1971**].
27. V. E. Henrich, P. A. Cox, *The Surface Science of Metal Oxides*, Cambridge University Press, [**1994**].
28. C. Wang, L. Yin, L. Zhang, D. Xiang, R. Gao, Metal oxide gas sensors: Sensitivity and influencing factors. *Sensors*, 10, [**2010**], 2088.
29. T. Seiyama, A. Kato, K. Fujiishi, M. Nagatani, A new detector for gaseous components using semiconductive thin films. *Anal. Chem.*, 34(11), [**1962**], 1502.
30. D. Chatterjee, S. Dasgupta, Visible light induced photocatalytic degradation of organic pollutants. *J. Photochem. Photobiol. C*, 6, [**2005**], 186.
31. F. Han, V.S.R. Kambala, M. Srinivasan, D. Rajarathnam, R. Naidu, Tailored titanium dioxide photocatalysts for the degradation of organic dyes in wastewater treatment: A review. *Appl. Catal. A Gen.*, 359, [**2009**], 25.
32. I. Chakraborty, Z. Guo, A. Bandyopadhyay, P. Sahoo. Physical modifications and algorithmic predictions behind further advancing two-dimensional water splitting photocatalyst: An overview, *Eng. Sci*, 20, [**2022**], 34–46.
33. F. Han, V.S.R. Kambala, M. Srinivasan, D. Rajarathnam, R. Naidu, Tailored titanium dioxide photocatalysts for the degradation of organic dyes in wastewater treatment: A review. *Appl. Catal. A*, 359, [**2009**], 25.
34. M. Janus, J. Choina, A.W. Morawski, Azo dyes decomposition on new nitrogen-modified anatase TiO2 with high adsorptivity, *J. Hazardous Mater.*, 166, [**2009**], 1.
35. J. Zhu, J. Xie, M. Chen, D. Jiang, D. Wu, Low temperature synthesis of anatase rare earth doped titania-silica photocatalyst and its photocatalytic activity under solar-light. *Colloids Surf. A*, 355, [**2010**], 178.
36. G. Wang, W. Lu, J. Li, J. Choi, Y. Jeong, S.Y. Choi, V-shaped tin oxide nanostructures featuring a broad photo current signal: An effective visible-light-driven photocatalyst. *Small*, 2, [**2006**], 1436.
37. S. Ghasemi, S. Rahimnejad, S.R. Setayesh, S. Rohani, M.R. Gholami, Transition metal ions effect on the properties and photocatalytic activity of nanocrystalline TiO2 prepared in anionic liquid. *J. Hazard. Mater.*, 172, [**2009**], 1573.
38. R. Qiu, D. Zhang, Y. Mo, L. Song, E. Brewer, X. Huang, Photocatalytic activity of polymer-modified ZnO under visible light irradiation. *J. Hazard. Mater.*, 156, [**2008**], 80.
39. B. Pare, S.B. Jonnalagadda, H. Tomar, P. Singh, V.W. Bhagwat, ZnO assisted photocatalytic degradation of acridine orange in aqueous solution using visible irradiation. *Desalination*, 232, [**2008**], 80.
40. S.H.S. Chan, T.Y. Wu, J.C. Juan, C.Y. Teh, Recent developments of metal oxide semiconductors as photocatalysts in advanced oxidation processes (AOPs) for treatment of dye wastewater. *J. Chem. Technol. Biotech.*, 86, [**2011**], 1130.
41. (a) R. Steim, F.R. Kogler, C.J. Brabec, Interface materials for organic solar cells. *J. Mater. Chem.*, 20, [**2010**], 2499. (b) S. Chen, J. R. Manders, S.-W. Tsang, F. So, Metal oxides for interface engineering in polymer solar cells, *J. Mater. Chem.*, 22, [**2012**], 24202.
42. S. Mitra, S. Ray, N. N. Ghosh, P. Hota, A. Mukherjee, A. Baguiand D. K. Maiti. Designed and synthesized de novo ANTPABA-PDI nanomaterial as an acceptor in inverted solar cell at ambient atmosphere. *Nanotechnol.*, 34, [**2023**], 315704.
43. M. Gra̎tzel, Photoelectrochemical cells. *Nature*, 414, [**2001**], 338.
44. N.K. Elumalai, C. Vijila, R. Jose, A. Shraf, S. Ramakrishna. Metal oxide semiconducting interfacial layers for photovoltaic and photocatalytic applications, *Mater. Renew. Sustain. Energy*, 4, [**2015**], 11.
45. C.W. Liu, W.T. Liu, M.H. Lee, W.S. Kuo, B.C. Hsu, Novel photodetector using MOS tunneling structures, *IEEE Electron Device Lett.*, 21, [**2000**], 307.

10

Progress in Two-dimensional Ferroelectrics and Potential Applications

Ateeb Naseer, Yogesh Singh Chauhan, Amit Agarwal, and Somnath Bhowmick*
*Correspondence: bsomnath@iitk.ac.in

10.1 Introduction

In recent years, there has been a growing interest in ferroelectric materials, driven by their potential applications in various technological fields. These materials have shown promise in high-speed, low-power field-effect transistors (FETs), non-volatile high-density memory devices, sensors, and more (1, 2). Ferroelectric materials exhibit spontaneous electric polarization that can be manipulated by applying an external electric field. Traditionally, bulk ferroelectrics have been extensively studied, but with the ongoing drive to miniaturize devices, the thickness of these materials has been reduced. However, a significant challenge arises as the material thickness approaches a critical value. At this point, the electric polarization tends to decrease or vanish due to the unscreened depolarization electrostatic field, surface reconstruction to minimize surface energy, diminished long-range Coulomb interaction, electron screening, and so on. To address this issue, researchers have focused on low-dimensional materials, especially atomically thin two-dimensional (2D) materials, with dangling bond-free interfaces, making them suitable for future-generation device applications (3–5).

10.2 State-of-the-Art

The recent progress in fabrication and characterization techniques has played a crucial role in facilitating the development of 2D ferroelectric technology. Chemical vapor deposition, physical vapor deposition, molecular beam epitaxy, mechanical exfoliation, and so on are standard fabrication techniques used to synthesize 2D ferroelectrics. Characterization of atomically thin ferroelectric materials remains a noteworthy challenge in experimental research. Some widely employed characterization techniques are piezo response force microscopy (PFM), second harmonic generation (SHG), scanning tunneling microscopy (STM), transmission electron microscopy (TEM), and so on. The ferroelectric nature of a material can be probed by examining the magnitude of the piezoelectric response using PFM. SHG microscopy uses second-order non-linear optical processes, and lack of

DOI: 10.1201/9781003439448-10

centrosymmetric naturally leads to the occurrence of SHG in ferroelectrics. The electronic properties of the materials can be acquired through the utilization of STM. The methodology entails conducting a scan utilizing a metal tip while maintaining a constant tunnel current. TEM operates on the fundamental principle of the interaction between high-energy electrons and individual atoms.

The initial experimental realization of 2D ferroelectricity was observed in the compound $CuInP_2S_6$, which exhibits a layered structure. The observation of room-temperature out-of-plane ferroelectricity in bulk $CuInP_2S_6$ for film thicknesses exceeding 100 nm was reported by Belianinov et al. (6). This discovery catalyzed further investigation into the properties of 2D ferroelectrics in their atomically thin state. In recent studies, researchers have reported the discovery of intrinsic room-temperature 2D ferroelectricity in various materials. Some notable examples of the experimentally realized 2D ferroelectrics are $CuInP_2S_6$, SnTe, α-In_2Se_3, β'-In_2Se_3, WTe_2, $2H\alpha$-In_2Se_3, d1T-$MoTe_2$, Bi_2O_2Se, and BA_2PbCl_4 (6–14).

Liu et al. confirmed the existence of out-of-plane polarization in a few-layer $CuInP_2S_6$ (7). The ferroelectric phase transition at a temperature of 320 K was observed through PFM and SHG techniques. STM was instrumental in obtaining the in-plane ferroelectricity in 2D SnTe (8) in 2016. Monolayer SnTe exhibited a distorted lattice structure at extremely low temperatures. Their analysis confirmed the presence of in-plane ferroelectricity, characterized by spontaneous domains and electric polarization in SnTe. Ferroelectricity has been observed in thin films of α-In_2Se_3 with a thickness as low as approximately three layers thick (14). Additionally, 2D ferroelectricity has been observed in thin films of WTe_2, with a thickness of 1.4 nm (two layers) (12), α-In_2Se_3 with a thickness of 1.2 nm (one to two layers) (15), and 1T $MoTe_2$ with a thickness of 0.6 nm (one layer) (16). The ferroelectric properties of α-In_2Se_3 and d1T-$MoTe_2$ have been demonstrated using TEM (9, 16). Besides experimental discovery, several ferroelectric and multiferroic materials have been predicted from ab initio calculations.

10.3 Classification of 2D Ferroelectric Materials

10.3.1 Intrinsic Ferroelectricity

Ferroelectricity originates from the lack of inversion symmetry of the crystal. Among 32 point groups, only 10 polar point groups can show ferroelectricity. A broad classification of such materials is given here:

2D non-vdW Ferroelectrics: These are thin films of well-known traditional ferroelectric materials. However, with reducing thickness, the ferroelectric properties vanish due to the unscreened depolarization field. The polarization can be maintained to some extent by screening the depolarization field or via external factors (strain, defects, etc.). Thin films of ferroelectric materials like perovskites ($BaTiO_3$, $SrTiO_3$, etc.) and HfO_2 have shown some ferroelectric retentivity almost to the 2D limits (17–19).

2D vdW Ferroelectrics: These new-generation 2D materials can overcome the limitations of unscreened depolarization in thin films of traditional ferroelectric materials. 2D layered materials with interlayer vdW coupling can be exfoliated to get atomically thin layers. Many 2D vdW ferroelectrics have been explored both experimentally and theoretically, like As, Sb, Bi, Te, $CuInP_2S_6$, SnTe, α-In_2Se_3, β'-In_2Se_3, WTe_2, 2Hα-In_2Se_3, d1T-$MoTe_2$, Bi_2O_2Se (6–12, 20). Generally, they have spontaneous polarization in either in-plane or out-of-plane direction (see Tables 10.1 and 10.2). Some vdW ferroelectric materials are reported to have both in-plane and out-of-plane polarization at room temperature, like α-In_2Se_3 (21). α-In_2Se_3 has been widely explored for logic and memory device applications.

2D Ferroelectric Metals: Bulk materials with a metallic nature lack ferroelectric properties due to the screening from the external field. However, 2D materials with a metallic character have been found to exhibit ferroelectric properties; these materials are referred to as 2D ferroelectric metals. CrN, $LiOsO_3$, $SrNbO_3$, and 1T-WTe_2 are all examples of 2D ferroelectric metals (12, 22, 23).

10.3.2 Extrinsic Ferroelectricity

The ever-increasing demand for scaled memory devices has prompted research into inducing ferroelectricity in non-ferroelectric materials using the following methods to break the crystal structure symmetry.

Doping: Even centrosymmetric 2D materials (like $CrBr_3$) can show ferroelectricity due to electronic doping-induced charge order (24).

Defect Engineering: Inducing ferroelectricity by breaking crystal symmetry through defect engineering has proven to be a successful technique. Out-of-plane polarization has been observed in MoS_2 and CrI_3, attributed to the vacancies of S and I atoms (25).

Composition Engineering: An ideal composite phase combines the best features of multiple phases. Composition engineering has proven to be beneficial for lowering the switching barrier by combining the metallic Tc phase of 2D ReS_2 with the ferroelectric Td phase of 2D ReS_2(26).

Surface Functionalization: Due to the very high surface-to-volume ratio, it is possible to induce ferroelectric properties via surface functionalization of 2D materials like graphene, germanene, silicene, stanene, and antimonene. For example, graphene has been reported to undergo hydroxylation, resulting in the formation of graphanol, demonstrating ferroelectric properties (27).

Strain Engineering: Strain-induced ferroelectricity originates from the softening of the polar phonon modes. Several studies showing polarization enhancement have been reported (28).

10.3.3 Multiferroics

Multiferroics are a distinct class of materials with the unique characteristic of simultaneously exhibiting two or more ferroic orders (like ferroelectricity, ferromagnetism, and ferroelasticity) within a single phase. Ferroelectricity arises due to the breaking of spatial inversion symmetry, while ferromagnetism arises due to the breaking of time-reversal symmetry. Rotational symmetry is broken in ferroelastic materials. Ferroelectricity, ferromagnetism, and ferroelasticity employ electric, magnetic, and stress fields for switching. The interrelationship among magnetoelectricity, magneto-elasticity, and piezoelectricity plays a crucial role in developing multiferroic devices with diverse functionalities. Multiferroic materials, possessing both ferroelectricity and ferromagnetism, exhibit a notable capacity for high-density data storage, accompanied by a minimal energy demand for both read and write operations. Some notable examples of 2D multiferroics are α-In$_2$Se$_3$, MX (M: Ge, Sn; X: S, Se), CrI$_3$, VOCl$_2$, Bi$_2$O$_2$X (X: S, Se, Te), MXenes, and so on (29–32).

Multiferroics can be categorized into two distinct classes: type-I and type-II. Type-I multiferroics are typically characterized by their high levels of spontaneous polarization and transition temperature, along with a relatively weak intercoupling between ferroelectricity and ferromagnetism. The type-I class can be further categorized into three distinct subclasses: lone-pair multiferroics, geometrically frustrated multiferroics, and charge-ordered multiferroics. Type-II multiferroics exhibit a pronounced interplay between the phenomena of ferroelectricity and ferromagnetism. The type-II class can be further classified into three distinct subclasses based on the underlying mechanism: spin-dependent (involving p- and d-orbital hybridization), exchange-striction, and spin current mechanism.

TABLE 10.1

The Out-of-Plane Spontaneous Polarization Values for Some Commonly Known 2D Materials

Material	P_s	Units	Ref.
t-MoS$_2$	0.23	$\mu C/cm^2$	[33]
t-MoSe$_2$	0.15	$\mu C/cm^2$	[33]
t-MoTe$_2$	0.10	$\mu C/cm^2$	[33]
WS$_2$	0.18	$\mu C/cm^2$	[33]
WSe$_2$	0.21	$\mu C/cm^2$	[33]
WTe$_2$	0.11	$\mu C/cm^2$	[33]
CuInP$_2$S$_6$	4.0	$\mu C/cm^2$	[34]
PbSe	5.0	$\mu C/cm^2$	[35]
PbS	5.5	$\mu C/cm^2$	[35]
α-In$_2$Se$_3$	11.0	pC/m	[36]
InSe	0.24	pC/m	[30]
GaSe	0.46	pC/m	[30]
BN	2.08	pC/m	[30]
SiC	6.17	pC/m	[30]
ZnO	8.22	pC/m	[30]
GaN	9.72	pC/m	[30]
AlN	10.29	pC/m	[30]
CuVP$_2$Se$_6$	0.65	pC/m	[37]
CuCrP$_2$Se$_6$	0.67	pC/m	[37]
CuVP$_2$S$_6$	0.78	pC/m	[37]
CuCrP$_2$S$_6$	0.79	pC/m	[37]
CrB$_2$	0.90	pC/m	[22]
AgBiP$_2$Se$_6$	1.2	pC/m	[38]

TABLE 10.2

The In-Plane Spontaneous Polarization Values for Some Commonly Known 2D Materials

Material	P_s	Units	Ref.
SiTe	42.0	$\mu C/cm^2$	[39]
SnTe	19.4	$\mu C/cm^2$	[39]
GeTe	32.8	$\mu C/cm^2$	[39]
$CrBr_3$	0.92	$\mu C/cm^2$	[24]
As	0.46	$10^{-10}C/m$	[20]
Sb	0.75	$10^{-10}C/m$	[20]
Bi	0.51	$10^{-10}C/m$	[20]
Te	1.02	$10^{-10}C/m$	[40]
SbN	7.81	$10^{-10}C/m$	[41]
BiP	5.35	$10^{-10}C/m$	[41]
GeS	5.06	$10^{-10}C/m$	[42]
GeSe	3.67	$10^{-10}C/m$	[42]
SnS	2.62	$10^{-10}C/m$	[42]
SnSe	1.51	$10^{-10}C/m$	[42]

10.4 Phenomenological Theory and *Ab Initio* Calculations

A thermodynamic theory for ferroelectric-paraelectric phase transition is described by the Landau-Ginzburg-Devonshire (LGD) phenomenological model (43). According to the LGD model, free energy (G) in terms of polarization (P) can be expressed as:

$$G = -EP + \sum_i \left[\frac{A}{2}P_i^2 + \frac{B}{4}P_i^4 + \frac{C}{6}P_i^6 \right],$$ (10.1)

where E is an applied electric field. The coefficients A, B, and C can be determined by fitting Equation 10.1 using the G and P values obtained from first principle calculations at zero applied fields. Change of energy as a function of polarization is calculated by systematically modifying the crystal structure from one ground state ($+P_S$) to another ($-P_S$) via the paraelectric state (P = 0), yielding the typical W-shaped double-well curve (Figure 10.1).

The electric field can be calculated from Equation 10.1 using E = dG/dP. The polarization P as a function of the applied electric field E takes the shape of a S-curve, as shown in Figure 10.1. The shape of the curve implies that around $P \approx 0$, the ferroelectric has a negative capacitance, which is proportional to the slope dP/dE. The value of the electric field at the boundary between the regions of positive and negative capacitance is denoted by E_c, which can be calculated using the following equation:

$$\left[\frac{dE}{dP} \right]_{E=E_c} = 0.$$ (10.2)

The desired value of E_c is of utmost importance for applications in memory devices. A smaller memory window is associated with a lower value of E_c, whereas a larger one

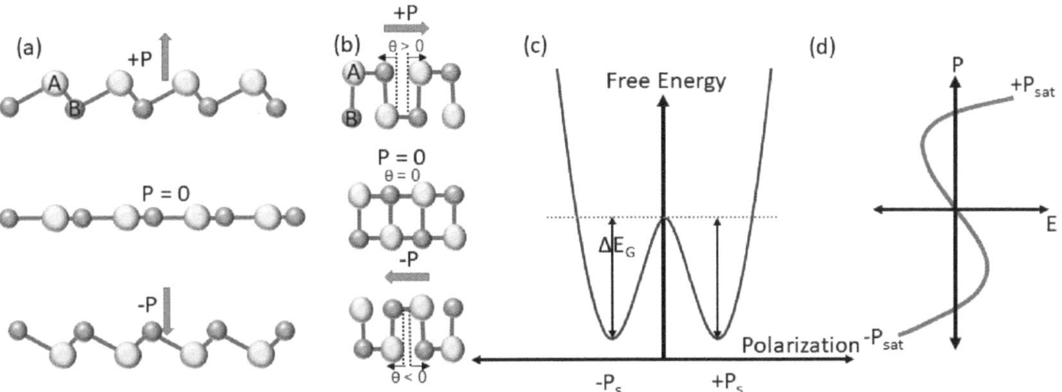

FIGURE 10.1
(a) out-of-plane ferroelectrics and (b) in-plane ferroelectrics. (c) Energy as a function of polarization. (d) The polarization versus field curve.

is associated with a higher value of E_c. However, a higher E_c necessitates a larger field to transition between states.

While it is tempting to convert the depth of the double-well curve (ΔE_G) to a temperature scale to estimate the para-to-ferroelectric phase transition temperature, it generally leads to an underestimation. A better estimation is obtained by adding a dipole–dipole interaction term to the free energy,

$$G = \sum_i \left[\frac{A}{2} P_i^2 + \frac{B}{4} P_i^4 + \frac{C}{6} P_i^6 \right] + \frac{D}{2} \sum_{i,j} \left(P_i - P_j \right)^2. \tag{10.3}$$

The coefficient D is determined by the differences in polarization between adjacent unit cells (42). The nearest dipole–dipole interactions are the primary determinants of phase transitions and the Curie temperature (T_C). A significant magnitude of dipole–dipole interactions D implies that the materials exhibit enhanced resistance to thermal fluctuations, thereby leading to elevated values of T_C (42). Critical temperature T_C can be estimated from $K_B T_C \simeq D \times P_s^2$, where K_B represents Boltzmann's constant.

Ferroelectric phase transition is a structural phase transition, giving rise to a spontaneous polarization in the crystal. Such a transition is generally related to soft phonon modes, whose frequency decreases anomalously near the transition point. Structure X, having softer or low-frequency phonon modes than structure Y, can have higher vibrational entropy (S) because of higher phonon occupancy. As a result, structure X has lower free energy (U – TS) than Y above some critical temperature, leading to a phase transformation from Y→X. Experimentally, soft modes can be identified by Raman and nuclear magnetic resonance spectroscopy, neutron scattering, and so on. Computationally, one can do a similar analysis using the phonon dispersion curves obtained from ab initio calculations using density functional perturbation theory.

The theory of vibrational free energy of ferroelectric materials revolves mainly around optical phonons. Since positive and negative ions are displaced in opposite directions, an optical mode creates a local electric field. When such a field gets stronger than the elastic restoring force (known as polarization catastrophe), the material undergoes a transition from the para-to-ferroelectric phase via a shift in the position of the ions. Equivalently, one can explain a ferro to paraelectric phase transition by considering the softening of an optical phonon mode. Generally, a transverse optical (TO) mode is of interest, as it has a lower frequency than the longitudinal optical mode (44). Experimentally, it has been observed that the phonon frequency of an optical mode vanishes at some point in the Brillouin zone, which is referred to as condensation or freezing of the phonon mode, as the transition temperature T_c is approached from below. The low-frequency TO mode results in a high value of static dielectric constant $\epsilon(0)$. According to the Lyddane-Sachs-Teller relation, the frequency of transverse and longitudinal phonon satisfies,

$$\frac{\omega_T^2}{\omega_L^2} = \frac{\epsilon(\infty)}{\epsilon(0)}. \tag{10.4}$$

Since the frequency of the TO mode decreases near the ferroelectric phase transition, the static dielectric constant $\epsilon(0)$ increases. Temperature dependence of $\epsilon(0) \propto 1/(T - T_c)$ and $\omega_T^2 \propto (T - T_c)$ has been observed experimentally, which further validates the role of soft TO modes behind ferroelectric phase transition.

For computational modeling of a ferroelectric material using the LGD phenomenological model, one must determine the free energy as a function of polarization from the density functional theory (DFT) based first-principles calculations. Since DFT can predict ground state electron density, one might be tempted to define bulk polarization as,

$$P_{cell} = \frac{1}{\Omega} \int_\Omega r\rho(r)\,dr, \tag{10.5}$$

where the integration is carried out over a unit cell of volume Ω, away from the surface. However, since the integral depends on the shape of the unit cell, the outcome is ambiguous in a solid, where the electronic charge density $\rho(\mathbf{r})$ is a continuous function in space. Let us discuss how this problem is addressed and how polarization is calculated in today's DFT packages, which use the modern theory of polarization based on the Berry-phase theory (45).

Electron density is expressed using a localized basis, known as the Wannier function $w_n(\mathbf{r})$, in a unit cell located at point $\mathbf{R}$ in space as,

$$w_n(r - R) = \frac{\Omega}{2\pi^3} \int_{BZ} dk\, e^{-ik \cdot R} \Psi_{nk}(r). \tag{10.6}$$

Here, n is the band index, $\Psi_{nk}(\mathbf{r}) = e^{-ik \cdot r} u_{nk}(\mathbf{r})$ are the Bloch functions, and $u_{nk}(\mathbf{r})$ has the periodicity of the Bravais lattice. Wannier center, defined as,

$$\bar{r}_n = \iota \frac{\Omega}{2\pi^3} \int_{BZ} dk\, e^{-ik \cdot R} u_{nk} | \frac{\partial u_{nk}}{\partial k}, \tag{10.7}$$

are taken as the "average position" of the electron, and the charge is assumed to be localized at that point. The total polarization is a sum of the contribution from all the ions (which are still treated as point charges) and electrons positioned at the Wannier centers of occupied Wannier functions,

$$P_{tot} = P_{ion} + P_{el} = \frac{1}{\Omega}\left(\sum_i q_i r_i + \sum_n^{occ} q_n \bar{r}_n\right). \tag{10.8}$$

Finally, spontaneous polarization is calculated as,

$$\Delta P = P_{tot}^{f} - P_{tot}^{0},\tag{10.9}$$

where 0 and f denote the initial (non-polar) and final (polar) structures, having ionic positions $\mathbf{r}_{i}^{0}$ and $\mathbf{r}_{i}^{f}$, and wavefunctions u_{nk}^{0} and u_{nk}^{f}. Since wavefunctions can be obtained from standard DFT codes, the previous method can be implemented to get the spontaneous polarization from first principle calculations.

Born effective charge is another useful quantity for the study of ferroelectric materials. Suppose an ion is displaced by ∂d_{β}, and as a result, polarization changes by an amount ∂P_{α}. Born effective charge $Z_{\alpha\beta}^{*}$ is defined as the ratio of the two quantities,

$$Z_{\alpha\beta}^{*} = \frac{\Omega}{e}\frac{\partial P_{\alpha}}{\partial d_{\beta}}.\tag{10.10}$$

Note that Born effective charge is a second-rank tensor. A displacement of an ion in the β direction changes polarization not only in the same direction but also in some direction α, which is perpendicular to β. An alternate definition of Born effective charge is the force ∂F_{α} induced on an ion (in the α direction) by an electric field ∂E_{β} (in the β direction),

$$Z_{\alpha\beta}^{*} = \frac{\Omega}{e}\frac{\partial F_{\alpha}}{\partial E_{\beta}}.\tag{10.11}$$

Compared to normal dielectric materials, an electric field exerts larger forces on the ions in ferroelectric materials. As a result, Born effective charges are anomalously high compared to the expected ionic charges in a ferroelectric crystal. Born effective charge tensor is a standard output of ab initio lattice dynamics calculations.

Another quantity of interest is the piezoelectric coefficient, which is a third-rank tensor

$$e_{\alpha\beta\gamma} = \frac{\partial P_{\alpha}}{\partial \varepsilon_{\beta\gamma}}.\tag{10.12}$$

The piezoelectric coefficient represents electromechanical coupling, which is measured in terms of change of polarization ∂P_{α} in the α direction by strain $\varepsilon_{\beta\gamma}$. Ab initio calculations are carried out to calculate the piezoelectric coefficient by applying a macroscopic strain (cell parameters changed) and measuring polarization using the Berry-phase method. Note that macroscopic strain also induces internal strain; that is, atomic coordinates relax to new equilibrium positions as the cell parameter is changed.

In summary, modern software packages for electronic structure calculation are capable of computing important parameters like free energy, spontaneous polarization, phonon dispersion, Born effective charge, and piezoelectric coefficient. One can use such material parameters to predict device characteristics.

10.5 Potential Applications

2D materials dominate the rapidly growing field of low-dimensional materials, showing remarkable promise for cutting-edge electronic, optoelectronic, spintronic, catalytic,

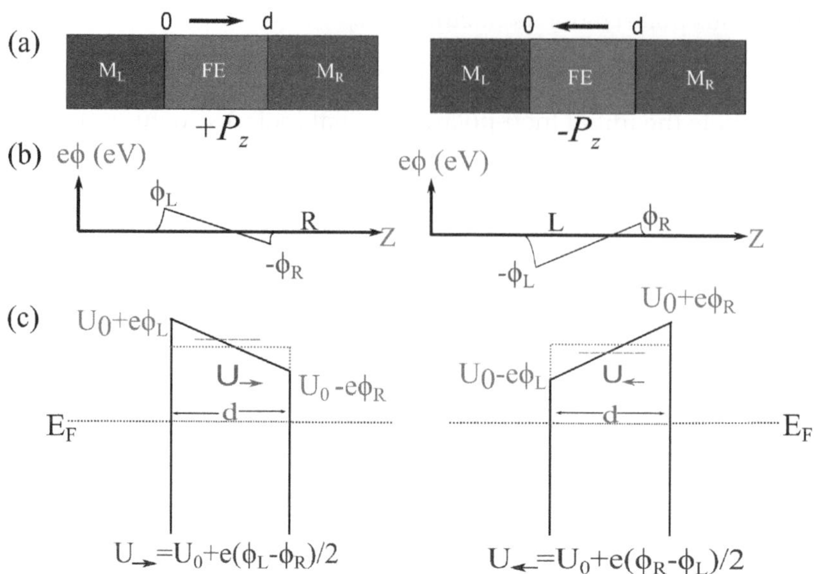

FIGURE 10.2
Working mechanism of FTJs.

energy-storage, biomedical, and sensing applications. Many desirable properties of 2D materials, such as high surface-to-volume ratio, tunability in bandgap and bandstructure, dangling bond-free interfaces, immunity to short channel effects, high mobility in ultra-thin form, and excellent mechanical flexibility, make them promising for future device applications.

10.5.1 Ferroelectric Tunnel Junctions

The field tunnel junction (FTJ) comprises a ferroelectric layer supporting switchable spontaneous polarization states and a high tunnel electroresistance ratio (TER) (see Figure 10.2). The high TER ratio is essential for incorporating tunnel junctions in non-volatile memory devices. The ferroelectric layer is between the two electrodes, heavily doped with p-type or n-type materials. Replacing a three-dimensional ferroelectric barrier in an FTJ with a 2D VdW material can decrease the FTJ size while maintaining the high TER. Kang et al. have predicted a 623% TER with B-doped and N-doped graphene/ BiP, comparable to 3D FTJs (46). Duan et al. achieved a TER of 1460% in the 2D In:SnSe/ SnSe/Sb:SnSe system with an in-plane ferroelectric polarization by dynamically controlling the width and height of the barrier during ferroelectric switching (47). Tunnel junctions exhibiting high TER values are advantageous for implementing ultra-fast and efficient non-volatile memory devices.

FTJs predominantly function through electronic carrier quantum tunneling across the ferroelectric layer. The devices comprise an asymmetrical and electrically adjustable tunnel barrier resulting from the polarization switching occurring in the ferroelectric layer. Additionally, a fixed tunneling barrier arises from the disparity in doping between the two electrodes and the ferroelectric layer. The fixed tunneling barrier is usually represented by a rectangular barrier (or potential U_0), about the Fermi level, referred to as E_F. The charges associated with spontaneous polarization are observed near the ferroelectric layer. Depending on their polarity, the surface charge carriers will either attract/repel the electrons, leading to a partial screening of the polarization charges. The mechanism establishes a depolarizing field and modifies the magnitude and shape of the potential profile. The charges located at the interfaces on the right and left sides will exhibit positive/negative and negative/positive characteristics, respectively, when polarization occurs in the +/- direction. This will result in the formation of an asymmetrical potential profile. We can express the electrostatic potential as,

$$\phi_{L/R} = (+/-)\frac{(\Gamma_{L/R}P_{sd})}{d + \epsilon_{FE}(\Gamma_L + \Gamma_R)}.$$

(10.13)

Here $\Gamma_{L/R}(= \delta_{L/R}/\epsilon_{L/R})$ is the ratio of the left/right electrode screening length and dielectric permittivity. P_s denotes the induced polarization at the ferroelectric-electrode interface, while d and ϵ_{FE} denote the height and dielectric constant of the ferroelectric layer, respectively. The asymmetrical potential produces two distinct tunneling conductances ($G_{\rightarrow}$ and $G_{\leftarrow}$), and hence, the finite tunneling electroresistance ratio, TER = $(G_{\rightarrow} - G_{\leftarrow})/G_{\leftarrow}$. The TER in the linear region can be approximated using the Wentzel–Kramers–Brillouin (WKB) approximation, ΔU (= $U_{\rightarrow} - U_{\leftarrow}$) $\ll U_0$. The TER ratio can be expressed as (48),

$$TER \approx \left[\frac{\sqrt{2m}\Delta U}{\hbar\sqrt{U_0}}\right] = exp\left[\frac{e}{\hbar}\sqrt{2m/U_0}\frac{P_s(\Gamma_L - \Gamma_R)d^2}{d + \epsilon_{FE}(\Gamma_L + \Gamma_R)}\right].$$

(10.14)

10.5.2 Ferroelectric Field-Effect Transistors

Ferroelectric field-effect transistors (Fe-FETs) are another popular application extensively investigated since their introduction in 1957. One-transistor (1T)-type Fe-FET structure comprises an integrated ferroelectric layer within the gate stack. Fe-FET can carry out both fast switching and high-density storage operations, thereby reducing the on-chip area for memory devices. The channel conductance of Fe-FETs can be controlled via switchable spontaneous polarization, allowing for non-destructive read-out operation. Metal-ferroelectric-semiconductor (MFS) devices suffer from the reactions and interdiffusions between ferroelectric and semiconductors. Metal ferroelectric-metal-insulator-semiconductor structures are susceptible to experiencing elevated levels of leakage currents. Consequently, there is a significant ongoing effort to enhance the operational performance of both configurations.

The operation of the Fe-FET is based on the principle that the presence of polarization-bound charges at the interface between the insulator and semiconductor significantly impacts the electrostatics. The shift in threshold voltage is achieved by reversing the polarity of the polarization charges, which is facilitated by applying an appropriate gate bias. Several 2D-based Fe-FETs have been demonstrated. Monolayer MoS_2-based Fe-FETs with P(VDF-TrFE) as ferroelectric in the gate stack have demonstrated a memory window of ~15 V for a sweep voltage of ±20 V along with a large ON/OFF ratio (~10^7) (49).

However, conventional Fe-FETs encounter depolarization field issues due to the potential drop across the interfacial dielectric and the occurrence of charge trapping at the interface between the ferroelectric material and the semiconductor (50). The effects mentioned earlier give rise to undesirable threshold voltage fluctuation, limiting their potential applications. Si et al. proposed that using ferroelectric materials as the semiconductor channel can effectively mitigate the depolarization field (21). The depolarization field can

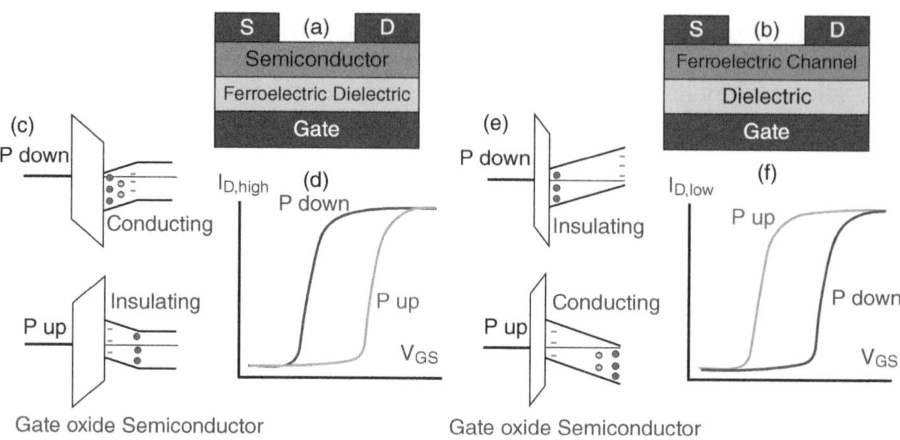

FIGURE 10.3
Working mechanism of FeS-FETs (21).

be screened by the mobile charges present in the semiconducting ferroelectric material. The proposed FeS-FET is based on the α-In$_2$Se$_3$ as a ferroelectric semiconductor channel (Figure 10.3). The polarization charges accumulate at the upper and lower surfaces of the ferroelectric semiconductor channel material, influencing the drain current. They additionally observed that the current hysteresis direction is sensitive to the strength of the electric field across the ferroelectric semiconductor, which, in turn, is affected by the oxide thickness of the bottom insulator.

For dielectrics with high oxide thickness, partial polarization occurs. Thus, in the polarization down state, the electrons are attracted toward the lower surface of the dielectric. Consequently, this attraction enhances the process of conduction. Dielectric materials with low equivalent oxide thickness can undergo complete polarization under the influence of an electric field. This polarization enables the accumulation of mobile charges at the upper surface, facilitating the conduction mechanism in the polarization upstate. Consequently, an anticlockwise hysteresis loop is observed.

10.5.3 Neuromorphic Computing

The von Neumann computer architecture has been responsible for handling computationally intensive mathematical algorithms. However, the current von Neumann architectures face significant computational challenges considering the exponential growth of computational data resulting from advancements in artificial intelligence, big data, machine learning, and the Internet of things. The challenges primarily arise from the physical separation between processing and storage units, which limits the speed of data read and write operations. Consequently, this leads to excessive power consumption.

Neuromorphic computing has emerged as a potential solution to overcome the limitations of conventional computing architectures, specifically the von Neumann bottleneck. This approach draws inspiration from the human brain's learning, memorization, and information-processing capabilities. The human brain exhibits a remarkable capacity to

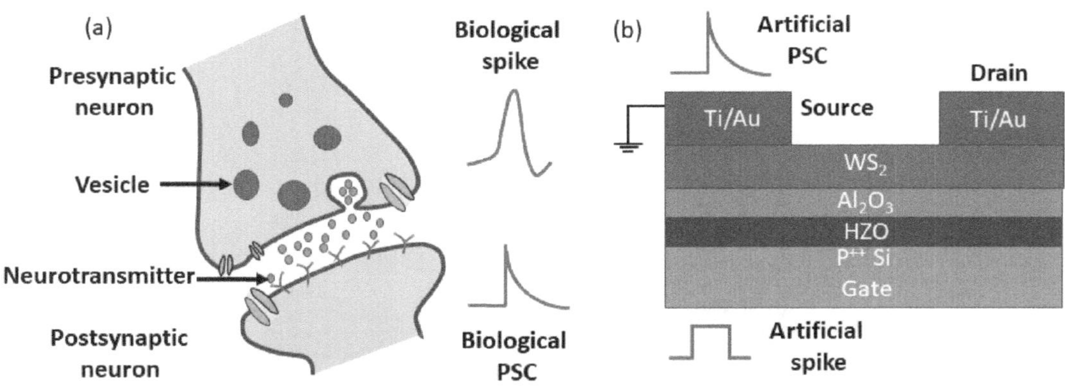

FIGURE 10.4
Working mechanism of a synaptic transistor (51).

process vast quantities of information simultaneously while maintaining a high level of energy efficiency, with power consumption levels below 20 W. By employing physically separated data processing and storage units, as well as addressing data-intensive tasks, neuromorphic computing aims to achieve high performance and energy efficiency in next-generation computing systems. Synaptic devices, serving as fundamental components within neuromorphic systems, possess the inherent capability to function as memory devices with highly accurate control of channel conductance. These devices can be seamlessly incorporated into artificial neural networks to facilitate in-memory computing, thereby amalgamating the functionalities of memory and computation. The modulation of ferroelectric polarization in Fe-FETs can be achieved by altering the proportion of up and down-polarized domains using pulsed electrostatic gating. This manipulation enables the emergence of conductance states with multiple levels, characterized by their non-volatile and history-dependent properties. The memory behavior exhibited by Fe-FETs allows their utilization as synaptic devices in a wide range of neuromorphic applications.

Chen et al. have successfully fabricated a 2D HZO/WS_2 Fe-FET device to emulate the plasticity observed in biological synapses (51). Within the context of a biological synapse, it is observed that neurotransmitter vesicles located within a presynaptic neuron can undergo diffusion across the synaptic cleft, ultimately reaching the postsynaptic neuron. Consequently, this process leads to initiating a postsynaptic current as a direct result of an electric stimulus (Figure 10.4). The synaptic weight can be modulated by applying a series of gate voltage pulses, resulting in the potentiation or depression of the postsynaptic current (drain-source current) released from the synaptic device. The device exhibits long-term synaptic plasticity due to the non-volatile polarization in HZO. Furthermore, the synaptic device utilizing the P(VDF-TrFE)/MoS_2 material combination has successfully exhibited a substantial ON/OFF ratio, exceeding 1,000 controllable states (52). The development of FeS-FETs utilizing two-dimensional ferroelectric semiconductors has also enabled the emulation of synaptic plasticity. α-In_2Se_3 is commonly employed in FeS-FETs because of the intercorrelated coupling between out-of-plane and in-plane polarization. This coupling implies that a planar or vertical electric field can manipulate out-of-plane and in-plane polarization, rendering α-In_2Se_3 appropriate for multi-terminal synaptic devices.

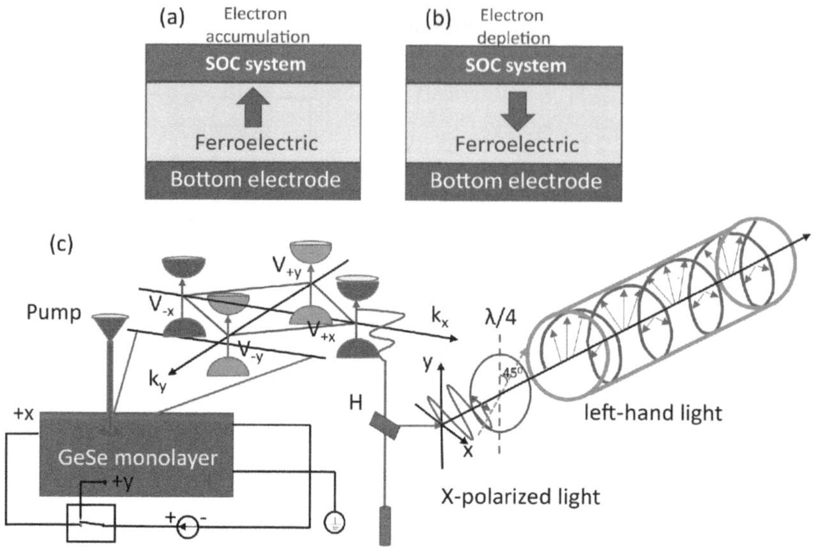

FIGURE 10.5
Applications of ferroelectrics in the domains of spintronics and valleytronics (53, 56).

10.5.4 Spintronics

Spin-based devices are based on spin polarization due to the segregation of electrons with up and down spins. The interaction between the ferroelectric polarization and spin-dependent phenomena in 2D ferroelectrics enables the realization of spin-based devices that can be electrically manipulated. The Rashba effect is a prominent spin-dependent phenomenon that can be effectively controlled by electrical means. This characteristic is particularly evident in 2D ferroelectrics, where an inherent electrical field is always present. The non-centrosymmetric nature of 2D ferroelectrics can result in non-trivial spin-orbit coupling effects, mainly when heavy atoms are present in the materials.

The switchable ferroelectric polarization can be effectively employed for the electrical manipulation of spin polarization direction, provided a correlation exists between the polarization and the Rashba effect. For example, Bruyer et al. predicted the coupling of the non-trivial Rashba effect and ferroelectric polarization in the monolayer of d1T-MoS$_2$ (33). The concept of ferroelectricity-controlled spin-charge conversion was introduced by Noël et al. This concept involves a Rashba spin-orbital coupling (SOC) system and a ferroelectric material (53). The switchable polarization in the adjacent ferroelectric material can electrically manipulate the Rashba SOC. The accumulation or depletion of electrons in the SOC material results from the varying band bending that occurs because of polarization switching (Figure 10.5 (a) and (b)).

10.5.5 Valleytronics

Some materials exhibit multiple valleys in their electronic structures, which serve as an additional degree of freedom of electrons. Early ideas primarily focused on the valley-dependent physical phenomena in 2D 2H-MoS$_2$ materials (54). The lack of inversion symmetry in the hexagonal 2H-MoS$_2$ monolayer gives rise to a valley contrasting orbital moment and Berry curvature. Consequently, this leads to opposing valley-dependent optical selection rules for circularly polarized light and the occurrence of the valley Hall effect. By applying external fields, valley polarization can be achieved by breaking the time-reversal symmetry. These fields serve to eliminate the degeneracy between the two valleys. In a subsequent study, Tong and co-workers predicted the occurrence of spontaneous valley polarization in buckled monolayers composed of group IV and group III-V binary materials (55). This phenomenon arises due to a robust exchange interaction, which functions as an internal magnetic field and breaks the time-inversion symmetry. Furthermore, it has been observed that the monolayer GeSe exhibits two valleys in the orthogonal ferroelectric phase, resulting in the broken inversion symmetry and thus generating different band gaps at V_x and V_y (Figure 10.5(c)). Shen et al. proposed an electrically tunable polarizer by utilizing the property of the distinctive optical selection rule (56).

10.6 Conclusion and Future Outlook

Conventional 3D ferroelectrics have been very effective in various domains and played a crucial role in advancing engineering technology. The growing need for smaller-footprint devices and better technology has generated an increased focus on studying 2D materials, including 2D ferroelectrics. 2D ferroelectric materials have garnered attention due to their potential applications in logic, memory, and optoelectronic devices that exhibit superior performance and require minimal energy consumption. Despite extensive predictions and investigations of various 2D materials, the investigation of inherent 2D ferroelectricity confined within a plane is still nascent, particularly in experimental exploration. In the pursuit of advancing 2D ferroelectrics, theoretical calculations commonly offer directional guidance for exploring and validating 2D ferroelectricity. Despite the difficulties in observing consistent and robust spontaneous polarization in ultrathin 2D materials, many methodologies are available to explore new 2D ferroelectric materials. These approaches encompass the investigation of fundamental physical properties and inherent mechanisms facilitated by rapid advancements in characterization and fabrication techniques. Investigating novel 2D ferroelectric materials remains a highly intriguing subject within academia and industry with potentially impactful possibilities.

Bibliography

[1] M. Hoffmann, F.P.G. Fengler, M. Herzig, T. Mittmann, B. Max, U. Schroeder, R. Negrea, P. Lucian, S. Slesazeck, T. Mikolajick, Unveiling the double-well energy landscape in a ferroelectric layer, *Nature* 565 (7740) (2019) 464–467.

[2] M. Rafiq, S.S. Parihar, Y.S. Chauhan, S. Sahay, Efficient implementation of max-pooling algorithm exploiting history-effect in ferroelectric-FinFETs, *IEEE Transactions on Electron Devices* 69 (11) (2022) 6446–6452.

[3] K. Nandan, A. Naseer, Y.S. Chauhan, Field-effect transistors based on two-dimensional materials (invited), *Transactions of the Indian National Academy of Engineering* 8 (1) (2023) 1–14.

[4] A. Naseer, K. Nandan, A. Agarwal, S. Bhowmick, Y.S. Chauhan, Di-metal chalcogenides: A new family of promising 2-D semiconductors for high-performance transistors, *IEEE Transactions on Electron Devices* 70 (5) (2023) 2445–2452.

[5] A. Naseer, K. Nandan, A. Agarwal, S. Bhowmick, Y.S. Chauhan, Performance evaluation of monolayer ZrS_3 transistors for next-generation computing, *IEEE Transactions on Electron Devices* 70 (10) (2023) 5435–5442.

[6] A. Belianinov, Q. He, A. Dziaugys, P. Maksymovych, E. Eliseev, A. Borisevich, A. Morozovska, J. Banys, Y. Vysochanskii, S.V. Kalinin, $CuInP_2S_6$ room temperature layered ferroelectric, *Nano Letters* 15 (6) (2015) 3808–3814.

[7] F. Liu, L. You, K.L. Seyler, X. Li, P. Yu, J. Lin, X. Wang, J. Zhou, H. Wang, H. He, S.T. Pantelides, W. Zhou, P. Sharma, X. Xu, P.M. Ajayan, J. Wang, Z. Liu, Room-temperature ferroelectricity in $CuInP_2S_6$ ultrathin flakes, *Nature Communications* 7 (1) (2016) 12357.

[8] K. Chang, J. Liu, H. Lin, N. Wang, K. Zhao, A. Zhang, F. Jin, Y. Zhong, X. Hu, W. Duan, Q. Zhang, L. Fu, Q.K. Xue, X. Chen, S.H. Ji, Discovery of robust in-plane ferroelectricity in atomic-thick SnTe, *Science* 353 (6296) (2016) 274–278.

[9] C. Cui, W.J. Hu, X. Yan, C. Addiego, W. Gao, Y. Wang, Z. Wang, L. Li, Y. Cheng, P. Li, X. Zhang, H.N. Alshareef, T. Wu, W. Zhu, X. Pan, L.J. Li, Intercorrelated in-plane and out-of-plane ferroelectricity in ultrathin two-dimensional layered semiconductor In_2Se_3, *Nano Letters* 18 (2) (2018) 1253–1258.

[10] S. Wan, Y. Li, W. Li, X. Mao, W. Zhu, H. Zeng, Room-temperature ferroelectricity and a switchable diode effect in two-dimensional $\alpha–In_2Se_3$ thin layers, *Nanoscale* 10 (2018) 14885–14892.

[11] S.M. Poh, S.J.R. Tan, H. Wang, P. Song, I.H. Abidi, X. Zhao, J. Dan, J. Chen, Z. Luo, S.J. Pennycook, A.H. Castro Neto, K.P. Loh, Molecular-beam epitaxy of two-dimensional In_2Se_3 and its giant electroresistance switching in ferroresistive memory junction, *Nano Letters* 18 (10) (2018) 6340–6346.

[12] Z. Fei, W. Zhao, T.A. Palomaki, B. Sun, M.K. Miller, Z. Zhao, J. Yan, X. Xu, D.H. Cobden, Ferroelectric switching of a two-dimensional metal, *Nature* 560 (7718) (2018) 336–339.

[13] W. Wang, Y. Meng, Y. Zhang, Z. Zhang, W. Wang, Z. Lai, P. Xie, D. Li, D. Chen, Q. Quan, D. Yin, C. Liu, Z. Yang, S. Yip, J.C. Ho, Electrically switchable polarization in Bi_2O_2Se ferroelectric semiconductors, *Advanced Materials* 35 (12) (2023) 2210854.

[14] C. Zheng, L. Yu, L. Zhu, J.L. Collins, D. Kim, Y. Lou, C. Xu, M. Li, Z. Wei, Y. Zhang, M.T. Edmonds, S. Li, J. Seidel, Y. Zhu, J.Z. Liu, W.X. Tang, M.S. Fuhrer, Room temperature in-plane ferroelectricity in van-der-Waals In_2Se_3, *Science Advances* 4 (7) (2018) eaar7720.

[15] F. Xue, W. Hu, K.C. Lee, L.S. Lu, J. Zhang, H.L. Tang, A. Han, W.T. Hsu, S. Tu, W.H. Chang, C.H. Lien, J.H. He, Z. Zhang, L.J. Li, X. Zhang, Room-temperature ferroelectricity in hexagonally layered α-In_2Se_3 nanoflakes down to the monolayer limit, *Advanced Functional Materials* 28 (50) (2018) 1803738.

[16] S. Yuan, X. Luo, H.L. Chan, C. Xiao, Y. Dai, M. Xie, J. Hao, Room-temperature ferroelectricity in $MoTe_2$ down to the atomic monolayer limit, *Nature Communications* 10 (1) (2019) 1775.

[17] J. Junquera, P. Ghosez, Critical thickness for ferroelectricity in perovskite ultrathin films, *Nature* 422 (6931) (2003) 506–509.

[18] D. Lee, H. Lu, Y. Gu, S.Y. Choi, S.D. Li, S. Ryu, T.R. Paudel, K. Song, E. Mikheev, S. Lee, S. Stemmer, D.A. Tenne, S.H. Oh, E.Y. Tsymbal, X. Wu, L.Q. Chen, A. Gruverman, C.B. Eom, Emergence of room-temperature ferroelectricity at reduced dimensions, *Science* 349 (6254) (2015) 1314–1317.

[19] H.J. Lee, M. Lee, K. Lee, J. Jo, H. Yang, Y. Kim, S.C. Chae, U. Waghmare, J.H. Lee, Scale-free ferroelectricity induced by flat phonon bands in HfO_2, *Science* 369 (6509) (2020) 1343–1347.

[20] C. Xiao, F. Wang, S.A. Yang, Y. Lu, Y. Feng, S. Zhang, Elemental ferroelectricity and antiferroelectricity in Group-V monolayer, *Advanced Functional Materials* 28 (17) (2018) 1707383.

[21] M. Si, A.K. Saha, S. Gao, G. Qiu, J. Qin, Y. Duan, J. Jian, C. Niu, H. Wang, W. Wu, S.K. Gupta, P.D. Ye, A ferroelectric semiconductor field-effect transistor, *Nature Electronics* 2 (12) (2019) 580–586.

[22] W. Luo, K. Xu, H. Xiang, Two-dimensional hyper-ferroelectric metals: A different route to ferromagnetic-ferroelectric multiferroics, *Physical Review B* 96 (2017) 235415.

[23] J. Lu, G. Chen, W. Luo, J. Iniguez, L. Bellaiche, H. Xiang, Ferroelectricity with asymmetric hysteresis in metallic $LiOsO_3$ ultrathin films, *Physical Review Letters* 122 (2019) 227601.

[24] C. Huang, Y. Du, H. Wu, H. Xiang, K. Deng, E. Kan, Prediction of intrinsic ferromagnetic ferroelectricity in a transition-metal halide monolayer, *Physical Review Letters* 120 (2018) 147601.

[25] Y. Zhao, L. Lin, Q. Zhou, Y. Li, S. Yuan, Q. Chen, S. Dong, J. Wang, Surface vacancy-induced switchable electric polarization and enhanced ferromagnetism in monolayer metal trihalides, *Nano Letters* 18 (5) (2018) 2943–2949.

[26] J. Zhang, S. Wu, Y. Shan, J. Guo, S. Yan, S. Xiao, C. Yang, J. Shen, J. Chen, L. Liu, X. Wu, Distorted monolayer ReS_2 with low-magnetic-field controlled magnetoelectricity, *ACS Nano* 13 (2) (2019) 2334–2340.

[27] G.B. Liu, D. Xiao, Y. Yao, X. Xu, W. Yao, Electronic structures and theoretical modelling of two-dimensional group-Vib transition metal dichalcogenides, *Chemical Society Reviews* 44 (2015) 2643–2663. doi:10.1039/C4CS00301B.

[28] A. Priydarshi, Y.S. Chauhan, S. Bhowmick, A. Agarwal, Strain-tunable in-plane ferroelectricity and lateral tunnel junction in monolayer group-IV monochalcogenides, *Journal of Applied Physics* 131 (3) (2022) 034101.

[29] H. Wang, X. Qian, Two-dimensional multiferroics in monolayer group-iv monochalcogenides, *2D Materials* 4 (1) (2017) 015042.

[30] L. Li, M. Wu, Binary compound bilayer and multilayer with vertical polarizations: Twodimensional ferroelectrics, multiferroics, and nanogenerators, *ACS Nano* 11 (6) (2017) 6382–6388.

[31] H. Ai, X. Song, S. Qi, W. Li, M. Zhao, Intrinsic multiferroicity in two-dimensional $VoCl_2$ monolayers, *Nanoscale* 11 (2019) 1103–1110.

[32] M. Wu, X.C. Zeng, Bismuth oxychalcogenides: A new class of ferroelectric/ferroelastic materials with ultra-high mobility, *Nano Letters* 17 (10) (2017) 6309–6314.

[33] E. Bruyer, D. Di Sante, P. Barone, A. Stroppa, M.H. Whangbo, S. Picozzi, Possibility of combining ferroelectricity and Rashba-like spin splitting in monolayers of the 1T-type transition-metal dichalcogenides MX_2(M=Mo,W;X=S,Se,Te), *Physical Review B* 94 (2016) 195402.

[34] L. You, Y. Zhang, S. Zhou, A. Chaturvedi, S.A. Morris, F. Liu, L. Chang, D. Ichinose, H. Funakubo, W. Hu, T. Wu, Z. Liu, S. Dong, J. Wang, Origin of giant negative piezoelectricity in a layered van-der-Waals ferroelectric, *Science Advances* 5 (4) (2019) eaav3780.

[35] P.Z. Hanakata, A.S. Rodin, H.S. Park, D.K. Campbell, A.H. Castro Neto, Strain-induced gauge and Rashba fields in ferroelectric Rashba lead chalcogenide PbX monolayers ($X = S, Se, Te$), *Physical Review B* 97 (2018) 235312.

[36] M. Soleimani, M. Pourfath, Ferroelectricity and phase transitions in In_2Se_3 van-der-Waals material, *Nanoscale* 12 (2020) 22688–22697.

[37] J. Qi, H. Wang, X. Chen, X. Qian, Two-dimensional multiferroic semiconductors with coexisting ferroelectricity and ferromagnetism, *Applied Physics Letters* 113 (4) (2018) 043102.

[38] B. Xu, H. Xiang, Y. Xia, K. Jiang, X. Wan, J. He, J. Yin, Z. Liu, Monolayer $AgBiP_2Se_6$: An atomically thin ferroelectric semiconductor with out-plane polarization, *Nanoscale* 9 (2017) 8427–8434.

[39] W. Wan, C. Liu, W. Xiao, Y. Yao, Promising ferroelectricity in 2D group-IV tellurides: A first-principles study, *Applied Physics Letters* 111 (13) (2017) 132904.

[40] Y. Wang, C. Xiao, M. Chen, C. Hua, J. Zou, C. Wu, J. Jiang, S.A. Yang, Y. Lu, W. Ji, Two-dimensional ferroelectricity and switchable spin-textures in ultra-thin elemental Te multilayers, *Materials Horizon* 5 (2018) 521–528.

[41] C. Liu, W. Wan, J. Ma, W. Guo, Y. Yao, Robust ferroelectricity in two-dimensional SbN and BiP, *Nanoscale* 10 (2018) 7984–7990.

[42] R. Fei, W. Kang, L. Yang, Ferroelectricity and phase transitions in monolayer group-iv monochalcogenides, *Physical Review Letters* 117 (2016) 097601.

[43] R. Cowley, Structural phase transitions I. Landau theory, *Advances in Physics* 29 (1) (1980) 1–110.

[44] L. Casella, A. Zaccone, Soft mode theory of ferroelectric phase transitions in the low-temperature phase, *Journal of Physics: Condensed Matter* 33 (16) (2021) 165401.

[45] R.D. King-Smith, D. Vanderbilt, Theory of polarization of crystalline solids, *Physical Review B* 47 (1993) 1651–1654.

[46] L. Kang, P. Jiang, N. Cao, H. Hao, X. Zheng, L. Zhang, Z. Zeng, Realizing giant tunneling electroresistance in two-dimensional graphene/BiP ferroelectric tunnel junction, *Nanoscale* 11 (2019) 16837–16843.

[47] X.W. Shen, Y.W. Fang, B.B. Tian, C.G. Duan, Two-dimensional ferroelectric tunnel junction: The case of monolayer In:SnSe/SnSe/Sb:SnSe homostructure, *ACS Applied Electronic Materials* 1 (7) (2019) 1133–1140.

[48] A. Sokolov, O. Bak, H. Lu, S. Li, E.Y. Tsymbal, A. Gruverman, Effect of epitaxial strain on tunneling electroresistance in ferroelectric tunnel junctions, *Nanotechnology* 26 (30) (2015) 305202.

[49] Y.T. Lee, D.K. H Wang, S. Im, High-performance MoS_2 nanosheet-based non-volatile memory transistor with a ferroelectric polymer and graphene source-drain electrode, *Journal of the Korean Physical Society* 67 (9) (2015) 1499–1503.

[50] E. Yurchuk, J. Mu¨ller, S. Mu¨ller, J. Paul, M. Pešić, R. van Bentum, U. Schroeder, T. Mikolajick, Charge-trapping phenomena in HfO_2-based FeFET-type non-volatile memories, *IEEE Transactions on Electron Devices* 63 (9) (2016) 3501–3507.

[51] L. Chen, L. Wang, Y. Peng, X. Feng, S. Sarkar, S. Li, B. Li, L. Liu, K. Han, X. Gong, J. Chen, Y. Liu, G. Han, K.W. Ang, A van-der-Waals synaptic transistor based on ferroelectric $Hf_{0.5}Zr_{0.5}O_2$ and 2D tungsten disulfide, *Advanced Electronic Materials* 6 (6) (2020) 2000057.

[52] B. Tian, L. Liu, M. Yan, J. Wang, Q. Zhao, N. Zhong, P. Xiang, L. Sun, H. Peng, H. Shen, T. Lin, B. Dkhil, X. Meng, J. Chu, X. Tang, C. Duan, A robust artificial synapse based on organic ferroelectric polymer, *Advanced Electronic Materials* 5 (1) (2019) 1800600.

[53] P. Noël, F. Trier, L.M. Vicente Arche, J. Bréhin, D.C. Vaz, V. Garcia, S. Fusil, A. Barthélémy, L. Vila, M. Bibes, J.P. Attané, Non-volatile electric control of spin–charge conversion in a $SrTiO_3$ Rashba system, *Nature* 580 (7804) (2020) 483–486.

[54] J.R. Schaibley, H. Yu, G. Clark, P. Rivera, J.S. Ross, K.L. Seyler, W. Yao, X. Xu, Valleytronics in 2D materials, *Nature Reviews Materials* 1 (11) (2016) 16055.

[55] W.Y. Tong, S.J. Gong, X. Wan, C.G. Duan, Concepts of ferrovalley material and anomalous valley hall effect, *Nature Communications* 7 (1) (2016) 13612.

[56] X.W. Shen, W.Y. Tong, S.J. Gong, C.G. Duan, Electrically tunable polarizer based on 2D orthorhombic ferrovalley materials, *2D Materials* 5 (1) (2017) 011001.

11

Logic Devices Based on 2D Semiconducting Materials

Muhammad Mubeen, Hasan Ubaid Ullah, and Yinghui Han*

Correspondence: hanyinghui@ucas.ac.cn

11.1 Introduction

Two-dimensional (2D) materials have received much attention in nanoelectronics due to their unique qualities, especially exceptional electrical and optical characteristics, and potential applications [1]. Various proof-of-concept devices constructed from 2D materials are being demonstrated, considering their research is still in its infancy, which highlights their significance toward advanced technology [2]. Particularly promising for logic circuits are 2D semiconducting materials (2D-SCM). It has excellent optical, electrical, metallic, semiconductor, photoacoustic, and photothermal properties, so many studies have applied them to logic electronics for biomedical, energy, and sensor applications [1–3]. Recent advances in the much-watched MoS_2 transistor, for example, demonstrate basic digital circuits operating at gigahertz frequencies. It is also a two-dimensional photocatalyst material in environmental remediation [4]. Two-dimensional transition metal disulfides (TMDs) are also materials of concern for large-scale applications in electronic devices [5], and their electrical transport processes, including doping, contact engineering, mobility enhancement, and standardized manufacturing of TMDS, directly affect their application prospects as memory for logic integrated computing [6]. The ferroelectric semiconducting α-In2Se3 channel can carry out non-volatile computer operations and logic operations, which leads to their possible application in logic devices. The van der Waals semiconductor/metal heterostructure is conducive to regulating the light response of 2D-SCM [7–9]. For example, the energy transfer process between single-layer lead sulfide-cadmium sulfide quantum dots (Q-dots) and single-layer tungsten disulfide (WS_2), based on the connection band alignment, utilizes the photoelectric response of the van der Waals heterostructure. To create single crystals of a rhombic 2D conjugated metal-organic framework with strong charge carriers and electrical conductivity mobility. Logically related semiconductor properties can be seen in 2D conjugated frameworks made of metals and organics. Rhombic 2D conjugated metal-organic skeleton single crystals with strong carriers and conductive mobility have been prepared, and mechanical and electrical properties of high mobility mono-lay InAs have been studied using first principles and their interfacial interactions with different metals have been analyzed [10–12].

The advantages of 2D-SCM are high electron mobility, low energy consumption, and strong adaptability to current technology. However, high-quality 2D-SCM synthesis

DOI: 10.1201/9781003439448-11

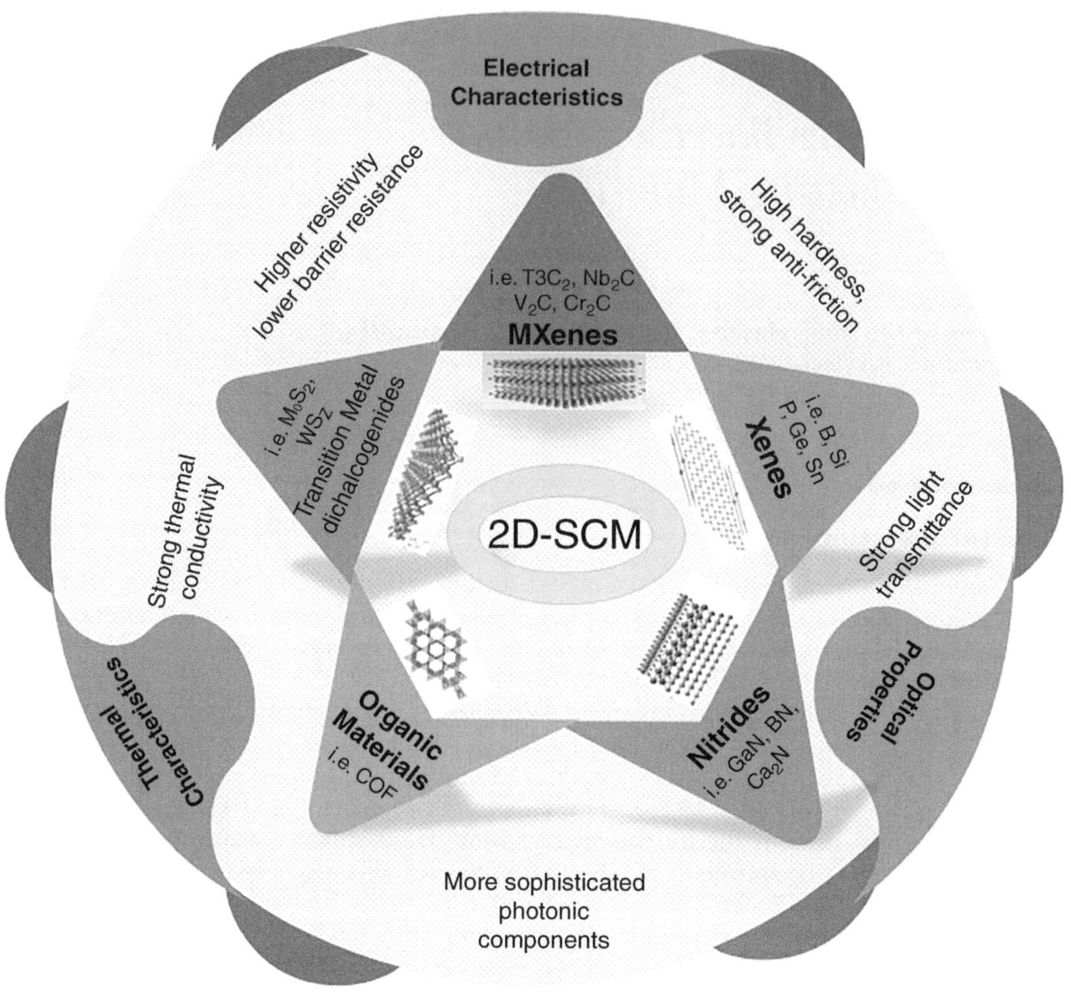

FIGURE 11.1
The schematic diagram of typical classes of 2D-SCMs and their optical, electrical, and thermal properties.

can be complex and requires a lot of research to improve the synthesis process. Some 2D-SCMs become unstable in the air, which limits their practical application. Many defects in 2D-SCM devices are related to the production process, from the choice of substrate to the technology used in synthesis. Strain engineering provides precise control of the optical and electronic properties of 2D-SCM. Surface-enhanced Raman spectroscopy can study the effect of dipoles on performance enhancement and reliability of various types of 2D-SCM materials by studying the correlation between the lattice arrangement of 2D-SCM and the unique characteristics of Raman enhancement. Contact engineering techniques applied in this area can offset the challenges associated with contact resistance by reducing the defect density at the substrate interface, resulting in optimal device functionality. **Figure 11.1** showcases typical 2D-SCMs and their optical, electrical, and thermal properties.

11.2 Logic Structure and Properties of 2D-SCM Devices

The logical structure of devices with 2D-SCM has great application potential in many fields and is now considered a fascinating research topic. Since independent lithography operations in the vertical channel direction are often much more difficult in 3D or 4D NAND flash memory, 2D NAND flash memory devices highlight the manufacturing advantages, and their storage transistors and series-select transistors are individually manufactured by parametric control using lithography and related manufacturing techniques. The program/erase duration for 2-D flash materials is only 20 ns. They display a reasonably large storage window above 50 V, which may be used to implement higher storage units than the TLC or QLC currently accessible [13]. Integrating 2D materials and silicon technology can solve the problem of traditional silicon-based electronic devices. Examples include silicon multi-bridge-channel field-effect transistor (MBCFET), whose stacked multilayer 2D nanosheets are separated by a general dielectric and gate, and fin field-effect transistor (FINFETs) with vertically grown 2D channel films that provide a three-sided gate surround. In addition to the silicon metal-oxide-semiconductor field-effect transistors (MOSFET) structure, two-dimensional DSFET, TFET, and NCFET are used to solve the energy-saving problem. The various functions provided by bipolar conductivity in 2D-SCM are currently being investigated for use in logic circuits and artificial neural networks. A schematic of an artificial neural network (ANN) architecture, shown in **Figure 11.2**, reveals how to connect neurons or computational primitives using synapses or memory components, and a schematic of a graphene mem-transistor that can be used to create artificial synapses to realize ANNs. Synthesis through various methods, engineering of atomic defects, investigation of emergent characteristics, and creation of devices for functional applications, focusing on the latest developments in the field of 2D-SCM. The unique applications of these devices, such as radio frequency switches and wearable biosensors, highlight their utility and adaptability. However, there are still some issues that need to be addressed, such as manufacturing methods, device performance enhancements, and integration using existing technologies [14, 15].

11.2.1 Logic Structure of 2D-SCM Devices

The fundamental principles associated with semiconductor physics and electronic device design are often the foundations of the logic structure of 2D-SCM electronic devices. Because of highly thick atomic layers, 2D-SCM like graphene, transition metal dichalcogenides (TMDCs), and black phosphorus offer unique features that enable the development of new device topologies and capabilities [1, 5, 6, 12, 13, 14].

11.2.2 Properties of 2D-SCM Devices

Among the most significant characteristics of 2D-SCM devices include thickness, tunable bandgap, high carrier movement, quantum confinement effects, flexibility and mechanical strength, optoelectronic properties and transparency, thermal characteristics, low-dimensional effects, energy efficiency, and chemical sensing [2, 4, 6, 9, 11, 13–16]. The exceptional qualities associated with their distinct structure have drawn much scientific interest in 2D materials, as shown in **Figure 11.3**. Strain Engineering has been widely used to modify physical properties, expanding its applications in flexible nanoelectronic and optoelectronic devices. Strain engineering is a crucial method for tailoring the lattice and electronic

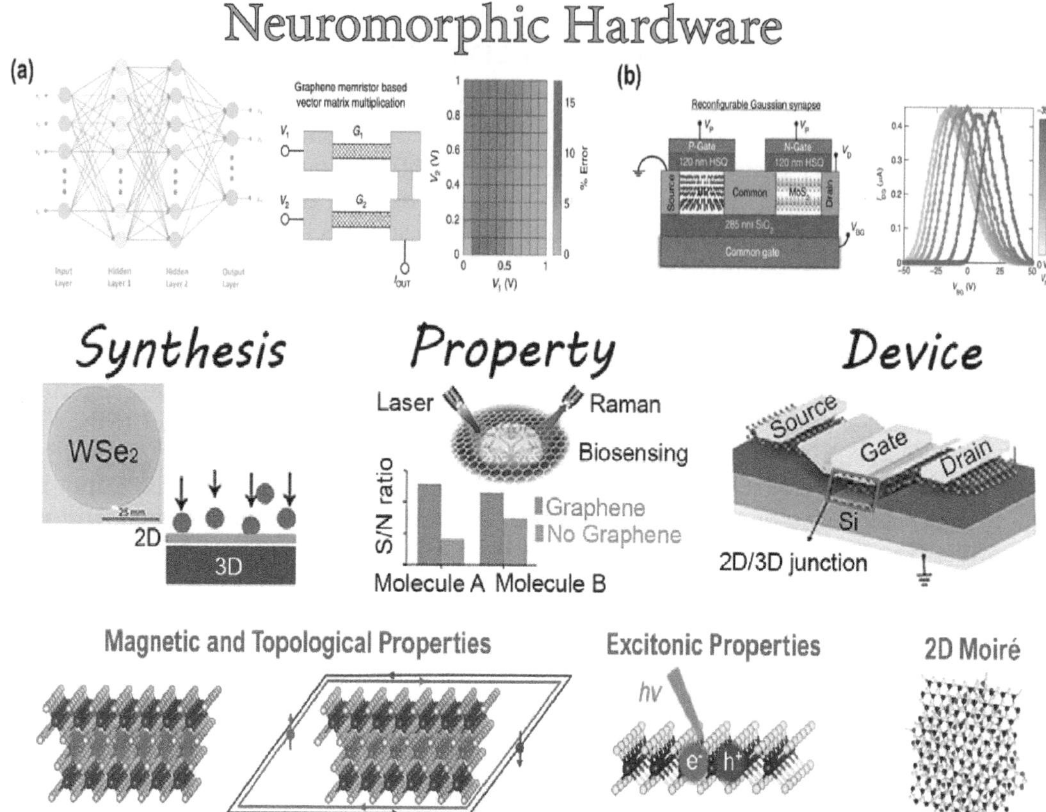

FIGURE 11.2

The schematic of smart sensing, security, and neuromorphic hardware for 2D-SCM. Redraw and adapted with permission [16], Copyright. The Authors, some rights reserved; exclusive licensee (American Chemical Society). Distributed under a Creative Commons Attribution License 4.0 (CC-BY-NC-ND).

structure of 2D materials. The techniques for applying strain to two-dimensional materials and each approach's benefits and drawbacks are discussed. We describe the phase transition of 2D materials and the consequences of strain on optical, electrical, and magnetic properties. In conclusion, we present the possible uses of strained 2D materials and anticipate the obstacles and opportunities they will present in real-world applications down the road [2].

11.3 Carrier Transport of 2D-SCM Logic Devices

2D-SCM logic circuits have great potential applications in high-performance electronics. Recent developments in transport carriers for 2D semiconductor logic devices emphasize transport characteristics, device design, and performance characteristics. The effective carrier transmission of 2D-SCM logic devices is an important problem. Inkjet printing processes that provide more capacity and cost-effectiveness in the production of electrical

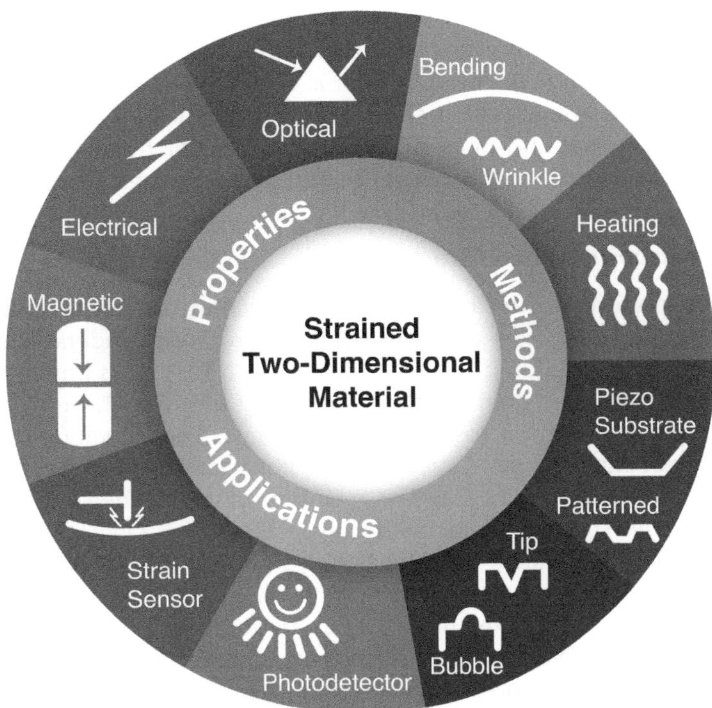

FIGURE 11.3
Special qualities of 2D-SCM associated with unique structures by strain engineering.

equipment is possible to manufacture scalable, affordable devices with enhanced on-chip transfer capabilities. By releasing a second metal layer on the MoS_2 channel, inkjet printing MoS_2 transistors with on-chip transmission advantages can be created. The efficiency of the device is improved by significantly reducing the optimal channel length and producing a carrier that is primarily on-chip. Integrating multiple materials (such as metals or insulators) into heterogeneous structures can improve carrier transport characteristics, or combining two-dimensional materials with other nanoscale elements can lead to new functions in logic circuits [17], even with reduced contact resistance, increased mobility, saturation current, and on–off ratios suitable for device logic applications. Thus, improving the integration of multiple materials (especially metals or insulators) into a heterogeneous structure can improve carrier transport characteristics, thereby improving device performance. 2D TMDs and metal contact can form a near-perfect van der Waals (vdW) interface without chemical interaction. Switching changes of synthetic synaptic electronic electrochemical random-access memory devices are controlled by a 2D material ion transport layer one atom thick. The adjustment of ion screening by a single layer of hexagonal boron nitride makes possible excellent non-volatile analog switches with good memory persistence and excellent durability [18, 19].

11.3.1 Scattering Mechanism

In recent years, significant advancements have been made in studying the 2D-SCM logic devices' scattering process. Because graphene spin valves have a long spin-diffusion length at room temperature, they have shown potential for spin logic circuits. In spintronic devices based on synthetic spin order, the artificial magnetic order in oxide superlattices can be regulated at an atomic scale. Zwitterionic surfactant-based label-free light-scattering sensors have demonstrated potential for detecting applications. All-optical logic gates are regularly hypothesized and demonstrated physically or through simulations. These platforms include metal surfaces, waveguides, and photonic crystals [20, 21].

11.3.1.1 Intrinsic Electron–Phonon Scattering

Intrinsic electron–phonon scattering is a crucial mechanism that significantly impacts materials' thermal and electrical characteristics. A profound understanding of phonon and electron interactions and their influence on carrier transport is imperative for advancing high-efficiency electronic devices. Many scholarly inquiries have delved into intrinsic electron–phonon scattering across various materials. These include layered materials like MoAlB and iron chalcogenides, metal halide perovskites, 2D-SCM, graphene heterostructures, and iron chalcogenides [22]. These studies have unveiled valuable insights into the behavior of carrier transport, elucidating aspects such as electron–phonon coupling, phonon-limited mobility, tunable electron-flexural phonon interaction, pressure-responsive photocarrier dynamics, modest electron mobility in 2D-SCM, electron redistribution facilitated by cesium mediation, the impact of excitation processes in carbon nanomaterials, computational approaches to thermal transport in low-dimensional materials, and the interaction between electrons and phonons within disordered crystals [23].

11.3.1.2 Photo-phonons on the Surface of an Oxide

Surface photo-phonons occurring on oxide 2D-SCM offer significant revelations regarding the behavior of semiconducting substances such as HOIPs. A comprehensive grasp of these behaviors is pivotal in unraveling the intricate interplay between photo-phonons and the surfaces of oxide 2D-SCM. The study of oxide 2D-SCM contributes a theoretical and empirical understanding of how light interacts with the surface photo-phonons of semiconductor materials [24].

11.3.1.3 Coulomb Heterogeneous Scattering

The significance of coulomb heterogeneous scattering and its effects on diverse systems such as molecular multiferroics, hydrogels, electromagnetic wave absorbers, electrolytes, nanoribbons, surfactants, two-dimensional materials, liquid metals, and biocatalysts have been extensively investigated. The influence of columbic interactions in controlling magnetoelectric coupling, proton conductivity, mineralization processes, electromagnetic wave absorption characteristics, ion transportation tendencies, charge transfer within molecular junctions, self-organization of surfactants and polymers, as well as heat transfer in materials with anisotropic arrangements. These revelations enhance comprehension of coulomb heterogeneous scattering phenomena and offer valuable insights for developing and improving materials with customized attributes [25].

11.3.1.4 Scattering Imperfection

The interplay between magnetic arrangement and charge conduction in a 2D magnetic semiconductor is explored. The modifiable electron conduction occurring within the magnetic phase of a 2D-SCM uncovers a robust interconnection between its magnetic structure and charge conduction. A thorough investigation of the magnetic ordering, specific to each layer down to the monolayer, is carried out through magneto transport analysis. Numerous aspects concerning scattering irregularities in semiconducting materials with a 2D configuration are examined. An exploration is conducted into phenomena like the interplay connecting magnetic order and charge conduction, the behavior of Raman scattering, the emergence of trap states due to imperfections, the diffusivity of excitons, dielectric features of stratified materials, dynamics of carrier relaxation, and surface customization. These findings collectively enhance our comprehension of the impact of imperfections on the electronic and optical traits of semiconducting materials with a 2D structure [26, 27].

11.3.1.5 Charge Trap and Phase Change of Metal-Insulator

The controllable manipulation of electron interactions within high-k dielectric films through gate tuning has a pronounced impact on confinement extent and electrostatic interactions. The groundwork for delving deeper into topological phase transitions and correlated occurrences in 2D semiconductor-metal-insulator systems. When the gate-to-film distance is minimized, there is an observable reduction in the spatial reach of the 2D long-range logarithmic interaction. Moreover, as lateral lengths surpass the film-gate separation dimension, a shift in this interaction becomes apparent, transitioning primarily into a dipolar or potentially exponential behavior [28].

11.3.2 Interface Engineering

Interface manipulation in 2D-SCM has garnered significant attention due to its potential to tailor their electronic and photonic performance. Recent investigations have unveiled fresh possibilities and challenges when exerting strain on 2D materials. These studies have underscored the reliance of strain-induced effects or strain-interaction applications on interfacial characteristics such as shear strength and adhesion. These insights provide valuable guidelines for strategically incorporating mechanical strains into ultrathin semiconductor devices. The synthesis of highly crystalline, semiconducting imine-based two-dimensional polymers has been demonstrated through interfacial synthesis. The optoelectronic traits of these materials have been probed using time-resolved terahertz spectroscopy, unveiling characteristics akin to p-type semiconductors characterized by notable hole mobility. Interface engineering is pivotal in fine-tuning the electronic and photonic elements of 2D-SCM. Tactics encompassing strain manipulation, interfacial synthesis, and chemical modification are effective strategies for developing high-performance devices rooted in 2D materials. The interfacial attributes are critical to the success of strain-related or strain-coupled applications of 2D materials, spanning shear strength, adhesion, and overall interfacial mechanics. The optimization of extrinsic transport elements, like electrode/semiconductor contacts and interfacial impurities, emerges as a critical consideration for augmenting carrier mobility in 2D transition metal chalcogenide semiconductors. Furthermore, the conceptualization and fabrication of heterostructures founded on integrating 2D materials with wide bandgap semiconductors open up novel prospects in optoelectronic applications [29, 30].

11.3.3 Mobility Model

Exploring mobility mechanisms within 2D-SCM holds paramount significance within organic electronics. Gaining insight into the intricate landscape of charge transport properties is pivotal in advancing high-performance electronic devices predicated on these materials. Notably, this inquiry delves into discotic liquid crystals, a unique class of molecules characterized by disk-like geometries that organize themselves into partially ordered phases, colloquially referred to as discotic liquid crystal phases. The extensive investigation of these materials is driven by their promising prospects in organic electronic applications. Various models have been formulated within this context to elucidate the charge mobility phenomena inherent to organic substances. The prevalent methodologies for quantifying charge mobility are also examined [31].

11.4 Electrical Contact and Doping of 2D-SCM

In the fields of semiconductor physics and electronics, electrical contact and doping are basic ideas. Due to the particular electrical characteristics of 2D-SCM, such as graphene, TMDCs, and other atomically thin materials, these principles are made much more fascinating. Effective electrical connections and careful doping level control are essential to create high-performance electronic and optoelectronic devices based on these materials.

The electrical contact is critical in influencing the overall device performance in 2D-SCM. Effective charge carrier injection and extraction need low-resistance, ohmic connections. To contact 2D-SCM, graphene is a viable contender that can be used to create purely 2D circuits [32]. Since 2D materials are atomically thin, problems like metal diffusion into them make producing stable and low-resistance connections challenging. Researchers are working to develop contact materials and deposition methods that retain a robust electrical connection while causing the least amount of damage to the 2D material's inherent features. Using materials that firmly adhere to the 2D material or using barrier layers are two possible strategies to stop metal diffusion.

Doping is essential for adjusting the material's conductivity and carrier concentration for diverse applications [33]. Doping significantly influences the electrical and optoelectronic characteristics of 2D-SCM because of their ultrathin nature and high surface-to-volume ratio. To prevent the added dopants from impairing the material's structural and electrical characteristics, the doping process in 2D materials must be carefully regulated. Reduced mobility, a non-uniform carrier distribution, and other undesirable outcomes may result from excessive doping. Researchers frequently experiment with various dopants and doping techniques to acquire exact control over the resultant electrical characteristics.

11.4.1 Contact Resistance

The contact resistance of 2D-SCM develops due to several variables, including the 2D material's electrical characteristics, the composition of the metal utilized for contact, and the effectiveness of the interface between the two materials. By choosing a suitable 2D metal, low semiconducting-metal contact resistance can be attained [34].

The atomically thin structure of 2D-SCM like graphene or TMDCs might make it challenging to create low-resistance electrical connections. Devices made of these materials, including transistors or sensors, may function less effectively overall due to contact resistance. High contact resistance can slow down machines and increase power consumption, causing inefficient carrier injection and extraction.

11.4.2 Phase Change Contact

Investigating phase change contacts within 2D-SCM constitutes a research domain to enhance the electrical connections between 2D materials and metals. Notable enhancements in contact resistance for 2D materials have been accomplished through various methods such as precise deposition of ultra-clean contact metals, doping, inducing phase changes through lithiation, introducing insulating layers between the metal and semiconductor (known as MIS), and engineering work functions. Nonetheless, achieving optimal contact resistance, particularly for monolayer (1L) 2D materials, remains significantly challenging, with values surpassing an order of magnitude. A crucial aspect of ameliorating contact resistance comprehends the transmission of charge carriers from the metal to the 2D materials.

Phase change materials have been under scrutiny due to their potential in energy storage and conversion applications, wherein the characteristics of interfaces and metal-semiconductor contacts play pivotal roles [35]. The capacity of 2D materials to transition between distinct crystal structures, each imbued with unique properties, is noteworthy.

11.4.3 Grid-Voltage Control Contact

Research is currently dedicated to enhancing the electrical connections between a grid electrode and 2D-SCM through grid-voltage control contact. This field involves deeply examining the interaction between the grid and the 2D material, focusing on both external and internal factors influencing the connection. A comprehensive comprehension of these factors is crucial to enhance these electrical contacts. The main hindrance to scaling 2D field-effect transistors (FETs) is the resistance encountered at the contact point, which poses a significant challenge. Overcoming this resistance is vital for achieving high-performance devices. Researchers have employed in situ transmission electron microscopy to scrutinize the contact attributes across various 2D materials [36]. By manipulating these attributes, the objective is to establish targeted electrical contacts.

Improving electrical contacts with 2D materials involves a significant aspect called ohmic contact engineering [37].

11.4.4 Tunneling Contact

Tunneling contacts are specialized electrical contacts used in 2D-SCM that rely on quantum tunneling events for carrier transport between the electrode and the 2D substance. Particles can cross an impassable potential energy barrier thanks to the quantum mechanical process known as quantum tunneling. Interest in tunneling contacts has grown as a solution to contact resistance issues and to facilitate effective carrier injection and extraction in 2D materials. Particularly with ultrathin materials like 2D-SCM, the energy barrier at the metal-semiconductor interface can cause excessive contact resistance and ineffective carrier injection/extraction in conventional connections. Using quantum tunneling effects, tunneling contacts provide a solution by enabling carriers to "tunnel" through the barrier without using thermal energy. The 2D material's and the electrode's energy levels are positioned in tunneling contacts to overlap their wavefunctions. Despite the barrier being energetically unfavorable according to classical physics, electrons can "tunnel" through it thanks to this overlap. The barrier height, thickness, effective mass, and energy of the carriers are all factors that affect how likely tunneling is to occur [38].

11.5 Logic and Memory Devices/Circuits Based on 2D-SCM

Due to their distinct electronic characteristics and possible uses in various electronic devices, including logic and memory devices, 2D-SCM have drawn much attention [38, 39]. While 2D-SCM have many benefits, it's crucial to remember that they also have drawbacks, such as contact resistance, uniform doping, and environmental stability. Research is now being done to overcome these obstacles and realize the full potential of 2D materials in logic and memory devices.

11.5.1 Logic Integrated Circuit

Designing and fabricating elements like transistors and linkages to create functional circuits is required to make logic integrated circuits (ICs) utilizing 2D-SCM. In 2012, atom-thick, two-dimensional materials, including graphene, were used to create the first integrated circuit components [40]. Transistors can be made based on the intrinsic properties of 2D materials, such as graphene and TMDCs. 2D-SCM can be used to manufacture many types of transistors, such as FETs and tunnel FETs (TFETs).

Creating working logic gates and more complicated circuits requires combining many transistors when designing logic ICs. As a result of their ultrathin structure and ability to scale, 2D materials are advantageous in this regard since they enable compact and densely packed circuits. After the transistors are created and integrated, software tools must be used to design and simulate the logic circuit. When designing with 2D materials, one must consider their distinctive electrical characteristics, mobility, and potential deviations because of flaws. Building logic ICs with 2D-SCM have the potential to result in high-performance, low-power electronic devices

11.5.1.1 Direct-Coupled FET Logic

FETs can act as the gate with built-in over-voltage protection and increased ON-current in 2D-SCM [41]. It adapts the complementary metal-oxide semiconductor (CMOS) and transistor-transistor logic families. In direct-coupled FET logic (DCFL), the logic gates are directly coupled, which means that no intermediary buffering is required because the output of one gate is immediately connected to the input of another. 2D-SCM offer the potential for efficient and low-power digital circuits based on DCFL. The advantages of 2D-SCM include low power consumption due to quick switching times and high carrier mobility, allowing operation at lower supply voltages. However, challenges like noise margin and signal integrity due to minor voltage swings must be addressed in their design. Additionally, the atomic thinness of 2D materials enables higher density and potentially faster switching rates, facilitating compact device designs and integration of multiple gates within a small area. DCFL implementation in 2D-SCM necessitates careful consideration of device design, circuit topology, and material parameters. Addressing issues like contact resistance, gate leakage, and unpredictability unique to 2D materials is challenging for scientists and engineers.

11.5.1.2 Complementary Metal-Oxide Semiconductor

While standard bulk semiconductors have traditionally served as the foundation for CMOS technology, there is growing interest in investigating the idea of adopting 2D-SCM to implement CMOS because of their distinct features. Logic circuits are constructed using MOSFETs in a conventional CMOS process. The channel portions of transistors can be built using 2D materials, such as graphene, TMDCs, and other atomically thin materials. To make complementary pairs, CMOS technology needs both n-type and p-type transistors. Both n-type and p-type transistors can be made in the case of 2D materials by choosing the proper materials and their doping. For instance, depending on their doping levels, some TMDCs may behave in an n-type or p-type manner.

Low-resistance contacts and interconnects must be made for the 2D material transistors for effective signal transmission. In 2D materials, contact resistance can be a significant problem, necessitating careful designing of interactions. Due to their ultrathin nature, 2D materials enable compact and dense circuit topologies. However, the circuit design stage must consider difficulties such as contact resistance, unpredictability, and consistent doping. Benefits, including high carrier mobility, low power consumption, and quick switching speeds, could be provided by 2D materials. These characteristics may help CMOS circuits become more scalable and perform better. Low-resistance contacts and interconnects must be made for the 2D material transistors for effective signal transmission.

Furthermore, it's crucial to provide homogeneous doping in 2D materials and address the environmental stability of these materials. 2D TMDCs that have undergone solution processing have shown extraordinary potential as components for CMOS devices in the future [42, 43].

11.5.2 Memory Circuit

High-speed operation, low power consumption, and small device designs are potential benefits of memory circuits based on 2D-SCM [44]. Some of these memory circuits' uses are data storage and cache memory in electronic devices. In **Figure 11.4**, it shows the

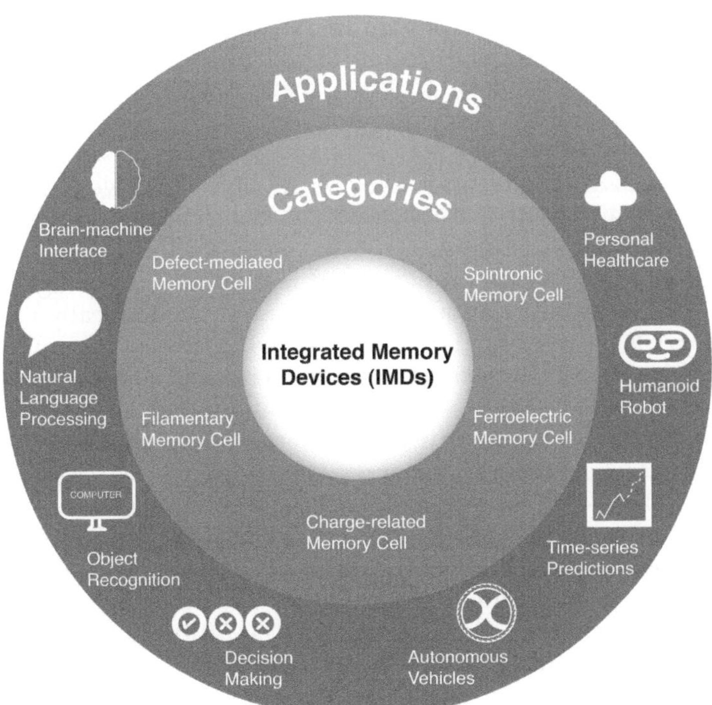

FIGURE 11.4
Applications of IMDs and memory taxonomy.

applications of IMDs and memory taxonomy. Depending on their operation, 2D materials-based IMDs can be divided into five groups: defect, filamentary, charge, ferroelectric, and spintronic integrated memory devices (IMDs). Potential applications are shown alongside their conventional structures. As the fundamental components of cutting-edge memory devices, 2D materials offer benefits for putting cutting-edge computing technologies into practice. On the other hand, these demand various functions, dense integration, and low energy consumption. Devices with atomic-thin 2D materials have high-density integration through highly scaled memory-device channels and efficient energy consumption. Together with their exceptional electrical and optical qualities, 2D materials' high surface-to-area ratio and wide exposed channel area in the horizontal plane enable the support of resistance-modulation and sensing functions. The absence of dangling boundaries and thickness-tunable band gaps in 2D materials allow for the customization of device architectures (e.g., heterostructure) for particular memory applications (e.g., sensing, processing, and storing activities).

11.5.3 Process Integration

Process integration in 2D-SCM entails several processes to construct functioning electronic devices with these atomically thin materials. The integration process can differ from conventional semiconductor production techniques because of its characteristics and difficulties. Choosing a suitable substrate is vital for integrating 2D materials effectively. Substrates should mechanically support, match lattice, and be chemically inert—examples: sapphire, silicon carbide, and silicon dioxide. Techniques like chemical vapor deposition (CVD), molecular beam epitaxy, and exfoliation create 2D materials based on application. CVD suits large sheets (e.g., graphene) or TMDCs. Some 2D materials grow on a growth substrate before being transferred to the target. Challenges due to 2D materials call for defect reduction, and quality control strategies. Post-processing (encapsulation, packaging) provides long-term stability. The transfer process can generally be used with 2D materials, regardless of their size or growth substrate type [45, 46]. A thorough understanding of the characteristics of the material and how it interacts with other components is necessary for process integration for 2D-SCM. Additionally, inventive fabrication methods address problems unique to these materials. Conclusion and Prospect

2D-SCM possess distinctive qualities that make them appealing for various electronic and photonic applications, which provide a fertile ground for investigating novel physical phenomena and advancing various technologies. Their atomically thin configuration grants these materials flexibility and transparency, which proves advantageous in applications involving flexible electronics, transparent displays, and wearable gadgets. Arranging multiple two-dimensional layers in a controlled manner to create van der Waals heterostructures leads to the emergence of fresh properties, opening up new possibilities for device engineering. Due to their efficient interaction with light, 2D-SCMs exhibit potential in optoelectronic roles like photodetectors, LEDs, and solar cells.

Nevertheless, these materials are associated with challenges in gaining widespread use. A key issue is the limited long-term stability of many remarkable 2D materials, impeding their practical applications. The limited carrier mobility and absorption rate of single-layer 2D materials also restrict their utility in electronics. Fabrication processes, encompassing material synthesis, transfer, patterning, and doping, significantly influence their performance and applications. Especially, the challenge lies in achieving large-scale material growth and seamless transfer integration methods. Ongoing research concentrates on devising approaches for the extensive synthesis of 2D materials and their seamless incorporation into devices and systems. As functional systems require the heterogeneous integration of diverse materials, the potential of 2D-SCM in surmounting miniaturization barriers and specific applications stands out. Future exploration must center on integrating 2D-SCM with other materials to craft high-performance electronic devices and systems.

Given all that, 2D-SCM possesses unique traits that suit various electronic and photonic applications. Yet, challenges like stability and carrier mobility persist. Advancements should target material integration, large-scale synthesis methods, and applications beyond electronics. The future outlook for 2D-SCM is promising, but certain aspects demand further research and development. Enhancing stability, improving carrier mobility, advancing large-scale synthesis techniques, delving into heterogeneous integration, and progressing manufacturing readiness constitute critical focal points.

Acknowledgments

This work was supported by the Key Funds for International Cooperation and Exchange of the National Natural Science of China (Grant No. 52320105003) and the Informatization Plan of the Chinese Academy of Sciences (Grant No. CAS-WX2023PY-0103).

References

1. Z. Xie, X. Meng, X. Li, W. Liang, W. Huang, K. Chen, J. Chen, C. Xing, M. Qiu, B. Zhang, G. Nie, N. Xie, X. Yan, H. Zhang, Two-dimensional borophene: properties, fabrication, and promising applications, *Research* 2020 (2020) 2624617.
2. S. Kang, D. Lee, J. Kim, A. Capasso, H. S. Kang, J. Park, C. Lee, G. Lee, 2D semiconducting materials for electronic and optoelectronic applications: potential and challenge, *2D Mater.* 7 (2020) 022003.
3. D. Lembke, S. Bertolazzi, A. Kis, Single-layer MoS_2 electronics, *Acc. Chem. Res.* 48 (2015) 100–110.
4. K. Khan, A. K. Tareen, M. Aslam, R. Sagar, B. Zhang, W. Huang, A. Mahmood, Nasir Mahmood, K. Khan, H. Zhang, Z. Guo, Recent progress, challenges, and prospects in two-dimensional photo-catalyst materials and environmental remediation, *Nano Micro Lett.* 12 (2020) 167.
5. X. Song, Z. Guo, Q. Zhang, P. Zhou, W. Bao, D. Zhang, Progress of large-scale synthesis and electronic device application of two-dimensional transition metal dichalcogenides, *Small* 13 (2017) 1700098.
6. J. Wang, F. Wang, Z. Wang, W. Huang, Y. Yao, Y. Wang, J. Yang, N. Li, L. Yin, R. Cheng, X. Zhan, C. Shan, J. He, Logic and in-memory computing achieved in a single ferroelectric semiconductor transistor, *Sci. Bull.* 66 (2021) 2288–2296.
7. M. Du, X. Cui, H. H. Yoon, S. Das, M. Uddin, L. Du, D. Li, Z. Sun, Switchable photoresponse mechanisms implemented in single van der Waals semiconductor/metal heterostructure, *ACS Nano* 16 (2022) 568–576.
8. S. Lim, D. Lee, H. Choi, Y. Choi, D. G. Lee, S. B. Cho, S. Ko, J. Choi, Y. Kim, T. Park, High-performance perovskite quantum dot solar cells enabled by incorporation with dimensionally engineered organic semiconductor, *Nano Micro Lett.* 14 (2022) 204.
9. L. Sporrer, G. Zhou, M. Wang, V. Balos, S. Revuelta, K. Jastrzembski, M. Löffler, P. Petkov, T. Heine, A. Kuc, E. Cánovas, Z. Huang, X. Feng, R. Dong, Near IR bandgap semiconducting 2D conjugated metal-organic framework with rhombic lattice and high mobility, *Angew. Chem. Int. Ed.* 62 (2023) e202300186.

10. K. Xu, W. Jiang, X. Gao, Z. Zhao, T. Low, W. Zhu, Optical control of ferroelectric switching and multifunctional devices based on van der Waals ferroelectric semiconductors, *Nanoscale* 12 (2020) 23488–23496.

11. J. Wu, C. Qiu, H. Fu, S. Chen, C. Zhang, Z. Dou, C. Tan, T. Tu, T. Li, Y. Zhang, Z. Zhang, L. M. Peng, P. Gao, B. Yan, H. Peng, Low residual carrier concentration and high mobility in 2D semiconducting Bi_2O_2Se, *Nano Lett.* 19 (2019) 197–202.

12] A. Bolotsky et al., Two dimensional material in biosensing and healthcare: from in vitro diagnostics to optogenetics and beyond, *ACS Nano* 13, no. 9 (Sep 2019) 9781–9810.

13. S. S. Kim, S. K. Yong, W. Kim, S. Kang, H. W. Park, K. J. Yoon, D. S. Sheen, S. Lee, C. S. Hwang, Review of semiconductor flash memory devices for material and process issues, *Adv. Mater.* (2022) 2200659.

14. S. Wang, X. Liu, P. Zhou, The road for 2D-SCM in the silicon age, *Adv. Mater.* 34 (2021) 2106886.

15. Y. Wang, T. Li, Y. Li, R. Yang, G. Zhang, 2D-materials-based wearable biosensor systems, *Biosens.-Basel* 12 (2022) 936.

16. Y. Lei et al., Graphene and beyond: recent advances in two-dimensional materials synthesis, properties, and devices, *ACS Nanosci. Au* 2, no. 6 (Dec 2022) 450–485.

17. S. K. Mondal, A. Biswas, J. R. Pradhan, S. Dasgupta, Inkjet-printed MoS_2 transistors with predominantly intraflake transport, *Small Methods* 5 (2021) 2100634.

18. G. Lee, S. Oh, J. Kim, J. Kim, Ambipolar charge transport in two-dimensional WS_2 metal–insulator–semiconductor and metal–insulator–semiconductor field-effect transistors, *ACS Appl. Mater. Interfaces* 12 (2020) 23127–23133.

19. R. D. Nikam, J. Lee, W. Choi, W. Banerjee, M. Kwak, M. Yadav, H. Hwang, Ionic sieving through one-atom-thick 2D material enables analog nonvolatile memory for neuromorphic computing, *Small* 17 (2021) 2103543.

20. P. Ghising, C. Biswas, Y. H. Lee, Graphene spin valves for spin logic devices, *Adv. Mater.* 35 (2023) 2209137.

21. C. Qian, X. Lin, J. Xu, Y. Sun, E. Li, B. Zhang, H. Chen, Performing optical logic operations by a diffractive neural network, *Light Sci. Appl.* 9 (2020) 59.

22. Z. Ou, B. Peng, W. Chu, Z. Li, C. Wang, Y. Zeng, H. Chen, Q. Wang, G. Dong, Y. Wu, R. Qiu, L. Ma, L. Zhang, X. Liu, T. Li, T. Yu, Z. Hu, Strong electron-phonon coupling mediates carrier transport in $BiFeO_3$, *Adv. Mater.* 10 (2023) 2301057.

23. Y. Zhou, Z. Fan, G. Qin, J. Yang, T. Quyang, M. Hu, Methodology perspective of computing thermal transport in low-dimensional materials and nanostructures: the old and the new, *ACS Omega* 3 (2018) 3278–3284.

24. B. T. Murti, A. D. Putri, S. Kanchi, M. I. Sabela, K. Bisetty, Inamuddin, A. M. Asiri, Light induced DNA-functionalized TiO_2 nanocrystalline interface: theoretical and experimental insights towards DNA damage detection, *J Photochem Photobiol.* 188 (2018) 159–176.

25. Y. Hu, S. Broderick, Z. Guo, A. T. N'Diaye, J. S. Bola, H. Malissa, C. Li, Q. Zhang, Y. Huang, Q. Jia, C. Boehme, Z. V. Vardeny, C. Zhou, S. Ren, Proton switching molecular magnetoelectricity, *Nat. Commun.* 12 (2021) 4602.

26. E. J. Telford, A. H. Dismukes, R. L. Dudley, R. A. Wiscons, K. Lee, D. G. Chica, M. E. Ziebel, M. Han, J. Yu, S. Shbani, A. Scheie, K. Watanabe, T. Taniguchi, D. Xiao, Y. Zhu, A. N. Pasupathy, C. Nuckolls, X. Zhu, C. R. Dean. X. Roy, Coupling between magnetic order and charge transport in a two-dimensional magnetic semiconductor, *Nat. Mater.* 21 (2022) 754–760.

27. S. Funk, M. Royo, I. Zardo, D. Rudolph, S. Morkötter, B. Mayer, J. Becker, A. Bechtold, S. Matich, M. Döblinger, M. Bichler, G. Koblmüller, J. J. Finley, G. Goldoni, G. Abstreiter, High mobility one- and two-dimensional electron systems in nanowire-based quantum heterostructures, *Nano Lett.* 13(2013) 6189–6196.

28. S. Kondovych, I. Luk'yanchuk, T. I. Baturina, V. M. Vinokur, Gate-tunable electron interaction in high-κ dielectric films, *Sci. Rep.* 7 (2017) 42770.

29. Z. Dai, L. Liu, Z. Zhang, Strain engineering of 2D Materials: issues and opportunities at the interface, *Adv. Mater.* 31 (2019) 1805417.

30. H. Sahabudeen, H. Qi, M. Ballabio, M. Položij, S. Olthof, R. Shivhare, Y. Jing, S. Park, K. Liu, T. Zhang, J. Ma, B. Rellinghaus, S. Mannsfeld, T. Heine, M. Bonn, E. Cánovas, Z. Zheng, U. Kaiser, R. Dong, X. Feng, Highly crystalline and semiconducting imine-based two-dimensional polymers enabled by interfacial synthesis, *Angew. Chem. Int. Ed.* 59 (2020) 6028–6036.

31. R. Termine, A. Golemme, Charge mobility in discotic liquid crystals, *Int. J. Mol. Sci.* 22 (2021) 877.

32. A. Allain, J. Kang, K. Banerjee, A. Kis, Electrical contacts to two-dimensional semiconductors, *Nat. Mater.* 14 (2015) 1195–1205.

33. H. Yoo, K. Heo, M. H. Ansari, S. Cho, Recent advances in electrical doping of 2D semiconductor materials: methods, analyses, and applications, *Nanomater.* 11(2021) 832.

34. R. Duflou, G. Pourtois, M. Houssa, A. Afzalian, Fundamentals of low-resistive 2D-semiconductor metal contacts: an ab-initio NEGF study, *NPJ 2D Mater. Appl.* 7 (2023) 38.

35. X. Chen, H. Yu, Y. Gao, L. Wang, G. Wang, The marriage of two-dimensional materials and phase change materials for energy storage, conversion and applications, *EnergyChem* 4 (2022) 100071.

36. L.-W. Wong, L. Huang, F. Zheng, Q. H. Thi, J. Zhao, Q. Deng, T. H. Ly, Site-specific electrical contacts with the two-dimensional materials, *Nat. Commun.* 11 (2020) 3982.

37. Y. Zheng, J. Gao, C. Han, W. Chen, Ohmic contact engineering for two-dimensional materials, *Cell Rep. Phys. Sci.* 2 (2021) 100298.

38. S. Kanungo, G. Ahmad, P. Sahatiya, A. Mukhopadhyay, S. Chattopadhyay, 2D materials-based nanoscale tunneling field effect transistors: current developments and future prospects, *NPJ 2D Mater. Appl.* 6 (2022) 83.

39. K. Zhu, S. Pazos, F. Aguirre, Y. Shen, Y. Yuan, W. Zheng, O. Alharbi, M. A. Villena, B. Fang, X. Li, A. Milozzi, M. Farronato, M. Muñoz-Rojo, T. Wang, R. Li, H. Fariborzi, J. B. Roldan, G. Benstetter, X. Zhang, H. N. Alshareef, T. Grasser, H. Wu, D. Lelmini, M. Lanza, Hybrid 2D-CMOS microchips for memristive applications, *Nature* 618 (2023) 57–62.

40. H. Wang, L. Yu, Y. Lee, Y. Shi, A. Hsu, M. L. Chin, L. Li, M. Dubey, J. Kong, T. Palacios, Integrated circuits based on bilayer MoS_2 transistors, *Nano Lett.* 12 (2012) 4674–4680.

41. Q. Qian, J. Lei, J. Wei, Z. Zhang, G. Tang, K. Zhong, Z. Zheng, K. J. Chen, 2D materials as semiconducting gate for field-effect transistors with inherent over-voltage protection and boosted ON-current, *NPJ 2D Mater. Appl.* 3 (2019) 24.

42. T. Zou, Y.-Y. Noh, Solution-processed 2D transition metal dichalcogenides: materials to CMOS electronics, *Acc. Mater. Res.* 4 (2023) 548–559.

43. D. Porotnikov, M. Zamkov, Progress and prospects of solution-processed 2D-SCM nanocrystals, *J. Phys. Chem. C* 124 (2020) 21895–21908.

44. J. Ma, H. Liu, N. Yang, J. Zou, S. Lin, Y. Zhang, X. Zhang, J. Guo, H. Wang, Circuit-level memory technologies and applications based on 2D materials, *Adv. Mater.* 34 (2022) 2202371.

45. A. Quellmalz, X. Wang, S. Sawallich, B. Jzlu, M. Otto, S. Wagnee, Z. Wang, M. Prechtl. O. Hartwig, S. Luo, G. S. Duesberg, M. C. Lemme, K. B. Gylfason, N. Roxhed, G. Steemme, F. Niklaus, Large-area integration of two-dimensional materials and their heterostructures by wafer bonding, *Nat. Commun.* 12 (2021) 917.

46. M. Caporali, A. Grüneis, S. Heun, T. Szkopek, Two-dimensional semiconductors: present and future challenges, *Phys. Status Solidi RRL* 14 (2020) 2000041.

12

Memristors Based on 2D Semiconducting Materials

Zhengyue Li and Huarui Sun*
Correspondence: huarui.sun@hit.edu.cn

12.1 Introduction

With the rapid development of the information age, Moore's law is going to reach its limit with traditional memory technologies based on complementary metal oxide semiconductor (CMOS) technology. Emerging non-volatile memories (NVRS) such as magnetic memories and phase-change memories (PCM) have aroused the interest of researchers. Compared with other NVRS, memristors have prominent performance on storage density, power consumption, read/write speed, erasable time, and data retention time [7]. In addition, memristors are easy to fabricate and scale down because of their sandwich structures. These properties make memristors one of the most promising devices for next-generation electronics. Besides non-volatile memory (NVM), memristors can be applied in non-volatile logic operation and neuromorphic computation. Non-volatile logic operations based on memristors can be used to solve the von Neumann bottleneck caused by the separation of computing and storage functions in traditional computer architecture and overcome the limit of information transmission rate. In neuromorphic computation, memristors are used to simulate biological synapses, with non-volatile synaptic weight, synaptic plasticity, nanoscale size, low power consumption, and large-scale integrability.

The specific definition of the memristor is ambiguous. Generally, resistive random access memory (RRAM), PCM, thermistors, gas discharge lamps, metal-insulator transition memory devices, spintronic memory devices as well as ion channels in living organisms can be classified as memristors [8], no matter whether they fit the mathematic model of memristors or not. In this chapter, we use the most common definition raised by Leon Chua: All two-terminal NVM devices based on resistance switching are memristors, regardless of the device material and physical operating mechanisms. Experimentally, a device with a pinched hysteresis loop I–V curve can be recognized as a memristor. A typical memristor's schematic structure is shown in Figure 12.1(a), and its I–V curve is shown in Figure 12.1(b, c, and d) [9].

Most memristors have a sandwich structure consisting of a top electrode (TE), an RS layer (or active/functional layer), and a bottom electrode (BE). The resistance can be switched between a high-resistance state (HRS) and a low-resistance state (LRS) by applying a voltage between the TE and BE. Early memristors were mostly comprised of metal-insulator-metal and most RS layer materials used were metal oxides (MOs) such as TiO_x, HfO_x, AlO_x, and TaO_x. Nevertheless, breakthroughs in performance are difficult on MO-based memristors because the RS behavior is not stable enough in MO materials. Besides, reducing the size is difficult in three-dimensional (3D) MO memristors. In recent years, with extensive

DOI: 10.1201/9781003439448-12

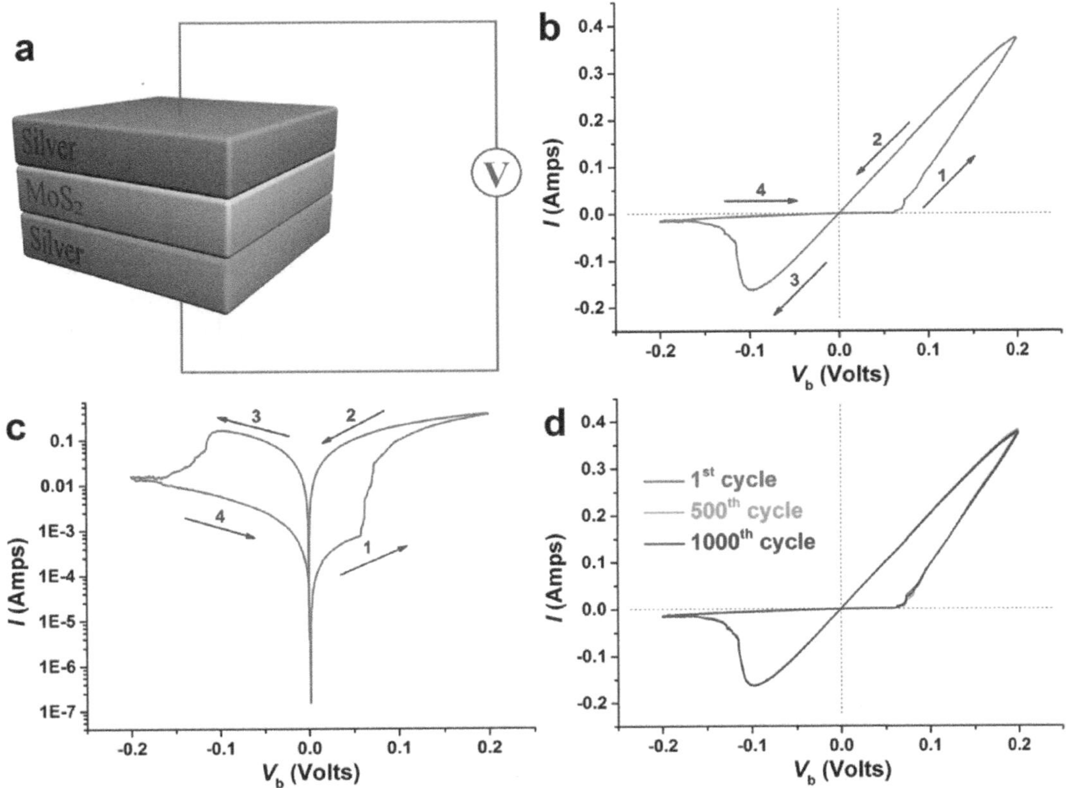

FIGURE 12.1

I–V characteristic of MoS_2 nanosheets based switch. (a) Schematic structure of Ag/MoS_2/Ag switch. (b) Typical I–V characteristic of Ag/MoS_2/Ag switch at room temperature. (c) Typical I–V characteristic of Ag/MoS_2/Ag switch in the logarithmic scale at room temperature. (d) Typical I–V characteristic of Ag/MoS_2/Ag switch at the 1st (red), 500th (green), and 1000th (blue) cycle at room temperature. Adapted with permission [9], Copyright 2017, American Chemical Society.

studies of 2D materials, researchers have started to use 2D materials to replace MOs as the RS layer. 2D materials have outstanding electrical, optical, mechanical, and thermal properties and scalability, which show great promise in new-generation device applications. A memristor based on 2D SnS was reported to exhibit a giant on/off ratio of 10^8, much higher than that of 3D memristors [10], and most 2D memristors have better stability than traditional memristors. This is because in traditional memristors a thin oxide layer has non-uniformity and random defects, which destroy the switching invariability and often cause large leakage currents and small windows, while a thick one has relatively high program voltages [7, 11]. RS behaviors have been observed in manifold 2D material-based memristors, such as MoS_2 [3, 4], h-BN, and even their monolayers. It is worth pointing out that most 2D materials exhibiting the RS behavior are semiconductors.

In this chapter, Section 12.2 introduces some typical 2D-SCMs and memristors based on them. Section 12.3 introduces common fabrication methods for 2D-SCM memristors. Section 12.4 introduces the main physical mechanisms for resistive switching (RS) in 2D-SCM memristors Section 12.5 introduces some basic applications of memristors. Section 12.6 gives a summary and outlook for 2D-SCM memristors.

12.2 2D-SCMs for Memristors

12.2.1 Metal Chalcogenides

12.2.1.1 Transition Metal Dichalcogenides

Transition metal dichalcogenides (TMDs), having the chemical formula of MX_2 (M is a transition metal atom such as Mo, W, and Hf; X is a chalcogen atom S, Se, or Te), are the most studied 2D-SCM materials. TMDs have a typical sandwich atomic structure, in which the M atoms are sandwiched between two layers of X atoms, tightly bound by covalent bonds. Most TMDs exist as 2H phase, which is semiconducting and thermodynamically stable, and have a large tunable bandgap and a high carrier mobility. Because of their excellent physical properties, TMDs have been widely used in electronic and optoelectronic devices.

MoS_2 is the most representative material in the TMD family, existing as molybdenite in nature with good stability [12]. MoS_2 can exhibit RS behavior from monolayer to several layers [3, 4, 9]. In principle, intrinsic MoS_2 does not show RS behavior, but dopants and lattice defects can bring RS into MoS_2, and, fortunately, defects are always introduced during the synthesis of MoS_2 nanosheets. The first batch of monolayer MoS_2 memristors showing NVRS behavior was reported by Sangwan et al. in 2015 [3]. The devices had switching ratios (R_{HRS}/R_{LRS}) up to 10^3 and could be switched between LRS and HRS repeatedly by grain boundary migration, which can further modulate the sulfur vacancy (V_S) concentration in adjacent areas, leading to resistance change.

However, the lateral structure is impractical for memristors as non-volatile switching was once considered to be not scalable to sub-nanometers owing to leakage currents. In 2018, Ge et al. reported NVRS behavior in monolayer TMD vertical devices for the first time [4]. The device structure of these memristors is shown in Figure 12.2(a, b, c, and d), and TEM characterization is shown in Figure 12.2(e and f). They used Au as noble TE and BE to avoid any influence of MOs and the devices showed NVRS behavior with a high on/off ratio above 10^4. No electro-forming process was observed, which is necessary for MO memristors to initialize a soft dielectric breakdown to form a conductive filament for RS. In 2021, Hus et al. studied the RS mechanism of $Au/MoS_2/Au$ memristors with a combination of scanning tunneling microscopy/scanning tunneling spectroscopy (STM/STS) [13]. It was revealed that sulfur vacancies, the most common lattice defects in MoS_2, have a one-to-one correlation between RS and metal adatom absorption. Because MoS_2's layered structure provides a sharp, clean interface with TE and BE, excessive leakage current is prevented even in the presence of vacancy defects.

In addition to defect-dominated TMD memristors, other types of MoS_2 memristors have also been designed by doping, intercalation, or other methods. Zhu et al. reported a planar structured $Au/Li_xMoS_2/Au$ memristor intercalated with a high concentration of Li^+ ions [14]. Li^+ ions were used to induce phase transitions ($2H \leftrightarrow 1T(1T')$) in MoS_2. Applying electric fields could drive Li^+ ions to redistribute, leading to the resistance change in the device. Based on the photoelectric characteristics of 2D MoS_2, Wang et al. prepared a MoS_2 nanosphere-based photomemristor [15]. The resistance state of the device could be controlled by the polarization of nanospheres. As oxygen ions have a lower migration barrier and higher mobility than intrinsic hydrogen and metal ions, oxidation methods were used to achieve ultrathin switching layers, reducing energy consumption and switching variations. Graphene/$MoS_{2-x}O_x$/graphene structures memristors with repeatable bipolar RS behavior were fabricated by Wang M et al. in 2018 [16]. In-situ HRTEM was used to observe high thermal stability in the $MoS_{2-x}O_x$ layer with the device heated up to 340°C.

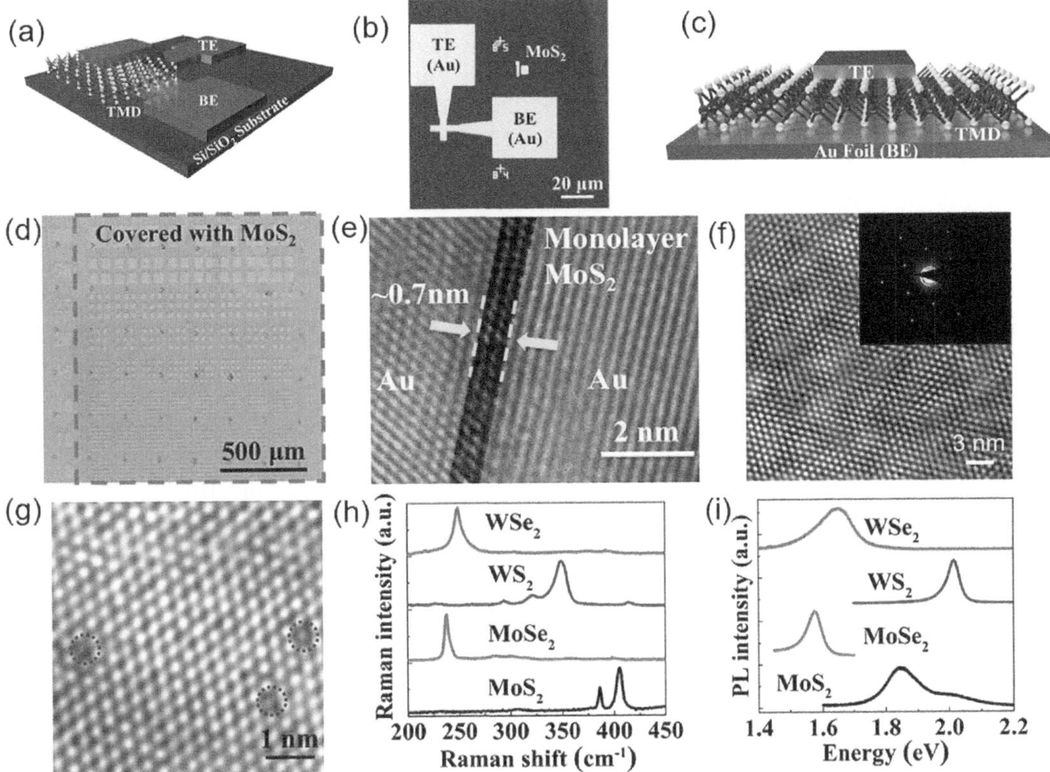

FIGURE 12.2
Schematic of the device and material characterization. (a, b) Schematic and optical image of metal–insulator–metal structures of TMD crossbar sandwich. (c) Schematic of TMD lithography-free and transfer-free sandwich based on MoS_2 grown on Au foil. (d) Optical image of fabricated MoS_2 litho-free devices with Au electrodes on Si/SiO_2 substrate. The dash box in panel (d) indicates the area covered with MoS_2. (e) TEM cross-section image of $Au/MoS_2/Au$ litho-free device revealing the atomically sharp and clean monolayer interface. (f) Atomic resolution TEM image of MoS_2 after transfer onto TEM grid. The hexagonal honeycomb structure has an in-plane lattice constant of ~0.31 nm indicating high-quality MoS_2. Inset: selective area electron diffraction (SAED) pattern captured from MoS_2 film, consisting of only one set of hexagonally arranged diffraction spots, suggestive of single-crystal growth. Adapted with permission [4], Copyright 2017, American Chemical Society.

WS_2, an analog of MoS_2 in the TMD family, has better thermal stability and resistance to oxidation than MoS_2. WS_2 was predicted to have the highest mobility due to the lightest effective mass, a high current ON/OFF ratio, a tunable bandgap up to 2.1eV, and a valance band splitting three times larger than MoS_2. A $Pd/WS_2/Pt$ memristor with ultralow power consumption down to femtojoules was fabricated by Yan et al. [11]. TEM showed that vacancy defects are the dominant factor of RS behavior in the memristor. Not only V_S but W vacancy (V_W) was also considered to play a role in RS behavior. Electron hopping between V_S and V_W is another factor that enhances the RS performance. The Joule heat generated by the voltage pulse promoted the movement of S ions and W ions, resulting in the increase of vacancies, which further produced a change in the vacancy gap and promoted electron hopping.

In addition to individual TMDs, van der Waals (vdW) heterostructure engineering is a good way to expand the properties of 2D materials [17]. Zhang et al. prepared an ultrathin memristor based on a 2D WS_2/MoS_2 heterostructure [18]. The device reached a high on/off ratio up to 10^4 and a clearly extended endurance with more than 120 switching cycles. The RS behavior was mainly caused by mild band modulation under applied voltages, eliminating material degradation during RS. These results showed that 2D WS_2/MoS_2 heterostructure memristors have a better memristive performance than those based on individual WS_2 or MoS_2 layers.

Recently, ReX_2 has attracted wide attention in the fields of microelectronic devices, photoelectric detection, energy storage, photocatalysis, and so on. ReS_2 and $ReSe_2$ are triclinic crystals with low symmetry, where the formation of Re_4 chains leads to extremely weak interlayer vdW coupling and strong structural anisotropy, which directly leads to changes in material intrinsic properties and novel nanophenomena. In 2021, Li et al. reported electron-beam-irradiated ReS_2-based memristors which had excellent switching uniformity [19]. The RS behavior was dominated by the migration of V_S without electro-forming. These ReS_2-based memristors had minimal temporal variation and were capable of simulating synaptic plasticity, presenting great potential for neuromorphic computing.

$MoTe_2$ was predicted to have the lowest energy difference between different phases in the TMD family [20], showing great potential for PCM-based memristors. Zhang et al. fabricated memristors based on $2H$-$MoTe_2$ and $Mo_{1-x}W_xTe_2$ in 2019 [21]. The RS behavior in these devices did not seriously depend on the material processing conditions. Applying electric fields could induce the devices to transit between the semiconducting $2H$ phase and the more conducting $2H_d$ phase (a transient state between the $2H$ phase and the metallic $1T'$ or T_d phase). They also observed the coexistence of T_d and $2H$ phases in $MoTe_2$ at room temperature. Moreover, according to thermodynamic principles, the addition of W made the phase transition in $Mo_{1-x}W_xTe_2$ devices easier, and their SET voltages were smaller than those of $MoTe_2$ devices.

12.2.1.2 *Post-transition Metal Chalcogenides*

Recently, post-transition metal chalcogenides (PTMCs) have become another kind of emerging 2D-SCM materials after TMDs. They have the general chemical formula MX, containing post-transition metal atoms (Ga, In, Sn, Tl, Pb, and Bi, sometimes containing Al and semimetal Ge, Sb, and Po) and chalcogenide atoms. The electronic and optoelectronic properties of PTMCs are quite different from those of TMDs. When the thickness of PTMCs is reduced from bulk to a few layers, their bandgaps increase substantially.

GaSe is a typical PTMC material with a direct bandgap for the bulk and a quasi-direct bandgap for the monolayer. Three-terminal lateral memristors based on 2D GaSe were reported by Yang et al., which were also called "memtransistors" because of their structural similarity to transistors [22, 23]. These memristors exhibited good NVRS behavior: after being exposed to air for one week, their on/off ratio increased to 5.3×10^5 and an ultralow threshold electric field of ~3.3×10^2 V cm^{-1} was reached. The 2D GaSe layer was intrinsically p-type, and the RS behavior was attributed to the migration of intrinsic Ga vacancies, which led to interface-type changing between the Schottky contact and Ohmic contact. The devices had an ultralow SET voltage as the Ga vacancy possesses a low migration energy.

12.2.1.3 *In_2Se_3*

Post-transition metal sesquisuifide $In_2Se_3 \cdot (2H)$ has become a material of attention in 2D-SCMs. Room-temperature in-plane and out-of-plane ferroelectricity in monolayer α-In_2Se_3 has been confirmed [24–26], while paraelectricity and antiferroelectricity exist in β and β′ phases, respectively. In addition, α-In_2Se_3 has an ultrahigh carrier mobility among all 2D ferroelectric materials. As for analogs of In_2Se_3, however, only weak ferroelectricity was observed in Sb_2S_3 [27, 28] and no ferroelectricity was confirmed in In_2Te_3 [29].

In terms of memristor behavior, ferroelectricity, and the low energy difference between different phases are beneficial characteristics. In 2021, ferroelectricity memristors based on In_2Se_3 were reported by Gabel and Gu [30], the RS behavior of which was regulated by applying an in-plane electric field. These devices exhibited a high on/off ratio up to 10^3 at certain biases and the LRS/HRS currents remained stable over 10^3 s. Based on the results, the multidomain formation and the associated energy barriers between domains result in RS behavior.

12.2.2 Other 2D-SCMs

12.2.2.1 Black Phosphorus

Black phosphorus (BP) has become one of the most popular 2D-SCMs due to its unique physical properties. BP possesses a direct bandgap at various thicknesses; when the thickness of BP is reduced from bulk to single layer, its bandgap increases from 0.3 to ~2.0 eV, which can bridge the energy gap between the zero band gap of graphene [31] and the larger band gap (1.5 eV to 2.5 eV) of many TMDs [32]. Moreover, BP has a high carrier mobility and internal quantum efficiency, with the best stability among all phosphorus allotropes.

In 2019, BP-based memristors with ultrahigh on/off ratio over 10^7 were reported by Wang et al. [5]. Self-assembly phosphorous oxide (PO_x) was obtained on the BP surface and served as the RS layer with BP. The introduction of a thin PO_x layer could increase the on/off ratio of the device, while the 2D BP layer provided a low leakage current. The RS behavior was dominated by the oxygen vacancy migration in the PO_x layer. Moreover, these memristors possessed a high data retention time of more than 10^4 s at room temperature and repeated programmable behavior. After repeated bending to 30° for 500 times, the devices retained RS behavior with an on/off ratio of about 3×10^2, showing great potential for wearable electronics and neuromorphic computing.

12.2.2.2 CuInP₂S₆

$CuInP_2S_6$ (CIPS) is another room-temperature strong ferroelectric 2D-SCM. It was predicted to be the only ferroelectric material that possesses four polarization states simultaneously [33]. Driven by a strong electric field, Cu ions can cross the layer spacing between adjacent layers and stabilize on the adjacent surface of the next layer. When all Cu ions in a layer pass through the layer spacing and stay on the surface of the adjacent layer, polarization conversion becomes opposite to the direction of the applied electric field. The defects caused by Cu ion interlayer migration can effectively reduce the potential barrier for interlayer migration of other Cu ions.

A multifunctional memristor based on 2D CIPS was reported by Liu et al. in 2023 [34]. By means of regulating the compliance current, the device could switch between volatile and non-volatile modes. When the compliance current was set under 10µA, the device would exhibit diode-like volatile behavior with a rectification ratio of 10^3. When the compliance current was set above 5mA, the device would switch to non-volatile mode with a stable on/off ratio up to 10^3. This mode switching could be attributed to an electrochemically conductive filament associated with the compliance current. Owing to these properties, this memristor was able to emulate short-term potentiation and long-term potentiation useful in artificial neural networks.

12.3 Fabrication of 2D-SCM Memristors

Fabrication of 2D memristors is relatively easy because of their simple structures compatible with CMOS processes, and a variety of technologies can be selected based on different requirements. According to the selected technology, the size of memristors can vary from nm to μm level. This section focuses on a brief introduction to the fabrication of conventional two-terminal sandwich structure memristors with one BE, one 2D-SCM RS layer, and one TE. Figure 12.3(a) shows a typical process of fabricating a sandwich structure 2D-SCM memristor. Figure 12.3(b) is the memristor array based on some memristor units.

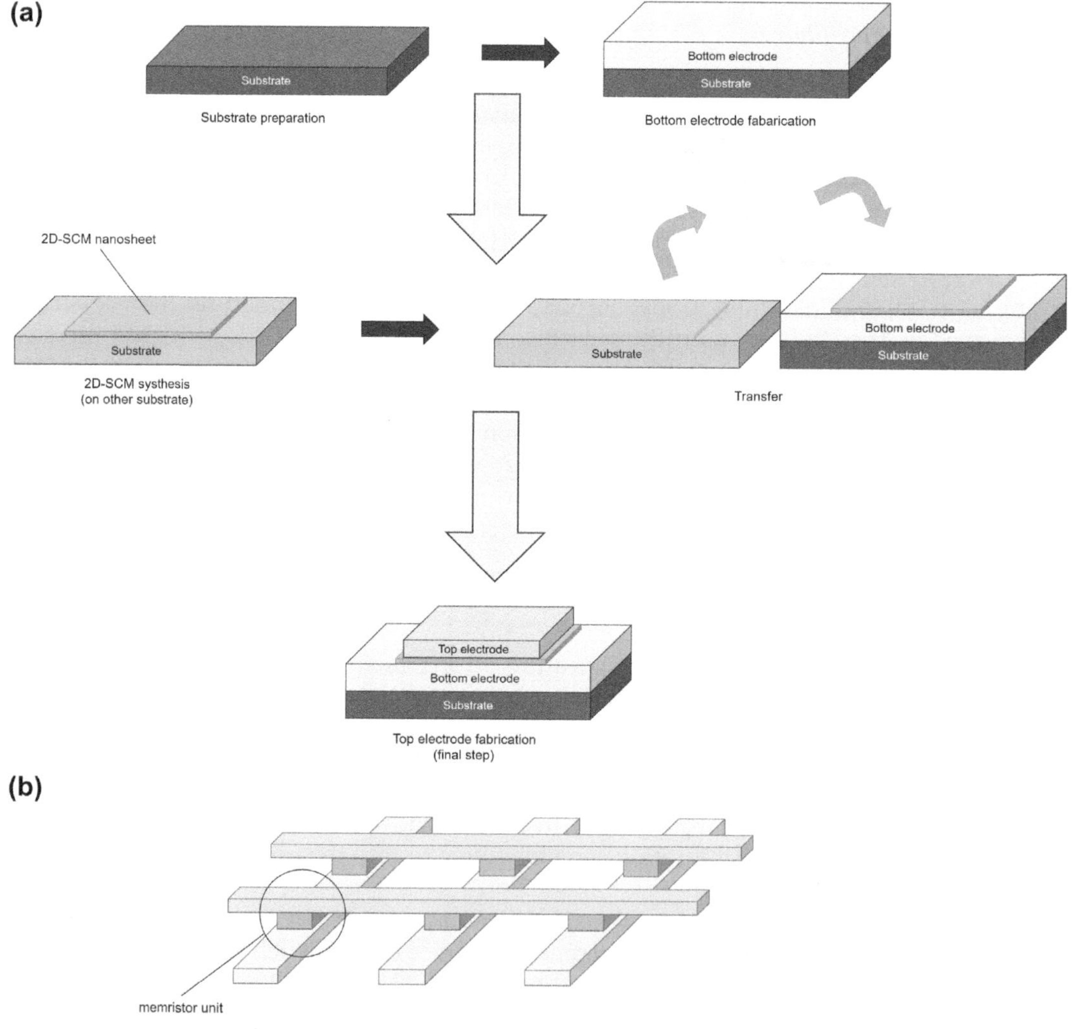

FIGURE 12.3

(a) Schematic of the fabrication process for a typical two-terminal 2D-SCM-based memristor, including substrate preparation, BE fabrication, 2D-SCM synthesis and transfer, and TE fabrication. (b) Memristor array with crossbar BE and TE.

12.3.1 Substrate and BE

Substrate selection is the first step to fabricate a memristor. Si/SiO_2 is usually chosen as a rigid substrate. In cases of manufacturing flexible electronics, flexible substrates such as ultrathin glass and polyimide films can also be selected. The selected substrate should not interact with the electrodes and RS layer during device operation. It is inevitable that thermal effects will have an impact on the performance, so the thermal conductivity of the substrate is often taken into account as an important parameter.

The materials used for BE in 2D memristors are generally metals with good electrical conductivity, commonly Au, Ag, and Cu. The selection of electrode materials needs to consider electrochemical effects on a case-by-case basis. The BE usually comes in two types: thin film and crossbar structures. The process of depositing a thin film electrode on a substrate has been well established, such as thermal evaporation, magnetron sputtering, electron beam deposition, and chemical vapor deposition (CVD). When memristor arrays need to be integrated, crossbar BEs should be prepared to form an electrode matrix as in Figure 12.3(b), requiring relatively complex processes such as photolithography or laser writer patterning.

12.3.2 RS Layer

Fabrication of 2D-SCM RS layers requires synthesis and transfer of 2D materials. It is difficult to prepare 2D flakes at the millimeter level, so the size of memristor cells is generally limited to the microns. Both top-down and bottom-up methods can be used to synthesize 2D flakes. This section briefly introduces the current progress in the synthesis and transfer of 2D-SCMs materials.

12.3.2.1 Synthesis

12.3.2.1.1 Mechanical Exfoliation

Mechanical exfoliation refers to the method of separating monolayer or few-layer 2D materials from a bulk by applying mechanical forces. Novoselov and Geim were the first to successfully separate monolayer graphene from graphite by mechanical exfoliation with Scotch tape [31], and the method was then widely adopted by researchers to obtain different kinds of monolayer or few-layer 2D materials. 2D materials obtained by this method have low yield and small area, but pure 2D crystals with different layer numbers can be obtained at one time, so they can still be used in the study of layer correlation for 2D memristors.

12.3.2.1.2 Chemical Vapor Deposition

CVD is widely used for preparing 2D materials with high quality and large scale. A CVD process usually involves exposing the substrate to one or more different precursors, which chemically react and/or decompose on the substrate surface to produce a film. Most memristors are manufactured based on the CVD method, which can synthesize materials through different routes, offering more possibilities. However, some problems exist in the CVD process for fabricating 2D-SCM memristors. Preparation of 2D-SCMs mainly uses solid phase precursors as raw materials, whose sublimation and diffusion processes are complicated and difficult to control, and material preparation reactions and side reactions coexist in the growth system. The precise control of process parameters and the improvement of reaction controllability of CVD remain great challenges in the synthesis of high-quality 2D materials. Moreover, the reaction temperature of the CVD process is usually close to 1000°C, which is not compatible with some substrates.

12.3.2.1.3 Physical Vapor Deposition

Physical vapor deposition (PVD) is a vacuum physical deposition method as opposed to reaction-based CVD. Compared with CVD, the PVD process has lower temperatures and less contamination. However, 2D compounds obtained using the PVD method usually have a large number of defects and exhibit poor electronic properties, and therefore PVD is generally suitable for the synthesis of simple substances such as BP.

12.3.2.1.4 Liquid-Phase Exfoliation

Liquid-phase exfoliation (LPE) is a solution treatment method for large-scale exfoliation of 2D materials. This method is similar to mechanical exfoliation, which utilizes the weak vdW force between layers to peel off the material. LPE is cost-effective, environmentally friendly, of high quality, and widely used in the industrial production of memristors. However, yield and quality are still not balanced for LPE. Traditional solvents are relatively pollution-free but hard to increase the yield, while improved solvents may cause more residues, influencing material properties.

12.3.2.2 Transfer

Some memristor substrates are not compatible with the synthesis process of 2D nanosheets, so 2D-SCMs are usually synthesized on other substrates such as sapphire, and then transferred to the memristor substrates. Dry transfer and wet transfer are two methods for the transfer process.

The dry transfer includes PDMS-assisted transfer, vdW interaction transfer, and non-atmospheric condition transfer [35]. PDMS-assisted transfer is one of the most common transfer methods in the laboratory. It can keep the intrinsic state of the material without introducing crystal defects, but PDMS may remain on the surface after transfer. The success rate of PDMS-assisted transfer relies on the flatness of the target substrate surface and the contact pressure. When the substrate surface has atomic flatness, the transfer success rate can reach 100%, while the transfer success rate decreases significantly in the case of rough substrates. VdW interaction transfer methods can preserve the properties of the material before transfer, not requiring contact with organic polymers, and reduce surface residues. When the material that needs to be transferred is unstable in air, such as BP, non-atmospheric condition transfer should be adopted.

Wet transfer is a common transfer method to construct large-area complex 2D heterojunctions. According to the type of media layer and peeling method, it can be divided into PMMA-assisted transfer, PLLA-assisted transfer, and so on. Wet transfer has a relatively low cost and high success rate, but impurities tend to be introduced and contaminate the material surface. PMMA-assisted transfer is the most commonly used wet transfer method, enabling large-scale transfer of 2D materials. Nevertheless, this method has a complex process and brings massive impurities, so numerous improved practical methods have been developed based on the original PMMA-assisted process. The PLLA-assisted transfer is a universal rapid transfer method, the whole process of which takes only a few minutes to complete, simultaneously avoiding the generation of crystal defects during sample transfer effectively.

12.3.3 Top Electrode

TE fabrication is the final and the most challenging process in memristor fabrication. Different from BE, TE should be placed exactly on the RS layer without contact with BE to avoid shorting, and therefore nano/micro process is required for TE patterning and deposition.

Similar to the case of other semiconductor devices, lithography is commonly used in the fabrication of memristor TE. Using a patterned lithography mask, TE can be deposited precisely on the RS layer. Then the fabrication of memristor is completed after removing the remaining photoresist. Residuals of photoresist may affect the performance of memristors, so other lithography-free processes are also adopted for TE fabrication. For example, a method that used SiO_2 to isolate BE and TE was reported by Datye et al. [36] E-beam evaporation was used to deposit a SiO_2 layer on the Ti/Au BE without covering the $MoTe_2$ RS layer, and then a Au TE was deposited on the $MoTe_2$ and SiO_2 directly by ALD.

12.4 Physical Mechanism for RS in 2D-SCM-Based Memristors

Understanding the RS mechanism in memristors is the key to designing memristors with high-performance or/and complex functions. 2D-SCMs themselves have a variety of RS mechanisms because of their semiconducting properties, coupled with quantum effects induced by the ultrathin thickness, which complicates the understanding of the underlying physics.

It should be noted that there are still controversies about the interpretation of memristor mechanism for the sake that RS behavior is extremely difficult to directly observe or measure inside the device when it occurs. Moreover, it is unrealistic to unify all the mechanisms, and the RS behavior of memristors often involves multiple mechanisms acting together, rather than a single mechanism. In this section, we divide the mechanisms into several classes, although the classification criteria used may be slightly different from those in other review papers.

12.4.1 Filamentary RS

Filamentary RS is the main physical mechanism for early memristive phenomena in MOs. The nature of this RS behavior is that metallic atoms are arranged like filaments inside the RS layer, forming a conductive channel. Filamentary RS that has been well recognized can be divided into the electrochemical mechanism (ECM), the thermochemical mechanism (TCM), and the valence change mechanism (VCM).

12.4.1.1 Electrochemical Mechanism

Memristors with an electrochemical mechanism (ECM) commonly have a structure of active electrode/solid electrolyte or dielectric material/inert electrode (Figure 12.4(a)). When an electric field is applied, the active layer will be oxidized and generate metal cations. The metal cations move along the direction of the electric field and are gradually reduced at the inert electrode, forming a metallic conductive filament in the RS layer. If the migration speed of metal cations in the RS layer is high, the cations will not be reduced

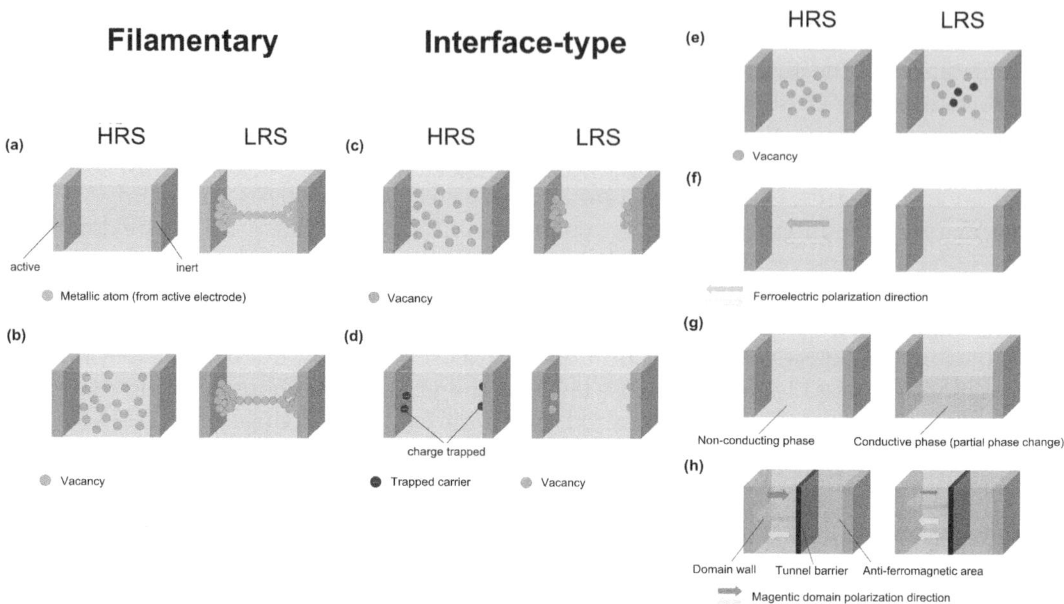

FIGURE 12.4
Schematic of resistive switching mechanisms. Filamentary resistive switching: (a) electrochemical mechanism and (b) valence change mechanism. Interface-type resistive switching: (c) valence change mechanism and (d) trapping and detrapping of charge. Other switching mechanisms: (e) space charge-limited current, (f) ferroelectric polarization, (g) phase-change memories, and (h) magnetic effects.

until they reach the inert electrode. In this case, a conductive filament grows from the inert electrode to the active layer, and vice versa. When the metallic filament completely forms, the device will switch from LRS to HRS. Under the action of a reverse electric field, the filament gradually breaks and the device returns to the original HRS.

In terms of 2D-SCMs, ECM is a rare mechanism because most 2D-SCMs are not solid electrolyte or dielectric material, and their ultrathin nature makes it hard to form a filament in the RS layer. Xu et al. reported ECM in vertical 2L-MoS$_2$ and 3L-MoS$_2$ memristors with Au BE and Cu TE [12]. They found that there was no RS behavior when the RS layer was monolayer MoS$_2$, where monolayer MoS$_2$ enables the current to pass through directly, leading to a short circuit. In the case of 3L-MoS$_2$, memristors needed an electro-forming process, while 2L-MoS$_2$ was electro-forming free. In contrast, Ge et al. reported memristive behavior [4] in Ag/1L-MoS$_2$/Au memristors which may be attributed to defect concentration, migration barrier difference, or other factors.

12.4.1.2 *Valence Change Mechanism*

VCM is a common RS mechanism in many 2D-SCM-based memristors. It can be divided into filamentary VCM and interface-type VCM. VCM is usually associated with vacancies and grain boundaries. Filamentary VCM is that, under the influence of an electric field, vacancies migrate in the RS layer, resulting in the change of valence states, and the arrangement of vacancies will form a nonmetallic conductive filament (Figure 12.4(c)). GaSe-based memristors reported by Yang et al. are typical VCM devices [22]. Ga vacancies continuously migrate and form a p-type conductive filament when a voltage is applied.

12.4.2 Interface-Type RS

Interface-type switching mechanism depends on interface effects between the electrode and the RS layer. In the case of 2D-SCM memristors, there is usually an interface contact resistance introduced by the Schottky barrier. The resistance state of the memristor is often changed by applying an electric field to adjust the barrier height. Moreover, the resistance of LRS is usually inversely proportional to the device area.

12.4.2.1 Valence Change Mechanism

In addition to forming a conductive filament, VCM can alter the interface between the RS layer and the electrode. It is generally believed that the migration of vacancies causes the change of the Schottky barrier height at the interface (Figure 12.4(d)) so that the transmission mechanism of interface current varies between Schottky and Ohmic conduction. Another possibility is that the increase of vacancy defects at the interface leads to the narrowing of the barrier, which results in direct tunneling or Fowler-Nordheim tunneling. The underlying mechanism can generally be inferred by the I–V curves. For example, Ge et al. reported I–V curves of $Au/1L\text{-}MoS_2/Au$ memristors at different temperatures [4]. They found that in LRS, the I–V curve is linear and the current decreases with temperature increasing, which is a typical metallic Ohmic conduction. In HRS, the I–V curve is nonlinear, and the current increases with the temperature increasing, which fits the Schottky emission model.

12.4.2.2 Trapping and Detrapping of Charge

Trapping and detrapping are usually associated with charge traps at the interface, which can be vacancy defects or dangling bonds on the surface of the material. When a carrier is trapped at the interface, the shape of the Schottky barrier is altered, leading to a resistance change (Figure 12.4(e)). The $W/1L\text{-}MoS_2/SiO_2/p\text{-}Si$ structure devices reported by He et al. were typical memristors based on trapping and detrapping of charge [37]. Dangling Si–O bonds at the interface of MoS_2/SiO_2 on p-Si serves as electron traps. When a forward bias is applied, electrons will be captured by the traps, leading to a higher barrier and the device switches to HRS. Electrons detrap when a reverse bias is applied and the device returns to LRS.

12.4.3 Other Switching Mechanisms

In addition to the broad categories discussed earlier, there are many other interesting RS mechanisms. When carriers are trapped by defects within the RS layer rather than traps at the interface, space-charge-limited currents occur and change the current transmission properties, realizing RS behavior (Figure 12.4(f)). Besides, other electronic effects such as ferroelectricity can realize memristive properties. As described in Section 12.2, the RS behavior based on ferroelectricity depends on the change of direction and/or intensity of ferroelectric polarization (Figure 12.4(g)).

Taking advantage of the low energy difference between different phases of 2D-TMDs, some researchers have induced phase transitions in various ways to achieve RS behavior (Figure 12.4(h)). For example, Li-ion intercalation $Au/Li_xMoS_2/Au$ memristors reported by Zhu et al. [14] were typical phase-change memristors.

Although often overlooked, magnetic phenomena caused by voltage changes can also be applied to memristors (Figure 12.4(i)). A multilevel memristor based on magnetic tunnel junction was reported by Lequeux et al. [38]. The memristive property is realized by controlling the motion of domain walls driven by the Slonczewski torque when voltage is applied.

There are still some other RS mechanisms not covered here, but, as 2D-SCMs themselves have a wide variety of properties, it is possible to design different types of memristors as long as there is sufficient understanding of the physical principles and the properties of the materials, not limited to those that have been reported.

12.5 Outlook

In this chapter, we mainly introduce some representative 2D-SCM-based memristors and explain their mechanisms. With excellent performance and rich physical mechanisms, 2D-SCM-based memristors are likely to be the focus of recent research in the field of memristors. However, both opportunities and challenges exist.

The state-of-the-art performance of 2D memristors has not reached a satisfactory stage. For example, most of the memristors lack long-term stability, which is essential for applications. Other performance indices such as switching ratio and power consumption also need to be improved. Even though some high-performance memristors have been reported, their industrial mass production is difficult to realize. To put 2D memristors in real applications, persistent efforts are needed to understand the fundamental physics and improve the fabrication process.

Moreover, as the 2D memristor is an emerging research topic, there are some misleading and inaccurate reports in this field. The study of memristors lacks a unified standard. The memristor test methods and standards vary from different groups and labs, which may generate inconsistent results. Different preparation processes may lead to differences in the properties of the same material, such as the defect type and density, which have often been overlooked in previous reports. Besides, the understanding of RS mechanisms is usually based on previous experience or knowledge in conventional memristors and may require in-depth verification for the 2D material system. There is still plenty of room for progress in the exploration of 2D-SCM memristors.

References

1. L. Chua, Memristor-the missing circuit element, *IEEE Transactions on Circuit Theory*, 18 (1971) 507–519.
2. D. Strukov, G. Snider, D. Stewart, R. Williams, The missing memristor found, *Nature*, 453 (2008) 80–83.
3. V. Sangwan, D. Jariwala, I. Kim, K. Chen, T. Marks, L. Lauhon, M. Hersam, Gate-tunable memristive phenomena mediated by grain boundaries in single-layer MoS_2, *Nature Nanotechnology*, 10 (2015) 403–406.
4. R. Ge, X. Wu, M. Kim, J. Shi, S. Sonde, L. Tao, Y. Zhang, J. Lee, D. Akinwande, Atomristor: Nonvolatile resistance switching in atomic sheets of transition metal dichalcogenides, *Nano Letters*, 18 (2017) 434–441.
5. Y. Wang, F. Wu, X. Liu, J. Lin, J. Chen, W. Wu, J. Wei, Y. Liu, Q. Liu, L. Liao, High on/off ratio black phosphorus based memristor with ultra-thin phosphorus oxide layer, *Applied Physics Letters*, 115 (2019) 193503.

6. W. Li, X. Zhang, J. Yang, S. Zhou, C. Song, P. Cheng, Y. Zhang, B. Feng, Z. Wang, Y. Lu, K. Wu, L. Chen, Emergence of ferroelectricity in a nonferroelectric monolayer, *Nature Communications*, 14 (2023) 2757.

7. M. Li, H. Liu, R. Zhao, F. Yang, M. Chen, Y. Zhuo, C. Zhou, H. Wang, Y. Lin, J. Yang, Imperfection-enabled memristive switching in van der Waals materials, *Nature Electronics*, 6 (2023) 491–505.

8. D. Liu, H. Cheng, X. Zhu, N. Wang, C. Zhang, Research progress of memristors and memristive mechanism, *Acta Physica Sinica*, 63 (2014) 187301.

9. P. Cheng, K. Sun, Y. Hu, Memristive behavior and ideal memristor of 1T phase MoS_2 nanosheets, *Nano Letters*, 16 (2015) 572–576.

10. X. Lu, Y. Zhang, N. Wang, S. Luo, K. Peng, L. Wang, H. Chen, W. Gao, X. Chen, Y. Bao, G. Liang, K. Loh, Exploring low power and ultrafast memristor on p-type van der Waals SnS, *Nano Letters*, 21 (2021) 8800–8807.

11. X. Yan, Q. Zhao, A. Chen, J. Zhao, Z. Zhou, J. Wang, H. Wang, L. Zhang, X. Li, Z. Xiao, K. Wang, C. Qin, G. Wang, Y. Pei, H. Li, D. Ren, J. Chen, Q. Liu, Vacancy-induced synaptic behavior in 2D WS_2 nanosheet-based memristor for low-power neuromorphic computing, *Small*, 15 (2019) 1901423.

12. R. Xu, H. Jang, M. Lee, D. Amanov, Y. Cho, H. Kim, S. Park, H. Shin, D. Ham, Vertical MoS_2 double-layer memristor with electrochemical metallization as an atomic-scale synapse with switching thresholds approaching 100 mV, *Nano Letters*, 19 (2019) 2411–2417.

13. S. Hus, R. Ge, P. Chen, L. Liang, G. Donnelly, W. Ko, F. Huang, M. Chiang, A. Li, D. Akinwande, Observation of single-defect memristor in an MoS_2 atomic sheet, *Nature Nanotechnology*, 16 (2020) 58–62.

14. X. Zhu, D. Li, X. Liang, W. Lu, Ionic modulation and ionic coupling effects in MoS_2 devices for neuromorphic computing, *Nature Materials*, 18 (2018) 141–148.

15. W. Wang, G. Panin, X. Fu, L. Zhang, P. Ilanchezhiyan, V. Pelenovich, D. Fu, T. Kang, MoS_2 memristor with photoresistive switching, *Scientific Reports*, 6 (2016) 31224.

16. M. Wang, S. Cai, C. Pan, C. Wang, X. Lian, Y. Zhuo, K. Xu, T. Cao, X. Pan, B. Wang, S. Liang, J. Yang, P. Wang, F. Miao, Robust memristors based on layered two-dimensional materials, *Nature Electronics*, 1 (2018) 130–136.

17. A. Geim, I. Grigorieva, Van der Waals heterostructures, *Nature*, 499 (2013) 419–425.

18. W. Zhang, H. Gao, C. Deng, T. Lv, S. Hu, H. Wu, S. Xue, Y. Tao, L. Deng, W. Xiong, An ultra-thin memristor based on a two-dimensional WS_2/MoS_2 heterojunction, *Nanoscale*, 13 (2021) 11497–11504.

19. S. Li, B. Li, X. Feng, L. Chen, Y. Li, L. Huang, X. Fong, K. Ang, Electron-beam-irradiated rhenium disulfide memristors with low variability for neuromorphic computing, *Npj 2D Materials and Applications*, 5 (2021) 1.

20. K. Duerloo, Y. Li, E. Reed, Structural phase transitions in two-dimensional Mo- and W-dichalcogenide monolayers, *Nature Communications*, 5 (2014) 4214.

21. F. Zhang, H. Zhang, S. Krylyuk, C. Milligan, Y. Zhu, D. Zemlyanov, L. Bendersky, B. Burton, A. Davydov, J. Appenzeller, Electric-field induced structural transition in vertical $MoTe_2$ and $Mo_{1-x}W_xTe_2$-based resistive memories, *Nature Materials*, 18 (2018) 55–61.

22. Y. Yang, H. Du, Q. Xue, X. Wei, Z. Yang, C. Xu, D. Lin, W. Jie, J. Hao, Three-terminal memtransistors based on two-dimensional layered gallium selenide nanosheets for potential low-power electronics applications, *Nano Energy*, 57 (2019) 566–573.

23. V. Sangwan, H. Lee, H. Bergeron, I. Balla, M. Beck, K. Chen, M. Hersam, Multi-terminal memtransistors from polycrystalline monolayer molybdenum disulfide, *Nature*, 554 (2018) 500–504.

24. Y. Zhou, D. Wu, Y. Zhu, Y. Cho, Q. He, X. Yang, K. Herrera, Z. Chu, Y. Han, M. Downer, H. Peng, K. Lai, Out-of-plane piezoelectricity and ferroelectricity in layered alpha-In_2Se_3 nanoflakes, *Nano Letters*, 17 (2017) 5508–5513.

25. F. Xue, W. Hu, K. Lee, L. Lu, J. Zhang, H. Tang, A. Han, W. Hsu, S. Tu, W. Chang, C. Lien, J. He, Z. Zhang, L. Li, X. Zhang, Room-temperature ferroelectricity in hexagonally layered α-In2Se3 nanoflakes down to the monolayer limit, *Advanced Functional Materials*, 28 (2018).

26. W. Han, X. Zheng, K. Yang, C. Tsang, F. Zheng, L. Wong, K. Lai, T. Yang, Q. Wei, M. Li, W. Io, F. Guo, Y. Cai, N. Wang, J. Hao, S. Lau, C. Lee, T. Ly, M. Yang, J. Zhao, Phase-controllable large-area two-dimensional In_2Se_3 and ferroelectric heterophase junction, *Nature Nanotechnology*, 18 (2022) 55–63.

27. J. Varghese, S. Barth, L. Keeney, R. Whatmore, J. Holmes, Nanoscale ferroelectric and piezo-electric properties of Sb_2S_3 nanowire arrays, *Nano Letters*, 12 (2012) 868–872.

28. L. Žigas, A. Audzijonis, J. Grigas, Origin of weak ferroelectricity in semiconductive Sb_2S_3 crystal, *Journal of Physics and Chemistry of Solids*, 101 (2017) 5–9.

29. D. Shao, J. Ding, G. Gurung, S. Zhang, E. Tsymbal, Interfacial crystal hall effect reversible by ferroelectric polarization, *Physical Review Applied*, 15 (2021) 024057.

30. M. Gabel, Y. Gu, Understanding microscopic operating mechanisms of a van der Waals planar ferroelectric memristor, *Advanced Functional Materials*, 31 (2021) 2009999.

31. K. Novoselov, A. Geim, S. Morozov, D. Jiang, M. Katsnelson, I. Grigorieva, S. Dubonos, A. Firsov, Two-dimensional gas of massless Dirac fermions in graphene, *Nature*, 438 (2005) 197–200.

32. X. Ling, H. Wang, S. Huang, F. Xia, M. Dresselhaus, The renaissance of black phosphorus, *Proceedings of the National Academy of Sciences of the United States of America*, 112 (2015) 4523–4530.

33. J. Brehm, S. Neumayer, L. Tao, A. O'Hara, M. Chyasnavichus, M. Susner, M. McGuire, S. Kalinin, S. Jesse, P. Ganesh, S. Pantelides, P. Maksymovych, N. Balke, Tunable quadruple-well ferroelectric van der Waals crystals, *Nature Materials*, 19 (2019) 43–48.

34. Y. Liu, Y. Wu, B. Wang, H. Chen, D. Yi, K. Liu, C. Nan, J. Ma, Versatile memristor implemented in van der Waals $CuInP_2S_6$, *Nano Research*, 16 (2023) 10191–10197.

35. J. Liao, J. Wu, C. Dang, L. Xie, Methods of transferring two-dimensional materials, *Acta Physica Sinica*, 70 (2021) 028201–028217.

36. I. Datye, M. Rojo, E. Yalon, S. Deshmukh, M. Mleczko, E. Pop, Localized heating and switching in $MoTe_2$-based resistive memory devices, *Nano Letters*, 20 (2020) 1461–1467.

37. H. He, R. Yang, W. Zhou, H. Huang, J. Xiong, L. Gan, T. Zhai, X. Guo, Photonic potentiation and electric habituation in ultrathin memristive synapses based on monolayer MoS_2, *Small*, 14 (2018) 1800079.

38. S. Lequeux, J. Sampaio, V. Cros, K. Yakushiji, A. Fukushima, R. Matsumoto, H. Kubota, S. Yuasa, J. Grollier, A magnetic synapse: multilevel spin-torque memristor with perpendicular anisotropy, *Scientific Reports*, 6 (2016) 31510.

13

Other Sensors Based on 2D Semiconducting Materials

Li Chen and Zheng Guo*

Corresponding author: zhguo@ahu.edu.cn

13.1 Introduction

2D SCMs, endowed with distinctive physicochemical attributes, have manifested profound potential for applications of sensor technology. With large surface areas, high surface reactivity, and exceptional electronic characteristics, these nanomaterials excel in the sensing of diverse targets, demonstrating both remarkable sensitivity and selectivity. Moreover, their nanoscopic scale facilitates rapid response times and commendable stability. Consequently, these 2D SCMs provide novel opportunities and avenues for improving sensor design and performance. In the following content, we explore the utilization of 2D SCMs in the field of electrochemical sensors, photoelectrochemical sensors, and chemiresistive gas sensors.

13.2 Electrochemical Sensor

Electrochemical sensors, extensively utilized for sensing and quantifying diverse chemical species, can be historically traced back to the early 19th century [1]. In 1833, Michael Faraday made his illustrious discovery of the fundamental laws governing electrochemical action, laying the groundwork for modern electrochemistry [1]. Subsequently, he introduced pivotal terminologies such as electrode, electrolyte, ions, anode, and cathode, which have become ubiquitous in contemporary scientific discourse [2]. In the early 20th century, electrochemical sensor concepts emerged as scientists explored using electrochemical principles for chemical sensing, though with limited practical applications. Rapidly advancing electrochemical sensors for sensing oxygen, hydrogen ions, and metal ions in various settings, driven by electronics and materials progress in the mid-20th century [3]. To date, the continued advancement of electrochemical sensors has garnered significant attention, largely owing to their distinctive advantages of high sensitivity, superior selectivity, low power consumption, and cost-effectiveness. Accordingly, electrochemical sensors have demonstrated versatile applicability, encompassing environmental monitoring, medical diagnostics, food safety, and numerous other fields. The working principle of the electrochemical sensor is to convert the chemical information generated by the reaction between the object being measured and the sensitive material into a quantifiable electrical signal. Electrochemical sensors can be divided into potential sensors, current sensors, and conductivity sensors according to various operational modes [4–5]. The conductivity sensors quantify chemical amounts via conductance, exhibiting high sensitivity but limited selectivity, thereby reducing its widespread application. No further expansion will be given in the following content.

Within the domain of potential sensors, the ion sensors (ion-selective electrodes) stand out as the most extensively studied and researched [6]. In general, ion-selective membranes facilitate the exchange of ions with those in the surrounding solution, generating two distinct liquid junction potentials at the interfaces [6]. This phenomenon constitutes a crucial aspect of the operation of ion sensors. MXenes are 2D nanomaterials with high aspect ratios and few atomic layers, which offer significant advantages for ion sensors. Ping et al. prepared two 2D MXene nanosheets, $Ti_3C_2T_x$ and Ti_2CT_x, for solid-contact ion-selective electrodes (**Figure 13.1(a)–(c)**) [7]. The incorporation of MXene coatings on the electrodes results in a substantial increase in double-layer capacitance, effectively facilitating the ion-to-electron transfer mechanism (**Figure 13.1(d)**). $Ti_3C_2T_x$ showed Nernstian responses (26.4 mV/decade) for Ca(II) concentrations between 10^{-1} and $10^{-5.5}$ M, with rapid response (<10 s) and low detection limits (0.79 μM) (**Figure 13.1(e)**).

Electrochemical current sensors surpass the potential sensor in wide usage and offer a superior solution for convenient, accurate, and safe current measurements. These sensors employ a three-electrode system to evaluate chemical quantities through the measurement of electrical currents. The recorded current signal exhibits a direct proportional relationship with the concentration of the redox material on the electrode, facilitating accurate determination of analyte concentrations and offering valuable insights into chemical processes. Examples of these sensing objects in the current sensor include ions, small biomolecules, and various chemical species. The hydrogen ion (H^+) sensor plays a pivotal role in diverse liquid chemical processes within industries such as chemical manufacturing, pharmaceuticals, and food production. It measures H^+ concentration in standard solutions, serving as a fundamental tool to ensure precision and efficiency in these critical applications. Sarkar et al. performed meticulous current measurements in response to varying pH levels, leading to the proposition of MoS_2 as a prospective candidate for a conventional H^+ sensor [8]. The pH sensing mechanism operates on the principle of protonation and deprotonation events involving the OH groups on the gate dielectric (**Figure 13.1(f)**), which are contingent upon the pH value of the surrounding electrolyte. Consequently, these dynamic interactions give rise to discernible changes in the dielectric surface charge, effectively enabling pH sensing with exceptional sensitivity and precision. Lower pH solutions protonate surface OH groups, resulting in positive surface charges on the dielectric. Conversely, higher pH solutions tend to deprotonate surface OH groups, generating negative charges (**Figure 13.1(g)**). Niu et al. introduce an innovative potentiometric pH sensor utilizing $Ti_3C_2T_x$ for non-invasive analysis of sweat pH in wearable applications [9]. The HF-$Ti_3C_2T_x$ sensor demonstrated notable sensitivities of -43.51 ± 0.53 mV pH^{-1} (pH 1–11) and -42.73 ± 0.61 mV pH^{-1} (pH 11–1), underscoring its potential for accurate and reliable pH monitoring in physiological environments. It provides accurate and continuous pH analysis, positioning the sensor as a valuable tool for comprehensive physiological monitoring and wearable health applications.

Researchers have significantly focused on developing sensing platforms tailored specifically for anion and heavy metal ion sensing in the current sensor field. Initial investigation into the application of current sensors in the field of anion sensing is essential for comprehensive understanding and advancement in 2D SCMs. The imperative nature of nitrite

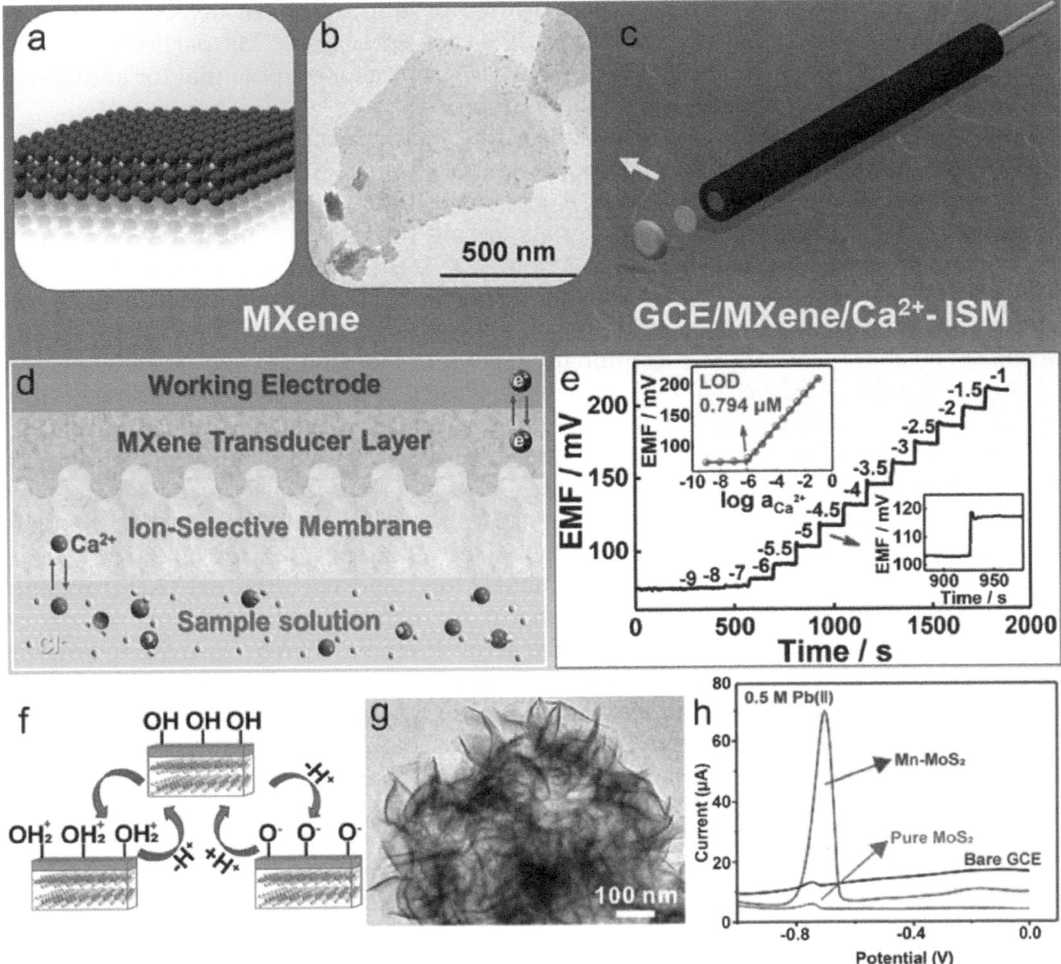

FIGURE 13.1

(a)–(b) Molecular structure diagram and its SEM diagram of $Ti_3C_2T_x$ nanosheets. (c) The physical Figure of glassy carbon electrode (GCE)/MXene/Ca^{2+}-ion-selective electrodes. (d) Schematic presentation of the application of MXene nanosheets as ion-selective electrodes for the detection of Ca^{2+}. (e) Dynamic potentiometric responses of the MXene-based Ca^{2+}-ion-selective electrodes. Adapted with permission [7]. Copyright (2019) Springer. (f) Illustration of the principle of pH sensing. Adapted with permission [8]. Copyright (2014) American Chemical Society. (g) Transmission electron microscopic image of $Mn-MoS_2$. (h) SWASV response of 0.5 mM Pb(II) on bare GCE, pure MoS_2, and $Mn-MoS_2$-modified GCE in a 0.1 M NaAc–HAc solution (Ph = 5.0). Adapted with permission [13]. Copyright (2018) The Royal Society of Chemistry.

sensing arises from its multifaceted hazards, spanning health implications, particularly concerning cancer-related risks, and adverse effects on potential threats to water quality and food safety. The integration of 2D SCMs, distinguished by their extensive specific surface area and rich availability of active sites, represents a sophisticated strategy for achieving superior electrochemical sensing of nitrite [10]. Notably, 2D SCMs exhibit exceptional sensing capabilities across various anionic species. For example, Yang et al. observed that MoS_2 exhibited remarkable catalytic activity in the redox reaction of SO_3^{2-}. The designed sensor demonstrated a wide dynamic linear range from 5.0×10^{-3} to 0.5 mM (r = 0.997, n = 15) for accurate quantification of SO_3^{2-}, accompanied by a low limit of detection (LOD) of 3.3×10^{-3} mM [11]. This research underscores the prowess of 2D SCMs, particularly MoS_2, in expanding the horizons of anion sensing and highlights their potential for addressing critical issues in environmental and health monitoring.

2D SCMs possess a high affinity for adsorbing heavy metal ions due to their active surface sites and unique chemical bonding characteristics. They exhibit tunable band structures and electron transport properties, enabling charge transfer and generating measurable electrochemical response signals during interaction with heavy metal ions. With their inherent chemical stability, 2D SCMs can sustain prolonged interactions with heavy metal ions, ensuring reliable sensing responses. 2D transition metal semiconductor sulfides, exemplified by MoS_2 and SnS_2, are characterized by their remarkable wealth of chalcogen atoms, serving as exceptionally favorable coordination sites for the selective binding of specific heavy metal ions. Consequently, these 2D SCMs find extensive utility in the adsorption of heavy metal ions. Lee et al. developed a novel vertically aligned MoS_2 nanofilm for in situ Pb^{2+} sensing. The sensor showed an excellent linear relationship with Pb^{2+} concentrations from 0 to 20 ppb with improved LOD was 0.3 ppb, using square wave anodic stripping voltammetry (SWASV) at -0.45 V vs Ag/AgCl [12]. Huang et al. present a remarkable achievement in ultrahigh sensitivity electrochemical sensing of Pb(II) by leveraging defect- and phase-engineered Mn-mediated MoS_2 nanosheets (**Figure 13.1(g) and (h)**). The research demonstrates that the strategic manipulation of Mn-MoS_2 enables potent chemical interactions between Pb(II) and the active S atoms, thereby facilitating efficient electron transfer and in situ catalytic redox reactions [13]. 2D SCMs also present compelling advantages for biomolecular sensing applications, encompassing notable attributes such as heightened sensitivity, rapid response times, molecular specificity, repeatability, and inherent stability [14–15]. These inherent capabilities make them indispensable tools in the fields of biomedical research, life sciences, and diagnostics, with the potential to drive the advancement of highly accurate, fast, and reliable biomolecular sensing technologies.

13.3 Photoelectrochemical Sensor

In 1960, Dewald was the first to propose the mechanism of photo-potential generation by semiconductor photoelectrodes and provided a comprehensive theoretical exposition on the field of photoelectrochemistry [16]. His pioneering work laid the foundation for a deeper understanding of the intricate processes governing photoelectrochemical phenomena in semiconductor materials [17]. In 1972, Fujishima and Honda made a significant discovery, demonstrating that TiO_2 possesses catalytic properties, particularly as a photoanode, enabling the photocatalytic decomposition of water under ultraviolet light irradiation [18]. This breakthrough marked a pivotal turning point in the field of chemistry, particularly in semiconductor photoelectrochemistry, sparking widespread interest and intensive research in this area. In the realm of both electrochemistry, which involves electron transfer and interfacial reactions, and photoelectric conversion with its intricate energy conversion processes, a significant approach has emerged [19]. This method revolves around establishing a quantitative relationship between the changes in photoelectric response and the analyte of interest. By carefully investigating how the analyte influences the sensing recognition process, particularly at the interface or during specific reactions, alterations in photocurrent or photovoltage are exploited to construct robust analytical methodologies. These innovative techniques have found widespread applications in various domains, such as biological and environmental analysis, offering precise and reliable solutions for scientific investigations. The photoelectrochemical sensor represents a cutting-edge device that harnesses the principles of photoelectrochemistry to discern substance concentrations with exceptional precision. It is the culmination of the fusion between photonics, chemical analysis, and computer technology, endowing it with elevated sensitivity, rapid response capabilities, and unparalleled accuracy. The working principle of a photoelectrochemical sensor is based on the photoelectrochemical effect, which refers to the phenomenon where semiconductor electrodes are excited by light, generating electron–hole pairs. These electron–hole pairs can move freely within the semiconductor electrode and produce a photocurrent under the influence of an external electric field. A semiconductor electrode, a liquid electrolyte, and reactive species generally make up the architecture of a photoelectrochemical sensor, as seen in **Figure 13.2**. Upon the introduction of reactive species into the electrolyte, these entities undergo adsorption onto the semiconductor electrode, initiating oxidation–reduction reactions and generating electron–hole pairs. Under illumination, excitation of conduction-band (CB) electrons toward the electrode prompts concurrent electron transfer from an electron donor species in the solution, resulting in an anodic photocurrent (**Figure 13.2**). Conversely, the migration of CB electrons toward an electron acceptor species in the solution elicits electron transfer from the electrode to recombine with the valence-band holes (h^+), leading to the formation of a cathodic photocurrent (**Figure 13.2**). In the photoelectrochemical process, changes in the anodic or cathodic photocurrent occur due to the capacity of the analyte to serve as an electron donor or acceptor, or its propensity to engage in chemical reactions with the electron donor or acceptor species at the semiconductor interface [20]. Exploiting these detectable variations in photocurrent enables both direct and indirect sensing of analytes, facilitating the development of sophisticated and precise analytical techniques with broad applications. The elucidated reaction mechanism highlights the pivotal role of the semiconductor electrode as the central component, helping to adsorb and catalyze the reactive species. The consequential fluctuations in photocurrent, directly correlating with the concentration of the reactive species, establish a robust and precise means for quantifying their concentrations via photocurrent measurements. This photoelectrochemical approach offers a highly accurate and sensitive method for quantitative analysis, showcasing the significance of the semiconductor electrode in facilitating reliable concentration determinations of the reactive species under investigation.

Currently, photoelectrochemical sensors are widely used in various fields such as environmental monitoring, life sciences, and industrial production. Due to their distinctive structure and fascinating electrical and optoelectronic capabilities, 2D SCMs have distinct benefits over conventional nanomaterials in improving charge kinetics [21]. Initially, by adjusting the number of layers or by combining them in synergistic ways with other nanomaterials, their band gap or edge location and light absorption may be precisely controlled. Additionally, the ultrathin nature of 2D semiconductors minimizes the migration distance of charge carriers, hence reducing the recombination of charge carriers. Lastly, the large specific surface area makes it easier for reactants to be adsorbed effectively. Furthermore, 2D nanomaterials greatly increase the size of the electrode/electrolyte contact, which encourages interfacial charge transfer and speeds up electrochemical processes. Most active sites are visible on the surface and actively involved in photocatalytic processes. Significantly, the photoconversion efficiency of these materials may be further improved by implementing sophisticated structural engineering, energy band engineering, and surface engineering techniques. Li et al. successfully synthesized 2D graphene-analogue

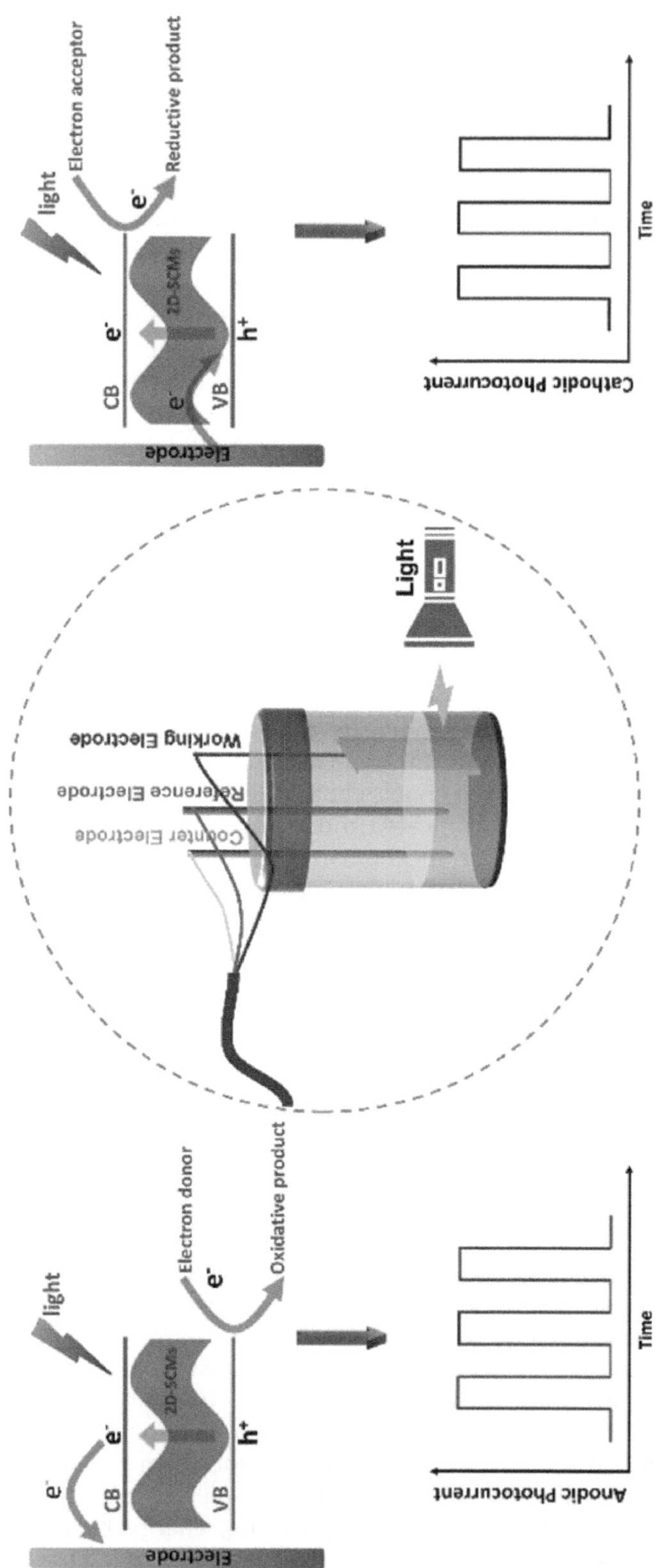

FIGURE 13.2
Diagram of the photoelectrochemical sensor with the traditional three-electrode system and the photocurrent generation mechanism.

carbon nitride (GA-C_3N_4) with 6–9 atomic thickness using a high-yield, large-scale thermal exfoliation technique (**Figure 13.3(a) and (b)**), which is suitable as a photoelectrochemical sensor for Cu^{2+} determination [22]. A Schottky heterostructure within GA-C_3N_4 nanosheets facilitates effective charge separation due to the continuous reduction of Cu^{2+} to Cu on their surface. The incorporation of Cu^{2+} as an electron acceptor leads to an amplified photocurrent response, demonstrating a direct correlation with the increasing concentration of Cu^{2+} (**Figure 13.3(c) and (d)**) [22]. Li et al. synthesized a novel graphene-like MoS_2/C_3N_4 (GL-MoS_2/C_3N_4) composite, enabling selective detection of trace Cu^{2+} in water. The 2D layered junction greatly increases contact area and enhances overall photoconversion efficiency by efficiently separating electron–hole pairs [23]. In addition, a 2D reduced graphene oxide (rGO)/MoS_2 material was constructed using a similar approach, enabling efficient photoelectrochemical detection of Hg^{2+} ions [24]. 2D semiconductor composites offer highly sensitive and efficient solutions for photoelectrochemical detection, with promising applications in analysis and detection, due to enhanced photoconversion efficiency, charge separation, large surface area, tunable properties, and high selectivity.

Aside from enabling the realization of photoelectrochemical sensing of heavy metal ions, 2D SCMs exhibit remarkable potential in facilitating the detection and manipulation of various small molecules. Luo et al. successfully fabricated a photoelectrochemical sensor by employing a single-layer nanoMoS_2-modified gold electrode (nanoMoS_2/GE). The significant enhancement in the photocurrent of nanoMoS_2/GE, achieved through the incorporation of dopamine, can be attributed to the improved generation efficiency of electron–hole pairs [25]. Hun et al. created a MoS_2 monolayer, a nanostructured p-type semiconductor, as the photoactive material. The photoelectrochemical response of this monolayer showed a significant decrease in the presence of methionine, with a logarithmic correlation to concentrations ranging from 0.1 nM to 1 M and a detection limit of 0.03 nM [26]. Zhang et al. successfully synthesized a sophisticated photoelectrochemical sensor by coupling g-C_3N_4 with CdS quantum dots, resulting in highly efficient photoactive species. The sensor exhibited a linear response to tetracycline within the concentration range of 10 to 250 nM, achieving an impressive detection limit (3S/N) of 5.3 nM [27]. Mahshid et al. developed a nanocomposite with MoS_2 nanosheets on nanohole-patterned TiO_2, enhanced with gold deposits. This design increased catalytic edge sites and improved optic-electrical coupling, resulting in excellent photoelectrochemical activities and superior performance compared to conventional systems [28].

Prior studies firmly establish the exceptional photoelectrochemical properties of 2D materials, showing their potential for enhancing sensing performance in diverse semiconductor-based photoactive materials. This is attributed to advantages such as ultrathin layers, excellent stability, large surface area, rapid electron mobility, and abundant surface-active sites. Despite significant progress, the ongoing development and application of 2D photoactive materials hold untapped potential, requiring further exploration to fully leverage their unique advantages in advancing photoelectrochemical sensors.

13.4 Semiconductor Gas Sensor

In 1962, Seiyama et al. achieved a pioneering breakthrough by developing an advanced gas component detector [29]. This detector capitalizes on the intricate correlation between gas adsorption and desorption phenomena, leading to a consequential modulation of electrical

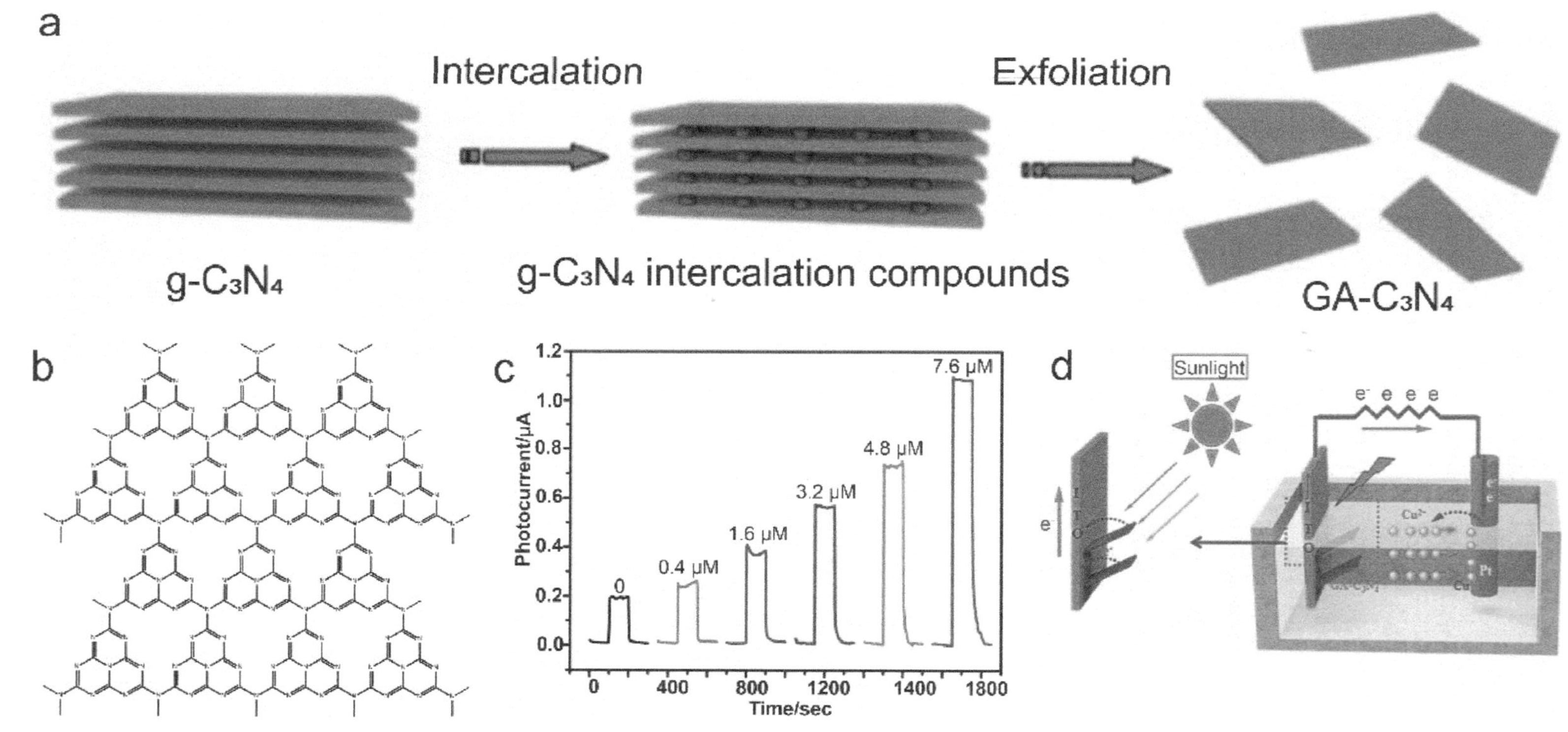

FIGURE 13.3

(a) Schematic representation of thermal exfoliation of the g-C$_3$N$_4$-based intercalation compound. (b) Perfect crystal structure of the graphene-analogue carbon nitride. (c) Photocurrent intensity of indium tin oxide/GA-C$_3$N$_4$ in the presence of different Cu^{2+} concentrations. (d) Schematic illustration of the exciton trapping mechanism and photoelectrochemistry for sensing of Cu^{2-}. Reproduced with permission [22]. Copyright 2014, The Royal Society of Chemistry.

conductivity within ZnO semiconductors. Subsequently, in the following decades, new materials, technologies, and application areas have been introduced, continuously improving and expanding the performance and application scope of semiconductor gas sensors. In recent years, there has been a surge in popularity surrounding the research on gas sensing mechanisms, owing to the proliferation of innovative characterization techniques and advancements in theoretical knowledge. The gas sensing mechanism of the semiconductor gas sensor primarily involves the selective adsorption and desorption of the target gas molecules at the gas-solid interface of the sensitive material [30]. This interaction leads to charge transfer processes, resulting in the alteration of the conductivity or resistance of the sensor, which serves as a quantitative indication of the presence and concentration of the target gas. Gas sensing mechanisms encompass various approaches, notably including Fermi level control, grain boundary barrier control, and electron depletion layer/hole accumulation layer theory [31–32]. Furthermore, an increasing number of researchers have directed their attention to exploring the intricate relationship between the sensitive materials and the target gas molecules of the sensor. This exploration includes the investigation of gas adsorption/desorption theory, resistance control mechanisms, and gas diffusion control theory [31]. These endeavors aim to gain deeper insights into the fundamental processes governing gas sensing and to enhance the performance of semiconductor gas sensors in various practical applications.

2D SCMs exhibit a multitude of distinct advantages when applied in semiconductor gas sensors. The schematic diagram of the prepared 2D-SCMs-based gas sensor devices is shown in **Figure 13.4**. The high surface-to-volume ratio of 2D SCMs, due to their atomic thinness, enhances interaction with gas molecules, boosting sensor sensitivity [33]. Manipulating the structure and composition allows for highly selective gas detection, altering electrical properties and enhancing sensitivity and selectivity. Superior carrier mobility and low surface energy contribute to rapid sensor response [34]. The single-atom thickness facilitates fast gas diffusion, reducing recovery time. Recent research has focused on applying 2D SCMs in gas sensors, emphasizing design, synthesis, and understanding of inherent gas sensing mechanisms.

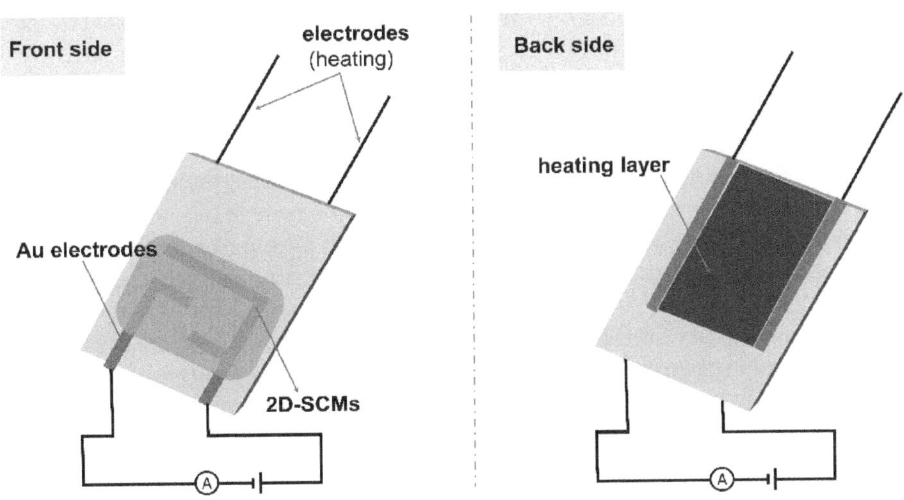

FIGURE 13.4
The schematic diagram of the prepared 2D-SCMs-based gas sensor devices.

Primarily, the progress of gas sensors fundamentally relies on the intricately orchestrated design and synthesis of sensitively tailored materials, possessing outstanding sensitivity and selectivity. 2D SCMs are defined as the semiconductor material exhibiting a planar structure, typically consisting of a few atomic layers, thereby possessing a reduced dimensionality. Presently, the primary types of 2D SCMs include graphene, sulfides (such as molybdenum disulfide and selenium disulfide), as well as phosphides, among others. To obtain these essential 2D SCMs, a variety of fundamental methodologies have been developed, including exfoliation, chemical vapor deposition (CVD), atomic layer deposition (ALD), electron beam evaporation, wet chemical synthesis, and selective etching [35]. Mechanical exfoliation is a common method for obtaining 2D SCMs by separating multilayer materials into thin flakes. Late et al. successfully achieved the synthesis of single-layer and few-layer $MoSe_2$ nanosheets at ambient temperature through the meticulous process of mechanical exfoliation from bulk crystals, thereby enabling the realization of a high-performance NH_3 gas sensor [36]. Liu et al. fabricated few-layered $MoTe_2$ flakes by mechanical exfoliation on a Si substrate. The resultant p-type $MoTe_2$ sensor demonstrated remarkable sensing capabilities toward NO_2 in ambient air, while exhibiting minimal response to H_2O, thus showcasing its immense potential for practical applications such as breath analysis and ambient NO_2 detection [37]. In contrast to the mechanical exfoliation technique, the aqueous exfoliation method offers several notable advantages including non-contact nature, facile scalability, enhanced efficiency, mild operational conditions, and environmental sustainability [38]. Deng et al. synthesized 1.5 nm thick 2D all-inorganic $NbWO_6$ perovskite nanosheets via liquid exfoliation (**Figure 13.5a**) [39]. The synthesized $NbWO_6$ nanosheets were initially dispersed in deionized water, and subsequently coated onto a ceramic tube with a gold electrode, leading to the fabrication of a miniaturized side-heated semiconductor gas sensor (**Figure 13.5b**). H_2S molecules act as a reducing gas, adsorbing onto the $NbWO_6$ nanosheets to be oxidized by pre-adsorbed oxygen species (e.g., O^-, O^{2-}, O_2^-), resulting in electron transfer from H_2S to $NbWO_6$. This process governs the gas sensing mechanism in the $NbWO_6$ nanosheet-based semiconductor gas sensor for H_2S detection (**Figure 13.5c**). To achieve superior quality 2D SCMs with attributes such as exceptional uniformity, precise thickness control, and homogeneous composition, a well-known approach as ALD has been developed for the fabrication of these materials. Lee et al. synthesized a 2D Al_2O_3/TiO_2 thin film heterostructure using ALD. The assembled sensors exhibited rapid detection of H_2 gas (<30 s at 300 K) without the need for heating modules, outperforming conventional H_2 gas sensors [40]. Kim et al. achieved remarkable sensing performance and superior gas selectivity for NO_2 gas by employing the ALD technique to directly grow 2D SnS_2 on a plastic substrate (**Figure 13.5(d) and (e)**) [41]. This innovative approach enables enhanced gas sensing capabilities, paving the way for potential advancements in gas sensor technology.

The physical method of exfoliation often consumes significant time and energy and can cause substrate damage. CVD exhibits significant advantages in the synthesis of 2D SCMs [42]. Its outstanding controllability, rapid growth kinetics, uniformity, resilience, material versatility, and functional adaptability render it an essential instrument in cutting-edge academic research and advanced industrial manufacturing. Initially, the discovery and synthesis of graphene were accomplished through the employment of the mechanical exfoliation method. However, the unsuitability of this method for facilitating large-scale production became evident. Subsequently, Koratkar et al. embarked on synthesizing graphene films via CVD to enable the detection of low concentrations of NO_2 and NH_3 in ambient air conditions, encompassing room temperature and atmospheric pressure [42]. In the field of room temperature gas sensors, notable advancements have

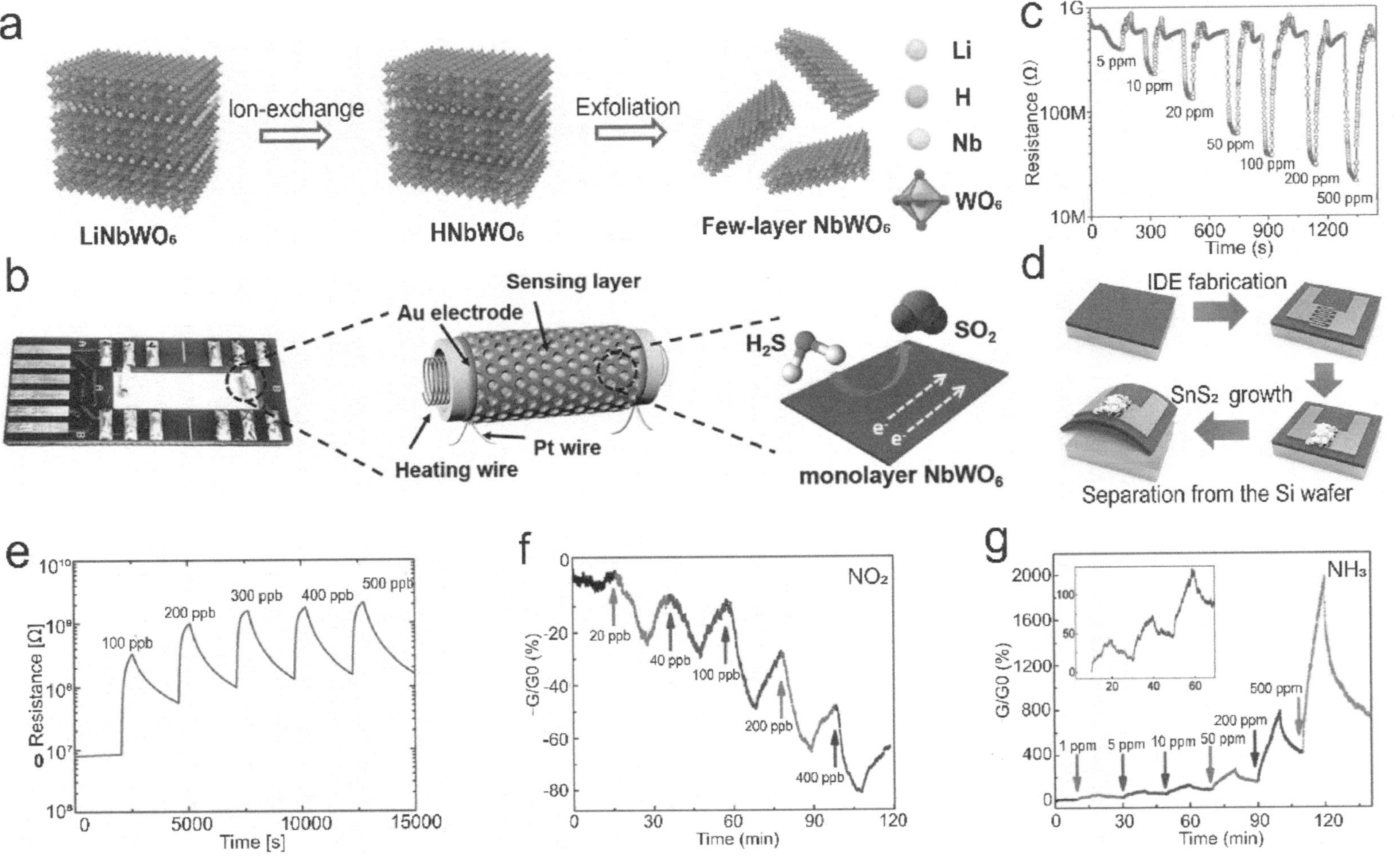

FIGURE 13.5

(a) The schematic illustration of the preparation process of the few-layer NbWO₆ nanosheets. (b) Optical photographs and a sketch of a side-heated NbWO₆ nanosheet-based gas sensor. (c) The response-recovery curve of the NbWO₆ sensor to H₂S with different concentrations. Reproduced with permission [39]. Copyright 2022, Wiley-VCH. (d) Schematic of the fabrication process for flexible SnS₂ gas sensor. (e) Variation in resistance of device with SnS₂ grown on Al₂O₃/PI/SiO₂/Si under pulses of NO₂. Reproduced with permission [41]. Copyright 2020, The Royal Society of Chemistry. (f–g) Real-time conductance change of MoS₂ with time after exposure to NO₂ and NH₃ under different concentrations. Reproduced with permission [43]. Copyright 2014, American Chemical Society.

been achieved in recent years through the ongoing progress of 2D SCMs fabricated via CVD. Zhou et al. successfully achieved a precise synthesis of monolayer MoS_2 using the Schottky-contacted CVD technique, resulting in remarkable current variations of 2~3 orders of magnitude when subjected to ultra-low concentrations of NO_2 and NH_3 (**Figure 13.5(f) and (g)**) [43]. In typical conditions, the robust binding between the target gas molecule and the reactive site at room temperature poses challenges in the dissociation of the target gas molecule, necessitating the application of thermal or optical energy to address this issue. Kumar et al. achieved enhanced performance in terms of rapid response and recovery kinetics, coupled with heightened sensitivity toward a concentration of 10 ppm NO_2 with the assistance of ultraviolet light [44]. In comparison to CVD, wet chemical methods offer distinct advantages, including ready availability of raw materials, user-friendly procedures, remarkable scalability, low-temperature preparation, and versatile applicability. Yang et al. fabricated a MoS_2/ZnO hetero-nanostructure 2D semiconductor material through a wet chemical method. The material displayed remarkable gas sensing capabilities, with an exceptional response of 3050% to a low concentration of 5 ppm NO_2 [45]. Regardless of the methodology employed, the fundamental goal of synthesizing 2D SCMs is to attain superior quality. This encompasses meticulous control over size and thickness parameters, optimization of heterostructure interfaces, realization of versatile material properties, and the establishment of cost-effective approaches for large-scale production.

The limited understanding of the fundamental sensing mechanism of 2D SCMs underscores the imperative for comprehensive research aimed at unraveling its intricate complexities. Normally, the mechanism of 2D semiconductor sensors involves the adsorption of target gas molecules on the surface of the material and the consequent charge transfer effects [46]. Target gas molecules adhere to the surface of 2D SCMs through either physical or chemical adsorption. Physical adsorption relies on weak van der Waals forces, while chemical adsorption involves stronger chemical bonds. This process alters the electronic structure and charge distribution, impacting the electrical properties of the sensor. Recent years have witnessed the utilization of advanced characterization techniques and theoretical calculations in the field of gas sensing to gain deeper insights into the underlying mechanisms. Cho et al. employed in situ photoluminescence characterization to confirm the charge transfer mechanism between the target gas molecules, specifically NO_2 and NH_3, and MoS_2 [47]. Advanced techniques like operando UV–Vis spectroscopy, XANES, and in situ ambient pressure X-ray photoelectron spectroscopy are increasingly used to investigate gas sensing mechanisms. In recent years, density functional theory (DFT) calculations and machine learning algorithms have played a crucial role in understanding intricate mechanistic aspects of gas-sensitive responses. DFT calculations offer precise characterization, modeling, energy optimization, computational efficiency, and high predictive capability [48]. They serve as a powerful tool for understanding sensor material properties and accelerating material development and optimization in sensor applications. Moreover, machine learning algorithms possess the capacity to autonomously acquire knowledge regarding the intricate relationship between sensor responses and gas characteristics, effectively extracting pertinent information and discernible patterns from extensive datasets. Through intricate analysis of sensor response patterns, these algorithms enable precise identification and classification of diverse gases, even amid the presence of interferences or overlapping response profiles.

Finally, optimizing 2D semiconductor sensor devices involves a systematic approach to enhance structural configuration and parameter settings for sensing applications using 2D SCMs. The goal is to maximize sensor effectiveness across diverse domains.

Achieving this optimization considers factors like materials, structure, interfaces, band structure, and signal processing. The focus lies on deploying 2D semiconductor sensors effectively in practical applications across fields such as energy, transportation, and security. Continuous technological advancements are expanding opportunities for gas sensor utilization in various domains [49–50].

13.5 Conclusion and Prospect

In this chapter, we present an exhaustive overview of the deployment of 2D SCMs in dominions such as electrochemical sensors, photoelectrochemical sensors, and chemiresistive gas sensors. Owing to their distinctive attributes and unparalleled sensing performance, 2D SCMs not only hold expansive potential within these sensor domains but also herald a promising future in enhancing sensor efficiency and transitioning into practical applications. Furthermore, beyond the sensor types elucidated previously, a plethora of sensors are pivoting their focus toward 2D SCMs, including but not limited to pulse and pressure sensors. Undoubtedly, there are challenges associated with integrating 2D semiconductor nanomaterials into sensor applications. These include the need to enhance their long-term stability within sensors, address interface issues when combining 2D materials with other substrates, and further optimize both the selectivity and sensitivity of the sensors. As research advances, we can anticipate the discovery of additional 2D SCMs exhibiting unique properties, thereby opening up new vistas for sensor applications. The amalgamation of 2D SCMs with other substances or technologies, including nanoparticles and quantum dots, holds the potential to yield multifunctional, high-performance sensors.

References

1. E. Bakker, M. Telting-Diaz, Electrochemical sensors, *Anal. Chem.* 74 (2002) 2781–2800.
2. G. Vinal, Electrochemical sources of electric power-I, *Electr. Eng.* 67 (1948) 354–357.
3. B. Privett, J. Shin, M. Schoenfisch, Electrochemical sensors, *Anal. Chem.* 82 (2010) 4723–4741.
4. H. Hamzah, S. Shafiee, A. Abdalla, B. Patel, 3D printable conductive materials for the fabrication of electrochemical sensors: a mini review, *Electrochem. Commun.* 96 (2018) 27–31.
5. L. Chen, J. Geng, Z. Guo, X. Huang, Polyoxometalates-based functional materials in chemiresistive gas sensors and electrochemical sensors, *Trends Anal. Chem.* 167 (2023) 117233.
6. M. Arnold, M. Meyerhoff, Ion-selective electrodes, *Anal. Chem.* 56 (1984) 20–48.
7. Y. Shao, Y. Yao, C. Jiang, F. Zhao, X. Liu, Y. Ying, J. Ping, Two-dimensional MXene nanosheets (types $Ti_3C_2T_x$ and Ti_2CT_x) as new ion-to-electron transducers in solid-contact calcium ion-selective electrodes, *Microchim. Acta* 186 (2019) 1–9.
8. D. Sarkar, W. Liu, X. Xie, A. Anselmo, S. Mitragotri, K. Banerjee, MoS_2 field-effect transistor for next-generation label-free biosensors, *ACS Nano* 8 (2014) 3992–4003.
9. R. Liang, L. Zhong, Y. Zhang, Y. Tang, M. Lai, T. Han, W. Wang, Y. Bao, Y. Ma, S. Gan, L. Niu, Directly using $Ti_3C_2T_x$ MXene for a solid-contact potentiometric pH sensor toward wearable sweat pH monitoring, *Membranes* 13 (2023) 376.
10. Z. Liu, V. Manikandan, A. Chen, Recent advances in nanomaterial-based electrochemical sensing of nitric oxide and nitrite for biomedical and food research, *Curr. Opin. Electroche.* 16 (2019) 127–133.

11. J. Yang, X. Xu, X. Mao, L. Jiang, X. Wang, An electrochemical sensor for determination of sulfite (SO_3^{2-}) in water based on molybdenum disulfide flakes/nafion modified electrode, *Int. J. Electrochem. Sci.* 15 (2020) 10304–10314.

12. J. Hwang, M. Islam, H. Choi, T. Ko, K. Rodriguez, H. Chung, Y. Jung, W. Lee, Improving electrochemical Pb^{2+} detection using a vertically aligned 2D MoS_2 nanofilm, *Anal. Chem.* 91 (2019) 11770–11777.

13. W. Zhou, S. Li, X. Xiao, S. Chen, J. Liu, X. Huang, Defect- and phase-engineering of Mn-mediated MoS_2 nanosheets for ultrahigh electrochemical sensing of heavy metal ions: chemical interaction-driven in situ catalytic redox reactions, *Chem. Commun.* 54 (2018) 9329–9332.

14. M. Ivanova, E. Grayfer, E. Plotnikova, L. Kibis, G. Darabdhara, P. Boruah, M. Das, V. Fedorov, Pt-decorated boron nitride nanosheets as artificial nanozyme for detection of dopamine, *ACS Appl. Mater. Interfaces* 11 (2019) 22102–22112.

15. Q. Lu, J. Deng, Y. Hou, H. Wang, H. Li, Y. Zhang, One-step electrochemical synthesis of ultrathin graphitic carbon nitride nanosheets and their application to the detection of uric acid, *Chem. Commun.* 51 (2015) 12251–12253.

16. J. Dewald, The charge distribution at the zinc oxide-electrolyte interface, *J. Phys. Chem. Solids* 14 (1960) 155–161.

17. H. Gerischer, Electrochemical behavior of semiconductors under illumination, *J. Electrochem. Soc.* 113 (1966) 1174.

18. A. Fujishima, K. Honda, Electrochemical photolysis of water at a semiconductor electrode, *Nature* 238 (1972) 37–38.

19. H. Gerischer, The impact of semiconductors on the concepts of electrochemistry, *Electrochim. Acta* 35 (1990) 1677–1699.

20. G. Wang, J. Xu, H. Chen, Progress in the studies of photoelectrochemical sensors, *Sci. China Ser. B* 52 (2009) 1789–1800.

21. W. Qian, S. Xu, X. Zhang, C. Li, W. Yang, C. Bowen, Y. Yang, Differences and similarities of photocatalysis and electrocatalysis in two-dimensional nanomaterials: strategies, traps, applications and challenges, *Nano-Micro Lett.* 13 (2021) 156.

22. H. Xu, J. Yan, X. She, L. Xu, J. Xia, Y. Xu, Y. Song, L. Huang, H. Li, Graphene-analogue carbon nitride: novel exfoliation synthesis and its application in photocatalysis and photoelectrochemical selective detection of trace amount of Cu^{2+}, *Nanoscale* 6 (2014) 1406–1415.

23. J. Yan, Z. Chen, H. Ji, Z. Liu, X. Wang, Y. Xu, X. She, L. Huang, L. Xu, H. Xu, H. Li, Construction of a 2D graphene-like MoS_2/C_3N_4 heterojunction with enhanced visible-light photocatalytic activity and photoelectrochemical activity, *Chem. Eur. J.* 22 (2016) 4764–4773.

24. X. Chen, R. Li, Y. Li, Y. Wang, F. Zhang, M. Zhang, Research of the multifunctional rGO/MoS_2 material in the sensing field: human breathing and Hg(II) pollution detection, *Mat. Sci. Semicon. Proc.* 138 (2022) 106268.

25. X. Hun, S. Wang, S. Wang, J. Zhao, X. Luo, A photoelectrochemical sensor for ultrasensitive dopamine detection based on single-layer $NanoMoS_2$ modified gold electrode, *Sensor. Actuator. B Chem.* 249 (2017) 83–89.

26. Y. Li, S. Mei, S. Liu, X. Hun, A photoelectrochemical sensing strategy based on single-layer MoS_2 modified electrode for methionine detection, *J. Pharmaceut. Biomed. Anal.* 165 (2019) 94–100.

27. Y. Liu, K. Yan, J. Zhang, Graphitic carbon nitride sensitized with CdS quantum dots for visible-light-driven photoelectrochemical sensing of tetracycline, *ACS Appl. Mater. Interfaces* 8 (2016) 28255–28264.

28. M. Jalali, R. S. Moakhar, T. Abdelfattah, E. Filine, S. S. Mahshid, S. Mahshid, Nanopattern-assisted direct growth of peony-like 3D MoS_2/Au composite for nonenzymatic photoelectrochemical sensing, *ACS Appl. Mater. Interfaces* 12 (2020) 7411–7422.

29. T. Seiyama, A. Kato, K. Fujiishi, M. Nagatani, A new detector for gaseous components using semiconductive thin films, *Anal. Chem.* 34 (1962) 1502–1513.

30. A. Dey, Semiconductor metal oxide gas sensors: a review, *Mater. Sci. Eng. B* 229 (2018) 206–217.

31. T. Li, W. Zeng, H. Long, Z. Wang, Nanosheet-assembled hierarchical SnO_2 nanostructures for efficient gas-sensing applications, *Sensor. Actuator. B Chem.* 231 (2016) 120–128.

32. L. Chen, L. Cai, J. Geng, C. Liu, Y. Wang, Z. Guo, Polyoxometalate-assisted in situ growth of ZnMoO$_4$ on ZnO nanofibers for the selective detection of ppb-level acetone, *Sensor. Actuator. B Chem.* 369 (2022) 132354.

33. Y. Kim, S. Lee, J. Song, K. Ko, W. Woo, S. Lee, M. Park, H. Lee, Z. Lee, H. Choi, W. H. Kim, J. Park, H. Kim, 2D transition metal dichalcogenide heterostructures for p- and n-type photovoltaic self-powered gas sensor, *Adv. Funct. Mater.* 30 (2020) 2003360.

34. Y. Kim, S. Kang, N. Oh, H. Lee, S. Lee, J. Park, H. Kim, Improved sensitivity in schottky contacted two-dimensional MoS$_2$ gas sensor, *ACS Appl. Mater. Interfaces* 11 (2019) 38902–38909.

35. X. Liu, T. Ma, N. Pinna, J. Zhang, Two-dimensional nanostructured materials for gas sensing, *Adv. Funct. Mater.* 27 (2017) 1702168.

36. D. Late, T. Doneux, M. Bougouma, Single-layer MoSe$_2$ based NH$_3$ gas sensor, *Appl. Phys. Lett.* 105 (2014) 233103.

37. E. Wu, Y. Xie, B. Yuan, H. Zhang, X. Hu, J. Liu, D. Zhang, Ultrasensitive and fully reversible NO$_2$ gas sensing based on p-Type MoTe$_2$ under ultraviolet illumination, *ACS Sens.* 3 (2018) 1719–1726.

38. Q. Liu, X. Li, Q. He, A. Khalil, D. Liu, T. Xiang, X. Wu, L. Song, Gram-scale aqueous synthesis of stable few-layered 1T-MoS$_2$: applications for visible-light-driven photocatalytic hydrogen evolution, *Small* 11 (2015) 5556–5564.

39. J. Wang, Y. Ren, H. Liu, Z. Li, X. Liu, Y. Deng, X. Fang, Ultrathin 2D NbWO$_6$ perovskite semiconductor based gas sensors with ultrahigh selectivity under low working temperature, *Adv. Mater.* 34 (2022) 2104958.

40. S. M. Kim, H. J. Kim, H. Jung, J. Park, T. Seok, Y. Choa, T. Park, S. Lee, High-performance, transparent thin film hydrogen gas sensor using 2D electron gas at interface of oxide thin film heterostructure grown by atomic layer deposition, *Adv. Funct. Mater.* 29 (2019) 1807760.

41. J. Pyeon, I. Baek, Y. Song, G. Kim, A. Cho, G. Lee, J. Han, T. Chung, C. Hwang, C. Kang, S. Kim, Highly sensitive flexible NO$_2$ sensor composed of vertically aligned 2D SnS$_2$ operating at room temperature, *J. Mater. Chem. C* 8 (2020) 11874–11881.

42. H. Huang, O. Tan, Y. Lee, T. Tran, M. Tse, X. Yao, Semiconductor gas sensor based on tin oxide nanorods prepared by plasma-enhanced chemical vapor deposition with post plasma treatment, *Appl. Phys. Lett.* 87 (2005) 163123.

43. B. Liu, L. Chen, G. Liu, A. Abbas, M. Fathi, C. Zhou, High-performance chemical sensing using schottky-contacted chemical vapor deposition grown monolayer MoS$_2$ transistors, *ACS Nano* 8 (2014) 5304–5314.

44. A. Agrawal, R. Kumar, S. Venkatesan, A. Zakhidov, G. Yang, J. Bao, M. Kumar, M. Kumar, Photoactivated mixed in-plane and edge-enriched p-type MoS$_2$ flake-based NO$_2$ sensor working at room temperature, *ACS Sens.* 3 (2018) 998–1004.

45. Y. Han, D. Huang, Y. Ma, G. He, J. Hu, J. Zhang, N. Hu, Y. Su, Z. Zhou, Y. Zhang, Z. Yang, Design of hetero-nanostructures on MoS$_2$ nanosheets to boost NO$_2$ room-temperature sensing, *ACS Appl. Mater. Interfaces* 10 (2018) 22640–22649.

46. A. Bag, N. Lee, Gas sensing with heterostructures based on two-dimensional nanostructured materials: a review, *J. Mater. Chem. C* 7 (2019) 13367–133683.

47. B. Cho, M. Hahm, M. Choi, J. Yoon, A. Kim, Y. Lee, S. Park, J. Kwon, C. Kim, M. Song, Y. Jeong, K. Nam, S. Lee, T. Yoo, C. Kang, B. Lee, H. Ko, P. Ajayan, D. Kim, Charge-transfer-based gas sensing using atomic-layer MoS$_2$, *Sci. Rep.* 5 (2015) 8052.

48. H. Tang, L. Sacco, S. Vollebregt, H. Ye, X. Fan, G. Zhang, Recent advances in 2D/nanostructured metal sulfide-based gas sensors: mechanisms, applications, and perspectives, *J. Mater. Chem. A* 8 (2020) 24943–24976.

49. Q. He, Z. Zeng, Z. Yin, H. Li, S. Wu, X. Huang, Fabrication of flexible MoS$_2$ thin-film transistor arrays for practical gas-sensing applications, *Small* 8 (2012) 2994–2999.

50. S. Dhall, B. Mehta, A. Tyagi, K. Sood, A review on environmental gas sensors: materials and technologies, *Sens. Int.* 2 (2021) 100116.

14

Telecommunications Devices Based on 2D Semiconducting Materials

Sunil Kumar Baburao Mane and Naghma Shaishta*

**Correspondence: nghmachem@gmail.com*

14.1 Introduction

The need for mixed data analysis particularly to the application operations at the organizational level is driven by the huge amounts of data such as artificial intelligence which define our daily existence. However, there are two significant issues from gadget to network. If transistor circuit elements keep on shrinking in the manner currently used, it is going to be more difficult to implement real-world changes to them as time passes because the short-channel consequences cannot be effectively managed. Systems in complicated contexts like the Internet of things and edge computing demand a variety of connections in particular contexts, boosting the need for systems that are integrated for computational tasks. However, the heterogeneous incorporation of the foundational components must adhere to strict constraints in operationally cohesive systems [1–4].

As a class of newly developed semiconductors, 2D semiconductors have novel qualities that indicate the potential to get around limitations in shrinking and application-specificity. First, it is advantageous to develop outstanding durability low-power electronic gadgets using 2D semiconductors with intact nanoscale architectures. Before the first monolayer MoS_2 transistors emerged, 2D semiconductors were unknown. These transistors showed extremely high on/off percentages, improved accessibility, and extremely low off-state currents. Second, the plentiful band design and the fundamentally effortless and versatile connection give the combination of various electrical activities a further level of independence, making it simple to develop more effective and specific uses through heterogeneous combination [5–8].

Because 2D semiconductors can carry out component architectures and functionalities that were formerly hard to achieve with other substances, various cutting-edge electronics, including ultrafast non-volatile flash, programmable logic, and logic-in-memory, were developed. In general, 2D semiconductors' rich physical properties might be exploited for certain electrical operations and show a lot of potential for all-in-one systems. A wide variety of disciplines are represented by 2D materials, which have undergone thorough reviews.

Because of their unusual optoelectronic characteristics as well as intriguing mechanical features resulting from their fundamentally thin dimensions, two-dimensional (2D) materials have been at the cutting edge of material studies recently. A novel and intriguing structure for layer-by-layer components and mixed gadget figuring has been made

DOI: 10.1201/9781003439448-14

possible by the segregation and production of a variety of fundamentally thin 2D materials. This structure allows for the discovery and customization of better or previously undiscovered characteristics and holds the potential for several novel innovations.

The investigations and uses that are the outcome of the natural restriction in out-of-plane movement in 2D materials are the topic of this unique attribute. As a result, these substances reveal themselves as a surface, setting them apart from other kinds of materials. Although numerous uses, such as electrical, optoelectronic, and photonic ones, remain in their beginnings, the traits resulting from this restriction require to be carefully tackled, and the special usage must be investigated. With an emphasis on a variety of subjects, this book chapter will address cutting-edge challenges, unsolved difficulties, and new insights into material characteristics and technological utilization of 2D materials, including electrical, telecommunication, and optical aspects.

14.2 Synthesis

A technique called chemical vapor deposition (CVD) is frequently used to create 2D semiconductor materials. 2D-heterojunctions can be created using CVD since it can give large-area, excellent, and properly managed layered development of 2D semiconductor materials. Mechanical exfoliation accompanied by transferring is frequently utilized for creating systems by assembling various 2D materials. Hydrothermal manufacturing, chemical exfoliation, thermal degradation, and electrochemical deposition are further potential formation processes. The preliminary colloidal method-based synthesis of cadmium selenide CdSe quasi 2D platelets with thicknesses of multiple atomic sheets and transverse widths up to dozens of nanometers occurred in 2008 [9]. Here, 2D materials include graphene, hexagonal boron nitride (h-BN), metal chalcogenides, phosphorene, and MXenes. **Figure 14.1** shows the traditional representative synthesis of 2D materials integration fabrication techniques for 2D materials.

14.3 Various 2D Materials with Their Application

14.3.1 Graphene

In 2004, Geim and Novoselov et al., who discovered graphene, a flat monolayer of carbon atoms organized in a 2D honeycomb lattice that has excellent thermal conductivity and transport of electrons, introduced the field [10]. One drawback of graphene is that it doesn't have a band gap, which makes it difficult to turn off field-effect transistors in digital technology. The importance of a 2D monolayer semiconductor is that it displays better piezoelectric coupling than bulk semiconductors that are more commonly used. This relationship might make implementations possible. Constructing nanoelectronics devices using graphene as an electrical conductor, h-BN as an electrical insulator, and a transition metal dichalcogenide as a semiconductor is one area of investigation that is being done.

Control and nonreciprocal passive elements, like isolators, circulators, switches, as well as power separators, are crucial in systems of communication. They perform a crucial

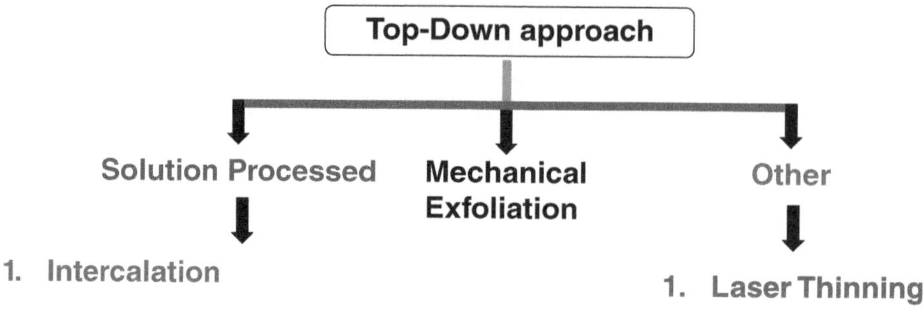

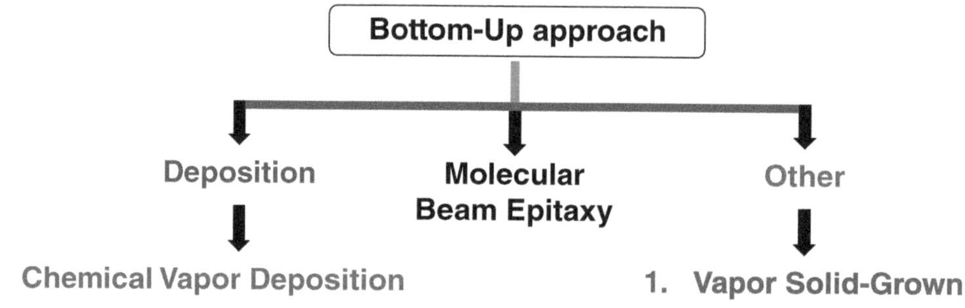

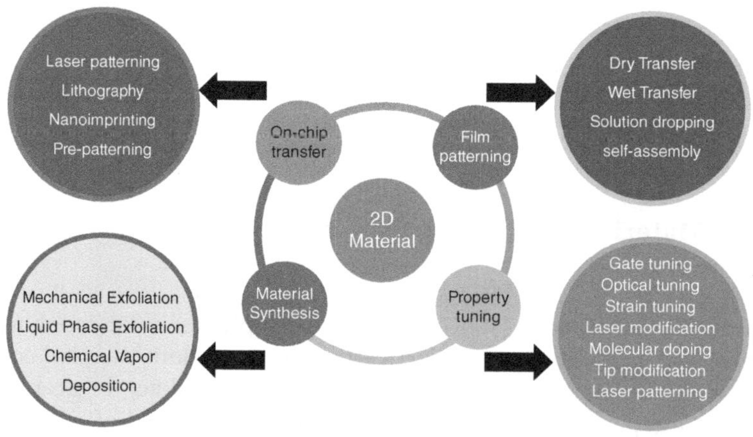

FIGURE 14.1
Conventional synthesis and integration techniques of 2D materials.

function in circuits that are integrated, either by shielding electromagnetic inputs from potential reflections from various circuit components or by transmitting electromagnetic signals to predetermined circuit components. These parts are typical ones used in microwave technology.

However, because the THz frequency zone is a complimentary area among electronics and photonics and currently lacks existing technological advances, it is regarded as the final unknown sector of non-ionizing radiations in the electromagnetic spectrum. Therefore, these parts with straightforward and tiny construction, good electrical properties, and adaptability in applicability are highly required for THz innovation as well.

An input signal is split among multiple outlets in power splitters. Although they typically offer a comparable ratio splitting, asymmetrical split connections are additionally employed in reality. For use in photonic integrated circuits, a wide range of power separators are currently being researched. There are various kinds of separators, such as those made from graphene and photonic crystals. These elements can be realized using magneto-optical resonators, linear waveguides, waveguide bends, Y-splitters, and other structures.

More specifically, two graphene-based power dividers with reciprocal and nonreciprocal features were proposed and evaluated. The magneto-optical leakage tensor and the graphene electrical conductivity tensor, accordingly, each define two materials with identical properties [11]. In contrast to those in metals, surface plasmon polaritons (SPP) in graphene have distinct characteristics, including a high modal gauge, significant field confinement in the THz location, moderate loss, ease of modification, or adaptability by electric and magnetic fields. For SPP-based optical nanodevice programs, graphene is an attractive material because of these characteristics. The well-defined promising application of graphene in flexible electronics such as organic light emitting diodes, organic solar cells, and transistors was illustrated in **Figure 14.2 [12]**.

In this regard, Nobre et al. reported the creation and research of an energy splitter using graphene that operates at two distinct THz domains [13]. A graphene ring resonator and eight graphene waveguides connected to the resonator make up the gadget. This arrangement of dielectric substrates contains the graphene layer. By dividing the signal input power and raising the number of transmission ports, the divider enters the biasing voltage (B.V.) phase when a bias voltage is applied to the graphene and the dielectric substrate setup. Alternatively, the equipment operates in the magneto-plasmon (M.P.) domain once the resonator is exposed to a DC magnetic field by dividing the input power similarly to how it would in the B.V. system, however, the transmitting and isolation outlets are distinct, resulting in origin safety. Utilizing dipole and quadrupole resonance topologies as an outcome of applied external electric and magnetic fields, correspondingly, the B.V. and M.P. phases operate. Surface plasmon-polariton waves that are traveling via the graphene waveguides are what excite these two phases. Both modes produce excellent results in delivery, isolation, and reflection and function as selector buttons.

The idea of a multi-input, multi-output (MIMO) antenna is examined by Yadav et al. in terahertz (THz) frequency bands [14]. All of the four sets of array elements for the two-port MIMO Yagi-Uda antenna that is being suggested are supplied in groups, with the corresponding ports retained on a semi-circular graphene circle. Using the vias to the graphene ring, the engineered elements are stimulated. The antenna's gain is 5.85 dB, and its directivity is 9.42 dBi, giving it a resonance frequency of 6.505 THz. In the operating passband, the isolation between the ports is kept to less than 15 dB. The higher-order TE61 mode's presence is confirmed by the suggested antenna performance. Additionally, the mean effective gain, envelop correlation coefficient, total active reflection coefficient, and diversity gain, which are all MIMO implementation variables, were also researched, and a fair level of concordance with the outcomes has been attained. The sensitivity of the small proportions and tailoring of the antenna performance are then analyzed using parametric analysis.

Using a graphene-graphene oxide hybrid waveguide, Ren et al. show how the phase reproduction of the star-16-quadrature-amplitude-modulation (star-16QAM) signal is

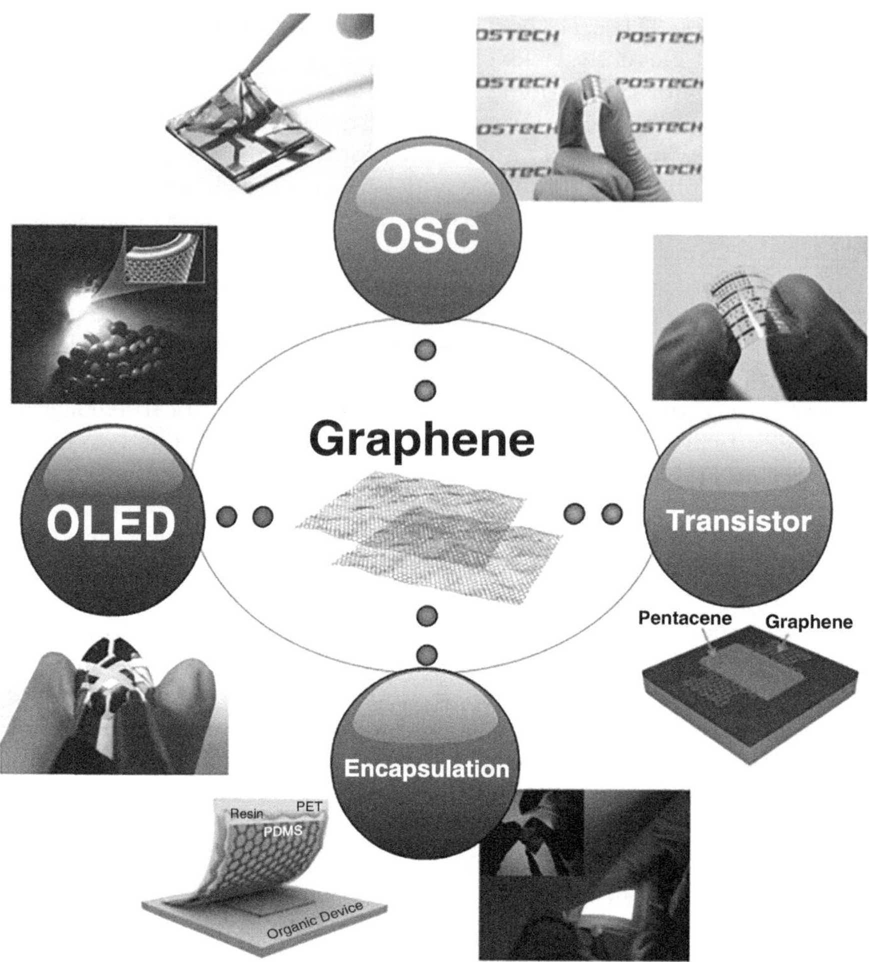

FIGURE 14.2
Application of graphene in the flexible electronics. Adapted with permission [12]. Copyright 2017, Elsevier Ltd.

accomplished [15]. The near-zero-dispersion point advances to the right with increasing applied electrochemistry, and graphene tunes the waveguide's multi-order dispersion features. By populating the waveguide with the high Kerr coefficient organic material graphene oxide, the waveguide's nonlinearity coefficient is 1.53×10^6/m/W and the dispersion is uniform in the ultra-wide wavelength range of 1310 to 1860 nm. Employing a dual-conjugate-pump degenerate four-wave mixing technique, the phase regeneration of the star-16QAM signal is accomplished, and the outcomes demonstrate that the error-vector magnitude is decreased from 39.25% to 2.14% and the phase noise is efficiently reduced.

14.3.2 Hexagonal Boron Nitride

The insulator with a high energy gap (5.97 eV) is called monolayer h-BN. However, because of its irregular outlines and openings, it may also be used as a semiconductor with improved conductivity. Because of its insulating ability, h-BN is frequently employed as a substrate and shield. Additionally, h-BN has a high thermal conductivity.

Promising applications for carbon nanodots (CNDs) in optoelectronic products include full-color displays and LEDs. In this regard, Wen et al. construct the CNDs/monolayer hexagonal boron nitride (ML-hBN)/sapphire heterostructures, where ML-hBN is appropriate for both CNDs and sapphire owing to the existence of the van der Waals (vdW) forces among ML-hBN and these materials and the CNDs may be altered chemically by affixing functionalities on the surface of the carbon cores of CNDs [16]. Characterization of the fundamental physical and chemical characteristics of CNDs and heterostructures **(Figure 14.3)**. The optical conductivity of CNDs/ML-hBN vdW heterostructures is also measured using terahertz time-domain spectroscopy, and the essential physical characteristics of these heterostructures are identified visually. These metrics' sensitivity to temperature is investigated. They discover that the existence of ML-hBN may significantly inhibit the clustering of dried CNDs on a sapphire base and that chemically altering the CNDs may regulate and change the electrical and optoelectronic features of the heterostructure. For the manufacture, insertion, and manipulation of CND-based vdW heterostructures for real-world gadget uses.

For use in medical diagnostics, piezoelectric nanomaterials (NMs) that can be remotely triggered by ultrasound are being explored. The association among the ultrasound and the piezoelectric intensity, as well as the statistical evaluation of piezoelectric effects in NMs, remains to be investigated. Mallick et al. showed how to make boron nitride nanoflakes using mechanochemical exfoliation, and they used the electrochemical approach to quantitatively assess how well the nanoflakes functioned as piezoelectric materials when exposed to ultrasonic waves [17]. In the electrochemical system, it was discovered how voltammetric charge, current, and voltage changed according to various acoustic pressures. The charge was increased by a net 49.54 C/mm^2 under 2.976 MPa to achieve a maximum of 69.29 C. Up to 597 pA/mm^2 of output current was observed, and the positive output voltage was adjusted from 600 to 450 mV. Furthermore, acoustic pressure led to a linear rise in piezoelectric efficiency.

14.3.3 MXenes

For the implementation of diverse functions such as battery backup, detectors, and photovoltaic cells, wire electrical conduction is a traditional approach for transferring components in electronic gadgets via the contact electrodes bridging the external circuit and internal operational unit. In wire equipment, the contact electrode, which is realized by metallic or semiconductive coatings, often requires an elevated conductivity. To effectively exchange carriers and reduce thermal energy loss, tiny Ohm contacts are typically needed at junctions among the contact electrode and the neighboring functional surface. To prevent the superfluous use of agitated transporters on their journey via the contacts of the equipment, the dose of functional material must also be kept to a minimum. All of these point to the need to enhance the wire electrical conduction technology to increase gadget efficiency [18, 19].

By stimulating components inside the functional system and relying on the penetrating energy field from distant outside influences, mobile electrical conduction is one of the most efficient ways to get over the contact electrode limitation. Antennas for Internet access, molecular vibrations, or energy harvesting may soak up the remote electromagnetic radiation energy. Radiant nuclear energy may be recognized and turned into electric energy using hybrid photovoltaic perovskites, and remote solar energy may be transformed into electrical energy. The atmospheric electric field (AEF) denotes enormous energy as a wireless energy diffused in the space environment. A single lightning flash in the atmosphere

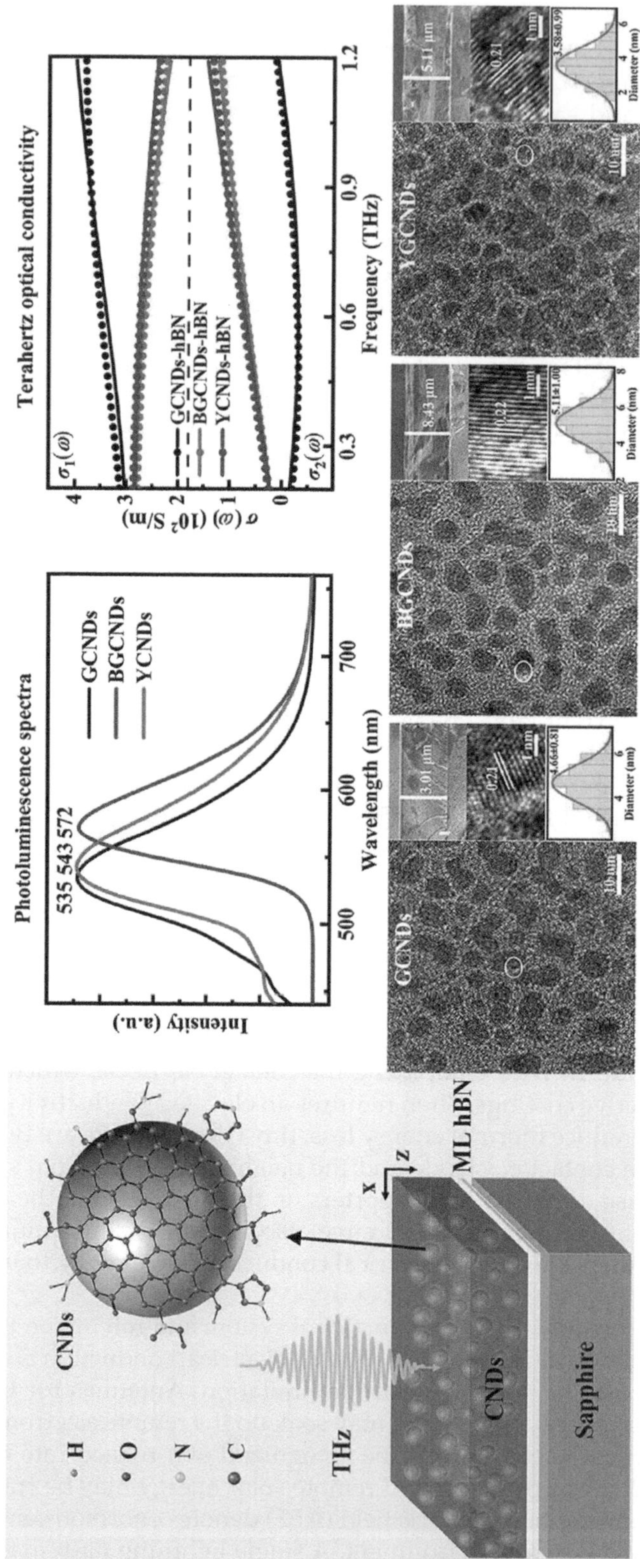

FIGURE 14.3
Graphical abstract representing the chemical alteration of h-BN for their possible THz optical characteristics. Adapted with permission [16]. Copyright 2023, Elsevier Ltd.

can cause the AEF to unleash hundreds of millions of joules of energy. Nevertheless, the unanticipated AEF is typically treated as contamination and protected by highly conductive materials like MXene and silver nanowires altered with ferroferric oxide [20, 21]. This AEF energy exhibits a significant output potential distinction that can be absorbed and used well.

Owing to the dispersive energy from numerous sources, electrostatic induction has been extensively utilized for collecting electric field energy in a variety of uses, including storing energy, the detection of atmospheric aerosol contamination, and tremors. With the use of electrostatic induction, the electret can catch electric field energy and convert it into electrical energy. Triboelectric nanogenerators are additionally able to gather mechanical energy. Electrostatic induction may also be employed to find electrical fields [22, 23]. The MXene family of innovative 2D materials has drawn a lot of interest because of their superior metallic electrical conductivity and hydrophilic surfaces with a wealth of functional groups. It has been used as an active compound in wire storage for energy, solar cells, detectors, tribology, and even electromagnetic interference screening due to its strong electrical conductivity. MXene was additionally used in an all-solid-state microsupercapacitor coupled with a triboelectric nanogenerator for distance sensing in AEF induction. In this section, it remains worthwhile to investigate the impact of energy storage in MXene-based materials on AEF induction as well as the use of vicinity sensing and AEF sensing in electrochemical systems [24].

In this context, Huang et al. explained that the conventional techniques for transferring carriers from external circuits into internal functional components in electronic equipment use wire electrical conduction; however, these visiting carriers always experience degradation along their journey pathways before adding to device efficiency [25]. To enable carrier use, a wireless electrostatic induction technique is devised for triggering native carriers only within MXene-based electrochemical supercapacitive devices. Findings from experiments and computer simulations show that native carriers are efficiently excited in porous MXene substance as well as in metallic foil substrate and ionic electrolyte in instrument areas influenced by AEF, providing a massive energy capacity of 541.6 F/g, which is significantly greater than the 258.5 F/g carried out by wire conductive charge. The inducing system is also investigated to sense wireless closeness trigger with a high sensitivity of 7.01 μA/m and high linearity of 97.9% in penetration depth ranging o to 20 cm, as well as to identify the AEF strength with the greatest sensitivity of 14.4 mA/v and high linearity of 95.2 %. For practical exploration, these intelligent features are also achieved with prototype apparatus. The MXenes serve as an intriguing connection to develop next-generation sensors by utilizing their use in modern chemistry. They possess excellent metallic conductivity, hydrophilicity, minimal diffusion obstacle, high mobility of ions characteristics, a substantial surface area, as well as simple integration. This article does a good job of highlighting the latest advances in the realm of MXenes and highlighting their crucial significance in analytical detection **(Figure 14.4)**. The final section has an emphasis on future prospects with inspired studies in the area of MXene-based detectors [26].

14.3.4 Metal Chalcogenides

A vast class of 2D materials known as metal chalcogenides naturally avoids the gapless property of graphene and makes it possible to absorb light and use it as a photodetector. Metal chalcogenides can be divided into transition metal and main group metal chalcogenides depending on the elements present; these categories will be briefly discussed to provide an outline of these materials. There are two subsets of transition metal chalcogenides

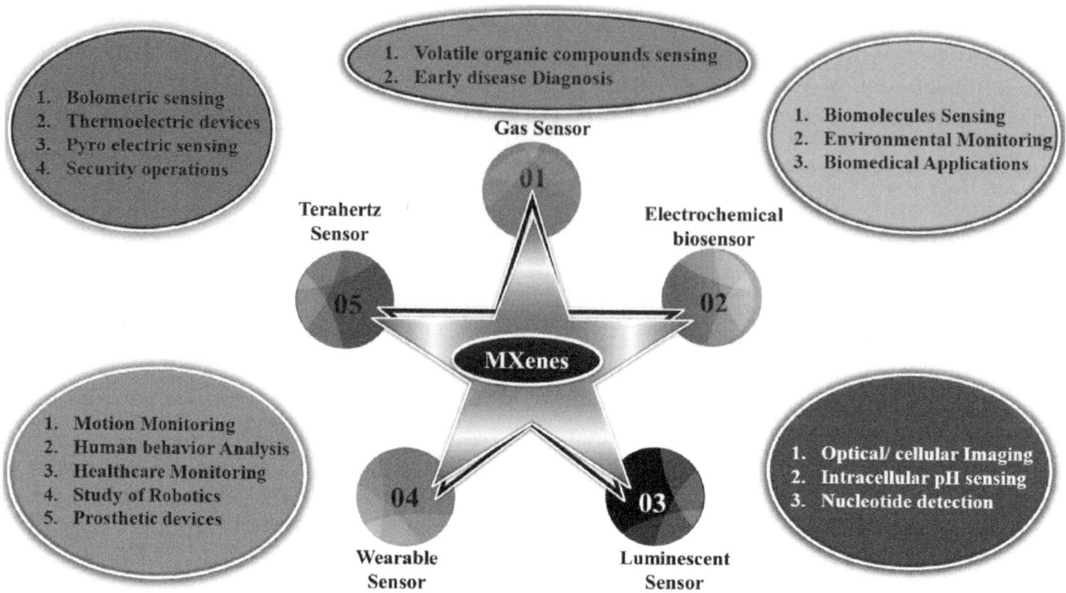

FIGURE 14.4
Application of mexene as the new technology for sensing and biosensing.

(TMCs): the renowned TMDs with a structure of MX_2 (M = Mo, W, as semiconductor; V, Nb, and Ta as metal) and the less studied transition metal trichalcogenides with the form of MX3 (M = Ti, Zr, Hf), where X denotes chalcogens (S, Se, Te). Following the finding of magnetism in its few-layer forms, these chalcogenides have drawn interest. They are joined by vdW forces; the TMCs have unusual optical, electrical, and catalytic characteristics.

For the creation of cutting-edge electrical and optoelectronic devices, it is crucial to increase our understanding of the characteristics of low-dimensional vdW materials, especially their responsiveness to environmental factors. A new family of low-dimensional materials with the potential for gate adaptability and photoreactivity are TMCs, which have adjustable optical band gaps and anisotropic conductivity. Field-effect transistor detectors for gases that are improved by light may be created using these characteristics.

Drozdowska et al. exhibited prototype nitrogen dioxide, ethanol, and acetone detectors made on zirconium trisulfide (ZrS_3). In photoinduced gas sensing, photoconductivity and photogating both play important roles [27]. The initial one predominates for blue (470 nm) and green (515 nm), while the second one does so for red (700 nm) irradiations. The findings imply that surface trap states cause the charge carriers to be trapped as well as scattered in the channel. Charge transfer and transport of carrier modulation control the gas detection, producing unique I-V characteristics under specific irradiation settings.

Voloshina et al. explained that Angle-resolved photoelectron spectroscopy is used to examine the electronic framework of the CoPS3, a broad bandgap vdW antiferromagnetic material that is much over the T_N transition temperature for this compound [28]. Spin-resolved band arrangement effectively explains the apparent dispersion of the electronic states in the valence band. This finding emphasizes the significance of local electron-electron correlations and shifts in explaining the electronic and magnetic characteristics and generates additional thoughts for research on the electronic composition of $CoPS_3$ and other comparable materials.

Naseer et al. explained that besides silicon-based supplementary metal-oxide semiconductor fabrication, transistors made of naturally occurring fundamentally thin 2-D semiconductors have demonstrated considerable promise for electronics [29]. They investigate a newly found 2-D family of group-11-chalcogen semiconductors M_2X (where M = Au and Ag and X = S, Se, Te) for transistor uses and evaluate the efficiency limit of transistors made of these semiconductors employing first-principles-based nanoscale gadget models. With gate lengths as small as 5 nm, M_2X transistor efficiency is positive and equivalent to that of potential 2-D transistors. The finest execution transistor is n-type, calm of Au_2S, with a large I_{ON}/I_{OFF} ratio and better electrostatic control for 5 nm gate length.

14.3.5 Phosphorene

Because of their special qualities and capacity to unveil intriguing novel events, 2D materials are the focus of substantial study endeavors. Phosphorene is a single layer of black phosphorus crystals that serves as a counterpart to graphene, and also it has a large surface-to-volume ratio. Due to its interesting properties, including strong optical absorption, high carrier accessibility, and numerous other alluring characteristics, phosphorene has earned a unique position within the group of 2D semiconducting materials with particular significance in electronic and optoelectronic uses. The outstanding mechanical, electrical, optical, and transportation capabilities of phosphorene are due to its anisotropic orthorhombic crystal form [30].

The present fossil fuel-dominated energy system must be transformed by utilizing cutting-edge methods for concurrently gathering universal, sustainable, and readily available solar energy using the photothermal phenomenon and effectively conserving the gathered thermal energy. The most hopeful strategy for achieving effective solar energy collecting and quick thermal energy preservation is the development of black phosphorene-based phase-change materials with optimum photothermal conversion rate as well as elevated latent heat. Cleaning superior black phosphorene nanosheets remains difficult, and there is still a need for an effective method for creating oriented black phosphorene structures that may optimize thermal conductivity increase.

Here, excellent black phosphorene nanosheets are created using an exfoliation technique that has been refined by Wang et al. [31]. The structure made of shipshape-compatible black phosphorene over a wide range is effectively constructed in the meantime by controlling the temperature gradient throughout freeze-casting, increasing the thermal conductivity of the poly(ethylene glycol) matrix up to 1.81 W/m/K at 20 vol% black phosphorene loading. The structure additionally provides the composite with a high latent heat of 103.91 J g-1 and outstanding phase-change material entrapment capacity. It is hoped that the research will develop the concept of juxtaposing architectures with tiny sheets to improve the controlled form of composites' thermal properties.

Rapid gene sequencing of proteins at individual molecule precision is still a formidable obstacle to overcome. At this point, Mittal et al. propose a remarkable biosensor that uses a comprehensible machine learning (ML) framework to swiftly recognize the essential characteristics of all 20 amino acids [32]. It is based on a solid-state 2D phosphorene nanoslit sensor. Using the XGBoost extrapolation approach, the transmission potential of each of the 20 amino acids can be accurately determined. The ML and density functional theory investigations that followed show that it is feasible to distinguish between specific amino acids with great accuracy and selectivity using information that transmits voltage and current signals. The phosphorene nanoslit technique may serve as a great choice for sequencing proteins with up to a 13-fold rise in transmission sensibility, according to a detailed comparison of findings to those from graphene nanoslit. The current study enables rapid screening of each of the 20 amino acids and may be expanded to additional biomolecules to determine the best course of treatment for various diseases.

14.3.6 2D-perovskite

When compared to their three-dimensional equivalents, efficient photodetectors (PDs) based on 2D perovskite single crystals (SCs) have been shown to be more stable. The sensitivity of 2D perovskite PDs is constrained in the ultraviolet and visible range because of their significant energy band gaps. In this regard, Xi et al. reported a thermally annealed Au nanoparticle (NP)-based broadband PD based on $(PEA)_2PbI_4$ (PEA = $C_6H_5(CH_2)_2NH^{3+}$) 2D perovskite SC is developed [33]. With the dark current remaining constant, the addition of Au NPs advances improvement in photocurrent throughout an extensive wavelength range.

Au NPs' improved induction of a localized electric field causes photocurrent amplification in the $(PEA)_2PbI4$'s light absorption spectrum. The photocurrent amplification coefficients typically exceed 1,000% beyond the $(PEA)2PbI4$ absorption range of 650–900 nm, which results from the hot holes created by Au NPs. It should be noted that the PD's reactive wavelength has been increased to include the telecommunication band; particularly, the plasmonic PD can transform light signals at 1,310 nm into reliable on-off electric signals. Such research encourages the creation of sensitive, broad-bandwidth perovskite photodetectors, offering an exciting approach to achieving photodetection in the telecommunication area.

Qiao et al. explained that for the forthcoming stage in developing photodetector with novel attributes, layered 2D perovskites offer more opportunities [34]. The outstanding dielectric efficiency of metal halide perovskite is hardly noticed in comparison to its superior optoelectronic features. This example shows a dual-purpose perovskite-based photovoltaic detector that can demultiplex two wavelengths. The extensive use of the photosensitive and dielectric characteristics of 2D perovskite in the black phosphorus/perovskite/MoS_2 organized photodetector enables the instrument to operate in several modes. Under 405 nm, the instrument exhibits a typical steady photoresponse, but at longer wavelengths of visible light, it exhibits a brief spike reply. Under 405 nm, the self-powered responsivity, rise/decay duration, and linear dynamic range can all exceed 100, 38 s/50 s, and 17.7 mA/W, correspondingly. The transient surge photocurrent with long wavelength illumination is shown to be linked to light intensity as well as interact with the typical photoresponse. It is possible to recognize two waveband-dependent signals and employ them to concurrently display additional information. To increase data transfer rates and secure interactions, this work presents a novel method for multispectral detection and demultiplexing. Additional multispectral photodetectors made of various layered 2D materials may be inspired by this working mode, particularly for optoelectronic applications of wide bandgap, strong dielectric photosensitive compounds. Several real-world uses associated with the advancement of photovoltaics depend on precise projections of the annual energy yield of solar energy systems beneath actual conditions of operation. Lipovsek et al. build and validate an illustration of energy yield for planar perovskite solar cells. The design incorporates the place and orientation of the instrument and relies on a mix of sophisticated optical modeling and in-depth optoelectrical evaluation [35]. Employing the framework, they first carry out a thorough sensitivity analysis and identify the crucial variables and their interactions that have the greatest impacts on the projected energy yield estimates. They demonstrate that the short-circuit current density's dependence on solar irradiance and the temperature-dependent relationships between the open-circuit voltage and fill factor rule the findings, while these relationships are jointly compensating.

Additionally, they examine how the use of external light control foils on the equipment affects its ability to produce energy. The research includes several surface characteristics, locations, and machine configurations. The findings demonstrate that even under realistic operating conditions, the general trends of the light operation effectiveness of textured foils are not significantly distinct from those noticed under conventional testing circumstances; the tetrahedral texture, which is considered to be ideal in all cases, yields the greatest energy yield gains, in the range of 7–11%.

14.4 Challenges

At the extent of the gadget, increasing the endurance—the number of instances a barrier may be switched requires additional research into how the underlying mechanism ages. To achieve massively connected device arrays that can mirror the hyper-connectivity and effectiveness of the brain, enhancing material homogeneity will be crucial. It is encouraging to note that several 2DMs have demonstrated memristive effects, and this collection is expected to expand in the years to come. To direct practical research and maximize the efficiency of memristive equipment, mathematical approaches are going to grow more and more important.

The massive chemical vapor deposition (CVD) development of phosphorene is hampered by issues such as a shortage of suitable platforms and intermediates. Due to phosphorene's strong reactivity with water as well as oxygen, creating phosphorene electronics presents a significant problem. As a result, it must be sufficiently safeguarded from deterioration by being sandwiched or encapsulated among various materials. The fundamental issue is that the efficiency of BP-based systems depends on the purity of the crystal lattice as well as the amount of sheets. Hazardous and impracticable solvents must be avoided; hence, the search for a novel exfoliation medium is necessary to develop eco-friendly solvents. Due to BP's rapid decomposition under atmospheric circumstances, studying it for the construction of devices grows increasingly difficult.

14.5 Conclusion

Upcoming mobile communications norms, such as 6G and higher, include stringent bandwidth and information speed demands that cannot be met by current innovations that have primarily relied on three-dimensional materials present in the environment. Networks for mobile interaction that cannot be implemented in the microwave and infrared bands may be implemented in the THz spectrum. The employing of graphene, transition metal dichalcogenides (TMDs), and perovskites has been encouraged as a result of the current revisiting of potential uses in the THz band of 2D materials. These have a chance to resolve certain long-standing problems regarding the invention of effective barriers for THz wave transmission and recognition. The research investigations of 2D materials with important electrical characteristics that may be employed in THz electronics to create efficient systems for potential mobile communications include graphene, TMDs, MXenes, and metal-organic frameworks. It was reported that the ball mill process was used to create BN NFs with an average length size of 0.1 to 0.2 m. The centrosymmetric BN MFs transitioned to non-centrosymmetric BN NFs, which suggests the existence of piezoelectric actions, according to the total microscopic and spectroscopic approaches.

Monolayer as well as multilayer 2D materials each exhibit distinctive physical characteristics in the mechanical, thermal, and electrical fields. The radio frequency (RF) front-end parts of transceivers and receivers, such as blenders, modulators, oscillators, toggles, and amplifiers needed for various signal modulators, are developed using these features. Because 2D materials possess substantial movement, high cut-off frequencies and large yields in analog and RF circuit construction are possible when they are used as active components in the channels of RF transistors. The construction of adaptable, compact, dependable patch-type RF monopole antennas with omnidirectional radiation for mobile telephones uses 2D materials, the upcoming evolution in intelligent materials for mobile communications. Due to its exceptional versatility, rigidity, electrical conductivity, and manufacturing comfort, MXenes, an unusual class of 2D materials, has already demonstrated great features across numerous wireless networks.

The electronic age of societies has demonstrated the enormous benefits of the 6G mobile communication system. By fusing aerial, shipping, and subterranean interactions, 6G makes wireless connectivity dependable, widespread, and nearly instantaneous. In addition, 5G and 6G interaction is becoming more and more necessary due to the advancement of modern technologies such as artificial intelligence, blockchain, millimeter wave interaction, non-orthogonal numerous availability, THz communication, and quantum interaction. A successful mobile messaging system requires being capable of meeting customer demands for increased capacity, speed, versatility, coverage, and overall excellence. The complexity of developing a 6G network is increasing along with the customer base and service region. Numerous investigations have recently applied ML to communication over wireless networks.

At the component stage, 2DMs outperform incumbent approaches in terms of efficiency. Furthermore, they claim to be simple to integrate with Si complementary metal-oxide semiconductor silicon architecture. They are therefore great options for greatly enhancing silicon chip performance. We are sure that 2DMs will play a bigger role in connected goods in the years to come. Achieving the necessary production capability levels, which may vary concerning the intended use, is the hurdle for 2D material-based heterogeneous circuits. Ultra-sharp smartphone photographs and an entirely novel category of incredibly environmentally friendly Internet of things detectors could be made possible by a new form of active pixel detector that makes use of a revolutionary 2D material.

References

1. H. Peng, C. Mao-Sheng, C. Wen-Qiang, Y. Jie. Developing MXenes from wireless communication to electromagnetic attenuation. *Nano-Micro Letters.* **2021**, 13, 115.
2. D. Yogita, H. Muruganandham, K. Manoj, J. Ankur, S. Debasish. Modified transition metal chalcogenides for high performance supercapacitors: Current trends and emerging opportunities. *Coordination Chemistry Reviews.* **2022**, 451, 214265.

3. L. Guohao, C. W. Brian, S. Fei, Y. Changqiang, W. Zhenjun, X. Xiuqiang, A. Babak, Z. Nan. 2D titanium carbide (MXene) based films: Expanding the frontier of functional film materials. *Advance Functional Material.* **2021**, 31, 2105043.

4. Y. Jean-Paul, N. N. Hassan, P. Benoit. The internet of modular robotic things: Issues, limitations, challenges, & solutions. *Internet of Things.* **2023**, 23, 100886.

5. H. Xiaohe, L. Chunsen, Z. Peng. 2D semiconductors for specific electronic applications: From device to system. *NPJ 2D Materials and Applications.* **2022**, 6, 51.

6. Y. Liu, D. Xidong, S. Hyeon-Jin, P. Seongjun, Y. Huang, D. Xiangfeng. Promises and prospects of two-dimensional transistors. *Nature,* **2021**, 591, 43–53.

7. L. Liu, L. Chunsen, J. Lilai, L. Jiayi, D. Yi, W. Shuiyuan, J. Yu-Gang, S. Ya-Bin, W. Jianlu, C. Shiyou, W. Z, David, Z. Peng. Ultrafast non-volatile flash memory based on van der Waals heterostructures. *Nature Nanotechnology.* **2021**, 16, 874–881.

8. M. M. Guilherme. Z. Yanfei, A. Avsar, W. Zhenyu, T. Mukesh R. Aleksandra, K. Andras. Logic-in-memory based on an atomically thin semiconductor. *Nature.* **2020**, 587, 72–77.

9. I. Sandrine, D. Benoit. Quasi 2D colloidal CdSe platelets with thicknesses controlled at the atomic level. *Journal of the American Chemical Society.* **2008**, 130, 16504–16505.

10. K. S. Novoselov. Electric field effect in atomically thin carbon films. *Science.* **2004**, 306, 666–669.

11. D. N. Elsheakh. Reconfigurable frequency and steerable beam of monopole antenna based on graphene pads. *International Journal of RF Microwave Computer-Aided Engineering.* **2020**, 30, e22156.

12. H. Tae-Hee, K. Hobeom, K. Sung-Joo, L. Tae-Woo. Graphene-based flexible electronic devices. *Materials Science and Engineering: R: Reports.* **2017**. 118, 1–43.

13. F. D. Nobre, S. D. Silva-Santos, M. W. B. da Silva, H. E. Hernandez-Figueroa, G. Melo, W. Castro. Graphene-based multifunctional signal divider in THz region. *Photonics and Nanostructures—Fundamentals and Applications.* **2023**, 54, 101115.

14. Y. Rajesh, G. Shailza, V. S. Pandey, K. Sandeep. Graphene based two-port MIMO yagi-uda antenna for THz applications. *Micro and Nanostructures.* **2023**, 181, 207616.

15. L. Ren, L. Xuefeng, W. Xiao, L. Hongjun. Phase regeneration of star-16QAM using phase-sensitive amplification in graphene silicon-based organic hybrid waveguide. *Optics Communications.* **2023**, 540, 129530.

16. H. Wen, B. Wang, X. Cheng, D. Song, H. Xiao, W. Xu, S. Lu. Fabrication and chemical modification of carbon nanodots/monolayer hexagonal boron nitride/substrate heterostructures and their terahertz optoelectronic properties. *Applied Surface Science.* **2023**, 629, 157441.

17. B. C. Mallick, H. Y. Chang, G. S. Chen. A new method for characterizing piezoelectric properties of hexagonal boron nitride nanoflakes activated by focused ultrasound. *Ultrasonics.* **2023**, 134, 107048.

18. G. Nafradi, E. Horvath, M. Kollar, A. Horvath, P. Andricevic, A. Sienkiewicz, L. Forro, B. Nafradi. Radiation detection and energy conversion in nuclear reactor environments by hybrid photovoltaic pervoskites. *Energy Conversion and Management.* **2020**, 205, 112423.

19. N. Wei, Q. Wang, Y. Ma, L. Ruan, W. Zeng, D. Liang, C. Xu, L. Huang, J. Zhao. Asymmetric supercapacitor for sensitive elastic-electrochemical stress sensor. *Journal of Power Sources.* **2018**, 402, 353–362.

20. N. Wei, Y. Li, Y. Tang, Y. Zhou, R. Ning, M. Tang, S. Lu, W Zeng, Y. Xiong. Resilient bismuthene-graphene architecture for multifunctional energy storage and wearable ionic-type capacitive pressure sensor device. *Journal of Colloid and Interface Science.* **2022**, 626, 23–34.

21. Y. Zhou, Q. Wang, X. Zhang, X. He, R. Ning, D. Rong, W. Zeng, N. Wei, Y. Xiong, S. Wang, T. Liao. Piezoionic transfer effect in topological borophene-bismuthene derivative micro-leaves for robust supercapacitive electronic skins. *Nano Energy.* **2022**, 104, 107970.

22. L. Huang, X. Zhou, R. Xue, P. Xu, S. Wang, C. Xu, W. Zeng, Y. Xiong, H. Sang, D. Liang. Low-temperature growing anatase TiO2/SnO2 multi-dimensional heterojunctions at MXene conductive network for high-efficient perovskite solar cells. *Nanomicro Letters.* **2020**, 31, 44.

23. H. J. Kim, S. Yoo, M. H. Chung, K. H. Yoo, J. Kim, H. Jeong. Power-free electrostatic collecting film development for purifying indoor air pollution. *Nano Energy.* **2019**. 65, 104034.

24. N. Wei, Y. Tang, Y. Li, Q. Wang, W. Zeng, D. Liang, R. Wu. Metal–organic frameworks-derived hollow octadecahedron nanocages for supercapacitors and wearable self-powered tactile stress sensor. *Applied Surface Science.* **2022**, 599, 153822.

25. L. Huang, Z. Dong, C. Wu, Y. Ma, L. Ruan, W Zeng, S. Wang, Y. Xiong, C. Xu, D. Liang. Wireless charging via atmospheric electrostatic induction in supercapacitive MXene: Huge energy-capture, sensitive electric-field- and proximity-sensing. *Electrochimica Acta.* **2023**, 460, 142597.
26. A. Sinha, Dhanjai, H. Zhao, Y. Huang, X. Lu, J. Chen, R. Jain. MXene: An emerging material for sensing and biosensing. *TrAC Trends in Analytical Chemistry.* **2018**, 105, 424–435.
27. K. Drozdowska, A. Rehman, S. Rumyantsev, M. Wurch, L. Bartels, A. Balandin, J. Smulko, G. Cywinski. Study of ZrS3-based field-effect transistors toward the understanding of the mechanisms of light-enhanced gas sensing by transition metal trichalcogenides. *Materials Today Communications.* **2023**, 34, 105379.
28. E. Voloshina, Y. Jin, Y. Dedkov. ARPES studies of the ground state electronic properties of the van der Waals transition metal trichalcogenide $CoPS_3$. *Chemical Physics Letters.* **2023**, 823, 140511.
29. A. Naseer, K. Nandan, A. Agarwal, S. Bhowmick, Y. S. Chauhan. Di-Metal chalcogenides: A new family of promising 2-D semiconductors for high-performance transistors. *IEEE Transactions on Electron Devices.* **2023**, 70, 2445–2452.
30. S. Derbali, O. Moudam. Insights into the application of 2D phosphorene in dye-sensitized solar cells. *Energy Technology.* **2023**, 11. https://doi.org/10.1002/ente.202300353.
31. Y. Wang, Y. Chen, W. Dai, Z. Zhang, X. Kong, M. Li, L. Li, P. Gong, H. Chen, X. Ruan, C. Jiao, T. Cai, W. Zhou, Z. Wang, K. Nishimura, C. T. Lin, N. Jiang, J. Yu. Anisotropic black phosphorene structural modulation for thermal storage and solar-thermal conversion. *Small.* **2023**, 19. https://doi.org/10.1002/smll.202303933.
32. S. Mittal, M. K. Jena, B. Pathak. Protein sequencing with artificial intelligence: Machine learning integrated phosphorene nanoslit. *Chemistry–A European Journal.* **2023**, 29. https://doi.org/10.1002/chem.202301667.
33. Y. Xi, X. Wang, T. Ji, G. Li, L. Shi, Y. Liu, W. Wang, J. Ma, S. Liu, Y. Hao, L. Xiao, Y. Cui. Plasmonic resonance enabling 2D perovskite single crystal to detect telecommunication light. *Advanced Optical Materials.* **2023**, 11, 2202423.
34. B. S. Qiao, S. Y. Wang, Z. H. Zhang, Z. D. Lian, Z. Y. Zheng, Z. P. Wei, L. Li, K. W. Ng, S. P. Wang, Z. B. Liu. Photosensitive dielectric 2D perovskite based photodetector for dual wavelength demultiplexing. *Advance Material.* **2023**, 35, 2300632.
35. B. Lipovsek, M. Jost, S. Tomsic, M. Topic. Energy yield of perovskite solar cells: Influence of location, orientation, and external light management. *Solar Energy Materials and Solar Cells.* **2022**, 234, 111421.

15

Chips Based on 2D Semiconducting Materials

Mamta Rani, Gurpreet Kaur, Manmeet Singh, and Daljit Kaur*
*Correspondence: daljitkhehra@gmail.com

15.1 Introduction to Semiconductor Chip Technology

Lee De Forest's discovery of the vacuum tube diode and triode in 1906 launched the modern electronic industry. In 1947, John Bardeen, William Shockley, and Walter Brattain at Bell Labs invented the transistor, which paved the way for solid-state devices. Although solid-state devices are smaller than vacuum tubes in size, full miniaturization could only be accomplished by Jack Kilby of Texas Instruments in 1959 by creating the integrated circuit (IC) by integrating the five components (transistors, diodes, and capacitors) on a Ge wafer. He created resistors using the inherent resistivity of Ge, but the devices were coupled by external wire. Robert Noyce created monolith IC by fabricating separate components on a single silicon wafer. Si is favored over Ge because it generates an oxide layer naturally. The size of ICs is smaller. Additionally, the distance that carriers must travel between components is cut down, increasing device speed, and lowering electrical losses. Feature size has decreased from a few tens of micrometers (μm) to nanometers (nm). As of 2021, 3 and 2 nm are under development, but further shrinking of device dimensions poses substantial technological obstacles. Chips, or ICs, can be classified into the following groups based on the density of integrated transistors, as shown in Table 15.1.

Chip manufacturing costs rise with device complexity. To counteract this cost, additional chips must be made. The industry must switch to the larger wafer size to achieve Moore's law cost reductions. Wafer diameters expanded from 50 mm in 1970 to 300 mm in 2000, and the industry is currently using 300 mm wafers, although the move to greater sizes has begun. Manufacturers including Samsung, Toshiba, and Micron employ 300 mm wafers for most chip fabrication. Doug Wong, senior member, of technical staff for Toshiba America Electronic Components, said Toshiba produces 100% on 300 mm and is considering switching to 450mm (18-inch) as shown in Figure 15.1. With 450 mm wafers, chipmakers will obtain 2.4 times more chips per wafer. When chipmakers switch to bigger wafers, production costs drop by 30% [1].

15.1.1 Primary Chips Manufactured by Semiconductor Companies

Classification of primary types of chips is either based on their functionality or IC employed. The four primary types of semiconductors are memory chips, microprocessors, standard chips like graphic process unit (GPU), and complex system-on-chips (SoCs), when categorized based on functionality. Chips can be classified as digital, analog, or mixed depending on the kind of IC they are used in memory chips, microprocessors, standard chips, and

DOI: 10.1201/9781003439448-15

TABLE 15.1

Transistor Density with Different Levels of Integration

Level	Device Density per Chip
Small-scale integration (SSI)	$<=50$
Medium-scale integration (MSI)	50–5K
Large-scale integration (LSI)	5–100K
Very-large-scale integration (VLSI)	10^5–10^6
Ultra-large-scale integration (ULSI)	10^6–10^8
Giga Scale integration (GLSI)	$>10^8$

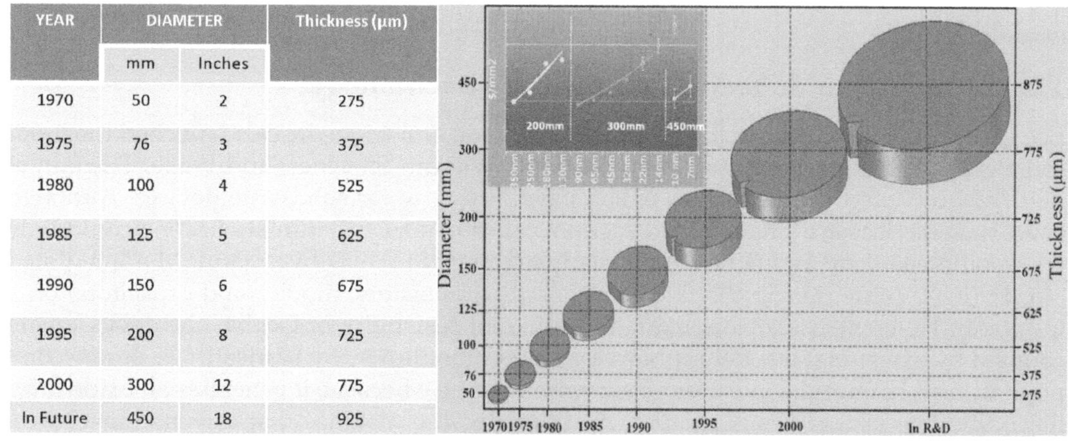

YEAR	DIAMETER		Thickness (µm)
	mm	Inches	
1970	50	2	275
1975	76	3	375
1980	100	4	525
1985	125	5	625
1990	150	6	675
1995	200	8	725
2000	300	12	775
In Future	450	18	925

FIGURE 15.1

Increasing trends of wafer size. Adapted with permission [2] Copyright the Authors, some rights reserved; exclusive licensee [Springer]. Distributed under a Creative Commons Attribution License 4.0 (CC BY) https://creativecommons.org/licenses/by/4.0/

complex SoC: a microcontroller chip, which typically combines the CPU with RAM, ROM, and input/output (I/O), has fewer capabilities than a SoC. The graphics, camera, audio, and video processing can all be integrated into a smartphone's SoC.

15.1.2 Silicon Wafer Manufacturing

There are various steps in silicon wafer manufacturing, which are described as follows:

(a) *Material manufacturing*: The starting material for Si wafer manufacture is called electronic-grade silicon (EGS). This is a single crystal ingot, which is shaped and cut into final wafers. EGS Should have impurity levels of the order of ppb with the desired doping levels so that it matches the chemical composition of the final wafers.

(b) *Conversion of sand to polycrystalline silicon*: To get the EGS, the starting material is metallurgical grade Si (MGS). The first step is the synthesis of MGS from Quarzite (SiO_2) or sand. Then the MGS is first converted to Polycrystalline grade silicon using Siemens Process. In this process, Si reacted with HCL gas to form trichlorosilane which is in gaseous form. A Si rod is used to nucleate the reduced Si obtained from silane gas.

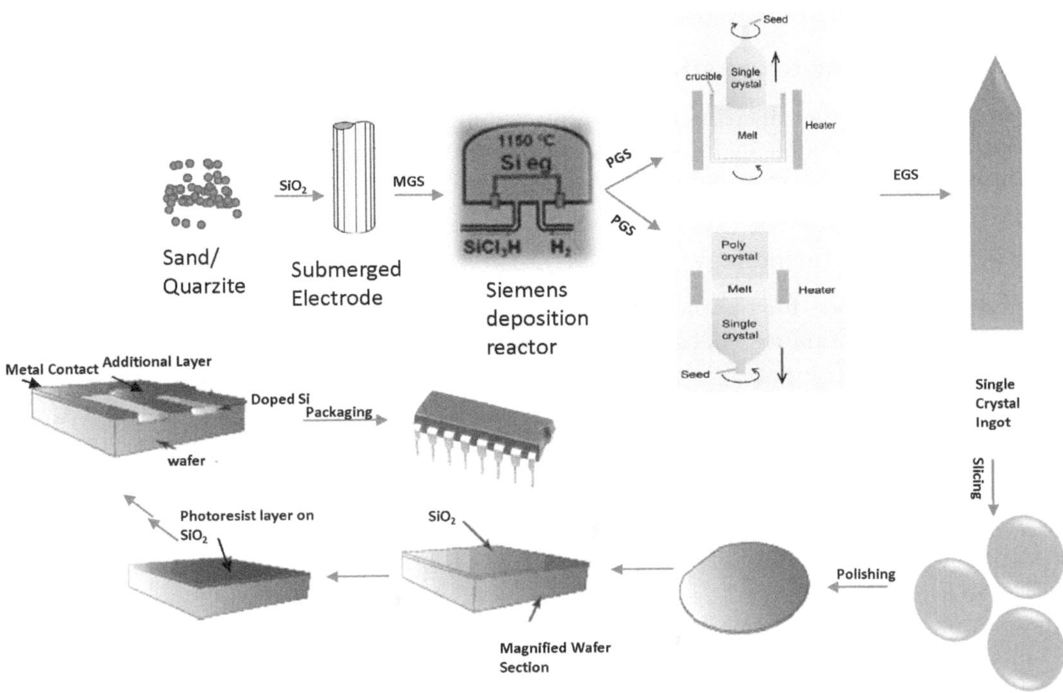

FIGURE 15.2
Various IC manufacturing steps starting from sand to packaging for final use.

(c) *Crystal growth by converting polycrystalline silicon to EGS*: There are two main techniques for converting Polycrystalline EGS into a single crystal ingot which is used to obtain wafers. This also involves the removal of impurities and doping of silicon if needed.

  (i) Czochralski technique (CZ)

  (ii) Float zone technique

(d) *Wafer manufacturing*: After a single crystal is obtained, this is further processed to produce wafers. Wafers are shaped and cut by using industrial-grade diamond-tipped saws. During shaping, the seed and tang ends of the ingot are removed and the surface of the ingot is grounded to get a uniform diameter across the length of the ingot. For further processing, resistivity and orientation are checked by using four probe techniques and X-ray diffraction, respectively. After this, one or more flats (primary flats and secondary flats) are ground along the length of the ingot. The primary flat, which acts as a visual reference to the orientation of the wafer, while the secondary flat is used for the identification of the wafer, dopant type, and orientation. After flats, before shipping, wafers are sliced to the required thickness, etched and polished, and subjected to final inspection [3].

(e) *IC fabrication steps*: Single crystal silicon wafer is the starting material for IC fabrication, and the product is a functioning chip as shown in Figure 15.2. The following are a set of intermediate processes used to create semiconductor devices and circuits: design and masking, wafer fabrication, testing and quality control, integration and packaging, assembly and final testing, soldering, and final quality control [4].

15.2 Moore's Law for Semiconductor Chip Technology

Moore's law states that the dimension of transistors and the integration density of transistors on ICs almost double every two years, postulated, and revised by American engineer, Gordon Moore.

15.2.1 Sub-10 nm Technology

The emerging trends for nanotechnology in electronics have explored the functionality of low dimensional nanostructures, owing to their unique structures and large surface-to-volume ratios. The sub-10 nm structures for the silicon (Si)-based IC manufacturing industry are contemplated by combining the most sophisticated lithography, etch, and film deposition processes. The specific surface area is much larger for sub-10 nm structures or nanoparticles compared to their bulk counterparts. In the modified architecture of field-effect transistors (FETs) from planar to fin FETs, the width of nano-fins has shrunken to 7 nm for the latest complementary metal-oxide semiconductor (CMOS) chips based on finfet technology. For FET density scaling, the pitch of Si-nano-fins is reduced from 60 to 34 nm to facilitate the latest chips with lower power consumption and higher performance. In 2002, Huang et al. reported the first lateral 3D GaN NW FETs, which demonstrated a D-mode (normally-on) FET based on a single GaN NW [5]. The improvised fabrication methods and advanced lithography techniques like extreme ultraviolet photolithography (EUL), block copolymer (BCP)-based directed self-assembly, electron-beam lithography, and focused ion beam helped in achieving nanostructured channels for the transistor.

15.2.2 Sub 5 nm Technologies

The attention toward advanced technology nodes with high speeds and low power as the critical parameters becomes important for applications like 5G mobile communications (in smartphones), artificial intelligence systems, high-performance computing, and advanced systems for autonomous drive (ADAS). The operation and manufacturing process for 5 nm technologies is claimed by a few semiconductor companies. But for the large-scale production of these chips along with the highest operating efficiency, manufacturing companies need to pay 100 million dollars for the design, fabrication, and packaging costs. This is the reason that many of semiconductor companies are now not leading-edge logic manufacturing companies. This number has come down from 20 in 2000 to only four in 2022 which are Intel, TSMC, Globalfoundries (GloFo), and Samsung. Recently IBM has sold their fabs to GloFo. Intel's 7 nm process, initially scheduled for 2017, was shifted to mid-2021, with a production ramp-up in late 2021 with the 12th CPU generation Alder Lake GPU. The first commercialized chip manufactured on a 5 nm technology node is the Apple A14 processor employed for next generation smartphones, and Qualcomm's Snapdragon SD875 embedding a 5G baseband chip combined with an Adreno GPU, for the smartphone market which is quite expensive [6].

The 5 nm technology is focused on the design rule compatible with the previous 7 nm, which is concerned with saving time and limiting the cost for re-design. The 5 nm technology enables more than 150 million transistors per millimeter square. Ten billion devices such as the 5G chipset of HiSilicon Kirin 990 would fit into an area of 8×8 mm^2. In comparison to the 7 nm process (N7), the 5 nm technology node boosts the power saving by 20–30%. For processing 5 nm technology, the design of MOSFETs or FETs is changed from planar to fin geometry known as FinFETs. The emerging follow-on device architectures after Fin-FET are gate-all-around (GAA) transistors as shown in Figure 15.4(a). The enhancement in logic density is achieved by employing contact over active gate and single diffusion breaks, as chips are power limited [7,8].

15.2.3 Sub 2 nm Technology

TSMC is expecting to start the production of ICs using its 2 nm technology by the end of 2024, about three years after the introduction of its 3 nm technology. The chips fabricated using TSMC's 2 nm technology will be introduced in consumer devices by around 2026 [6]. In the year 2021, Intel announced that its 20Å technology would use its own version of nanosheet FET (NSFET) named ribbon FET, which would be launched by 2024 [9]. The research center Interuniversity Microelectronics Centre (IMEC) has developed more advanced GAA forms, such as nanowire FET (lateral and vertical NWFET), nanosheet FET, and fork-sheet FET, as shown in Figure 15.4(b). The complementary FET (CFET) geometry is targeted for 2 nm and beyond. The CFET is a complex version of GAA, while

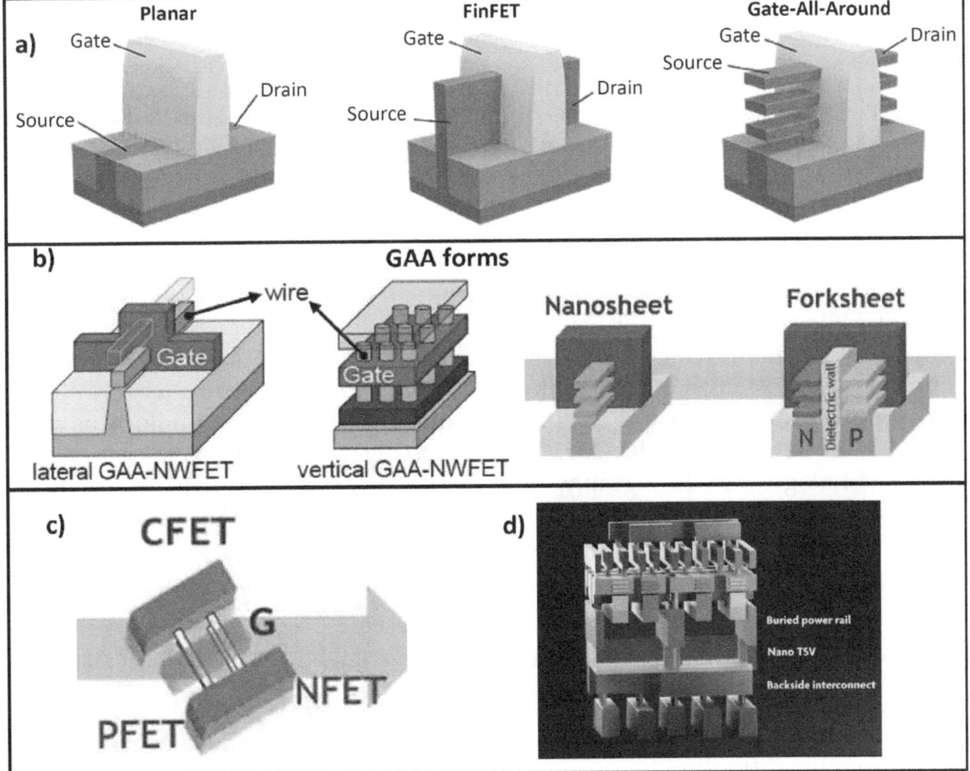

FIGURE 15.3
(a) Design of planar, Fin-FET, and gate-all-around (GAA) transistor geometries. (b) Different forms of gate-all-around (GAA) transistor geometries. (c) Complementary FET (CFET) and (d) buried power rail (BPR), nano through silicon vias (TSV), and backside interconnect schematic. Adapted from IMEC Knowledge Center (a) [10], (b) Technology Focus of Semiconductor Today [11], (c) and (d) Tomshardware Technology News Webpost [12]. Exclusive licensee (IMEC). Distributed under a Creative Commons Attribution License 4.0 (CC BY) https://creativecommons.org/licenses/by/4.0/

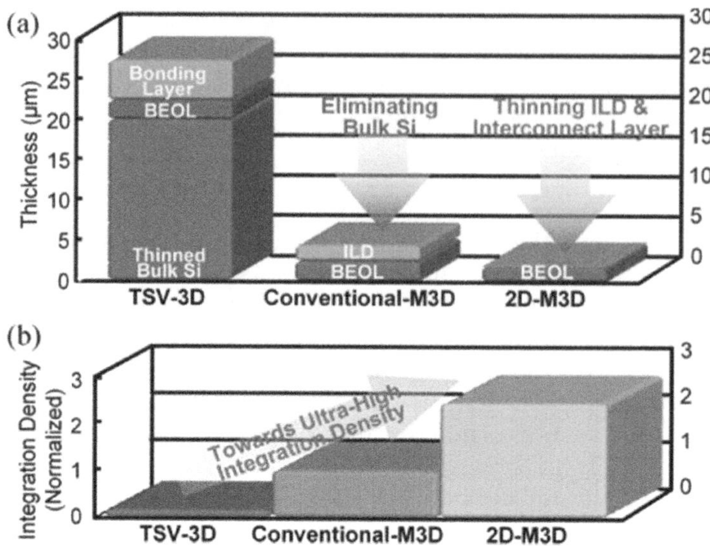

FIGURE 15.4

(a) Estimated tier thickness and (b) predicted integration density for TSV-based 3D integration (TSV-3D), conventional monolithic-3D (conventional-M3D), and proposed monolithic-3D integration with 2D materials (2D-M3D). The integration densities are normalized with respect to the integration density of conventional M3D. Adapted with permission [15], Copyright The Authors, some rights reserved; exclusive licensee (IEEE). Distributed under a Creative Commons Attribution License 4.0 (CC BY) https://creativecommons.org/licenses/by/4.0/

conventional GAAs can stack several p-type wires in one device and n-type wires on the other. The CFETs stack both n-FET and p-FETs together to reduce the area as shown in Figure 15.4 (c). The main characteristics of the technology node 2 nm are explored using the educational tool Microwind, by making the applicability of nanosheet (NS) FET with the buried power rails (BPR) and the through-silicon-vias (TSV) technology as shown in Figure 15.4(d). Extreme ultraviolet (EUV) lithography was expected to play a crucial role in sub-2 nm technology. EUV lithography uses very short wavelengths (10–30 nm) of light to etch extremely small features on semiconductor wafers, enabling the creation of sub-2 nm structures. The most sophisticated electron-beam lithography (EBL) systems can achieve 1 nm scale spot size and even down to the Angstrom scale.

15.2.4 Roadmap to 1 nm and sub 1 nm

The various architectures of FETs with channels made up of a new class of layered or 2D semiconducting materials (SCMs) mainly transition metal dichalcogenide (TMDC) MoS_2, are attracting attention because of unprecedented increases in integration density and enabling 1 nm or sub 1 nm technology. On the architectural level, research efforts in the direction of improving the structures of transistors in front-end-of-line, contacts in the middle-of-line, and wiring in the back-end-of-line (BEOL) are being carried out. IMEC shared its sub-"1 nm" process and transistor roadmap till 2036 with partnership from industry giants like Intel, ASML, TSMC, and Samsung at its Future Summit event in Antwerp, Belgium, in May 2022. The fourth-generation EUV tool with 0.33 nm numerical aperture (NA) will be replaced by the fifth-generation EUV tool with 0.55 nm NA that will enhance the transistor density from ~500 MTr/mm^2 (million transistors per square mm) up to ~1000 MTr/mm^2. They envisioned that silicon would be replaced by atom-thick new materials and 2D atomic channels as well as interconnects with robust electrical properties. As the GAA/nanosheet FET changes to atomic channel CFET, the nodes will reduce from 5 nm to sub 1 nm. The design of the standard chip was changed on three methodologies viz. system technology co-optimization, back power distribution network, and CMOS 2.0 with 3D System-on-chip design [12]. These methodologies enable the segregation of the power delivery network to the data transmission network, 3D stacking of memories and caches onto processing cores, and improving the wiring network at the BEOL.

15.3 Chips Based on 2D Semiconducting Materials

2D semiconducting materials (SCMs) are layered materials that can be grown as a one-atom thick layer like graphene. These 2D SCMs are flexible and transparent and have high mobilities at low power which make them promising candidates for high-performance ultrathin device heterostructures with increased integration of electronic devices onto a single wafer or chip. The main advantage of incorporating 2D materials onto a chip is increasing the device miniaturization with excellent electronic functioning to meet the future expectation of chip feature size to sub 1 nm using one-atom thick layer of 2D SCM in GAA or planar or vertical heterostructure stacking architectures. Wu et al. reported the sidewall MoS$_2$ transistor with the edge of graphene as a gate having a length of less than 1 nm [13], while Liu et al. reported MoS$_2$-based vertical transistors with MoS$_2$ channel thickness of 0.65 nm [14]. Jiang et al. analyzed that Moore's law can be extended well beyond the foreseeable roadmap of transistor scaling, and beyond-Moore heterogeneous integration can be realized by monolithic integration of 2D materials in 3D-ICs [15].

Atomically thin vertical dimensions of 2D semiconducting channel materials, graphene interconnects and carefully designed inter-tier electrostatics with graphene shielding for enhanced heat dissipation can lead to aggressive scaling of tier thickness from several μm (25–50 μm) to sub-μm. This can increase tenfold higher integration density with respect to through-silicon-via-based (TSV-based) 3D integration and >150% integration density improvement with respect to conventional (monolithic 3 dimensional) M3D integration (shown in Figure 15.5), with possibility for further improvements via incorporation of appropriate low-dielectric inter-layer dielectric materials.

15.3.1 Prospective 2D Semiconducting Materials for Chips Manufacturing

There are thousands of layered 2D materials which exhibit metallic, semiconducting, and insulating behavior. Out of all the 2D materials, the following 2D SCMs have been classified based on chemical constituents as following: elemental semiconductors, monochalcogenides, dichalcogenides, trichalcogenides, phosphides, iodides, arsenides, and oxides. Various synthesis methods have been explored to deposit 2D layered materials like physical vapor deposition, molecular beam epitaxy, atomic layer deposition, chemical vapor deposition (CVD), metal–organic chemical vapor deposition (MOCVD), suphurization/selenization, molten droplet method, chemical bath, hydrothermal, and post-etching/annealing.

The most successful synthesis method with good crystalline quality of monolayers is achieved by CVD or MOCVD. The bandgaps of most 2D SCMs lie in the range of 1–2.5 eV, while the bandgap of their bulk counterparts is indirect, which changed to a direct bandgap in monolayers. This change in the bandstructure and bandgap also affects its electrical properties like high on/off switching ratios, carrier mobilities, and electron or hole effective masses.

Out of these hundreds of 2D SCMs, the 12 mainly used 2D SCMs for transistor research can be classified based on their carrier types as follows:

(a) *N-type*: The n-type 2D semiconductors have electrons as the majority carriers. The n-type 2D semiconductors have been explored more than p-type SCs, as the n-type can be easily realized due to the strong electron doping from interfacial charge impurities and intrinsic structural defects. The common n-type 2D semiconductors are MoS_2, $MoTe_2$, InSe, SnS_2, Bi_2O_2Se, $PtTe_2$, and $SnSe_2$ [16]. The origin of n-type behavior is attributed to Fermi-level pinning near the conduction band and variation of Fermi-level energy over the surface due to structural defects such as sulfur vacancies, along with impurities (like rhenium and gold) and non-stoichiometric MoS_2 with a variable Mo/S ratio from ~1:1.8 to 1:2.3 [17,18].

(b) *P-type*: The p-type semiconductor has holes as the majority carriers. Black phosphorus (BP) also known as phospherene is a p-type 2D semiconductor because of its unique band structure (valence band having significant dispersion in the k-space compared to that of the conduction band), the presence of intrinsic defects like vacancies, interstitials, and grain boundaries, environmental sensitivity to oxygen and moisture which lead to the formation of electron-trapping surface states, thus reducing electron mobility and promoting hole conduction and small hole carrier effective mass (m^*_h = 0.15–0.20 m_0) of single- to few-layer BP. BP has shown the highest carrier (hole) mobility of 1000 $cm^2V^{-1}s^{-1}$ with a direct bandgap of 0.3 eV [19]. The other 2D reliable p-type 2D semiconductors with holes as majority carriers are still scarce and are yet to be explored experimentally. The potential candidates for p-type 2D semiconductors are germanene, selenene, tellurene, b-arsenene, SnS, GaSe, $MoTe_2$, $h\text{-}TiO_2$, $\beta\text{-}TeO_2$, Cu_2O, GeAs, GeP, AsP, $CsSnI_3$, and so on [20].

(c) Ambipolar: Ambipolar SCMs are those which possess unique carrier dynamics; that is, both electron and hole mobility can be tuned by gate control or electrostatic field. Ambipolar conduction in 2D materials can be changed to unipolar p-type/n-type conduction by affecting the behavior of charge carriers with the following different strategies: contact engineering, chemical doping, surface charge transfer doping, electrostatic gating, and dielectric engineering in transistors and electronic devices [21]. The commonly used ambipolar 2D SCMs are WSe_2, WS_2, $MoTe_2$, $MoSe_2$, $ReSe_2$, and so on. The indirect (in) and direct (dr) bandgaps of various 2D SCMs along with their semiconducting characteristics have been tabulated in **Table 15.2**.

Apart from 2D semiconducting materials, 2D graphene is generally used in place of metal contact/interconnect/metallic layer due to its zero bandgap and large momentum relaxation time. Hexagonal boron nitride (h-BN) has insulating electrical behavior (bandgap >5.0 eV) with low-dielectric constant and high thermal conductivity, which make it a potential 2D insulating dielectric candidate as front-end and back-end dielectric and passivation layer in 2D devices for enhancing their device performance.

TABLE 15.2

Characteristics of Some Semiconductors

Material	Bandgap (1L)/(2L) (eV)	Bandgap (Bulk 3D) (eV)
MoS_2	1.80 (1L)	1.29 (in), 1.78 (dr)
$MoSe_2$	1.57 (1L)	1.10 (in), 1.42 (dr)
WS_2	2.03 (1L)	1.32 (in),
WSe_2	1.67 (1L)	1.21 (in)
$SnSe_2$	1.69 (1L), 1.51 (2L), 1.37 (3L), 1.26 (4L)	1.07 (in), 1.84 (dr)
GaSe	3.50 (1 TL), 3.00 (2 TL), 2.30 (3 L)	2.1 (in)
SnS_2	2.41 (1L), 2.34 (2L), 2.29 (3L), 2.22 (4L)	2.18 (in), 2.61 (dr)
InSe	2.1 (1L), 1.9 (2L)	1.4 (in), 1.28–1.29 (dr)
$MoTe_2$	1.1 (1L, dr)	0.9 (in)
$ReSe_2$	1.34 (1L)	1.2 (in)
$PdSe_2$	1.3 (1L, in)	0
$PtSe_2$	1.8 (1L, in), 0.6 (2L)	0
PtS_2	1.68 (1L, in), (2L, in)	0.25 (in)
$PtTe_2$	0.4 (1L, in)	0
HfS_2	1.2	2.85 (in), 3.6 (dr)
$HfSe_2$	1.1 (20L)	1.13 (in)
Phosphorene	1.88	0.3
Silicene	0.1–0.4	0.3
Arsenene	1.5–1.7	0.3
Telluerene	0.79 (IL, dr)	0.325
AsP	2.52 (IL, in)	1.55 (dr)
$ZrSe_2$	0.439 (IL)	1.2 (dr)
GeS	2.34 eV (IL, in)	1.65 (in)

15.3.2 Electronic Devices and Functionalities of 2D SCMs

The integrated chips have thousands of components, which are mainly transistors and junction devices. There are a large number of reports for 2D NMOS and PMOS FETs, junction devices (metal–semiconductor and semiconductor–semiconductor (VdW heterostructure)), and inverters of which some reports with significant outcomes have been summarized in Table 15.3. However, the reports on the fabrication of complete IC architecture or microprocessors made up of 2D semiconductors are very few [22]. Watcher et al. have reported a microprocessor of 115 transistors made up of 2D MoS_2. The circuit is made of NMOS inverters where both pull-up and pull-down networks are built by n-type E-mode FETs. They have fabricated 18 devices per wafer, with FET channels made with bilayer MoS_2 films grown on large-area using CVD on a silicon wafer with 280-nm-thick silicon dioxide employing gate-first technology [23]. Two Ti/Au metal layers were used to interconnect the transistors and Al_2O_3 was used as gate oxide. The 1-bit implementation was successfully achieved with the device executing user-defined programs stored in external memory, performing logical operations, and communicating with the periphery. Wang et al. have fabricated ICs with two to 12 transistors placed side by side on a single sheet of bilayer MoS_2, which comprise an inverter, a NAND gate, a static random-access memory, and a five-stage ring oscillator based on a direct-coupled transistor logic technology. Both enhancement-mode and depletion-mode transistors were fabricated by using a gate of metals with different work functions [24]. Lin et al. have fabricated phase-pure, defect-free

TABLE 15.3

Various Devices and Their Characteristics

Device	Architecture	Device Characteristics	Ref.
Field-effect transistors (FETs)	MoS_2 FET on SiO_2/p^+–Si substrate	mobility of ~9 $cm^2\ V^{-1}\ s^{-1}$ from single domain and of 0.5–3.0 $cm^2\ V^{-1}\ s^{-1}$ from large-scale sheet on glass Among 16 MoS_2-based FETs, 13 devices successfully work (yield was more than 80%) Inverter (NOT gate) has a high-voltage gain of over 12 at a supply voltage of 5 V, with a 60 µs switching speed	Kwon et al. [26]
	MoS_2 FET with a 1 nm wide gate electrode made of a single-walled carbon nanotube	Subthreshold swing (SS) of ~65 mV/decade and large current ON/OFF ratio of ~10^6	Desai et al. [27]
	MoS_2 sidewall transistor made of atom-thin MoS2 channel and edge of graphene as gate electrode	On/off ratios up to 1.02×10^5 and subthreshold swing (SS) values down to $117\,mV\,dec^{-1}$	Wu et al. [13]
	Vertical FET with MoS_2 stacked in between source (metal) and drain (graphene) with 0.65 nm channel thickness	With on/off ratios of 26 for a channel length of $0.65\,nm$ and 10^3 for a length of $3.60\,nm$	Liu et al. [28]
	Y_2O_3 on MoS_2 as a buffer layer before HfO_2 dielectric for top-dated transistor	A low interface defect density down to $(2.3 \pm 0.8) \times 10^{12}\ cm^{-2}\ eV^{-1}$ is achieved and MoS_2 FETs exhibit a large on-state current (526 µA/µm)	Zou et al. [29]
	Few-layer $NiPS_3$ FET with gold contacts	n-type with on/off ratios of ~10^3–10^5 and mobility ~0.5–1 cm^2/Vs	Jenjeti et al. [30]
	Unipolar *n*-type $MoTe_2$ transistors with Ag contacts and AlOx encapsulation	Achieved highest saturation current (>400 µA/µm at 80 K and >200 µA/µm at 300 K) and relatively low contact resistance (1.2 to 2 kΩ·µm from 80 to 300 K)	Mleczko et al. [31]
Integrated inverter	Enhancement-mode FET made of large-scale CVD-grown single-layer MoS_2	Wide range of V_{dd} values, 2 to 6 V with a voltage gain of 45 at an operating voltage of 5V	Yu et al. [32]

TABLE 15.3 *(Continued)*

Various Devices and Their Characteristics

Device	Architecture	Device Characteristics	Ref.
Inverter	Large-scale WSe_2 flake with Pt as a contact for n-type FET and substitutional doping of potassium on contact regions for p-type FET	n-type and p-type FET with on/off current ratio $>10^4$ and complementary logic inverter with DC voltage gain > 12	Tosun et al. [33]
NMOS inverter	MoS_2 and graphene are severed as the transistor channel and contact electrode	Carrier mobility of 100 cm^2/Vs and a high on/off current ratio of $\sim 10^8$ has inverter gain $\sim$ 12 within an operating voltage of 3 V	Kim et al. [34]
NMOS logic inverter	CVD-grown ReS_2 film with graphene as electrodes	NMOS inverter in depletion mode operates at a supply voltage of 1 V with voltage gain > 3.5	Dathbun et al. [35]
PMOS inverter	2H $MoTe_2$ as channels and 1T $MoTe_2$ as contacts and interconnects	Inverter has complete rail-to-rail signal inversion with voltage gain >35 with an operating voltage of −6 V	Zhang et al. [36]
Complementary inverter	By transferring p-type WSe_2 and n-type MoS_2 FETs on the sapphire substrate, gated by ion gel	Reported the highest voltage gain of 110 as shown in **Figure 15.6**	Pu et al. [37]
CMOS inverter	By combination of CVD-grown MoS_2 n-type FET and a Si-NM p-type FET on a flexible plastic substrate Concurrent growth of both p-type WSe_2 and n-type $MoSe_2$ FETs on the same wafer	The logic inverters with a maximum DC voltage gain of 16 reproducible and high-voltage gain of 23	Das et al. [38] Chiu et al. [39]
CMOS inverter	Both n- and p-type MoS_2 FETs by directly growing MoS_2 film on different types of source/drain contact electrodes with fin-shaped FETs	Fin-FET has an ON/OFF ratio $>10^6$ with a low OFF current of a few pA; The high-voltage gain of an inverter of over 20	Lan et al. [40]
CMOS complementary logic inverter	p- and n-doped wafer-scale $PtSe_2$ film	A voltage gain of about 1 under the supply voltage of 3 V	Xu et al. [41]
Ionotronic memrister for 6-bit storage	CVD-grown monolayer WS_2 with Ag contacts as source and drain	Reverse hysteresis with memory windows larger than 25 V and extinction ratio greater than 10^6, stable retention, and endurance greater than 100 sweep cycles and 400 pulse cycles, in addition to 6-bit pulse-programmable memory	Mallik et al. [42]

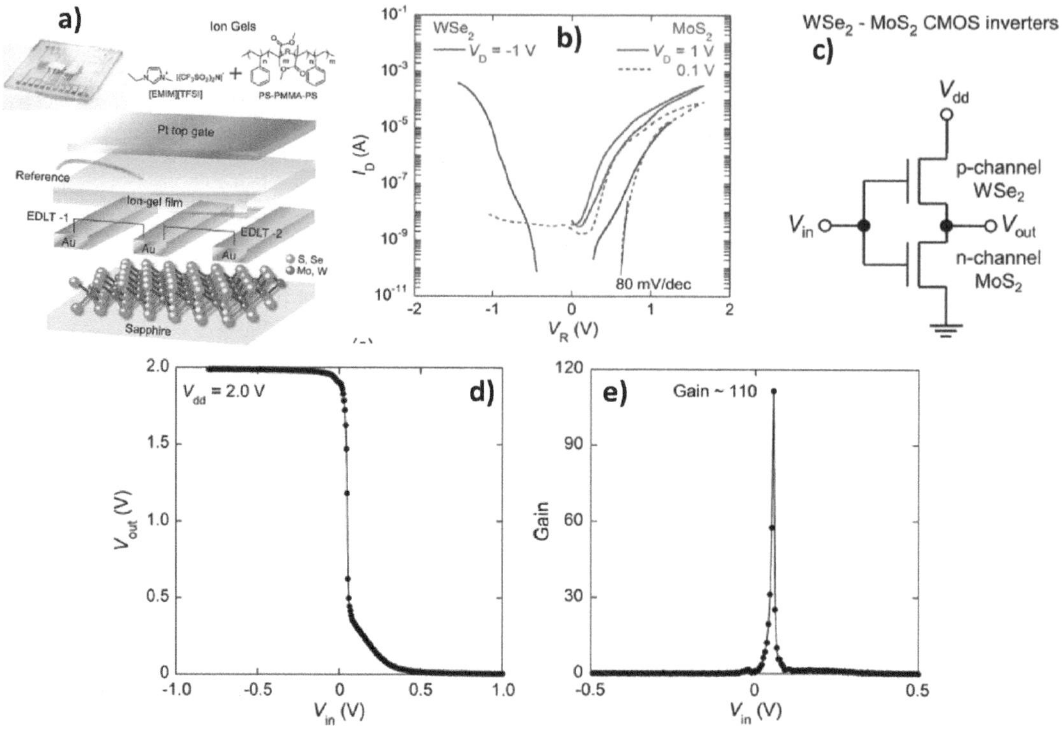

FIGURE 15.5

(a) High-gain monolayer WSe$_2$–MoS$_2$ CMOS inverters on a rigid sapphire substrate. (b) Transfer characteristics of the monolayer WSe$_2$ EDLTs (gray) and monolayer MoS$_2$ EDLTs (green). (c) The equivalent circuit diagram of a CMOS inverter with a monolayer WSe$_2$ p-channel transistor connected in series with a monolayer MoS$_2$ n-channel transistor. (d) V$_{in}$ – V$_{out}$ characteristics of a fabricated WSe$_2$–MoS$_2$ CMOS inverter at V$_{dd}$ of 2.0 V. (e) The voltage gain is plotted versus V$_{in}$. The maximum gain is above 110 with a low off-state voltage. Adapted with permission [37].

solution-processed wafer-scale MoS$_2$ thin films and then successfully created transistors, logic inverters, and integrated computational circuits like logic NAND, NOR, and XOR gates by integrating 3, 3, 5, and 11 thin-film transistors, respectively, from solution-processable 2D semiconductor thin films [25]. The complexity of integrated logic circuits requires the use of defect-free thin-film transistors of 2D SCMs.

15.3.3 Growth and Performance Challenges of 2D Semiconducting Chips

It can be concluded that 2D SCMs are the best candidates for future sub 1 nm chip technology as they have robust electrical properties at atomic scale or single monolayer with one atom thickness range. But there are some challenging issues which have been reported in recent years. These issues can be divided into two subcategories:

15.3.3.1 Challenges Related to the Growth as Well as Incorporation of 2D SCMs on Semiconductor Chips

The different challenges during the growth and incorporation of layered SCM monolayers in a heterostructure or wafer stack are as follows:

(i) *Scalability*: Wafer-scale synthesis methods for growing large-area high-quality 2D layered materials are required to be designed and tried for reproducible results.

(ii) *Low-temperature growth*: As concluded from the previous synthesis methods, the deposition of 2D SCMs requires high growth temperature (> 600°C) in facile CVD/MOCVD and other techniques. These temperatures cannot be applied in chip manufacturing steps as the Si wafer, other deposited materials, and device heterostructure degrade at a temperature higher than 450°C. So, generally, the transfer technique is employed to place a 2D monolayer or its heterostructure on the wafer.

(iii) Precise control on growth of 2D monolayers: there are many 2D SCMs whose precise growth of monolayer or few layers with uniformity on suitable substrates for the semiconductor industry has not been explored. or achieved due to lattice mismatch and sensitivity to environmental impurities like oxygen and moisture. This leads to the requirement of expensive ultra-high vacuum chambers and related equipment, whose laboratory small-scale research becomes unaffordable.

(iv) *Interface defects and lattice mismatch*: The adherence on the substrates and deposition of stable monolayers of 2D materials due to lattice mismatch with substrate orientation can be challenging. The interface defects can also increase during the stacking of metal–semiconductor layers in a transistor. The metallic contacts can be replaced by metallic 2D materials like graphene whose lattice mismatch is quite less comparable to other contact metals.

(v) Stability and environment sensitivity: These materials are prone to degrade with time due to exposure to moisture and air, hence proper encapsulation or passivation layer coating is essential to overcome these issues. The long-term stability of 2D SCM-based electronic devices is also a challenging issue as some of these materials like can undergo phase transitions with time.

15.3.3.2 Transistor or Electronic Device Performance-Based Challenges

(i) *Contact resistance*: The major issue which restricts the transistor performance made up of monolayer of MoS_2, WS_2, and WSe_2, is the large contact resistance due to the large Schottky barrier (SB) at the metal 2D semiconductor interface. The Schottky barrier height (SBH) at the metal–semiconductor (M–S) interface is determined by the metalwork function and electron affinity of the semiconductor. The band alignment is also affected by the interfacial defect states, which could induce Fermi-level pinning and dominate the SBH. Such a large SB is fundamentally limited by the gap-state pinning effects. There are two major types of gap-state pinning effects: defect-induced gap states and metal-induced gap states (MIGS) [43].

Although photodetector and rectifying M–S junction require Schottky contact and large SBH, FETs and other electronic circuits require low contact for improved device performance. The type of metal contacts and their deposition method also affect the contact resistance of the metals with a high work function, Φ (Au with $\Phi = 5.1$ eV, Pt and Pd) form low SB contacts with BP, while metals with a low work function (Al with $\Phi = 4.1$ eV) result in a large SB for holes. Graphene, a semimetal with a gate-tunable work function, is a promising alternative to traditional bulk metal electrodes for 2D semiconductors [20]. The transport property of the device can be tuned effectively by the graphene gate which has shown large gate tenability with small leakage currents and can also reach the Schottky-Mott limit. The insertion of an oxide tunneling layer (like Ta_2O_5) between MoS_2 and metal contact can also reduce the SB [44]. Recently it was found that ultralow contact resistance between Bi semimetal and 2D monolayer can overcome this issue. TSMC currently uses W (Tungsten) interconnects and Intel uses Co (cobalt) interconnects, which can now be replaced by Bi that can be easily deposited by helium ion beam lithography.

(ii) *Short-channel effect*: Due to ultrathin channel in one-atom thick 2D layer, the possibility of short-channel effect (SCE) in FETs can occur in which the channel length is comparable to depletion layer widths of source and junction leading to poorer device turn off behavior. To avoid the SCEs in 2D-SCMs-based transistors the high dielectric insulating gate layer thinner than 1 nm, that is, of a 2D layered insulator like h-BN is required.

(iii) *Thermal issues/heat dissipation*: The increase in transistor integration density and very small size compete against the thermal ceiling as there is a limit on how much heat can be dissipated by a given surface area. As, it gets very smaller, the ways to cool the chip or dissipate its heat are necessary. To overcome this issue, research in the cooling as well as packaging of chips has started and the use of microfluidic channels or material layers that increase the surface area for transference of heat is required. The 2D-SCMs-based transistors have a thermionic constraint of a subthreshold swing (SS) of 60mV decade^{-1} at room temperature. Hua et al. have put forth the Ag atomic threshold switching FET (ATS-FET) by inserting the Ag layer [45]. The other strategies used to obtain lower SS are band-to-band tunneling, impact ionization, nanoelectromechanical switching, Dirac source, negative capacitance, and negative differential resistance. Arrighi et al. addressed this issue by showing that a high thermal interface conductance G ~70 MW m^{-2} K^{-1} of MoS$_2$ with another layer which possess large thermal conductivity like hBN (450 Wm^{-1} K^{-1}) can help dissipate heat efficiently [46].

15.4 Conclusion

This book chapter has summarized the history, development, and current state-of-the-art technological advancement in the field of ICs/chips. The importance of 2D layered SCMs is requisite for the scaling down of the node process of chip technology to sub 1 nm by incorporating these materials. The prospective 2D SCMs, along with their bandgaps and growth techniques, has been summarized. The growth and performance challenges of 2D-SCMs-based transistors, logic, and ICs have been put forth, along with a review of literature having important outcomes like high-voltage gain and low SS values.

References

1. More Chip Production Moves to 300mm Wafers | DigiKey. (n.d.). www.digikey.in/en/articles/more-chip-production-moves-to-300mm-wafers.
2. P. Pal, V. Swarnalatha, A.V.N. Rao, A.K. Pandey, H. Tanaka, K. Sato, High speed silicon wet anisotropic etching for applications in bulk micromachining: a review, *Micro Nano Syst. Lett.* 2021 91. 9 (2021) 1–59.

3. S.M. Sze, *VLSI technology*, McGraw-Hill (1983). https://books.google.com/books/about/VLSI_Technology.html?id=IBpTAAAAMAAJ.

4. P. Swaminathan, *Semiconductor materials, devices and fabrication* (2016). www.amazon.in/Semiconductor-Materials-Fabrication-Parasuraman-Swaminathan-ebook/dp/B07BMMR96P.

5. Y. Huang, X. Duan, Y. Cui, C.M. Lieber, Gallium nitride nanowire nanodevices, *Nano Lett.* 2 (2002) 101–104.

6. The 2020 Mac Mini Unleashed: Putting Apple Silicon M1 to the Test. (n.d.). www.anandtech.com/show/16252/mac-mini-apple-m1-tested.

7. E. Sicard, L. Trojman, E. Sicard, L. Trojman, I. Finfet, E. Sicard, *Introducing 5-nm FinFET technology in microwind* (2021). To cite this version: HAL Id: hal-03254444. https://hal.science/hal-03254444/document.

8. E. Sicard, Introducing 14-nm FinFET technology in microwind. *Proc. 49th Des. Autom. Conf.—DAC* 12. 637371 (2017) 37. To cite this version: HAL Id: hal-0154117.

9. Intel, Annual Reports: Intel Corporation (INTC), *Intel reports* (2021). www.intc.com/filings-reports/annual-reports.

10. From FinFETs to Gate-All-Around (n.d.). https://semiengineering.com/from-finfets-to-gate-all-around.

11. Z. Tokei, Logic technology scaling options for 2nm and beyond. 16 (2021) 98–100. https://www.imec-int.com/en/articles/logic-technology-scaling-options-2nm-and-beyond

12. 4 Technologies That Sum Up IMEC'S Sub-1 nm Silicon Roadmap—EDN (n.d.). www.edn.com/4-technologies-that-sum-up-imecs-sub-1-nm-silicon-roadmap.

13. F. Wu, H. Tian, Y. Shen, Z. Hou, J. Ren, G. Gou, Y. Sun, Y. Yang, T.L. Ren, Vertical MoS2 transistors with sub-1-nm gate lengths, *Nature* 7900. 603 (2022) 259–264.

14. L. Liu, L. Kong, Q. Li, C. He, L. Ren, Q. Tao, X. Yang, J. Lin, B. Zhao, Z. Li, Y. Chen, W. Li, W. Song, Z. Lu, G. Li, S. Li, X. Duan, A. Pan, L. Liao, Y. Liu, Transferred van der Waals metal electrodes for sub-1-nm MoS2 vertical transistors, *Nat. Electron.* 2021 45. 4 (2021) 342–347.

15. J. Jiang, K. Parto, W. Cao, K. Banerjee, Ultimate monolithic-3D integration with 2D materials: Rationale, prospects, and challenges, *IEEE J. Electron Devices Soc.* 7 (2019) 878–887.

16. X. Huang, C. Liu, P. Zhou, 2D semiconductors for specific electronic applications: from device to system, *NPJ 2D Mater. Appl.* 6 (2022). https://doi.org/10.1038/s41699-022-00327-3.

17. M. Chhowalla, D. Jena, H. Zhang, Two-dimensional semiconductors for transistors, *Nat. Rev. Mater.* 1 (2016) 1–15.

18. S. McDonnell, R. Addou, C. Buie, R.M. Wallace, C.L. Hinkle, Defect-dominated doping and contact resistance in MoS2, *ACS Nano.* 8 (2014) 2880–2888. https://doi.org/10.1021/NN500044Q/SUPPL_FILE/NN500044Q_SI_001.PDF.

19. M. Akhtar, G. Anderson, R. Zhao, A. Alruqi, J.E. Mroczkowska, G. Sumanasekera, J.B. Jasinski, Recent advances in synthesis, properties, and applications of phosphorene, *NPJ 2D Mater. Appl.* 1 (2017) 1–12.

20. Y. Xiong, D. Xu, Y. Feng, G. Zhang, P. Lin, X. Chen, P-type 2D semiconductors for future electronics, *Adv. Mater.* 35 (2023) 2206939.

21. M.I. Beddiar, X. Zhang, B. Liu, Z. Zhang, Y. Zhang, Ambipolar-to-unipolar conversion in ultrathin 2D semiconductors, *Small Struct.* 3 (2022).

22. H. Tang, H. Zhang, X. Chen, Y. Wang, X. Zhang, P. Cai, W. Bao, Recent progress in devices and circuits based on wafer-scale transition metal dichalcogenides, *Sci. China Inf. Sci.* 62 (2019) 1–19.

23. S. Wachter, D.K. Polyushkin, O. Bethge, T. Mueller, A microprocessor based on a two-dimensional semiconductor, *Nat. Commun.* 8 (2017) 1–6.

24. H. Wang, L. Yu, Y.H. Lee, Y. Shi, A. Hsu, M.L. Chin, L.J. Li, M. Dubey, J. Kong, T. Palacios, Integrated circuits based on bilayer MoS 2 transistors, *Nano Lett.* 12 (2012) 4674–4680.

25. Z. Lin, Y. Liu, U. Halim, M. Ding, Y. Liu, Y. Wang, C. Jia, P. Chen, X. Duan, C. Wang, F. Song, M. Li, C. Wan, Y. Huang, X. Duan, Solution-processable 2D semiconductors for high-performance large-area electronics, *Nature* 562 (2018) 254–258.

26. H. Kwon, P.J. Jeon, J.S. Kim, T.Y. Kim, H. Yun, S.W. Lee, T. Lee, S. Im, Large scale MoS2 nanosheet logic circuits integrated by photolithography on glass, *2D Mater.* 3 (2016).

27. S.B. Desai, S.R. Madhvapathy, A.B. Sachid, J.P. Llinas, Q. Wang, G.H. Ahn, G. Pitner, M.J. Kim, J. Bokor, C. Hu, H.S.P. Wong, A. Javey, MoS2 transistors with 1-nanometer gate lengths, *Science (80-.)* 354 (2016) 99–102.

28. J. Zhang, F. Gao, P.A. Hu, A vertical transistor with a sub-1-nm channel, *Nat. Electron.* 2021 45. 4 (2021) 325.

29. X. Zou, J. Wang, C.H. Chiu, Y. Wu, X. Xiao, C. Jiang, W.W. Wu, L. Mai, T. Chen, J. Li, J.C. Ho, L. Liao, Interface engineering for high-performance top-gated MoS2 field-effect transistors, *Adv. Mater.* 26 (2014) 6255–6261.

30. R.N. Jenjeti, R. Kumar, M.P. Austeria, S. Sampath, Field effect transistor based on layered NiPS3, *Sci. Rep.* 8 (2018) 1–9.

31. M.J. Mleczko, A.C. Yu, C.M. Smyth, V. Chen, Y.C. Shin, Contact engineering high performance n-type MoTe 2 transistors, *Nano Lett.* 19. 9 (2019) 6352–6362.

32. L. Yu, D. El-Damak, S. Ha, X. Ling, Y. Lin, A. Zubair, Y. Zhang, Y.H. Lee, J. Kong, A. Chandrakasan, T. Palacios, Enhancement-mode single-layer CVD MoS2 FET technology for digital electronics, *Tech. Dig.—Int. Electron Devices Meet. IEDM.* 2016 February (2015) 32.3.1–32.3.4.

33. M. Tosun, S. Chuang, H. Fang, A.B. Sachid, M. Hettick, Y. Lin, Y. Zeng, A. Javey, High-gain inverters based on WSe 2 complementary field-effect transistors, *ACS Nano.* 8. 5 (2014) 4948–4953.

34. T. Kim, S. Fan, S. Lee, M.K. Joo, Y.H. Lee, High-mobility junction field-effect transistor via graphene/MoS2 heterointerface, *Sci. Reports* 2020 101. 10 (2020) 1–8.

35. A. Dathbun, Y. Kim, S. Kim, Y. Yoo, M.S. Kang, C. Lee, J.H. Cho, Large-area CVD-grown sub-2 V ReS2 transistors and logic gates, *Nano Lett.* 17 (2017) 2999–3005.

36. Q. Zhang, X.F. Wang, S.H. Shen, Q. Lu, X. Liu, H. Li, J. Zheng, C.P. Yu, X. Zhong, L. Gu, T.L. Ren, L. Jiao, Simultaneous synthesis and integration of two-dimensional electronic components, *Nat. Electron.* 2019 24. 2 (2019) 164–170.

37. J. Pu, K. Funahashi, C.H. Chen, M.Y. Li, L.J. Li, T. Takenobu, Highly flexible and high-performance complementary inverters of large-area transition metal dichalcogenide monolayers, *Adv. Mater.* 28 (2016) 4111–4119.

38. T. Das, X. Chen, H. Jang, I.K. Oh, H. Kim, J.H. Ahn, Highly flexible hybrid CMOS inverter based on Si nanomembrane and molybdenum disulfide, *Small.* 12 (2016) 5720–5727.

39. M.H. Chiu, H.L. Tang, C.C. Tseng, Y. Han, A. Aljarb, J.K. Huang, Y. Wan, J.H. Fu, X. Zhang, W.H. Chang, D.A. Muller, T. Takenobu, V. Tung, L.J. Li, Metal-guided selective growth of 2D materials: demonstration of a bottom-up CMOS inverter, *Adv. Mater.* 31 (2019) 1900861.

40. Y.W. Lan, P.C. Chen, Y.Y. Lin, M.Y. Li, L.J. Li, Y.L. Tu, F.L. Yang, M.C. Chen, K.S. Li, Scalable fabrication of a complementary logic inverter based on MoS2 fin-shaped field effect transistors, *Nanoscale Horizons* 4 (2019) 683–688.

41. H. Xu, H. Zhang, Y. Liu, S. Zhang, Y. Sun, Z. Guo, Y. Sheng, X. Wang, C. Luo, X. Wu, J. Wang, W. Hu, Z. Xu, Q. Sun, P. Zhou, J. Shi, Z. Sun, D.W. Zhang, W. Bao, Controlled doping of wafer-scale PtSe2 films for device application, *Adv. Funct. Mater.* 29 (2019) 1805614.

42. S.K. Mallik, R. Padhan, M.C. Sahu, G.K. Pradhan, P.K. Sahoo, S.P. Dash, S. Sahoo, Ionotronic WS2 memtransistors for 6-bit storage and neuromorphic adaptation at high temperature, *NPJ 2D Mater. Appl.* 7 (2023) 1–12.

43. Y. Lin, P.C. Shen, C. Su, A.S. Chou, T. Wu, C.C. Cheng, J.H. Park, M.H. Chiu, A.Y. Lu, H.L. Tang, M.M. Tavakoli, G. Pitner, X. Ji, C. McGahan, X. Wang, Z. Cai, N. Mao, J. Wang, Y. Wang, W. Tisdale, X. Ling, K.E. Aidala, V. Tung, J. Li, A. Zettl, C.I. Wu, J. Guo, H. Wang, J. Bokor, T. Palacios, L.J. Li, J. Kong, Contact engineering for high-performance N-type 2D semiconductor transistors, *Tech. Dig.—Int. Electron Devices Meet. IEDM.* 2021 December (2021) 37.2.1–37.2.4.

44. S. Lee, A. Tang, S. Aloni, H.S. Philip Wong, Statistical study on the Schottky barrier reduction of tunneling contacts to CVD synthesized MoS2, *Nano Lett.* 16 (2016) 276–281.

45. Q. Hua, G. Gao, C. Jiang, J. Yu, J. Sun, T. Zhang, B. Gao, W. Cheng, R. Liang, H. Qian, W. Hu, Q. Sun, Z.L. Wang, H. Wu, Atomic threshold-switching enabled MoS2 transistors towards ultralow-power electronics, *Nat. Commun.* 2020 111. 11 (2020) 1–10.

46. A. Arrighi, E. del Corro, D.N. Urrios, M. V Costache, J.F. Sierra, K. Watanabe, T. Taniguchi, J.A. Garrido, S.O. Valenzuela, C.M.S. Torres, M. Sledzinska, Heat dissipation in few-layer MoS2 and MoS2/hBN heterostructure, *2D Mater.* 9 (2021) 15005.

16

2D Semiconductors for Electrochemical Energy Applications

Yash Desai, Anuj Kumar, and Ram K. Gupta*

Corresponding author: ramguptamsu@gmail.com (Ram K. Gupta)

16.1 Introduction

Global energy demands have been escalating enormously, making it crucial to search for novel materials for energy applications, wherefore sustainable energy systems development is of great importance. Population growth, per capita consumption, economic development, and supply in remote locations has resulted in an exponential and continual rise in energy prices [1]. Rechargeable batteries, supercapacitors, solar cells, fuel cells, and photovoltaic cells are examples of sustainable energy sources and storage systems that benefit from modern material investigation and creative technical solutions. One of the core and most prospering themes in today's scientific activity is the discovery and creation of unique materials that are useful in various energy devices. Energy systems are growing exponentially because of alternate energy resources, interest in technical developments, demand, prices, and environmental effects. Traditional energy generation is based on fossil fuels; however, this has been steadily replaced by contemporary creative technology, with an emphasis on renewable resources such as solar and wind. While most of the current development is centered on improving the basic science and engineering of materials for sustainable energy is crucial to addressing the present challenges.

Even though many self-sustaining concepts stand longer, they demand the development of high-performance compounds designed to resolve certain challenges, such as dependency on highly expensive and scarce noble metals and deprived long-term stability. Numerous investigations have been conducted by researchers to discover different materials that may be used in revolutionary energy gadgets but to get the best possible capability of the energy devices, it becomes crucial to incorporate nanomaterials into the synthesis. Materials whose basic unit is composed of nanoscale and one dimension is at least three-dimensional. These materials have been popularly studied by studied while in the field of energy storage due to the structural, mechanical, and electrochemical properties they offer. For example, 2D nanomaterial graphene can be modified architecturally such as phase transformation, intercalation, metal hybridization, and hierarchical structuration whose schematics are represented in **Figure 16.1** [2]. These structural modifications are made to enhance the electrochemical properties of the electrode materials during the construction of supercapacitors. As discussed, 2D semiconductors can be modified structurally and used in several energy applications.

DOI: 10.1201/9781003439448-16

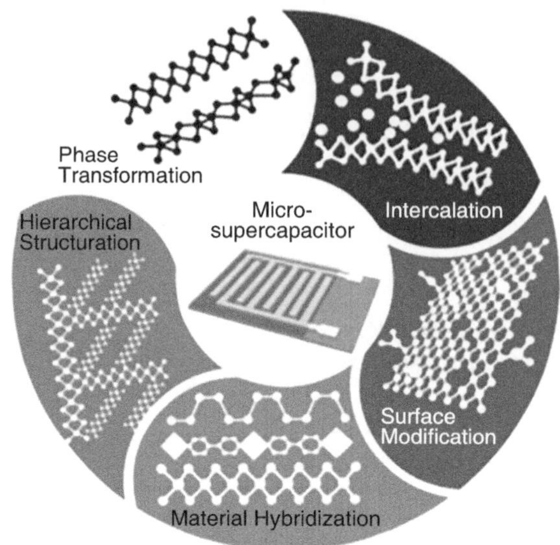

FIGURE 16.1
Engineering 2D architectural modifications of nano-graphene structure to construct high-performance micro-supercapacitors. Reprinted with permission [2]. Copyright 2018, John Wiley and Sons.

2D graphene has been used as a precursor and substrate for electrode materials in several electrochemical devices such as batteries, supercapacitors, fuel cells, and even solar cells which are discussed in further parts of the chapter [3]. Different 2D materials have different behavioral patterns of conductions upon which their applications are classified. MXene-based materials have been extensively used as part of supercapacitors and batteries in the form of electrode materials [4].

Accompanied by MXene are MOFs that have been added to electrode materials to manufacture high-quality supercapacitors. On the other side, fuel cells have different issues with their performance such as liquid electrolyte hazard; therefore, 2D semiconductors have been looked into to synthesize solid electrolytes to construct solid-state fuel cells [5]. These electrolytes also have lower operating temperatures compared to conventional liquid electrolyte-based fuel cells. One such example of synthesis was studied through synthesizing $LSTCrCeO_3$-based perovskite electrolyte materials which works through the generation of surface defects and surface O-vacancies at the junction which increases the ionic conductivity. The respective electrolyte material was synthesized through sol–gel technology as discussed in **Figure 16.2** [5]. Perovskite 2D materials have also been used in solar cells to gain high conversion efficiency, which indirectly promotes the usage of these solar cells as an alternative to conventional electrochemical devices [6]. Heterostructures are also used to fabricate several components such as anode, cathode, or electrolyte materials, and insulators are doped to create impurities that can act as semiconducting materials. Such 2D semiconducting materials have been successfully implemented in high efficiency offering fuel cells and supercapacitors [7]. Synthesizing such materials will be beneficial in manufacturing high-quality energy storage devices and the same will be discussed in further parts of the chapter.

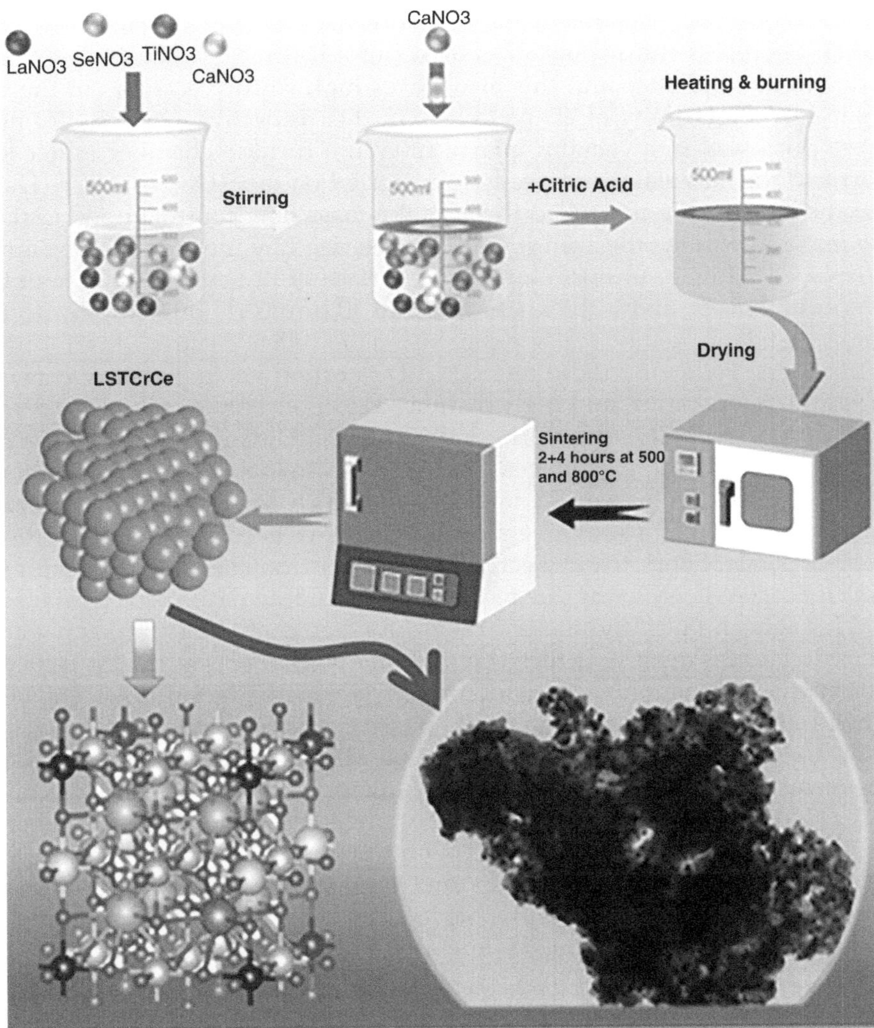

FIGURE 16.2

Synthesis route of LSTCrCeO$_3$ solid electrolyte for a fuel cell using sol–gel technology. Reprinted with permission [5]. Copyright 2023, American Chemical Society.

16.2 Synthesis and Characterization of 2D Semiconductors

The development of semiconducting materials using 2D materials has critical issues regarding van der Waals contact, power consumption, and interface instability, which generally leads to low charge-carrier mobility. Therefore, several efforts have been made to overcome all such drawbacks in the synthesis and device engineering of 2D semiconducting electronic components [8]. Chemical vapor deposition (CVD) has been extensively for the synthesis of 2D semiconductors. One such study was recently published by Kim et al. [9], where they used CVD and synthesized Bi$_2$S$_3$-type semiconducting materials. Bi$_2$O$_3$ was directly grown on a SiO_2 substrate using Bi_2O_3 powder and H_2S gas. However, in the conventional CVD process sulfur powder is used but it becomes difficult to provide steady density due to its solid state and this also leaves high sulfur vacancies in the structure resulting in low crystallinity of 2D materials. In this respective synthesis 2D multilayer flakes were processed in a vacuum quartz tube in a furnace chamber under low pressure 7 torr and the material was directly transported on the SiO_2/Si surface. The density issues for the sulfur were controlled through the mass flow controller and CVD system. Once the material was synthesized it was characterized by the atomic force microscope which displayed a highly uniform and clean surface of Bi_2S_3 crystals without chemical residue, critical defects, and a thickness of about 10 nm. CVD offers optimized properties of high crystalline quality, scalable dimensions, excellent electronic performance, and tunable thickness. Following these merits, CVD is expensive and energy consuming as it requires high temperature and high vacuum conditions. Other than the CVD process colloidal/solution synthesis of uniform and ultrathin 2D nanocrystals has been a great point of research as it has advantages. By using the colloidal synthesis method Son et al. [10] synthesized CdSe-based 2D nanocrystals which have a potential application in various electronic devices. NCs for semiconductors have been extensively studied due to their shape-dependent optical and electrical characteristics. The main reason for selecting cadmium chalcogenide NCs was due to their adoption of two different crystal structures cubic and hexagonal. But to synthesize 2D NC for semiconductors hexagonal structures are preferred in excess quantity which can amplify the charge through the structure.

Solution synthesis is mainly carried out through two methods, top-down and bottom-up approaches. In the top-down method liquid phase exfoliation, chemical ion-intercalation exfoliation, and electrochemical ion-intercalation exfoliation. Whereas in the bottom-up approach metal ion/chalcogen precursors, organic ligands, and solvents are mixed in a flask under N_2 protection, and the 2D material starts to grow at elevated temperature in a way following the classical LaMer theory: with the increase in reaction time, precursors undergo nucleation process, growth and then Oswald ripening [11]. Once the synthesis is complete transmission electron microscopic images are captured which can show $In_{1.8}Se_3$ nanocrystals formed upon injection of selenourea and nanosheets formed thereupon injection of a short aminonitrile [12]. The synthesized crystal structure exhibited fast and high photoresponse over the full visible range, which makes them very appealing for device application in layered structures with other 2D materials, for example by deposition over graphene as photodetectors at low cost of manufacturing 2D semiconductors.

16.3 Electrochemical Energy Applications of 2D Semiconductors

16.3.1 2D Semiconductors for Supercapacitors

Supercapacitors have been explored lately due to the energy storage demand and to address energy and environmental concerns. For such reasons, the selection of electrode materials has been very crucial to improving the electrochemical properties of supercapacitor devices such as energy and power density and cyclic stability. Electrode materials not only deliver electrochemical performance but should also be able to provide good thermal and chemical stability. Apart from electrode materials supercapacitors also require other constituents in the form of electrolyte materials, current collectors, separators, and sealants, which indirectly contribute to the electrochemical properties of the supercapacitors.

2D materials or composites for an electrode in supercapacitors such as MXenes, metalorganic frameworks (MOFs), covalent organic frameworks (COFs), and many others have been explored in the respective portion of the chapter [13].

MXenes materials were first explored in 2011, which were 2D inorganic compounds made of transition metals nitrides, carbides, and carbonitrides, that form multilayer precursors. These materials are promising candidate in the field of energy storage, especially as electrodes for supercapacitors, due to their enhanced combination of hydrophobicity and metallic conductivity [14]. Using the MXenes-MnO_2/Ti_3-C_2T_x Ar composite researchers fabricated an aqueous pseudocapacitor which gave out a higher specific capacitance of 212 F/g compared to pure $Ti_3C_2T_x$ Ar. Following this, the respective composite when used as electrode material, showed a higher capacity retention of 88% after 10,000 cycles compared to the symmetric capacitor fabricated with pure MnO_2-based electrodes [15]. The second type of nanomaterials is MOFs, which are hybrid materials with a crystalline structure and high porosity.

Despite all the superior electrochemical properties of MOFs, there are still major limitations that are derived due to the high crystallinity, such as short cyclic stability at higher charge-discharge rates. To overcome such deficiencies Salunke et al. worked on enhancing the surface area of the metal oxides through the thermal decomposition of MOF precursors in the presence of N_2 gas before annealing the precursors in air, which inhibits the release of volatile gases, thus collapsing the framework and improving electrical conductivity, the same synthesis process has been explained in **Figure 16.3** [16]. A similar study was conducted by He et al. [3], where the composite material showed a specific capacitance of 339 F/g at 0.5 A/g in 1M Na_2SO_4 electrolyte [3]. Apart from MXenes and MOFs, COFs are other promising materials for energy harvesting and storage that have been attracted for use as supercapacitor electrode materials. The COF network is formed by an organic framework joined by covalent bonds. The COFs are tunable to the molecular design and can possess various functionality making them suitable for electronic, and conductive equipment [17]. COFs are commonly used as electrode material for supercapacitors due to their superior pseudocapacitance. This pseudocapacitance is due to the redox mechanism supplied by the existing functionality present in the COF backbone. Using COFs Kim et al.

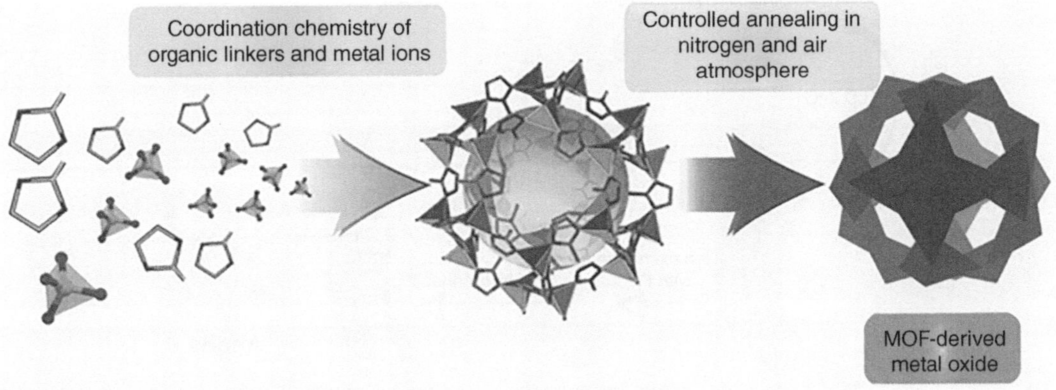

FIGURE 16.3

Illustration representing the fabrication of porous metal oxide nanostructures with large surface area and high porosity from MOFs composed of metal ions and organic linkers as precursors through two-step annealing in N_2 and air, respectively. Reprinted with permission [16]. Copyright 2017, American Chemical Society.

[18] synthesized N-doped carbon through the carbonization of COFs using azine linkage to form a 2D network in the carbonized micropores delivered the specific capacitance of 234 F/g at 1.0 A/g [18]. In another study, researchers coated Ni nanowires with COFs and a high specific capacitance of 314 F/g at 50 A/g and capacitance retention of 74% at 2 A/g, accompanied by fast charged discharge derived from high current density [19].

Metal nitrides have gained a lot of attention from scientists as electrode material for supercapacitors due to their unique high conductivity, high chemical stability, electronic structure, and astonishing mechanical properties such as ductile behavior and high hardness. 2D transition metal nitrides can be synthesized through different compounds such as tetramethyl-disiloxane, and transition metal oxides (TMOs), by various synthesis methods and delamination processes. For example, Zhu et al. [20] synthesized two metal nitrides using different metals on vertical-aligned graphene sheets. The Fe_2N nanoparticles and TiN porous layers have been used as electrodes for supercapacitors and the same is shown in **Figure 16.4** as schematical fabrication of the material. Wherefore, the Fe_2N and TiN acted as anode and cathode material respectively, which showed a specific capacitance of 58 F/g at 4 A/g and remained stable even after the 20,000 cycles. The device exhibited a high volumetric energy density of 0.55 mWh/cm at a power density of about 220 mWh/cm at 8 A/g [20].

In the past decade, supercapacitor devices have been advanced quickly by incorporating various materials with different approaches and configurations to provide excellent electrochemical properties. The combination of novel 2D materials in supercapacitor electrodes has exhibited improved capacitive properties, power density, and energy density.

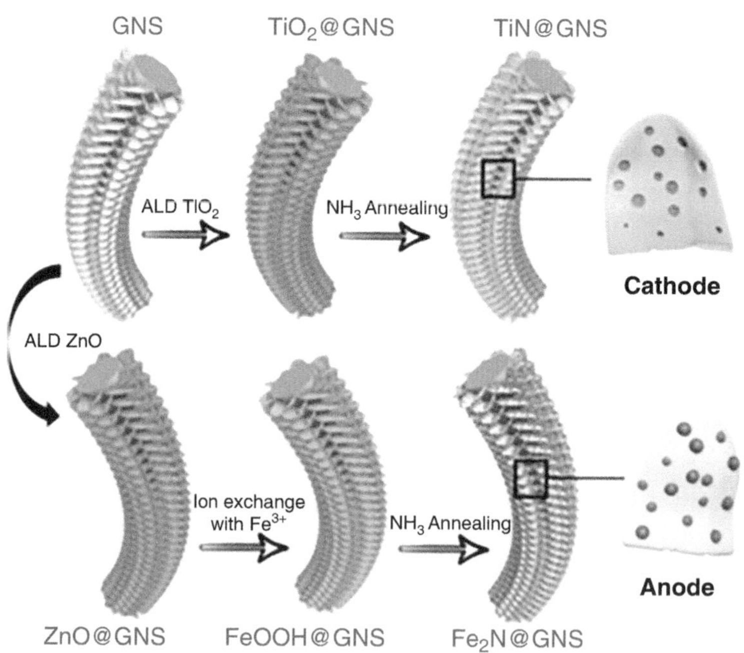

FIGURE 16.4

Explanation of the development of metal nitride as anode and cathode electrodes for supercapacitors. Reprinted with permission [20]. Copyright 2015, John Wiley and Sons.

In this section, we discussed 2D nanomaterials such as MXenes, MOFs, COFs, and metal nitrides used for manufacturing supercapacitor electrodes and reinforcing the electrochemical, chemical, and thermal properties of the electronic devices [21].

16.3.2 2D Semiconductors for Batteries

The development of 2D materials has promoted the emergence of new energy fields. To overcome people's needs, electrochemical devices such as lithium-ion batteries (LIBs), sodium-ion batteries (SIBs), and magnesium-ion batteries (MIBs). Among all these batteries LIBs have been widely applied in the smart electronic market with the most mature system. Still, researchers are looking into exploring high-performance LIBs with high capacity, long life, and high rate. One such study was performed by Zhu et al. [22], where they used 2D black phosphorus (BP) as an anode material for LIBs. The process used in the preparation of the material was mechanical exfoliation, this traditional method is to peel BP repeatedly with a scotch tape and the residuals on the scotch tape are transferred to SiO_2 substrate by certain general solvents like acetone, isopropyl alcohol, and methanol.

The interlayer distance of BP is larger compared to graphene which makes intercalation of ions easier, resulting in better ionic conductivity. BP also provides an ultrafast diffusion channel for Li^+, and Na^+, giving a theoretical specific capacity of 2596 mAh/g and a working voltage range of 0.4–1.2 V in LIBs [22]. Other than LIBs, NIB battery systems have gained popularity accompanied by a surge of research in two-dimensional nanostructures. Researchers demonstrated strain-engineered Na-ion batteries and storage in 2D MoS_2 as anode material, which enhanced the operating performance of Na-ion batteries [23]. One more application of transition metal sulfide was encountered by Qi et al., where they used Rhenium disulfide (ReS_2) rGO 2D semiconductor composite synthesized through a one-pot hydrothermal method. The composite showed an interconnected, hierarchical, and porous architecture. The composite acted as anode material for LIBs. It delivered a large initial capacity of 918 mAh/g at 0.2 C. Additionally, the composite also exhibited much better electrochemical cycling stability and rate capability than virgin ReS_2. This was only possible due to the composites' unique structure which allowed electrolyte interaction, ionic diffusion, and electron transfer [24]. MIBs have also incorporated two-dimensional carbon allotropes such as twin T-graphene, as the TTG possesses a wide band gap of 2.70 eV and the bulk TTG also exhibits great potential as an anode material of magnesium batteries with a theoretical capacity of 556 mAh/g, and diffusion energy barrier of 0.96 eV [25]. Much more work is to be done to improve the efficiency of the Mg-ion batteries built with such 2D semiconductors which will lower the dependence on LIB as a quality energy storage device.

Several 2D materials have been used to engineer batteries, MXenes are one of a kind and taken into consideration while constructing Li-ion and Na-ion batteries. Due to MXenes exceptional mechanical strength, hydrophilicity, and great dispersion quality compared to other 2D materials, it is the perfect material for the synthesis of films and membranes used as active anode material with precise morphology and tunable nanochannels. Such tunable qualities enable outstanding ion storage and electron transport. One more astonishing quality of MXenes was mechanical strength in terms of flexibility making them ideal for flexible, portable, and highly integrated devices [26]. Xu et al. [27] synthesized single-layer MXenes and net-shaped electrodes assembled through heterojunction of $MoSe_2/Ti_3AlC_2$ as shown in **Figure 16.5**. The electrochemical performance of the net electrode was also investigated. Two peaks at 0.78 V and 0.35 V in the first cycle correspond to the cathodic peak where sodium-ion inserts in the intercalation of $MoSe_2$ and makes a $NaMoSe_2$ alloy

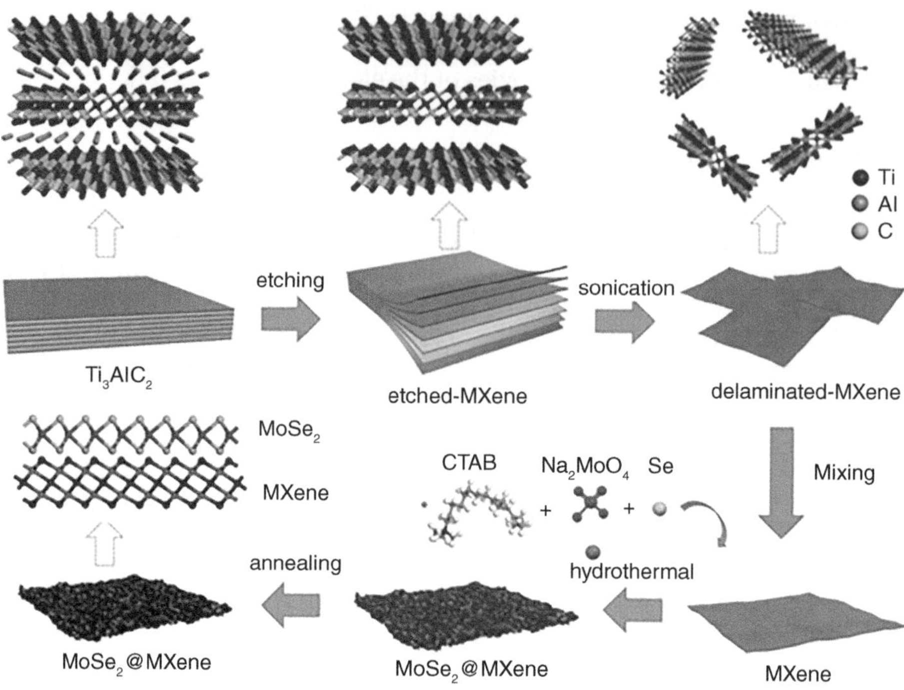

FIGURE 16.5

Schematical representation of the synthesis of single-layer MXenes and MoSe₂/MXenes heterojunction. Reprinted with permission [27] Copyright 2020, Elsevier.

at 0.78V. Whereas, for 0.35V peak the alloy decomposes to metal Mo and Na$_2$Se. The other two cycles gave out an anodic and cathodic peak at 1.7V and 1.37V respectively. Overall, as a potential anode 2D heterojunction material for Na$^+$ ion batteries, it exhibited excellent specific capacity, prolonged cycle life at high reversible capacities of 434 mAh/g at 1 A/g for 200 cycles, and an outstanding rate capability. The electrode material also showed a columbic efficiency of 99.8% making it a promising anode material for SIBs and elevating the development of transition metal-based dichalcogenides 2D MXenes materials for high-performance batteries [22].

A similar type of study on SIBs was done by Zhang et al. [28] where they designed cobalt sulfide and Ti$_3$C$_2$ MXene flake composite with heterolayered architecture, in which CoS nanoparticles joined with MXene flakes. The role of MXene was to provide a conductive network and CoS/MXene electrode delivered a high reversible capacity of 267 mAh/g after 1700 cycles at 2 A/g and ultra-low capacity decay rate of only 0.0072% per cycle accompanied by excellent rate performance of 272 mAh/g at 5 A/g [28]. Such advanced 2D semiconducting nanomaterials have been very useful in the construction of high-performance, efficient, and mechanically stable batteries.

16.3.3 2D Semiconductors for Fuel Cells

The gross energy usage on earth is increasing on an exponential scale due to the rise in the global population growth. There has been significant improvement in energy storage devices and many storage devices are available in the market. Out of all, solid-oxide fuel

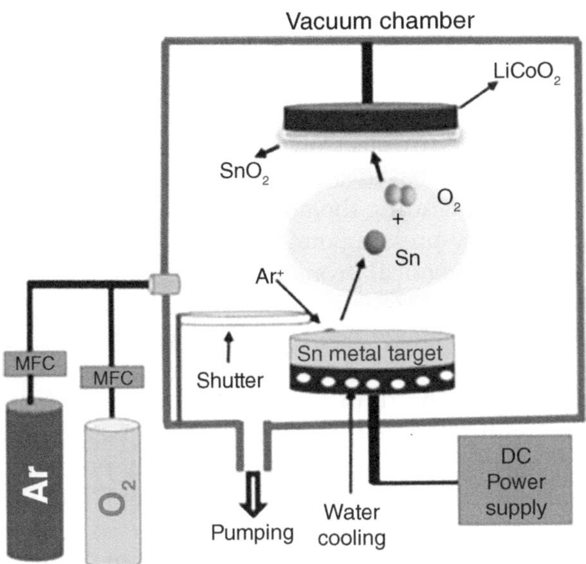

FIGURE 16.6
Schematic diagram of SnO$_2$ thin-film formation on LiCoO$_2$ pellet substrate by DC magnetron sputtering. Reprinted with permission [30]. Copyright 2022, American Chemical Society.

cells have been in great popularity due to their potential to provide power more efficiently with low or no pollutant emissions. Conventional solid oxide fuel cell (SOFC) mainly consists of three components: electrolyte, cathode, and anode. The electrolyte is the most important part of the cell as it prevents short circuits and facilitates ion mobility. However, due to liquid electrolyte usage, there are many limitations such as safety hazards, and low ionic mobility. For such reasons, single-layer fuel cells or electrolyte-layer-free fuel cells are gaining attention in research [29]. Following the single-layer fuel cell idea, Ganesh et al. designed a semiconductor junction fuel cell using LiCoO$_2$/SnO$_2$ 2D semiconducting material. Therefore, SnO$_2$ nano-powder was synthesized using the co-precipitation method. Once the SnO$_2$ nanoparticle was synthesized, LiCoO$_2$ powder and the nanoparticle were fabricated in the form of thin-film p–n junction fuel cells by DC magnetron sputtering as shown in **Figure 16.6**. Researchers also fabricated bulk planar fuel cells using the dry-pressing method. This new generation of solid-oxide fuel cells is made in three different configurations with improved material functionalities and fuel cell performance. The X-ray photoelectron spectroscopy revealed transferred electrons from LiCoO$_2$ to SnO$_2$ attesting the formation of the p–n junction. The bulk planar, bulk heterojunction and thin-film planar p–n junction fuel cell devices exhibit maximum power densities of 0.61, 0.82, and 0.30 W/cm^2, respectively. As being electrolyte-free, the materials prevented passing on electrons to the internal device and avoided circuit problems.

As discussed about drawbacks of commercial electrolyte materials, researchers went ahead in developing solid-oxide fuel cells, but the operation temperature of respective cells is above 800°C. Following this researcher went ahead and designed a new electrolyte that delivers high ionic conductivity and long durability as well as a lower operating temperature of >600°C. Shah et al. [5], designed LaSrTCrCeO$_3$ synthesized through the sol–gel technique as described and used the 2D semiconductor material as an electrolyte for low-temperature-ceramic fuel cells. The novel perovskite semiconductor fuel cell showed a performance of 1031 mW/cm^2 and good ionic conductivity of 0.16 S/cm at a low working temperature of 520°C. Furthermore, the electrolyte showed a good fuel cell performance of 270 and 340 mW/cm^2 at even low temperatures of 370 and 330°C, respectively, [5]. Following such trends to synthesize electrolyte materials for fuel cells, some researchers recently derived heterostructure electrolytes from a-Al$_2$O$_3$ for solid-state fuel cells. The composite electrolyte was made based on CeO$_2$/amorphous alumina produced through the solid mixing method and blending them to form a suspension solution. Once the composite material was ready, a dry-pressing method was used to fabricate the solid-oxide fuel cell. While testing the electrochemical properties of the material, the electrolyte exhibited high ionic conductivity of up to 0.127 S/cm and achieved a maximum power density of 1017 mW/cm^2 with an open circuit voltage of 1.14 V at a low operating temperature of 550°C. The study also suggested that a potential energy barrier was formed at the heterointerfaces caused by the wide band gap of the alumina insulator material which played a crucial role in restraining electron conduction. But with the material being a metal oxide composite was left with too many oxygen vacancies which eventually promoted ionic transport and ended up being beneficial in terms of fuel cell performance [7]. In this part of the chapter, we looked into the recent advancement which took place in terms of fuel cell development, where we observed that scientists are more interested in developing solid-oxide or ceramic-based fuel cells which sometimes is also linked with developing solid electrolyte materials that have so far been very beneficial in term of inhabiting the short circuit issues and enhancing the fuel cell performance by enabling high ionic conductivity at subsequent lower operating temperatures. We also saw fuel cells have the potential to overcome the ever-growing energy demands by providing efficient power generation technology with exceptional energy efficiency, fuel flexibility, and low or no pollution emissions.

16.3.4 2D Semiconductors for Solar Cells

Solar cells, a device that converts solar radiation directly into electricity, are renewable energy conversion technology. Since their development in the early 1950s, improved conversion efficiency has been researched. Considering the photoelectric conversion mechanism, these cells can be divided into two main categories, conventional solar cells and excitonic solar cells (XSCs) [7]. These solar cells have been a point of interest for scientists due to the increase in energy consumption and after all deriving new and sustainable techniques that transform and store energy. For such reasons, they require novel materials that can provide enhanced photochemical properties. Moreover, plentiful 2D materials possess exotic electronic and optoelectronic properties, making them interesting donor and acceptor materials in producing high-efficiency fuel cells. Thus, the development of suitable 2D donor and acceptor materials with high carrier mobility and good light-harvesting performance is significant for the development of XSCs. Following these trends, Wang et al. [31] designed two 2D structures, named α-and β Sb$_2$TeSe$_2$ monolayers by exfoliation process. Once the synthesis process was completed, the researchers studied the electronic, structural, and optical properties of the 2D semiconducting material through the density functional theory function calculations which attested that the two monolayers are dynamically, thermally, mechanically, and chemically stable. The monolayers have a considerable band gap of 1.1 eV to absorb light in the full range of the solar spectrum and reflect high carrier mobility up to 300 cm^2/Vs. Additionally, the moderate band gap of the monolayers makes them favorable donor materials to construct 2D heterojunction XSCs. Lastly, the designed α-Sb$_2$TeSe$_2$/HfS$_2$ and β-Sb$_2$TeSe$_2$/BiOI heterojunction XSCs gave out high power conversion efficiency (PCE) up to 22.5 and 20.3% accordingly, making them potential candidates for photovoltaic applications [31].

Two-dimensional Janus structures such as MoSSe and graphene have been of popular interest due to their novel properties and interesting behaviors. The material has a high-half metallic gap of 0.98 eV, a curie temperature of which can be up to 292 K. Due to such favorable edge designs of materials such as Sc_2CHCl, Sc_2COHCl, and many more, these monolayers are promising contenders for photocatalytic applications. The material also possesses large efficient absorption in the visible and UV light regions. The proposed materials, when developed as XSC by composing a Se_2COHCl/InS heterostructure device achieved a PCE of 21.04%. Such Janus MXenes should be looked into to derive a facile configuration to get more efficient solar cells [32]. Solar cells are also made by the conjunction of 3D and 2D perovskite materials. Referring to the previous statement Jiang Et al derived perovskite solar cells which consisted of Ruddlesden-Popper 2D perovskites as conducting material and 3D perovskite material as an active precursor material. These 2D materials provide alternately connected bulky organic cations and conductive inorganic slabs which subsequently provide high stability to the solar cells. The precursor solution of mixed-ion 2D-3D perovskite $(FA_{0.85}MA_{0.15})Pb$ $(I_{0.85}Br_{0.15})_3$ was prepared through the solution deposition method and spin-coated on fluorine-doped tin oxide/compact TiO_2/mesoporous TiO_2 substrate as shown in **Figure 16.7a**. As observed from **Figure 16.7b**, the 2D perovskite was deposited through the spin coating on the 3D perovskite and illustrates the device architecture. Once the device was designed, photovoltaic tests were performed where the device exhibited PCE of 18.9% with an open circuit voltage (OCV) of 1.06V and short circuit current density of 23.8mA cm^{-2} at a fill factor of 0.75. In a further study, the cell performed even better giving PCE of 21.6% with OCV of 1.10 V, corresponding with a fill factor of 0.81 [33]. The results suggested that the materials had higher stability and efficiency compared to other 2D materials used to design the solar cells and could be used for obtaining high-performance devices. A similar study was performed by Jang et al, where they designed polymer solar cell using 2D/3D halide junction cells via solid phase in-plane growth. The applied solid phase in-plane growth can be demonstrated as 2D material applied on the 3D material as discussed in the previous research. However, the current research was done in the absence of liquid state precursors that provided better control of the thickness of the

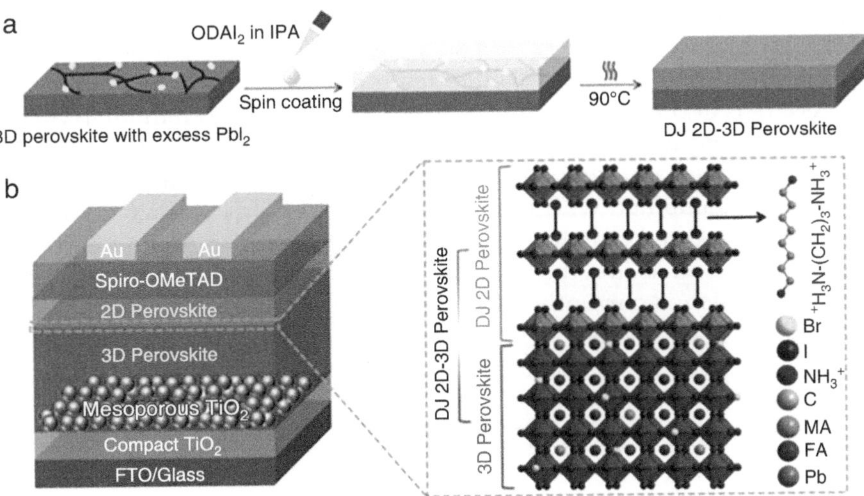

FIGURE 16.7

(a) Schematics representing fabrication of 2D–3D perovskite. **(b)** Diagram illustrates the 2D–3D perovskite and device architecture. Reprinted with permission [33]. Copyright 2020, Elsevier.

2D material upon the substrate. However, the entire design enhanced the overall stability and performance of the device. In the synthesis procedure 2D solid film was stacked on top of the 3D film to make direct contact. Heat and pressure are then applied perpendicular to each surface to enable 2D perovskite from the solid 2D film to the top of the 3D film. Initially, 2D seeds are formed on the surface which is eventually transformed from 3D film to a new 2D film layer [34].

The development of a 2D/3D halide junction was successful. The perovskite solar cell gave out a PCE of 24.59% at OCV of 1.185V and the encapsulated device retained 94% efficiency after 1056 h under the damp heat test at 85°C temperature and 85% relative humidity). Followed by 98% efficiency after 1620h under full sun illumination [34]. The construction of such efficient devices through 2D materials will improve the transformation of solar energy into electricity. We hope such technologies will help in providing different alternatives to batteries and other conventional electrochemical devices and induce direct energy to support humankind.

16.4 Conclusion and Outlook

In this chapter, the basics of electrochemical energy storage devices to advanced nanomaterials which are currently being studied to manufacture high-quality energy storage devices are provided. The introduction gave a brief idea regarding the importance of energy storage devices and the inclusion of 2D semiconducting nanocomposites with their respective roles in developing such devices. Once the materials are found they need to be designed and manufactured considering the type of application they will be used in. For such reason there is a requirement of characterization, so there was a glance at synthesis and characterization techniques of various types of 2D semiconductors which included sol–gel, exfoliation, CVD, and several other synthesis techniques, and the same was explained in the second part with some elaborated examples. Real-time application of 2D semiconducting nanomaterial was discussed in the third part which explained the use of semiconducting material for the fabrication of hi-tech electrochemical energy storage devices and solar cells. We also saw through the explained research that materials performed well, attesting to the capabilities of such advanced materials. As a part of the outlook, this chapter glorified the 2D nanomaterials and their application by presenting the past and current research being carried out by the scientists. This chapter also provided suggestions for future research projects and launching novel material combined with high-performance energy storage devices. From the chapter, it is also observed that 2D materials have endless exploration and the research accomplished until now is just the beginning. They will be the most appropriate alternative for metal which is currently used and could also provide energy application in flexible or wearable electronics.

References

1. J.P. Paraknowitsch, A. Thomas, Doping carbons beyond nitrogen: An overview of advanced heteroatom doped carbons with boron, sulphur and phosphorus for energy applications, *Energy Environ. Sci.* 6 (2013) 2839–2855.

2. Y. Da, J. Liu, L. Zhou, X. Zhu, X. Chen, L. Fu, Engineering 2D architectures toward high-performance micro-supercapacitors, *Adv. Mater.* 31 (2019) 1–28.

3. K. Zhao, Z. Xu, Z. He, G. Ye, Q. Gan, Z. Zhou, S. Liu, Vertically aligned MnO_2 nanosheets coupled with carbon nanosheets derived from Mn-MOF nanosheets for supercapacitor electrodes, *J. Mater. Sci.* 53 (2018) 13111–13125.

4. S. Jariwala, Y. Desai, R.K. Gupta, Pseudocapacitive Materials for 3D Printed Batteries. In: R.K. Gupta, (eds) *Pseudocapacitors Engineering Materials*. Springer, Cham (2024). https://doi.org/10.1007/978-3-031-45430-1_21

5. M.A.K.Y. Shah, Y. Lu, N. Mushtaq, M. Yousaf, N. Akbar, N. Arshad, M.S. Irshad, B. Zhu, Designing a novel semiconductor electrolyte ($LaSrTiCrCeO3$) with enhanced ionic caonduction for low-temperature ceramic fuel cells, *ACS Appl. Energy Mater.* 6 (2023) 2832–2844.

6. M.A. Green, A. Ho-Baillie, H.J. Snaith, The emergence of perovskite solar cells, *Nat. Photonics.* 8 (2014) 506–514.

7. Y. Zhang, D. Zhu, X. Jia, J. Liu, X. Li, Y. Ouyang, Z. Li, X. Gao, C. Zhu, Novel n-i $CeO2/a-Al2O3$ heterostructure electrolyte derived from the insulator a-Al2O3 for fuel cells, *ACS Appl. Mater. Interfaces.* 15 (2023) 2419–2428.

8. Y. Kim, E. Jeong, M. Joe, C. Lee, Synthesis of 2D semiconducting single crystalline $Bi2S3$ for high performance electronics, *Phys. Chem. Chem. Phys.* 23 (2021) 26806–26812.

9. I.J.T. Jensen, A. Ali, P. Zeller, M. Amati, M. Schrade, P.E. Vullum, M.B. Muñiz, P. Bisht, T. Taniguchi, K. Watanabe, B.R. Mehta, L. Gregoratti, B.D. Belle, Direct observation of charge transfer between NOx and monolayer $MoS2$ by operando scanning photoelectron microscopy, *ACS Appl. Nano Mater.* 4 (2021) 3319–3324.

10. J.S. Son, J.H. Yu, S.G. Kwon, J. Lee, J. Joo, T. Hyeon, Colloidal synthesis of ultrathin two-dimensional semiconductor nanocrystals, *Adv. Mater.* 23 (2011) 3214–3219.

11. J. Polte, Fundamental growth principles of colloidal metal nanoparticles—a new perspective, *CrystEngComm.* 17 (2015) 6809–6830.

12. G. Almeida, S. Dogan, G. Bertoni, C. Giannini, R. Gaspari, S. Perissinotto, R. Krahne, S. Ghosh, L. Manna, Colloidal monolayer β-In2Se3 nanosheets with high photoresponsivity, *J. Am. Chem. Soc.* 139 (2017) 3005–3011.

13. H. Zhang, Ultrathin two-dimensional nanomaterials, *ACS Nano.* 9 (2015) 9451–9469.

14. M. Ghidiu, M.R. Lukatskaya, M.Q. Zhao, Y. Gogotsi, M.W. Barsoum, Conductive two-dimensional titanium carbide "clay" with high volumetric capacitance, *Nature.* 516 (2015) 78–81.

15. R.B. Rakhi, B. Ahmed, D. Anjum, H.N. Alshareef, Direct chemical synthesis of $MnO2$ nanowhiskers on transition-metal carbide surfaces for supercapacitor applications, *ACS Appl. Mater. Interfaces.* 8 (2016) 18806–18814.

16. R.R. Salunkhe, Y. V Kaneti, Y. Yamauchi, Metal-organic framework-derived nanoporous metal oxides toward supercapacitor applications: Progress and prospects, *ACS Nano.* 11 (2017) 5293–5308.

17. S. Dalapati, M. Addicoat, S. Jin, T. Sakurai, J. Gao, H. Xu, S. Irle, S. Seki, D. Jiang, Rational design of crystalline supermicroporous covalent organic frameworks with triangular topologies, *Nat. Commun.* 6 (2015) 1–8.

18. G. Kim, J. Yang, N. Nakashima, T. Shiraki, Highly microporous nitrogen-doped carbon synthesized from azine-linked covalent organic framework and its supercapacitor function, *Chem.—A Eur. J.* 23 (2017) 17504–17510.

19. Y. Han, N. Hu, S. Liu, Z. Hou, J. Liu, X. Hua, Z. Yang, L. Wei, L. Wang, H. Wei, Nanocoating covalent organic frameworks on nickel nanowires for greatly enhanced-performance supercapacitors, *Nanotechnology.* 28 (2017).

20. C. Zhu, P. Yang, D. Chao, X. Wang, X. Zhang, S. Chen, B.K. Tay, H. Huang, H. Zhang, W. Mai, H.J. Fan, All metal nitrides solid-state asymmetric supercapacitors, *Adv. Mater.* 27 (2015) 4566–4571.

21. X. Hong, J. Fu, Y. Liu, S. Li, X. Wang, W. Dong, S. Yang, Recent progress on graphene/polyaniline composites for high-performance supercapacitors, *Materials (Basel).* 12 (2019) 1451.

22. S.C. Dhanabalan, J.S. Ponraj, Z. Guo, S. Li, Q. Bao, H. Zhang, Emerging trends in phosphorene fabrication towards next generation devices, *Adv. Sci.* 4 (2017).

23. J. Hao, J. Zheng, F. Ling, Y. Chen, H. Jing, T. Zhou, L. Fang, M. Zhou, Strain-engineered two-dimensional $MoS2$ as anode material for performance enhancement of Li/Na-ion batteries, *Sci. Rep.* 8 (2018) 1–9.

24. F. Qi, Y. Chen, B. Zheng, J. He, Q. Li, X. Wang, J. Lin, J. Zhou, B. Yu, P. Li, W. Zhang, Hierarchical architecture of ReS 2/rGO composites with enhanced electrochemical properties for lithium-ion batteries, *Appl. Surf. Sci.* 413 (2017) 123–128.

25. J. Lin, X. Chen, B. Zhang, C. Tan, Q. Lin, X. Wang, Two dimensional twin T-graphene: Monolayer for visible-light photocatalytic water splitting and bulk for anode material of magnesium batteries, *RSC Adv.* 12 (2022) 30349–30358.

26. N. Iqbal, U. Ghani, W. Liao, X. He, Y. Lu, Z. Wang, T. Li, Synergistically engineered 2D MXenes for metal-ion/Li-S batteries: Progress and outlook, *Mater. Today Adv.* 16 (2022).

27. E. Xu, Y. Zhang, H. Wang, Z. Zhu, J. Quan, Y. Chang, P. Li, D. Yu, Y. Jiang, Ultrafast kinetics net electrode assembled via MoSe2/MXene heterojunction for high-performance sodium-ion batteries, *Chem. Eng. J.* 385 (2020) 123839.

28. Y. Zhang, R. Zhan, Q. Xu, H. Liu, M. Tao, Y. Luo, S. Bao, C. Li, M. Xu, Circuit board-like CoS/MXene composite with superior performance for sodium storage, *Chem. Eng. J.* 357 (2019) 220–225.

29. B. Zhu, Y. Mi, C. Xia, B. Wang, J.-S. Kim, P. Lund, T. Li, A nanoscale perspective on solid oxide and semiconductor membrane fuel cells: materials and technology, *Energy Mater.* 1 (2022) 100002.

30. K.S. Ganesh, L. Fan, B. Wang, P. Jeevan Kumar, B. Zhu, Built-in electric field for efficient charge separation and ionic transport in LiCoO2/SnO2 semiconductor junction fuel cells, *ACS Appl. Energy Mater.* 5 (2022) 12513–12522.

31. C. Wang, Y. Jing, X. Zhou, Y.F. Li, Sb2TeSe2 Monolayers: Promising 2D semiconductors for highly efficient excitonic solar cells, *ACS Omega.* 6 (2021) 20590–20597.

32. Y. Zhang, B. Sa, N. Miao, J. Zhou, Z. Sun, Computational mining of Janus Sc2C-based MXenes for spintronic, photocatalytic, and solar cell applications, *J. Mater. Chem. A.* 9 (2021) 10882–10892.

33. X. Jiang, J. Zhang, S. Ahmad, D. Tu, X. Liu, G. Jia, X. Guo, C. Li, Dion-Jacobson 2D-3D perovskite solar cells with improved efficiency and stability, *Nano Energy.* 75 (2020) 104892.

34. Y.W. Jang, S. Lee, K.M. Yeom, K. Jeong, K. Choi, M. Choi, J.H. Noh, Intact 2D/3D halide junction perovskite solar cells via solid-phase in-plane growth, *Nat. Energy.* 6 (2021) 63–71.

17

Solar-Rechargeable Energy Conversion/Storage Systems: Overview of 2D Nanomaterials

Mahmoud Adel Hamza*, Mahmoud Moussa, Ayat N. El-Shazly, and Mohammed Salah

**Correspondence: mahmoud.gharib@adelaide.edu.au*

17.1 Introduction

Due to the explosive increase in the world population, the energy demand has explosively increased for both domestic and industrial uses. Unfortunately, our world is mainly dependent on fossil fuels for the generation of electricity which has a severe impact on the environment due to carbon emissions which are the main source of air pollution and global warming [1]. The UN's Sustainable Development Goals (SDGs) focused on these challenges, especially SDG 7 concerning ensuring access to affordable and clean energy for all, and SDG 13 concerning taking actions to combat climate change [2]. Hence, there is an urgent need to make a paradigm transition from fossil fuels to green renewable energy resources [3]. This stimulates the ongoing advance of cutting-edge energy conversion and storage systems using green energy [4]. Solar energy is a permanently accessible eco-friendly source of energy that could be efficiently utilized in photocatalytic energy conversion and storage applications. The earth's surface receives ~ 100,000 TW of solar power per hour which is much greater than the world's energy consumption per year (<20 TW/year) [5]. Therefore, it is of pivotal significance to maximize the efficient utilization of permanently accessible solar energy that can be converted/stored in the form of electrical energy. This could be achieved by converting solar power into electricity using solar cells or solar fuels (e.g., green hydrogen fuel) [6,7]. However, some stumbling blocks prevent the broad use of solar energy. Among them is its intermittent nature which makes it unusable sometimes such as during the night and in the winter [8,9]. This motivated researchers to find ways to convert and then store solar energy to reuse it in times of need [10].

Solar-rechargeable energy systems such as solar cells, solar supercapacitors, and photo-rechargeable batteries have emerged as outstanding alternatives to traditional storage systems [4]. The common element of these systems is the utilization of a semiconductor nanomaterial (e.g., TiO_2) as the photoactive electrode. TiO_2 is one of the most common semiconductors; however, its wide bandgap energy limits its activity under UV irradiation which represents *ca.* 4% of the solar irradiation [1,11]. Thus, the utilization of visible-light-drive semiconductors is a promising approach to efficiently manipulate a broader range of solar energy. Owing to their high surface area and tunable optical/electronics properties, 2D semiconductor nanomaterials including metal-organic frameworks (MOFs), metal sulfides (e.g., MoS_2 and $ZnIn_2S_4$), and metal-free semiconductors (e.g., $g\text{-}C_3N_4$) have

DOI: 10.1201/9781003439448-17

demonstrated outstanding performance as photocatalysts and photoactive materials in these hybrid energy/storage systems (**Figure 17.1a**). **Figure 17.1b** shows a comparison of crystal structures of different families of 2D semiconductor materials and the corresponding bandgap energies. The broad ranges of bandgaps displayed by 2D semiconductors enhances their potential utilization in numerous can be exploited in a various optoelectronics/photonic applications including photovoltaics, light-emitting diodes (LEDs), fiber-optics communication, and thermal imaging [12].

Hence, this chapter will offer an overview of the utilization of solar energy in energy storage and conversion systems. It will be focused on three main devices: solar cells, photo-supercapacitors, and rechargeable batteries as summarized in **Figure 17.1**. This chapter is limited to just giving the main concept behind these devices, and some examples of recent works of some 2D nanomaterials will be demonstrated. This chapter is not a critical review; however, key reviews have been cited to guide the readers once they require detailed information and a summary of the recent advances and the current state of the art. Nevertheless, it is believed that the general ideas discussed here will be valuable to the readers and new scholars interested in the field of energy conversion and storage.

17.2 Solar Cells

Solar cells (SCs) or photovoltaics (PV) can be considered green, efficient, and successful technology to directly convert solar energy into electric power. Recently, organic-/perovskite-based thin-film solar cells significantly achieved great enhancement in power conversion efficiency exceeding 18% and 25% for organic SCs and perovskite SCs, respectively [13]. Along with the high-power conversion efficiencies, perovskites/organics thin-film SCs are characterized by low-cost large-scale roll-to-roll (R2R) processing compatibility, mechanical flexibility, lightweight, and semi-transparency [14]. Thus, perovskites/organic SCs are placed at the focal point of attention of researchers for developing SC technologies. In this context, 2D nanomaterials can be considered ideal contact layer materials in thin-film SCs. This could be ascribed to the unique features of 2D nanomaterials including the tunable electronic structure, high optical transparency, and high charge-carrier mobility [15].

Various 2D nanomaterials have been employed as photoelectrodes, electron transport layer (ETL), hole transport layer (HTL), and additives in active layers for improving the efficiency of SCs as shown in **Figure 17.2a**. Some main families of these 2D nanomaterials have been employed in SC applications such as graphene and graphene-like elemental 2D nanomaterials (antimonene, black phosphorous, bismuthine, borophene, etc.), MOFs, metal oxides, transition metal dichalcogenides (such as WSe_2 and MoS_2) [16], metal frees semiconductors (e.g., g-C_3N_4, and covalent organic frameworks (COFs)), MXenes, and polymers. Some review articles summarized the utilization of 2D nanomaterials in SCs [17]. To adhere to the context of the book, we will focus on 2D semiconducting nanomaterials.

MoS_2 is one of the most employed 2D nanomaterials in various solar energy applications. For instance, Jiang et al. [18] fabricated an organic perovskite solar cell (PSC) containing a modified Spiro-OMeTAD layer using the flower-like 2H semiconducting phase of MoS_2 as HTL as shown in **Figure 17.2b(i)–(iii)**. There is a remarkable reduction in the intensity of the photoluminescence (PL) peak of the MoS_2-doped sample as shown in **Figure 17.2b(iv)**. This PL quenching reveals that the MoS_2-doped Spiro-OMeTAD exhibits an enhanced hole extraction feature upon being used as an HTL in PSCs. Therefore, the PSC fabricated

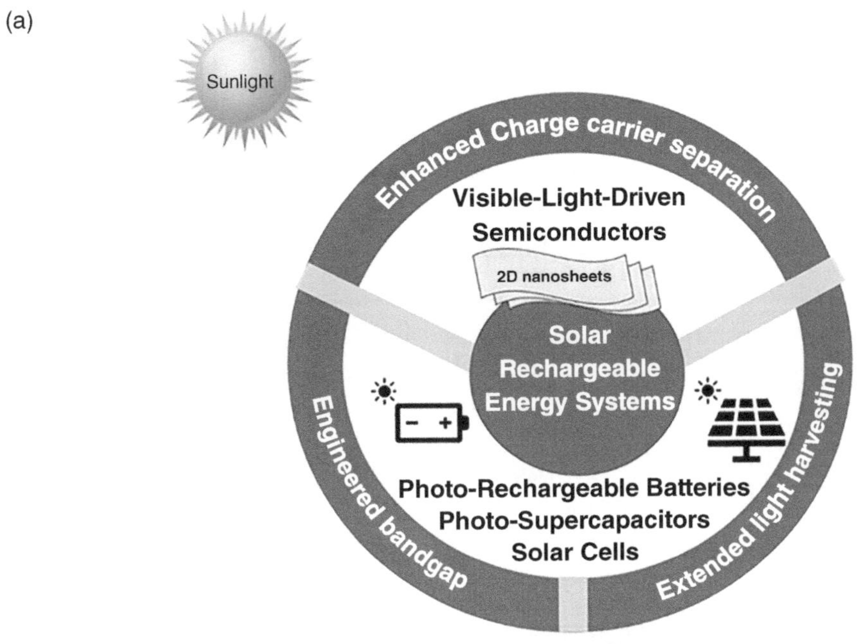

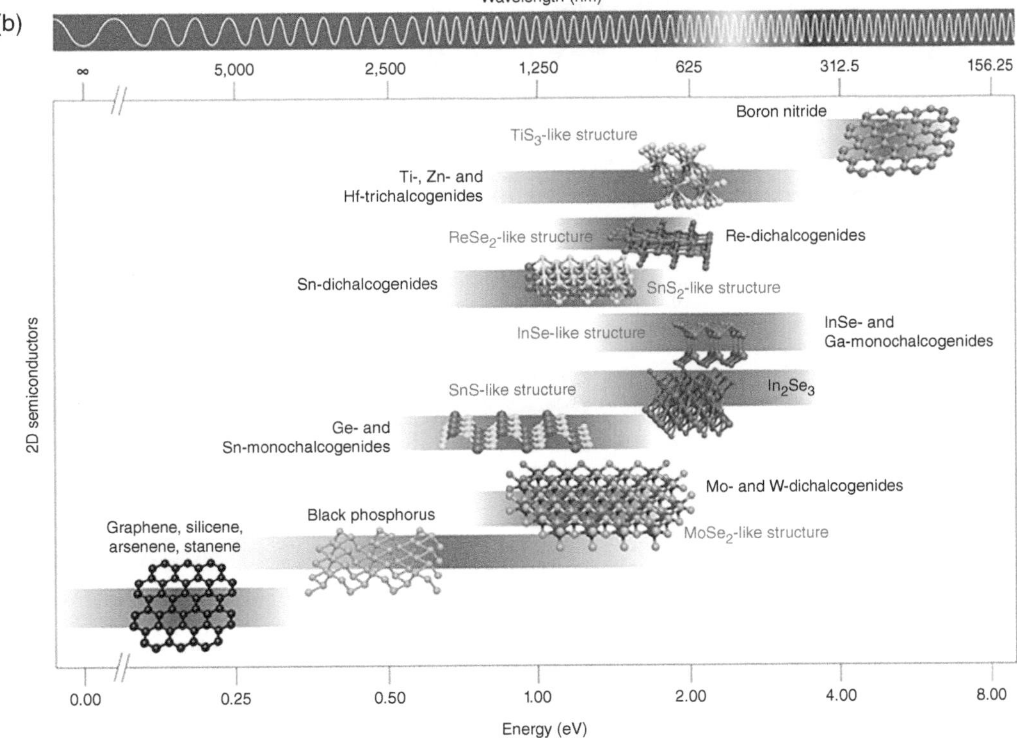

FIGURE 17.1

(a) Schematic diagram of the context of the chapter for demonstrating the concept of solar-rechargeable systems and (b) schematic comparison of crystal structures of different families of 2D semiconductor materials. Adapted with permission [12], Copyright (2016), Nature Publishing Group.

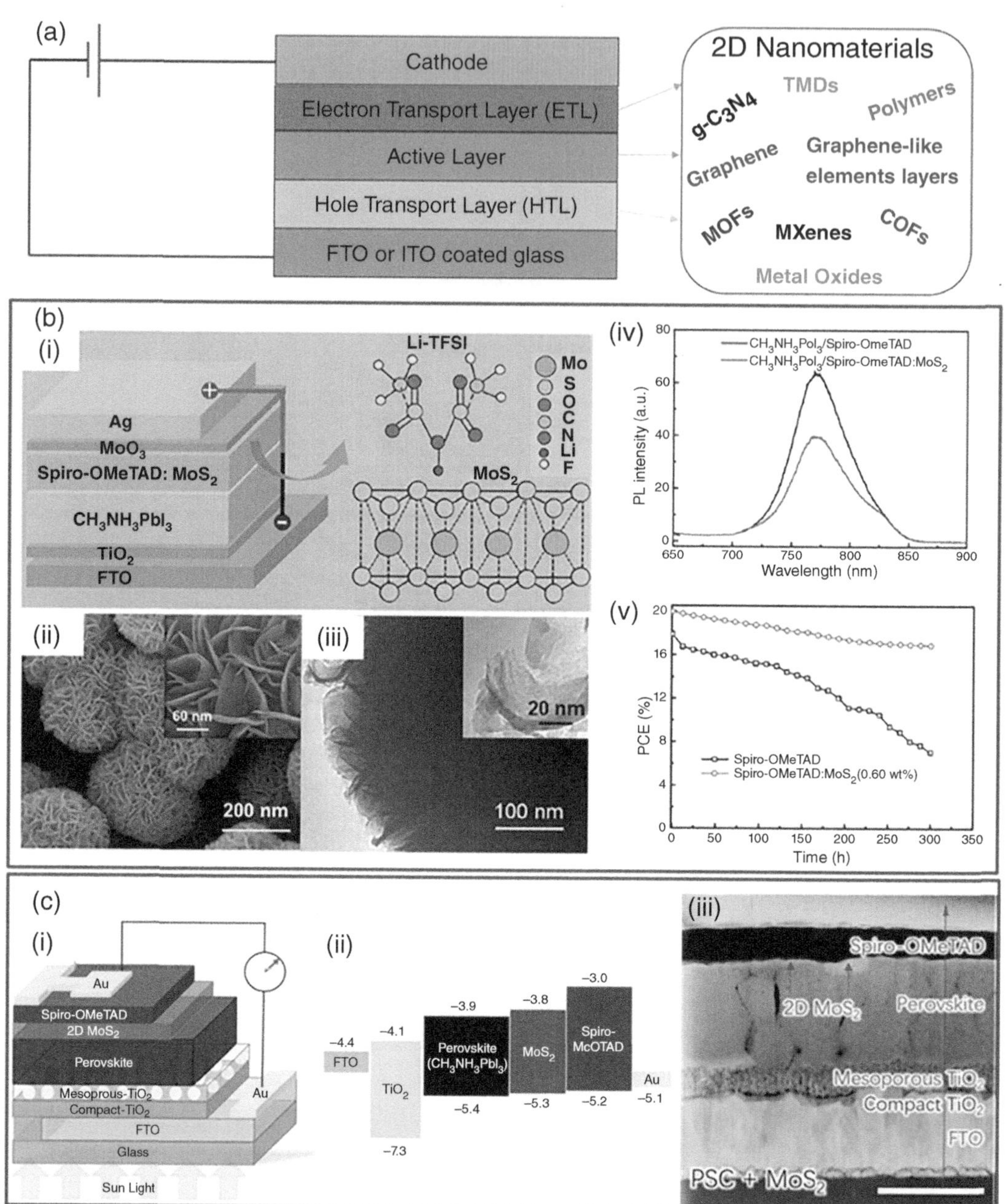

FIGURE 17.2

(a) Schematic representation of the primary layers required for fabrication of solar cells and opportunities of 2D nanomaterials, (b)(i) a sketch of the organic PSC structure, and the molecular structures of MoS_2 and Li-TFSI, (ii) SEM and (iii) TEM images of pure MoS_2, (iv) PL spectra of $CH_3NH_3PbI_3$ films coated with Spiro-OMeTAD and Spiro-OMeTAD: MoS_2 film, and (v) the recorded temporal PCE degradation in the fabricated PSCs without encapsulation while being exposed to air at room temperature in the dark. Adapted with permission [18], Copyright (2019), Royal Society of Chemistry, and (c)(i) a Sketch of the PSC demonstrating the employed 2D MoS_2 nanoflakes as a buffer layer between the Spiro-OMeTAD HTL and perovskite layer, (ii) band energy diagram of the whole structure, and (iii) cross-section TEM image of the PCS. Adapted with permission [19] under Creative Commons Attribution (CC BY) license 4.0, Copyright (2020), Nature Publishing Group.

without encapsulation displayed outstanding stability by retaining about 85% of the initial photoconversion efficiency (PCE) after 300 h exposure to air compared to just 30% of the initial PCE was retained in case of the reference device as demonstrated in **Figure 17.2b(v)**. The incorporation of MoS_2 showed crucial roles in developing the film stability of Spiro-OMeTAD and the hole mobility. Similarly, Liang et al. [18] investigated the behavior of 2D metal chalcogenides (MoS_2 and $MoSe_2$) as a buffer layer between the HTL and the perovskite layer to develop the stability of the organometallic-halide PSC as demonstrated in **Figure 17.2c(i)–(iii)**. The as-fabricated solar cell achieved a higher PCE of 14.9% with greater lifetime stability compared to the reference PSC. After 1 h, the PCE of the PSC containing MoS_2 as a buffer layer retained about 93% of the initial value, while only 78% of the initial PCE was retained in the case of the reference PSC. It was found that MoS_2 and $MoSe_2$ can act both as an additional HTL and a protective layer that efficiently enhances the PCE of the PCSs.

The fabrication of heterojunctions (type I, type II, Z-scheme, Schottky junction, p–n junctions, etc.) could enhance the charge separation and significantly diminish the e-h recombination [1]. For instance, TiO_2 is known as an efficient ETL; hence, the fabrication of heterojunction is expected to achieve high efficiency of electron transport and consequently, the PCE will be increased. Xie et al. [20] investigated the in-*situ* synthesis of g-C_3N_4 that was introduced into TiO_2-based PSC using a simple glass-assisted annealing method. The as-fabricated solar cell (**Figure 17.3a(i)**) employed bare TiO_2 (TO) or TiO_2/g-C_3N_4 (TOCN) as ETL. It was found that the utilization of TOCN as ETL achieved the highest PCE of 20.46%, which is ca. 20% improvement compared to the reference cell (17.18%) as estimated from the recorded current density–voltage (J–V) curves displayed in **Figure 17.3a(ii)**. The incident photon-to-current efficiency (IPCE) spectra of fabricated PSCs are shown in **Figure 17.3a(iii)**. The PSC based on the TOCN ETL displayed enhanced IPCE values in the wavelength range of 455–770 nm compared to that PSC based on the TO ETL. This was accompanied by an enhancement in the measured short circuit current density (J_{sc}) from 22.83 to 24.44 mA cm^{-2}. This was ascribed to the facilitated fast migration of electrons and the significantly reduced charge recombination at the interface in the case of TOCN ETL. Electrochemical impedance spectroscopy was utilized to obtain the Nyquist impedance spectra and the corresponding equivalent circuits of the fabricated PSCs as displayed in **Figure 17.3a(iv)**. It was found that each spectrum is composed of one hemicircle which was attributed to the recombination resistance (R_{rec}) and chemical capacitance (C_{rec}) at the ETL/perovskite/HTL interfaces. The TOCN-based PSC has a lower series resistance (Rs) but a higher R_{rec} compared to TO-based PSC. This confirmed the enhanced charge transfer and the suppressed charge recombination at the TOCN/perovskite layer/ spiro-OMeTAD interfaces; this was consistent with the improved Jsc value. This enhanced charge separation is illustrated in the proposed band structure of the whole system displayed in **Figure 17.3a(v)**.

Lin et al. [21] inspected the electronic structures of the g-C_3N_4/WTe_2 van der Waal (vdW) heterostructure and their solar cell performances *via* first-principles calculations. The effect of the vdW interactions on the electronic structure of the g-C_3N_4/WTe_2 heterostructure was investigated by changing the vdW gap (d_{layer}). The effect of d_{layer} on estimating the bandgap, band alignment, and work function of g-C_3N_4/WTe_2 heterostructure with different d_{layer} was estimated. Different functional density functional theory (DFT) methods showed a similar trend that the E_g decreases by decreasing the d_{layer} (**Figure 17.3b(i)**). Likewise, the work function and band alignment showed an analogous trend of bandgap change with altering the d_{layer}. As the d_{layer} decreased, the conduction band minimum shifted downward continuously, and the valence band maximum shifted upward continuously; hence, the band gap was reduced and vice versa (**Figure 17.3b(ii)**). The g-C_3N_4/WTe_2 heterostructure exhibited a bandgap energy of 1.24 eV with a type II band alignment (**Figure 17.3b(iii)**). Remarkably, the bandgap energy of the g-C_3N_4/WTe_2 could be increased to 1.44 eV by increasing the vdW gap to enhance and extend its absorption capabilities to a more visible light range. It was found that the reduced band alignment difference achieved by the tunable vdW gap achieved enhancement of the PCE up to 17.68%.

Wen et al. [22] employed DFT calculations to investigate various promising heterostructures composed of M_2CO_2 (M = Hf,Zr) and MoX_2(X = S,Se,Te) for solar energy applications. They concluded that M_2CO_2/MoX_2 heterostructure exhibited the intrinsic features of type

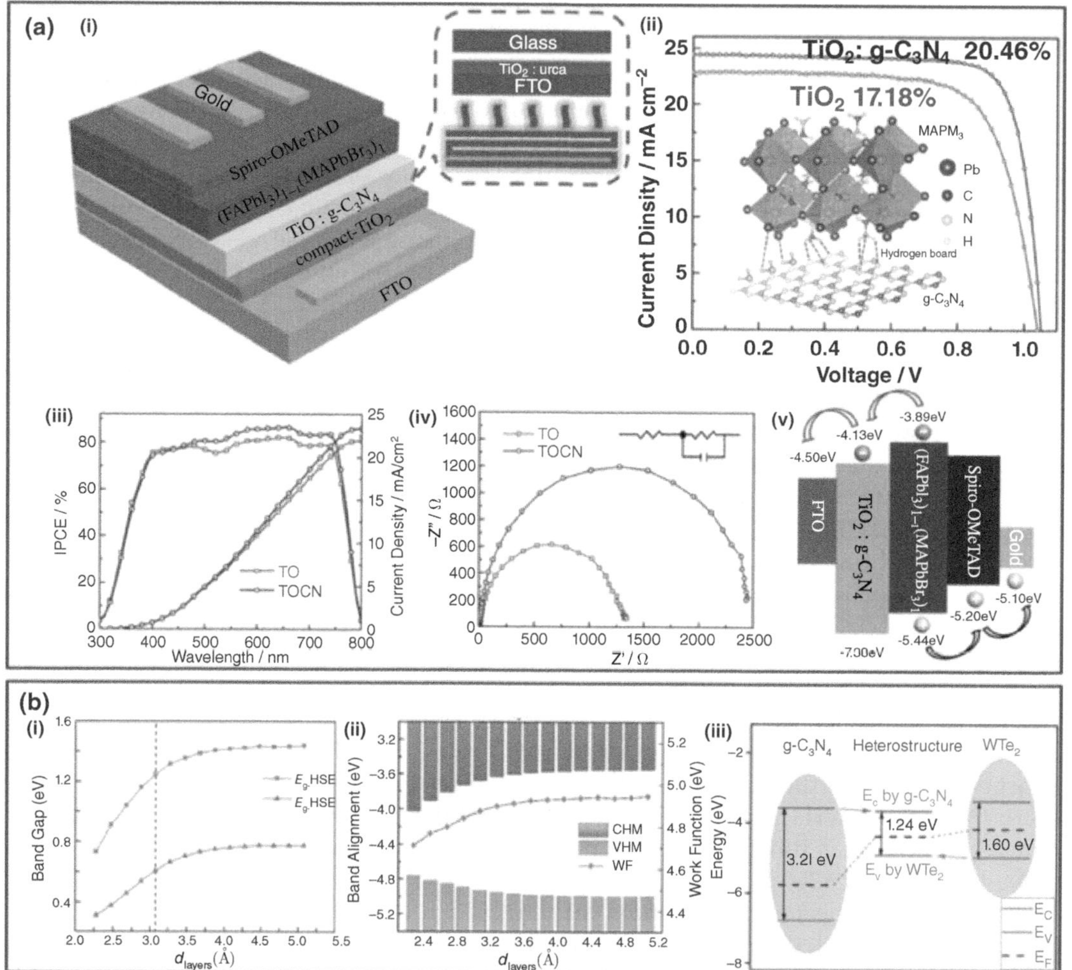

FIGURE 17.3

(a)(i) Schematic illustration of a fabricated PCS using TiO_2/g-C_3N_4 (TOCN) as electron transport layer (ETL), (ii) the recorded J–V curves, (iii) IPCE spectra, and (iv) Nyquist impedance spectra and equivalent circuit (inset) of the PSC TOCN ETL compared to the bare TiO_2 ETL (TO), and (v) the band energy diagram of the whole structure of the fabricated solar cells using TO and TOCN ETLs. Adapted with permission [23], Copyright (2021), Elsevier, and (b)(i) bandgap, (ii) work function and band alignments of g-C_3N_4/WTe_2 heterostructure as a function of van der Waal (vdW) gap, and (iii) the band alignment diagrams for isolated g-C_3N_4, WTe_2 monolayer and heterostructure interface. Adapted with permission [21], Copyright (2022), the Royal Society of Chemistry.

II band structures that efficiently diminish the electron–hole recombination and improve the photocatalytic behavior. M_2CO_2/MoX_2 heterostructures exhibited more suitable bandgaps compared to the monolayers of M_2CO_2 and MoX_2. This resulted in the enhancement of the light-harvesting capabilities in the UV–visible light ranges.

17.3 Optical Supercapacitors

Supercapacitors are one of the most promising options for energy storage owing to quick charge/discharge rates, a long cycle life, low maintenance requirements, and a strong commitment to environmental sustainability and safety [24,25]. Supercapacitors fundamentally distinguish themselves from conventional capacitors by storing electrical energy through two distinct mechanisms based on used active materials: electrical double layers (carbon-based materials such as activated carbon, graphene, and CNTs) [26] and rapid faradaic processes (conducting polymers, metal oxides, etc.) [27]. As a result, supercapacitor applications are constrained to the storage of exclusively electrochemical energy. **Figure 17.4(a)** demonstrates a simplified plot known as the Ragone plot that is commonly utilized to compare the power and energy capabilities of energy storage systems. It was found that fuel cells are high-energy systems, whereas supercapacitors are high-power systems, and batteries exhibit intermediate power/energy features.

However, recent developments have focused on the synergistic coupling of diverse energy sources. One notable example is the integration of SCs with supercapacitors, resulting in what is known as optical/solar/photo-supercapacitors, where solar energy conversion is directly harnessed for energy storage [28]. An innovative approach by Xu et al. [29] involved combining organometallic PSCs with polypyrrole-based supercapacitors to

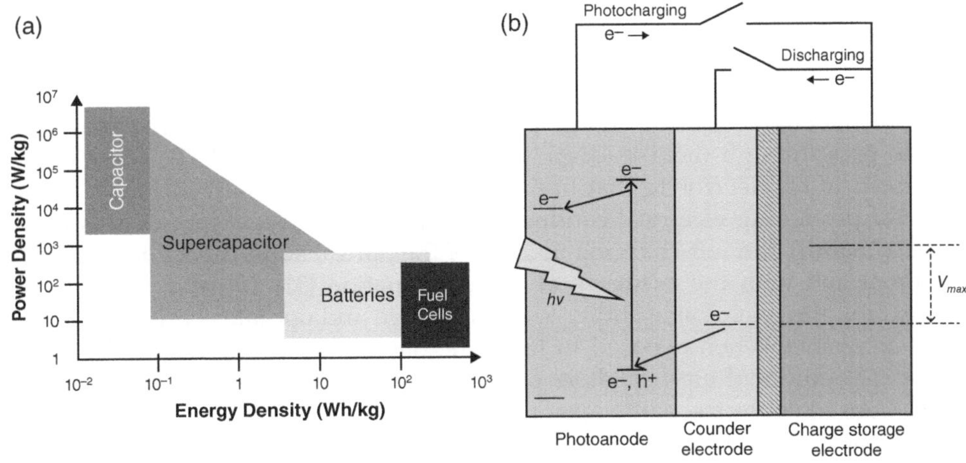

FIGURE 17.4

(a) Simplified Ragone plot representing the distribution of energy density and power density for electrochemical energy storage systems. Adapted with permission [31], Copyright (2023), Springer, and (b) schematic illustration of the setup of a solar-rechargeable energy storage system. Adapted with permission [5], Copyright (2015), Wiley.

efficiently store converted solar energy. On a parallel front, Chien et al. [30] introduced a power system that incorporates organic photovoltaic cells connected in series alongside supercapacitors, serving dual roles in energy conversion and storage. It is worth noting that many of these hybrid cells represent a departure from the traditional approach, where energy conversion and storage cells are separately integrated.

The general concept of photo-rechargeable energy systems including photo-supercapacitors and photo-rechargeable batteries is demonstrated in **Figure 17.4(b)** [5]. A photoactive storage system consists of a photoanode, a counter electrode, and a charge storage electrode. Upon exposing the photoanode to light irradiation, the photons with an energy exceeding the bandgap energy of the photoactive material can excite it and form electron–hole pairs. The electrons transfer to the charge storage electrode to be stored there. In parallel, the electrons of the counter electrode can counterbalance the holes at the photoanode, and the device will be charged. During the discharge process, the electrons transfer from the charge storage electrode to the counter electrode. A membrane is used as a separator to separate the counter electrode from the charge storage electrode, in addition to providing the counterbalance by the cations. The mechanisms of energy storage in supercapacitors and batteries are different. However, there are some mutual electrochemical similarities such as the occurrence of the energy-providing processes at the phase boundaries of the electrode/electrolyte interfaces and the separation of transport of electrons and ions.

Hence, the primary challenge posed by these devices lies in the complex process of integrating two or more distinct cells. Consequently, there is a strong desire to directly incorporate both light energy conversion and storage components within a compact electrochemical cell "optical supercapacitor" and the design of transparent conductive electrode materials [32]. Numerous benefits are provided by the architectural layout of these integrated devices, including a novel method for utilizing renewable energy sources, and meeting the growing demand for effective and sustainable power solutions, a decrease in energy waste, continuous self-generated energy storage for autonomous device applications, an increase in the lifespan of supercapacitors, streamlined device connections, and lower overall costs [33].

It is important to be aware of the basic elements and operations of the optical supercapacitors to get a thorough knowledge of this mechanism [33]. Optical supercapacitors consist of two active electrode materials, serving as the core energy storage sites that are essential for the charge and discharge processes. They are mostly transparent to allow photons to pass through them as displayed in **Figure 17.5(a)(i)** [32,33]. These electrodes are contained in an electrolyte that has anions and cations—charged particles that are essential for promoting electrical conduction within the device. Moreover, as displayed in **Figure 17.5(a)(ii)**, the self-charging activity of the optical supercapacitor was evaluated both without and with the existence of UV illumination [33]. Notably, optical supercapacitors exhibit the remarkable ability to self-recharge and operate continuously if a light source is accessible. When exposed to light, the photon-absorbing material initiates the generation of additional electron–hole pairs, effectively replenishing the stored energy reservoir. Consequently, the bandgaps of various semiconductors, hole-electron recombination, and the caliber of various interfaces are factors that affect overall photoconversion and storage efficiency. It is essential to optimize these variables to increase optical supercapacitor efficiency.

A semitransparent, and flexible solid-state optical supercapacitor was developed by $Ti_3C_2T_x$ MXene all-solution-processed. The transparent MXene film was used as both the top electrode of the photovoltaics cell and the bottom electrode of the supercapacitor units [34]. This kind of all-solution-processed optical supercapacitor built at low temperatures

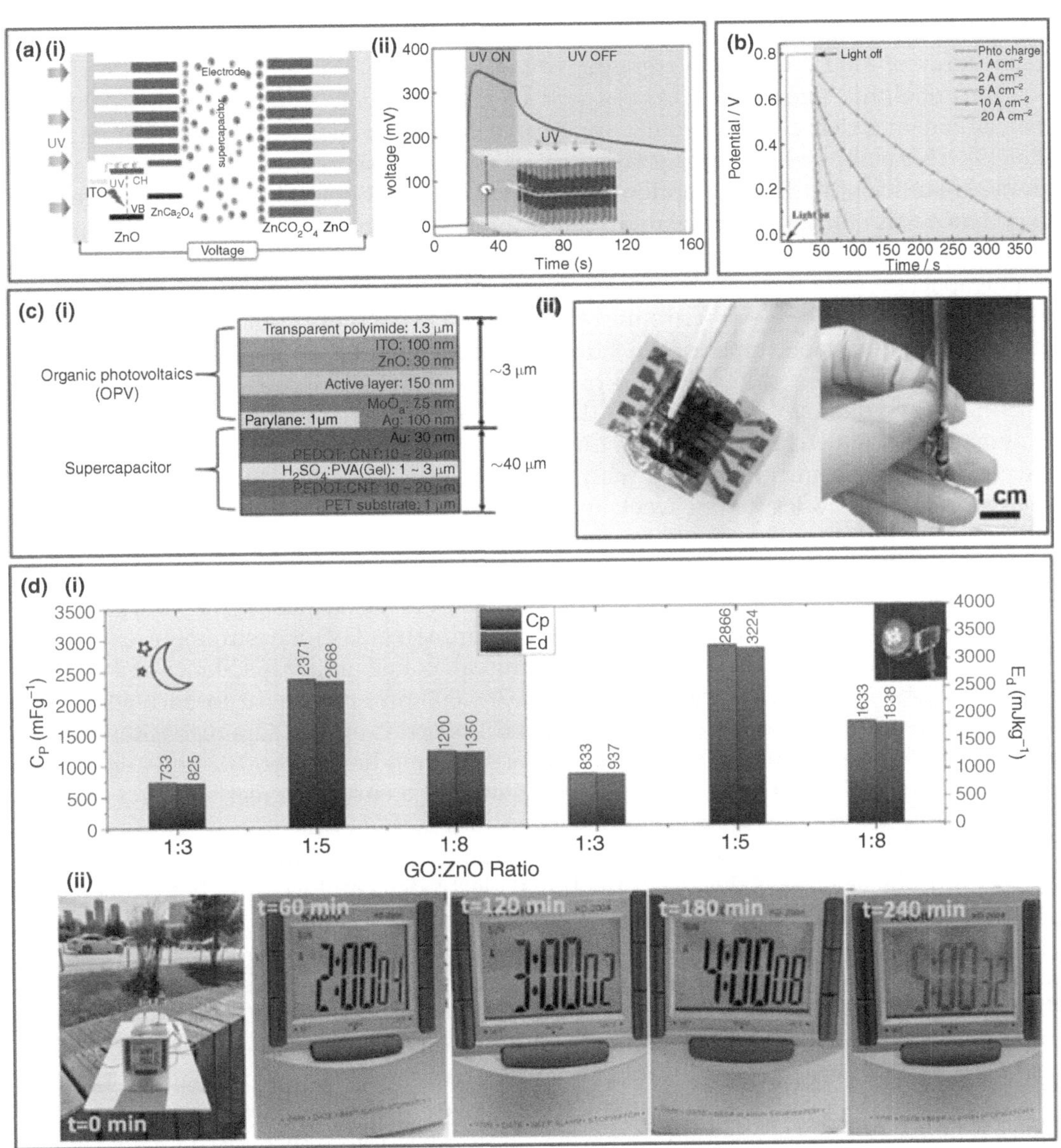

FIGURE 17.5

(a)(i) Schematic showing the operation mechanism of the optical supercapacitor under UV illumination and (ii) self-charging process of an optical supercapacitor under UV illumination. Adapted with permission [33], Copyright (2019), the American Chemical Society. (b) Photo-charging and discharge response of the MXene optical supercapacitor during sunlight-simulated illumination. Adapted with permission [34], Copyright (2020), Royal Society of Chemistry. (c) Photographs of the as-fabricated ultra-thin optical supercapacitor (left) and the device wrapped onto a rod. Adapted with permission [36] Copyright (2020), Wiley. (d)(i) Calculated Cp and Ed values of the PSC devices at dark and under UV illumination and (ii) outdoor experiment of two cells of photo-supercapacitor charged at daylight and no bias. Adapted with permission [37], copyright (2023), Elsevier.

provided high optical transparency of over 33.5% at a wavelength of 550 nm, great flexibility, and excellent storage efficiency of 88%. **Figure 17.5(b)** illustrates the charging and discharging behavior of this MXene optical supercapacitor, the device fully charged within a charging time of 2 s (yellow section) under the illumination of AM 1.5 and spent a longer discharge time of more than 350 s after turning off the light source, delivering a volumetric capacitance of 410 F cm^{-3} at a current density of 1 A cm^{-3}.

The monolithic integration process was used to develop compact, lightweight, low-power-consumption electrochromic supercapacitors with energy conversion photovoltaics [35]. This self-powered optical supercapacitor was designed to produce photocurrent under insufficient sunlight or indoor artificial lights (time for fully charging under LED irradiance was ≈91 s), providing an all-day operation device. The storage part of this integrated optical supercapacitor affords dual functions as electrochromic cell displays and supercapacitors, and the photovoltaic was fabricated to generate energy matches with the power required for the electrochromic cell operation to get highly efficient optical supercapacitor with areal capacitance of ≈26.2 mF cm^{-2} at current density of 0.04 mA cm^{-2}.

A flexible and ultra-thin (≈50 µm) optical supercapacitor was assembled by combining an organic photovoltaic cell (3 µm thickness) with another thinner carbon nanotube/polymer-based supercapacitor (40 µm thickness) (**Figure 17.5(c)**) [30]. This device provided excellent electrochemical stability with 96% efficiency retention after operation for 100 charge/discharge cycles for one week and presented impressive mechanical robustness of 94.66% efficiency retention after 5000 times of bending.

3D carbon foam-MoS$_2$ optical micro-supercapacitor was developed by Kumar et al. [38] The presence of optically sensitive MoS$_2$ semiconductors facilitated the charge generation for efficient energy storage. The cyclic voltammetry (CV) measurement exhibited a larger CV area under light illumination, and the calculated areal capacitance at a scan rate of 5 mVs^{-1} showed an enhancement of ~46% (λ ~ 600 nm) compared to dark conditions. Furthermore, under the same illumination conditions, this optical supercapacitor exposed a nearly 80% increase in the discharge time at a current density of 0.026 mA cm^{-2} and a small IR (~0.076 V), the calculated areal capacitance at a current density of 0.013 mA cm^{-2} was 7.24 mF cm^{-2} with an improvement in capacitance of ~145%.

In the same manner, Altaf et al. [37] fabricated graphene oxide/zinc oxide (GO/ZnO) nanocomposite as photoactive dual-functional electrodes employed in photo-supercapacitors. They optimized the critical ratio of GO/ZnO composition. The PL spectra and electron paramagnetic resonance analysis suggested that the nanocomposite with the 1:5 ratio exhibited a very low carbon-related defect concentration with respect to the other investigated ratios. The GO: ZnO composite was tested as solid-state PSC devices and it was found that the specific capacitance of the optimum nanocomposite reached 6612 mFg^{-1} after UV irradiation achieving 2.7-fold enhancement as demonstrated in **Figure 17.5(d)(i)**. The fabricated device displayed superior stability over 30,000 cycles with 99.6% capacitance retention and 100% columbic efficiency. Also, it was found that the calculated energy density and power density were 6.3 Whkg^{-1} and 625 Wkg^{-1}, respectively at 2.5 V operating voltage under UV irradiation. Notably, the outdoor experiments were evaluated after connecting two photo-supercapacitor cells in series and keeping them in daylight for just 10 seconds and interestingly the clock worked for more than 4 hours (**Figure 17.5(d)(ii)**).

17.4 Photo-Rechargeable Batteries

Currently, batteries have drawn a significant amount of interest in the past few decades for energy storage in the world. Batteries are used in numerous gadgets including electronic cameras, mobile devices, power tools, and laptops. Thus, the development of low-cost high-performance rechargeable batteries with high energy density has become an urgent demand [39]. Nanoscale electrode design with a large electrochemically active surface and short ion diffusion paths can facilitate the device reaction kinetics, which significantly enhances energy and power density and cycle life. The direct combination of photoactive materials with storing devices has been seen as one of the most attractive trends. This is attributed to the expected benefits such as enhanced efficiency and reduced volume-to-weight ratio when compared to conventional systems which consist of two separated devices [40].

In non-integrated systems, a photovoltaic cell is utilized to work as a solar energy harvester [41] with an external electrochemical device (rechargeable battery or supercapacitor) to store electricity externally in the form of chemical energy [42]. These two separate systems are connected by an external electrical wire, which usually suffers from increased Ohmic resistance due to the long distance, low integration levels between the two systems, and high energy losses [40]. To avoid such losses, integrated photo-rechargeable energy storage systems have been adopted such as photo-capacitors, photo batteries (PBATs), and redox flow batteries (RFBs) [5,43]. PBATs will be the center of focus in this section.

Integrated PBATs can be categorized based on the number of electrodes (i.e., two or three) [10]. A schematic diagram of two-electrode and three-electrode PBATs configurations is shown **in Figure 17.6(a)(i) and (ii).** A photoelectrode (PE), a counter electrode (CE)

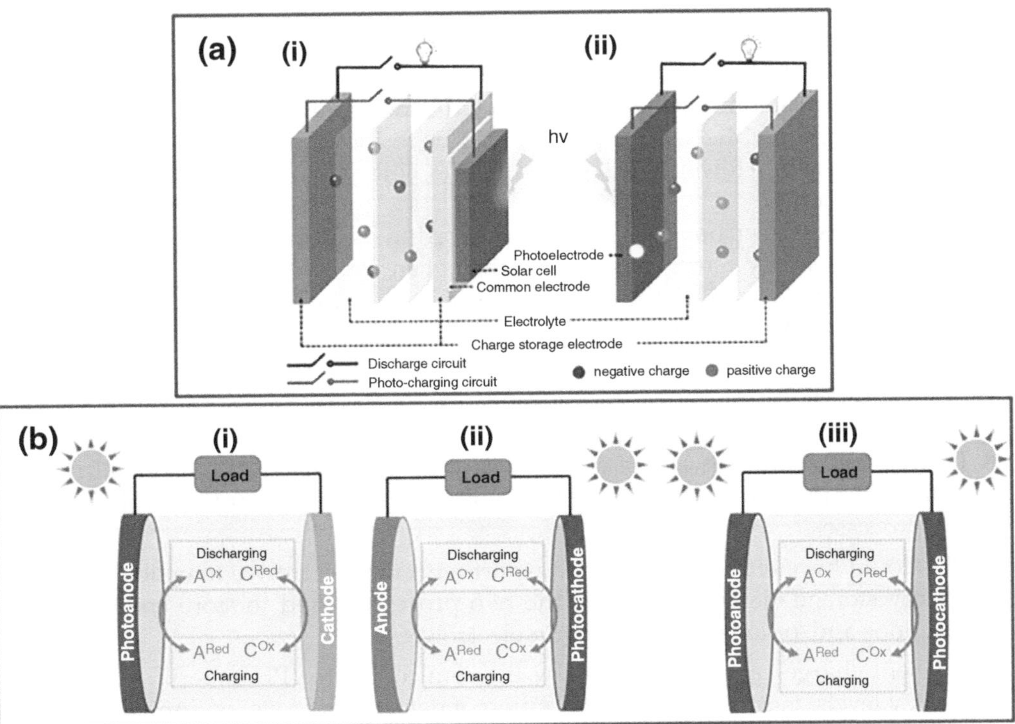

FIGURE 17.6

(a) Battery configuration of integrated PBATs: (i) three-electrode PBAT and (ii) two-electrode PBAT. Adapted with permission [46], Copyright (2020), Wiley. (b) Schematic of photo-responsive battery systems with two-electrode configuration. (i) The photoanode-cathode design, (ii) the anode-photocathode design, and (iii) the tandem design (the photoanode-photocathode design). Adapted with permission [45] Copyright (2019), Wiley that serves as a redox electron transfer surface, and an energy storage electrode (anode) are the main typical components of the three-electrode PBAT [44]. Under illumination, the photoactive material in PE is excited and forms e-h pairs. Assuming the voltage generated has been enough to activate the electrochemical charging process inside the battery, the electrons travel to the anode to be either stored or used to decrease the dynamic species [40]. On the other hand, holes in the PE are counterbalanced *via* electrons at the CE or by redox species which are reduced at the CE. An external bias must be applied when the voltage has not been enough to reduce the active species. In such cases, the device is described as a photo-assisted rechargeable battery [5]. It aids in reducing the overpotential; however, these devices cannot be charged solely using solar energy [40]. An integrated two-electrode PBAT can directly convert/store solar energy. It includes a photoanode in addition to a photocathode where one or both of them should participate in the energy conversion process as well as the energy storage process [45]. Any traditional rechargeable battery usually consists of cathodes, anodes, separators, and electrolytes [8,42]. Therefore, the configuration of the PBAT should fully take into consideration the compatibility of such parts. Based on the role of each of the two photoelectrodes, two-electrode PBATs are categorized into three designs as shown in **Figure 17.6(b)(i)–(iii)**, which represents the photoanode-cathode design, the anode-photocathode design, and the tandem electrode design, respectively [45].

PE represents the main parameter that controls the performance of PBATs and usually its total efficiency. Thus, designing and selecting the materials used as PE is critical for the efficient conversion and storage of solar energy. Some of the drawbacks impeding the enhanced performance of PBATs and related to PE include, but are not limited to, low light absorption, high charge recombination rate, poor charge transportation, and weak photo and chemical stability [46]. In this aspect, the focus of this section is centered on the photoactive materials utilized in PBATs. PE can be in the form of a photovoltaic cell or a simple semiconductor. However, integrated PBATs with photovoltaic cells are outside the scope of this section.

In integrated two-electrode PBATs, PE is referred to as bifunctional as it has two functions: solar light energy harvesting and storage chemically. As reported earlier [40], the following conditions represent the essential requirements for the semiconductor PE to get enhanced performance within PBATs:

- PE should have an optimal bandgap of 1–2 eV to ensure efficient light absorption within a broad range of the spectrum.

- PE material should be stable in terms of photo-, thermo-, and chemical performance within the range of the working voltage during the charging and discharging processes.

- PE should have efficient charge transport with the electrolyte to enable excellent energy storage via one or two of the two processes used to store energy: redox reactions and/or ions transfer within electrolytes.

- Efficient charge-carrier transport by the semiconductor PE material through its energy levels, morphology, concentration of defects, and level of doping.

Jia et al. [47] developed a bifunctional photo-assisted $Li–O_2$ system using siloxene nanosheets due to its superior light harvesting, and good semiconductor characteristics, in addition to the low rate of recombination. The utilization of the bifunctional photocatalyst into $Li–O_2$ batteries enabled a very low charging potential of 1.90 V and a very high discharge of 3.51 V. These results have achieved 185% round-trip efficiency, an extended cycle life with efficiency retention of 92% after 100 cycles, and a high reversible capacity of 1170 mAh g^{-1}. Another study by Li et al. [48] used SiC/RGO to act as a photo-electro-catalytic hybrid cathode in a photo-assisted Li–CO$_2$ system to overcome the polarization issue common in conventional Li–CO$_2$ batteries. RGO can transfer high-energy electrons generated on SiC because of absorbing photons, which prevents the recombination of pho-togenerated electrons. The discharge voltage of the battery was found to be 2.77 V, which was found to be close to the theoretical potential for the reaction $4Li+3CO_2{\rightarrow}2Li_2CO_3+C$ (i.e., 2.80 V). Additionally, upon charging, the holes generated on the photoelectrode promoted the Li$_2$CO$_3$ decomposition. This PBAT revealed a promising energy efficiency of 84.4%. Liu et al. [49] utilized a Li–O$_2$ battery composed of a cathode and a photoelectrode made of g-C$_3$N$_4$ on carbon paper. This was found to possess a very low charge voltage (1.96 V), causing a negative overpotential in addition to a promising charge/discharge cycling performance.

17.5 Conclusion

To sum up, solar-rechargeable energy systems such as SCs, solar supercapacitors, and photo-rechargeable batteries have emerged as outstanding alternatives to traditional storage systems. The employment of semiconducting 2D nanomaterials in these systems showed promising performance. Tuning the bandgap energy, enhancing the charge sepa-ration, and extending the optical absorbance to the visible region are the main challenges to improving the photocatalytic performance of these 2D nanomaterials. These can be achieved by various strategies and the fabrication of heterojunctions was found to be one of the most promising strategies to enhance the optical and photocatalytic performance of semiconductors. There should be a focus on addressing key technical challenges of the solar chargeable systems in parallel with fundamental research and exploring uncharted territory with novel materials and inventive device configurations. Finally, the future of solar energy systems seems incredibly bright and is rife with potential for growth, innova-tion, and beneficial societal effects.

References

1. A.N. El-Shazly, M.A. Hamza, A.E. Shalan, Major trends and mechanistic insights for the devel-opment of TiO$_2$-based nanocomposites for visible-light-driven photocatalytic hydrogen pro-duction, in: A.E. Shalan, A.S. Hamdy Makhlouf, S. Lanceros-Méndez (Eds.), *Adv. Nanocomposite Mater. Environ. Energy Harvest. Appl. Eng. Mater.* Springer International Publishing, Cham (2022) 771–794.
2. F. Fuso Nerini, J. Tomei, L.S. To, I. Bisaga, P. Parikh, M. Black, A. Borrion, C. Spataru, V. Castán Broto, G. Anandarajah, B. Milligan, Y. Mulugetta, Mapping synergies and trade-offs between energy and the sustainable development goals, *Nat. Energy.* 3 (2018) 10–15.
3. M.A. Hamza, A.N. El-shazly, S.A. Tolba, N.K. Allam, Novel Bi-based photocatalysts with unprecedented visible light-driven hydrogen production rate: Experimental and DFT insights, *Chem. Eng. J.* 384 (2020) 123351.

4. N. Flores-Diaz, F. De Rossi, A. Das, M. Deepa, F. Brunetti, M. Freitag, Progress of photocapacitors, *Chem. Rev.* 123 (2023) 9327–9355.

5. D. Schmidt, M.D. Hager, U.S. Schubert, Photo-rechargeable electric energy storage systems, *Adv. Energy Mater.* 6 (2016) 1500369.

6. A.N. El-Shazly, A.H. Hegazy, E.T. El Shenawy, M.A. Hamza, N.K. Allam, Novel facet-engineered multi-doped TiO_2 mesocrystals with unprecedented visible light photocatalytic hydrogen production, *Sol. Energy Mater. Sol. Cells.* 220 (2021) 110825.

7. A.N. El-Shazly, M.A. Hamza, N.K. Allam, Enhanced photoelectrochemical water splitting via engineered surface defects of $BiPO_4$ nanorod photoanodes, *Int. J. Hydrogen Energy.* 46 (2021) 23214–23224.

8. M. Salah, T. Pathirana, E. Alvarez De Eulate, C. Hall, B. Stoehr, S. Rudd, R. Kerr, M. Fabretto, Physical vapor deposition cluster arrival energy enhances the electrochemical performance of silicon thin-film anodes for Li-ion batteries, *ACS Appl. Energy Mater.* 4 (2021) 12243–12256.

9. A. Paolella, A. Vijh, A. Guerfi, K. Zaghib, C. Faure, Review—Li-ion photo-batteries: Challenges and opportunities, *J. Electrochem. Soc.* 167 (2020) 120545.

10. L. Wang, L. Wen, Y. Tong, S. Wang, X. Hou, X. An, S.X. Dou, J. Liang, Photo-rechargeable batteries and supercapacitors: Critical roles of carbon-based functional materials, *Carbon Energy.* 3 (2021) 225–252.

11. M.A. Hamza, S.A. Rizk, E.-E.M. Ezz-Elregal, S.A.A. El-Rahman, S.K. Ramadan, Z.M. Abou-Gamra, Photosensitization of TiO_2 microspheres by novel Quinazoline-derivative as visible-light-harvesting antenna for enhanced Rhodamine B photodegradation, *Sci. Rep.* 13 (2023) 12929.

12. A. Castellanos-Gomez, Why all the fuss about 2D semiconductors? *Nat. Photonics.* 10 (2016) 202–204.

13. U.K. Aryal, M. Ahmadpour, V. Turkovic, H.G. Rubahn, A. Di Carlo, M. Madsen, 2D materials for organic and perovskite photovoltaics, *Nano Energy.* 94 (2022) 106833.

14. G. Li, R. Zhu, Y. Yang, Polymer solar cells, *Nat. Photonics.* 6 (2012) 153–161.

15. U.K. Aryal, M. Ahmadpour, V. Turkovic, H.G. Rubahn, A. Di Carlo, M. Madsen, 2D materials for organic and perovskite photovoltaics, *Nano Energy.* 94 (2022) 106833.

16. X. He, Y. Iwamoto, T. Kaneko, T. Kato, Fabrication of near-invisible solar cell with monolayer WS2, *Sci. Rep.* 12 (2022) 1–8.

17. L. Shenghuang, F. Nianqing, B. Qiaoliang, Recent advances in the applications of elemental two-dimensional materials and their derivatives serving as transport layers in solar cell, *Chem. J. Chinese Univ.* 42 (2021) 412–431.

18. L.-L. Jiang, Z.-K. Wang, M. Li, C.-H. Li, P.-F. Fang, L.-S. Liao, Flower-like MoS_2 nanocrystals: a powerful sorbent of Li^+ in the Spiro OMeTAD layer for highly efficient and stable perovskite solar cells, *J. Mater. Chem. A.* 7 (2019) 3655–3663.

19. M. Liang, A. Ali, A. Belaidi, M.I. Hossain, O. Ronan, C. Downing, N. Tabet, S. Sanvito, F. El-Mellouhi, V. Nicolosi, Improving stability of organometallic-halide perovskite solar cells using exfoliation two-dimensional molybdenum chalcogenides, *NPJ 2D Mater. Appl.* 4 (2020) 40.

20. F. Xie, G. Dong, K. Wu, Y. Li, M. Wei, S. Du, In situ synthesis of g-C_3N_4 by glass-assisted annealing route to boost the efficiency of perovskite solar cells, *J. Colloid Interface Sci.* 591 (2021) 326–333.

21. P. Lin, N. Xu, X. Tan, X. Yang, R. Xiong, C. Wen, B. Wu, Q. Lin, B. Sa, The interlayer coupling modulation of a g-C_3N_4/WTe_2 heterostructure for solar cell applications, *RSC Adv.* 12 (2022) 998–1004.

22. J. Wen, Q. Cai, R. Xiong, Z. Cui, Y. Zhang, Z. He, J. Liu, M. Lin, C. Wen, B. Wu, B. Sa, Promising M_2CO_2/MoX_2(M=Hf,Zr;X=S,Se,Te) heterostructures for multifunctional solar energy applications, *Molecules.* 28 (2023).

23. F. Xie, G. Dong, K. Wu, Y. Li, M. Wei, S. Du, In situ synthesis of g-C_3N_4 by glass-assisted annealing route to boost the efficiency of perovskite solar cells, *J. Colloid Interface Sci.* 591 (2021) 326–333.

24. M.M. Taha, L.G. Ghanem, M.A. Hamza, N.K. Allam, Highly stable supercapacitor devices based on three-dimensional bioderived carbon encapsulated g-C_3N_4 nanosheets, *ACS Appl. Energy Mater.* 4 (2021) 10344–10355.

25. L.G. Ghanem, M.A. Hamza, M.M. Taha, N.K. Allam, Symmetric supercapacitor devices based on pristine g-C_3N_4 mesoporous nanosheets with exceptional stability and wide operating voltage window, *J. Energy Storage.* 52 (2022).

26. J. Zhang, J. Luo, Z. Guo, Z. Liu, C. Duan, S. Dou, Q. Yuan, P. Liu, K. Ji, C. Zeng, J. Xu, W. Liu, Y. Chen, W. Hu, Ultrafast manufacturing of ultrafine structure to achieve an energy density of over 120 Wh kg^{-1} in supercapacitors, *Adv. Energy Mater.* 13 (2023) 1–9.

27. Lichchhavi, A. Kanwade, P.M. Shirage, A review on synergy of transition metal oxide nanostructured materials: Effective and coherent choice for supercapacitor electrodes, *J. Energy Storage.* 55 (2022) 105692.

28. C.H. Ng, H.N. Lim, S. Hayase, I. Harrison, A. Pandikumar, N.M. Huang, Potential active materials for photo-supercapacitor: A review, *J. Power Sources.* 296 (2015) 169–185.

29. X. Xu, S. Li, H. Zhang, Y. Shen, S.M. Zakeeruddin, M. Graetzel, Y.B. Cheng, M. Wang, A power pack based on organometallic perovskite solar cell and supercapacitor, *ACS Nano.* 9 (2015) 1782–1787.

30. C.T. Chien, P. Hiralal, D.Y. Wang, I.S. Huang, C.C. Chen, C.W. Chen, G.A.J. Amaratunga, Graphene-based integrated photovoltaic energy harvesting/storage device, *Small.* 11 (2015) 2929–2937.

31. I.S. Ismail, M.F.H. Othman, N.A. Rashidi, S. Yusup, Recent progress on production technologies of food waste–based biochar and its fabrication method as electrode materials in energy storage application, *Biomass Convers. Biorefinery.* 13 (2023) 14341–14357.

32. S. Kim, S. Lee, Transparent supercapacitors: From optical theories to optoelectronics applications, *Energy Environ. Mater.* 3 (2020) 265–285.

33. B.D. Boruah, A. Misra, Voltage generation in optically sensitive supercapacitor for enhanced performance, *ACS Appl. Energy Mater.* 2 (2019) 278–286.

34. L. Qin, J. Jiang, Q. Tao, C. Wang, I. Persson, M. Fahlman, P.O.A. Persson, L. Hou, J. Rosen, F. Zhang, A flexible semitransparent photovoltaic supercapacitor based on water-processed MXene electrodes, *J. Mater. Chem. A.* 8 (2020) 5467–5475.

35. J. Cho, T.Y. Yun, H.Y. Noh, S.H. Baek, M. Nam, B. Kim, H.C. Moon, D.H. Ko, Semitransparent energy-storing functional photovoltaics monolithically integrated with electrochromic supercapacitors, *Adv. Funct. Mater.* 30 (2020).

36. R. Liu, M. Takakuwa, A. Li, D. Inoue, D. Hashizume, K. Yu, S. Umezu, K. Fukuda, T. Someya, An efficient ultra-flexible photo-charging system integrating organic photovoltaics and supercapacitors, *Adv. Energy Mater.* 10 (2020) 1–8.

37. C.T. Altaf, T.O. Colak, A.M. Rostas, M. Mihet, M.D. Lazar, I. Iatsunskyi, E. Coy, I.D. Yildirim, F.B. Misirlioglu, E. Erdem, M. Sankir, N.D. Sankir, GO/ZnO-based all-solid-state photo-supercapacitors: Effect of GO:ZnO ratio on composite properties and device performance, *J. Energy Storage.* 68 (2023) 107694.

38. S. Kumar, A. Mukherjee, S. Telpande, A. Das Mahapatra, P. Kumar, A. Misra, Role of the electrode-edge in optically sensitive three-dimensional carbon foam-MoS_2 based high-performance micro-supercapacitors, *J. Mater. Chem. A.* 11 (2023) 4963–4976.

39. E. Riyanto, T. Kristiantoro, E. Martides, Dedi, B. Prawara, D. Mulyadi, Suprapto, Lithium-ion battery performance improvement using two-dimensional materials, *Mater. Today Proc.* 87 (2023) 164–171.

40. C. Rodríguez-Seco, Y.S. Wang, K. Zaghib, D. Ma, Photoactive nanomaterials enabled integrated photo-rechargeable batteries, *Nanophotonics.* 11 (2022) 1443–1484.

41. M.W. Aljibory, H.T. Hashim, W.N. Abbas, A review of solar energy harvesting utilising a photovoltaic–thermoelectric integrated hybrid system, *IOP Conf. Ser. Mater. Sci. Eng.* 1067 (2021) 012115.

42. M. Salah, T. Pathirana, E.A. de Eulate, C. Hall, R. Kerr, M. Fabretto, Effect of vinylene carbonate electrolyte additive and battery cycling protocol on the electrochemical and cyclability performance of silicon thin-film anodes, *J. Energy Storage.* 46 (2022) 103868.

43. M. Skunik-Nuckowska, P. Rączka, J. Lubera, A.A. Mroziewicz, S. Dyjak, P.J. Kulesza, I. Plebankiewicz, K.A. Bogdanowicz, A. Iwan, Iodide electrolyte-based hybrid supercapacitor for compact photo-rechargeable energy storage system utilising silicon solar cells, *Energies.* 14 (2021).

44. H. Xue, H. Gong, Y. Yamauchi, T. Sasaki, R. Ma, Photo-enhanced rechargeable high-energy-density metal batteries for solar energy conversion and storage, *Nano Res. Energy.* 1 (2022) 1–8.
45. Z. Fang, X. Hu, D. Yu, Integrated photo-responsive batteries for solar energy harnessing: Recent advances, challenges, and opportunities, *Chempluschem.* 85 (2020) 600–612.
46. Q. Zeng, Y. Lai, L. Jiang, F. Liu, X. Hao, L. Wang, M.A. Green, Integrated photorechargeable energy storage system: Next-generation power source driving the future, *Adv. Energy Mater.* 10 (2020).
47. C. Jia, F. Zhang, L. She, Q. Li, X. He, J. Sun, Z. Lei, Z. Liu, Ultra-large sized siloxene nanosheets as bifunctional photocatalyst for a Li-O_2 battery with superior round-trip efficiency and extra-long durability, *Angew. Chemie Int. Ed.* 60 (2021) 11257–11261.
48. Z. Li, M.-L. Li, X.-X. Wang, D.-H. Guan, W.-Q. Liu, J.-J. Xu, In situ fabricated photo-electro-catalytic hybrid cathode for light-assisted lithium-CO_2 batteries, *J. Mater. Chem. A.* 8 (2020) 14799–14806.
49. Y. Liu, N. Li, K. Liao, Q. Li, M. Ishida, H. Zhou, Lowering the charge voltage of Li–O_2 batteries via an unmediated photoelectrochemical oxidation approach, *J. Mater. Chem. A.* 4 (2016) 12411–12415.

18

Energy Generation Devices Based on 2D-Semiconducting Materials

Ramesh Sivasamy*, Marutheeswaran Srinivasan, Edgar Mosquera Vargas, and Konrad Szaciłowski

**Correspondence: rameshsiva_chem@yahoo.com*

18.1 Introduction

Nanomaterials are a fascinating class of materials distinguished by size and can be systematically categorized depending on their dimensions. They are an essential part of the vast subject of nanoscience and nanotechnology. Two-dimensional (2D) nanomaterials, a new family of materials with atomic- to nanoscale thicknesses, have emerged because of the discovery of graphene.[1] Unlike their bulk counterparts, 2D nanomaterials exhibit distinct properties. Because of their unique physical and chemical properties, 2D nanomaterials have considerable potential for usage in various sectors, including electronics, biology, energy, and sensing.[2,3] With graphene, whose impressive physical, optical, and electrical characteristics have drawn researchers' interest, various ultrathin 2D nanomaterials have been explored that have layered structures yet display unique and fascinating features. Researchers found roughly 5,619 layered 2D nanomaterials out of 108,423 distinct, experimentally known 3D materials in the database.[4] These layered materials include transition metal dichalcogenides (TMDs),[5] graphitic carbon nitride (g-C3N4),[6] Metal Xenes (MXenes),[7] Nonmetal Xenes,[8,9] layered metal oxides,[10] layered double hydroxides (LDHs)[11] and many other 2D nanosheets.[8,12–14]

Let's start with graphene, as an exemplary material, possesses remarkable attributes such as ultrahigh carrier mobility at room temperature and broad-spectrum light absorption despite its lack of a bandgap, making it clearly superior to semiconducting silicon.[15] This absence of a bandgap in graphene leads to significantly low on/off ratios in field-effect transistor devices because it cannot completely shut off the off-state current. It results in imperceptible photocurrents in photodetectors. To overcome the limitations imposed by graphene's zero gap energy band structure, many 2D layered materials are found to be semiconductors with wide-range band gaps.[16]

Recently, metal chalcogenide layer materials comprising semiconductors with optical bandgap (1.6–2.0 eV) have been considered promising for the next generation of high-performance electronics, energy generation, and optoelectronic devices.[17] Here are several reasons 2D metal chalcogenides (TMC) are ideal for these applications: One of the most compelling features of 2D metal chalcogenides is their stability and tunable bandgap. Researchers can precisely control the electronic bandgap by varying the type of metal, the chalcogen element, or the layer thickness.[18] For example, MoS_2, $MoSe_2$, WS_2, and WSe_2

DOI: 10.1201/9781003439448-18

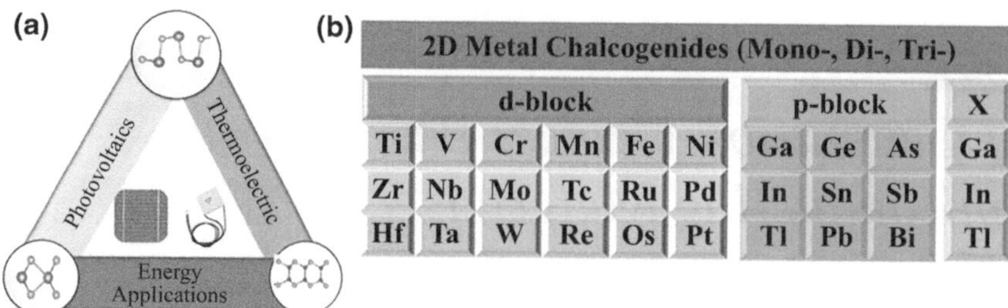

FIGURE 18.1
(a) Schematic layout of the 2D SMCs energy application and (b) possible combinations of the metal chalcogenide layers.

were shown to undergo bandgap transition (indirect-to-direct) when their thickness was reduced to a single layer.[19] High electron carrier mobility is another important characteristic of electrical devices, and several 2D metal chalcogenides exhibit high carrier mobility.[20] 2D metal chalcogenides also exhibit strong light-matter interaction due to their 2D nature and direct bandgap properties. This makes them excellent candidates for optoelectronic devices, such as photodetectors, light-emitting diodes, and solar cells.[18] Their ability to absorb and emit light efficiently can lead to highly efficient devices.[21] Additionally, TMDs possess more flexible and high mechanical strength. This is especially useful for flexible and wearable electronics since they cannot be made from conventional bulk materials. Finally, 2D metal chalcogenides can be easily synthesized through various methods, including chemical vapor deposition, liquid-phase exfoliation, and mechanical exfoliation methods.[22, 23] All the above properties, such as stability, high carrier mobility, efficient light absorption, and quantum solid confinement, contribute to energy-efficient devices, which are increasingly important in modern electronics. The schematic layout of the 2D SMCs energy application and possible combinations of the metal chalcogenide layers are shown in **Figure 18.1(a) and (b)**.

This chapter comprehensively overviews the latest advancements in 2D semiconductor metal chalcogenide nanomaterials on elucidating the practical utility in energy applications, such as thermoelectric devices and photovoltaic systems. The central theme of this chapter revolves around strategies and methodologies aimed at enhancing the efficiency of these nanomaterials when integrated into real-world devices.

18.1.1 Energy Applications Using Chalcogenides

18.1.1.1 Photovoltaics

Two-dimensional metal chalcogenide semiconductor materials are promising materials for utilization in solar cell applications because of their distinctive qualities and the flexibility to adjust their characteristics. Solar energy has gained significant prominence as a renewable energy option to effectively address the escalating global energy requirements and combat the adverse effects of climate change. Photovoltaic (PV) solar cells are semiconductors that transform incident solar radiation into electrical power. The advancement of tools that enhance the effective utilization of solar energy is paramount. Solar cells are of utmost importance in decarbonizing energy infrastructure, mitigating carbon footprints, and making progress toward establishing a low-carbon society. Solar cells make a substantial contribution toward achieving these aims, as they generate clean, ecologically friendly, and sustainable power and possess additional benefits.

Materials chemists, for their potential to exhibit photovoltaic properties under solar light illumination, have found various materials. The ability of these materials to absorb light within the visible spectrum is limited because of their wide bandgap, which surpasses 3.2 eV. The Earth's surface receives around 43% of ultraviolet-visible (UV–Vis) radiation, and within this portion, only a tiny fraction of 5% is efficiently utilized for solar energy conversion. In contrast, around 50% of near-infrared (NIR) light penetrates the Earth's surface, significantly enhancing the efficacy of solar power collection. In 2021, the global proportion of electricity generated from solar modules was a modest 4.5%, although coal-based sources made up a significantly more significant portion of 37% [1]. Therefore, exploring new photo-responsive materials that exhibit improved efficiency in boosting solar energy utilization, particularly within the UV–Vis band, is crucial. PV technology has been steadily developing over the past two decades by improving the power conversion efficiencies (PCEs) of solar cells.

The developments have currently reached the fourth-generation solar cell. The first generation of the photovoltaic cell consists of monocrystalline silicon and gallium arsenide. The efficiency was achieved by 15 to 24% with the bandgap of the material being 1.1 eV. The projected life period of the system is 25 years. Notable advantages of the system include its stability, high performance, and extended service life. There are several limitations associated with the subject at hand. Firstly, the manufacturing cost is significantly high. Additionally, the material exhibits heightened sensitivity to temperature variations. Furthermore, there is an issue with absorption, and the material is prone to loss. The second generation encompasses the advancement of initial-generation photovoltaic cell technology, alongside the progress made in thin-film photovoltaic cell technology using materials such as microcrystalline and amorphous silicon, copper indium gallium selenide (CIGS; efficiency 20%), and cadmium telluride/sulfide (CdTe/CdS) cells.

The third generation encompasses technologies such as nanocrystalline films, quantum dots, dye-sensitized, and organic polymer solar cells (5–20%). Organic materials (OSC; 11–17%), Perovskites cells (PSC; 25%), dye-sensitized cells (DSSC; 5–20%), Quantum dots cells (QD; 11–17 %), and multi-junction cells (20.3%). Currently, the fourth generation includes the low flexibility or low cost of thin-film polymers along with the durability of innovative inorganic-organic hybrid nanostructures such as metal oxides and metal nanoparticles or organic-based nanomaterials such as graphene, carbon nanotubes, and graphene derivatives. These devices, which are frequently referred to as nano photovoltaics, have the potential to become the photovoltaic industry's most promising future. Creating a broad spectrum of novel active semiconductors that exhibit satisfactory and consistent photovoltaic efficiency poses a significant and daunting obstacle. It can employ various approaches to enhance the efficiency of solar cell materials, including bandgap engineering, surface area enhancement, and carrier mobility improvement. Chalcogenide materials, which comprise a diverse array of compounds incorporating elements from the chalcogenide group, have notable benefits in effectively capturing solar energy. This prompted the scientific community to undertake comprehensive experiments into different photovoltaic (PV) materials and structures to achieve high PCE while maintaining low costs. Remarkable achievements have been witnessed globally in recent decades, indicative of substantial advancement. It is worth mentioning that quadruple junction solar cells have achieved an exceptional PCE of 44.7%, setting a record in the field [2]. There is anticipated to be a decline in the cost of solar power in the following years, leading to a rise in the annual share of photovoltaic (PV) panels in the worldwide electricity composition. The global photovoltaic (PV) business, which holds a market worth $100 billion, is expected to experience significant expansion due to emerging market opportunities across diverse sectors, including residential, commercial, wearable, and industrial applications.

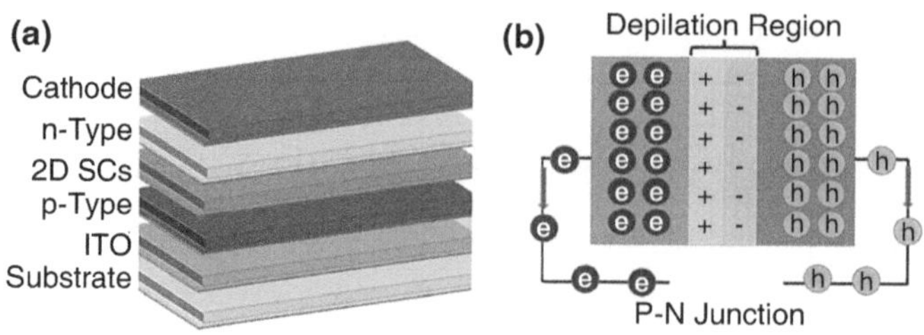

FIGURE 18.2

(a) Schematic layout of the several layers of the solar cell and (b) illustration of the p–n junction formation.

18.2 Photovoltaics Device Construction

The different layers of metal chalcogenide-based photovoltaic devices provide unique purposes critical to the device's overall functionality. The precise materials and later layouts might vary depending on the type of metal chalcogenide and the required device performance. Here, we have outlined the different layers of devices and their construction and functions, as shown in **Figure 18.2**. These layers collaborate to form an effective photovoltaic device.

A multitude of layers encompass the substrate, which functions as the initial layer, providing a sturdy base for the device's structural integrity. Typically, it is fabricated using glass or a comparable transparent substance that facilitates light transmission to the active layers. The subsequent layer corresponds to the transparent conductive oxide (TCO) layer. Indium tin oxide and fluorine-doped tin oxide are widely utilized as TCO material layers. The third layer is commonly employed as the front electrode, is the hole transfer layer (p-type), and is positioned atop the substrate. Its primary function is to facilitate the transmission of positively charged holes generated within the absorber layer to the front electrode. The gadget incorporates PEDOT: PSS, spiro-OMeTAD, or CuSCN to facilitate light transmission and establish electrical connectivity. Utilizing the fourth or window layer is a common practice to enhance electrical contact between the TCO and the subsequent layers. It can also serve as a diffusion barrier, effectively impeding any undesirable chemical interactions between different layers. Frequently, it comprises a transparent material, such as cadmium sulfide (CdS) or zinc oxide (ZnO).The fifth layer of a device is commonly referred to as the electron transport layer (n-type), which serves as the central component responsible for the device's functionality. The electronic transport layer (ETL) facilitates the movement of negatively charged electrons from the absorber layer to the back electrode, often composed of a metal chalcogenide such as cadmium telluride (CdTe) or CIGS. The layer mentioned earlier handles the absorption of light energy and the subsequent generation of electron–hole pairs, initiating the photovoltaic phenomenon. The sixth layer, the cathode or counter-electrode buffer layer, is commonly utilized to enhance charge extraction efficiency from the absorber layer to subsequent layers. It facilitates the mitigation of energy dissipation and improves the overall operational efficiency of the gadget. The choice of materials for back electrodes includes platinum (Pt) or carbon-based compounds. The rear electrode, often known as the cathode, captures the electrons transported by the ETL and finalizes the electrical circuit.

TABLE 18.1

List of the Various Transition Metal Chalcogenides-Based Pv Solar Cells and Their Successive Layers

Layers PV Based On	Electron Transport Layer (ETL)	Hole Transport Layer (HTL)	Active Layer	Counter-Electrode	Ref.
MoS_2	TiO_2, ZnS, Al_2O_3	MoO3, Spiro-OMeTAD, PEDOT:PSS	MoS_2	Pt	[6]
CdTe	ZnO, CdS, TiO2, SnO2	PEDOT:PSS, CuSCN, MoO3, Spiro-OMeTAD	CdTe	Pt, C, Au	[7]
CIGS		PEDOT:PSS, MoO3, Spiro-OMeTAD, PTAA	CIGS	Pt, C	[8]
CuInS2		Spiro-OMeTAD, PTAA	$CuInS_2$		[8, 9]
CZTS		PEDOT:PSS, CuSCN, MoO3, Spiro-OMeTAD	CuZnSnS		[10, 11]
Cu2S–CuS		PEDOT: PSS, CuSCN	Cu2S–CuS		[12]
Bi2S3			Bi2S3		[10]
In2S3			In2S3		
PbS–PbSe	TiO2, ZnO, PbSe, CuSCN	PEDOT:PSS, MoO3, Spiro-OMeTAD	PbS/PbSe		[10]
WS2–Ag2S	TiO2, ZnO	MoO3, Spiro-OMeTAD, PEDOT:PSS	WS_2–Ag2S		[13]
Fe/Ni-S2/Se2			Fe/Ni–S2/Se2		[14, 15]
ZnSe–ZnTe	ZnO, TiO2, CdS, MoO3	PEDOT:PSS, CuSCN, Spiro-OMeTAD	ZnSe–ZnTe		[5]
SnS–SnSe			SnS–SnSe		[4]

We have seen various semiconductor phases in transition metal chalcogenides (TMCs), such as MoS_2, CrS2, WS_2, TiS2, $MoSe_2$, $CrSe_2$, WSe_2, TiSe2, and various combinations [3]. According to the source, TMCs comprise a diverse range of macroscopic crystal structures manifest in separate crystalline phases, namely 1T, 2H, and 3R. Roughly 66% of these materials exhibit layered architectures [4]. In particular, it is noteworthy that MoS_2 shows a mechanical strength surpassing steel's by approximately 30%. It is seen that MoS_2 is susceptible to rupture after undergoing a deformation of 1%. The material under investigation has shown remarkable strength. The manufacturing process of photovoltaic devices has witnessed the replacement of platinum (Pt) counter electrodes by molybdenum disulfide (v) [5]. A list of the various transition metal chalcogenides-based PV solar cells and their successive layers are shown in **Table 18.1**.

18.3 Performance Metrics Associated with Photovoltaic Systems

The efficiency (η) of a system measures its ability to convert inputs into desired outputs. Solar panels with 2D metal chalcogenides measure the efficiency of converting incident solar energy into electricity statistically. Usually, efficiency is measured as $\eta = P_{in}/P_{out} \times 100\%$, where Pout is the PV module's power output. Solar power incident on PV module is a pin. Metal chalcogenides' bandgap may impair efficiency, hence this characteristic is optimized. Short-circuit current (Isc) is the factor that denotes higher current output likely when PV cell terminals are linked in a short-circuit configuration. The absorption characteristics and electronic properties and the electrical properties of metal chalcogenide materials play an important role in defining the Isc. Open-circuit voltage (Voc) is the maximum potential difference that a photovoltaic (PV) cell may produce when its terminals are detached in a circuit with no current flow. This measurement is performed in accordance with predetermined test conditions to ensure uniformity and comparability. Fill factor (FF) is the ratio between the maximum power point (P_{max}) and the product of the open-circuit voltage (Voc) and the short-circuit current (Isc). A greater FF indicates increased performance of the device. Maximum Power Point (P_{max}) refers to the highest level of electrical power that may be generated by a photovoltaic (PV) system. This optimal power output is achieved at the specific point when the multiplication of current and voltage is maximized. The aforementioned point is commonly referred to as the maximum power point. Maximum Power Point (Pmax) refers to the maximum amount of electrical power that a photovoltaic (PV) system is capable of producing. This ideal power output is reached at the precise moment when the product of current and voltage is at its maximum. The degradation rate is a metric that quantifies the yearly decline in performance, offering valuable information regarding the enduring dependability of photovoltaic systems based on metal chalcogenides [16, 17].

Stability and reliability measures evaluate the capacity of photovoltaic (PV) devices to sustain their performance over an extended period, encompassing their resistance to external elements such as humidity, temperature, and ultraviolet (UV) radiation. In recent years, technical advancements in solar energy have significantly progressed, surpassing the first stages witnessed in the 19th century. However, the average efficiency of solar panels remains at approximately 15–20%. This implies that about 80–85% of the incident raw energy originating from our primary celestial body is dissipated. In addition, it is noteworthy that silicon solar cells, which represent the predominant photovoltaic technology in use, possess a theoretical efficiency limit of approximately 35%. The reported single and multi-junction photovoltaic materials and their efficiency in all four generations are reported in **Table 18.2**.

Table 18.2 A List of Transition Metal Chalcogenide PVs and Their Efficiencies

Single-Junction				Multiple-Junction		
S. No.	Materials	Efficiency %	Ref.	Materials	Efficiency %	Ref.
1	Si	26.8	[8]	InGaP/GaAs/InGaAs	37.9	[13]
2	GaAs	29.1	[10]	GaInP/GaAs	32.8	[18, 19]
3	InP	24.2	[13]	GaInP/GaInAsP/S	35.9	[20]
4	CIGS	23.3	[15]	GaInP/GaAs/Si	35.9	[21]
5	CIGSSe	20.3	[22]	GaAsP/Si	23.4	[23]
6	CdTe	21.0	[24]	GaInP/GaInAs/Ge; Si	34.5	[25]
7	CZTSSe	12.1	[24]	Perovskite/Si	33.7	[25]
8	Perovskite	24.3	[26]	Perovskite/perovskite	28.2	[27]
9	GaInP	22.0	[28]	GaInP/GaAs/GaInAs	37.8	[29]
10	Organic	15.2	[30]	4J Minimodule	41.4	[19]

18.4 Challenges in Chalcogenide Photovoltaics

The materials that were discussed earlier have been shown to have outstanding performance in a variety of different applications. However, current research and development efforts are focusing on a number of different aspects to increase their effectiveness and investigate potential that has not yet been exploited. The following is a list of some of the most prominent fields in which there has been recent progress in research.

- *Materials toxicity*: Hazardous elements such as cadmium and lead can be found in certain metal chalcogenides. Researchers are now looking into less hazardous materials or procedures to address potential environmental and human health concerns. Non-toxic substances such as copper zinc tin sulfide (CZTS) or copper zinc tin selenide, for example (CZTSe)

- *Efficiency and performance*: The goal of achieving PCEs comparable to popular technologies such as silicon-based solar cells has proven to be a significant challenge. One potential solution for increasing efficiencies is to focus on improving crystal quality, optimizing film depositing processes, and increasing charge carrier mobility. Bandgap engineering, tandem cell architectures, and sophisticated light-trapping methods are also being investigated.

- *Stability and reliability*: Certain metal chalcogenides deteriorate and have a shorter lifespan when exposed to environmental factors such as moisture, temperature, and light. These issues may be improved by encapsulation techniques, protective coatings, and a better understanding of degradation processes.

- *Material homogeneity and defects*: Material defects, compositional non-uniformities, and grain boundaries can all affect charge transit and recombination within a material.

- *Advanced material engineering*: Modifying the composition and structure of metal chalcogenides can improve their electrical and optical properties. By adjusting the band gaps by the solar spectrum, this strategy has the potential to improve solar absorption.

- *Interface engineering*: Interface engineering entails carefully designing interfaces between various materials to maximize charge carrier capture and reduce recombination. This encourages efficient charge transport. Incorporating charge-selective layers, such as buffer layers or electron/hole transport layers, is frequently used in engineering to improve device efficiency.

18.5 Emerging Progress and Current Trends in Chalcogenide Photovoltaics Research

Recent advances and shifting research trends in metal chalcogenide photovoltaics have had a significant impact on the field of solar energy conversion. The investigation of tandem and multifunction solar cells, which integrate metal chalcogenide solar cells with other photovoltaic technologies to improve overall efficiency, is a significant area of research. Combining distinct elements or components enables the efficient absorption of light over a broader spectrum of solar wavelengths, thereby enhancing the device's overall performance. Exploration of hybrid architectures, which includes the seamless integration of metal chalcogenides and perovskite materials, is an equally intriguing area of research. The primary objective of this investigation is to increase light absorption and optimize the separation of charges within perovskite layers, with the potential to increase efficiency [7–9].

In the case of $BaZrO_3$, bandgap engineering via anion alloying produces an insulating material with a bandgap of approximately 5eV. In contrast, BaZrS3 has a bandgap of approximately 1.8 eV. Similar to chalcogenide perovskites, the bandgaps of these materials can be controlled by incorporating different anions and cations during synthesis. The substitution of titanium (Ti) for zirconium (Zr) and the addition of selenium (Se) to sulfur (S) have produced encouraging results. For instance, the bandgap of the chalcogenide perovskite $BaZr_{0.75}Ti_{0.25}S_3$ was reduced to 1.43 eV. This value is close to the optimal value, also referred to as the Shockley-Queisser limit. In contrast, the bandgap of $BaZr(S_{0.6}Se_{0.4})_3$ was 180 meV smaller than that of $BaZrS_3$ [6, 12].

In photovoltaic devices, it is common to remove photogenerated electrons selectively via the ETL. The ETL prevents the movement of electron vacancies via a large Schottky barrier at the perovskite material-ETL boundary. The heterojunction tunneling mechanism inhibits electron mobility, thereby making hole extraction easier. Consequently, insufficient charge transfer and charge flow obstruction will result in charge recombination at the contact, leading to energy loss. Recombination at the interface involves the nonradiative recombination of charge carriers. This process is brought about by contrasting energy levels, surface imperfections, and carrier migration toward the interface. The development of more effective solutions for defect passivation has been facilitated by a greater comprehension of defects and their subsequent effect on the operation of devices. By strategically employing surface treatments and interface engineering techniques, researchers have successfully lowered recombination rates and increased charge carrier lifetimes [10,11,14,15,30,32].

Innovative techniques for characterizing materials and electronics, such as high-resolution imaging and spectroscopy, have yielded crucial insights into their fundamental properties and intricate operational mechanisms. Utilizing techniques such as Kelvin probe microscopy, terahertz spectroscopy, and time-resolved photoluminescence. The emergence of flexible and transparent substrates has facilitated the development of flexible and partially transparent metal chalcogenide solar cells. This technological development permits the seamless incorporation of these devices onto various surfaces and objects, thereby expanding their applications. This effort demonstrates innovative light-absorption techniques for thin-film devices. These include studying plasmonic nanoparticles, nanostructured surfaces, and photonic crystals to improve light absorption. Machine learning and computer modeling accelerate the discovery of new materials and the optimization of devices. This collaboration forecasts material properties and identifies the most effective experimental methods. To increase the production of metal chalcogenide solar cells, academics are researching scalable fabrication methods. Among these methods are solution-based and roll-to-roll production.

A strong commitment to sustainability has been the driving force behind the implementation of eco-friendly synthesis and recycling processes. This commitment was characterized by a significant emphasis on substituting rare and hazardous substances with abundantly available earthly resources. Using metal chalcogenide solar cells in architectural components such as glass, windows, and facades generates electricity while improving aesthetics, thereby promoting novel architectural concepts. Surpassing this threshold instills a sense of assurance that high-performance, cost-effective photovoltaic systems can be successfully introduced to the market. There is increasing competition among materials scientists as seen by recent developments reported by LONGI, a prominent Chinese corporation that holds a significant market share in global solar panel production, recently unveiled its latest innovation in the form of a tandem solar panel with an impressive efficiency rating of 33.5%. A research group affiliated with King Abdullah University of Science and Technology in Saudi Arabia. According to the professor, their research team has successfully attained an efficiency rate of 33.7%. Although this situation is certainly exciting, it is essential to remember that we are still in the beginning stages. Researchers have been addressing this issue for a considerable amount of time. A National Renewable Energy Laboratory team developed a panel with an efficiency of 47%. This model cannot be widely adopted due to its high price. A recent study conducted by scientists at the Fraunhofer Institute for Solar Energy Systems ISE demonstrates that the efficiency of the most advanced four-junction solar cell to date has significantly increased. This enhancement was realized through the application of a novel antireflection coating. Specifically, the solar cell's efficiency was increased from 46.1% to 47.6%. This achievement represents a significant global milestone, as there is currently no solar cell that surpasses its efficiency on a global scale [33].

18.5.1 Thermoelectric Chalcogenides

18.5.1.1 Thermoelectricity and Its Importance

Over the past few years, the combustion of fossil fuels has exerted a notable influence on the environment and climatic circumstances. Carbon dioxide emission has attained an unprecedented level in the past decade, which raises significant concerns regarding future implications. Moreover, multiple datasets suggest that the energy utilized in overall consumption constitutes a mere 40%, dissipating the remaining energy into the environment as thermal energy. Scientific results have supported the urgent execution of coordinated measures to mitigate environmental catastrophes. The escalating global need for power, coupled with the depletion of fossil fuel supplies and mounting concerns regarding climate change, has spurred substantial research on alternative energy sources and techniques of energy conversion. There exists a finite set of alternative strategies for addressing the energy challenge, among which is the utilization of recovered energy, specifically in the context of wasted heat. Thermoelectric modules serve as a form of energy conversion technology [34].

To convert heat into usable electricity, the thermoelectric generator uses a solid-state setup. **Figure 18.3** shows that it can also be used as a solid-state refrigerator. All of these methods rely on the Seebeck and Peltier phenomena, which can create electricity from a simple temperature differential within a material. There aren't any mechanical forces or chemical processes involved, therefore they don't break down, they're small and portable, they last a long time, and they produce good results. Therefore, it is of paramount importance to material chemists that they discover thermoelectric materials that are not only effective but also helpful to the environment and financially viable. It has taken almost 150 years for the thermoelectric concept to be widely accepted, but only recently has an understanding of electron and phonon transport events provided a firm basis for it. An illustration (**Figure 18.3(b)**) illustrating the configuration of thermoelectric modules designed for power generation [35].

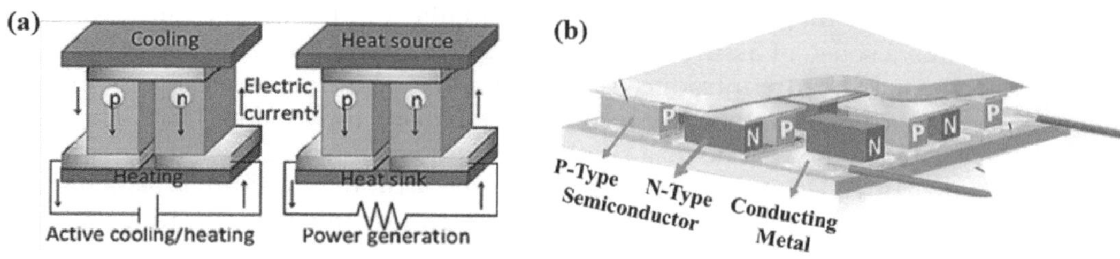

FIGURE 18.3

(a) Schematic representation of the Seebeck and Peltier effect. (b) Thermoelectric modules for power generation. Adapted with permission [35], Copyright: The Authors, some rights reserved; exclusive licensee (International Research Journal of Engineering and Technology). Distributed under a Creative Commons Attribution License 4.0 (CC BY) https://creativecommons.org/licenses/by/4.0/

18.5.1.2 Thermoelectric Properties: Thermoelectric Figure of Merit

The efficiency of converting heat to electricity in thermoelectric materials is evaluated using the dimensionless figure of merit, zT. This figure of merit is calculated using the equation $zT = S^2\sigma T/\kappa$, where S represents the Seebeck coefficient, σ represents the electrical conductivity, T represents the temperature, and κ represents the thermal conductivity. To get an adequate power output, a substantial temperature difference is imperative, denoted as $\Delta T = Th–Tc$. The good thermoelectric material must exhibit good efficiency η seen across a limited temperature difference ΔT. The highest efficiency η across a finite temperature difference ΔT. We can readily estimate this value using S, ρ, and κ.

Numerous factors influence the success of thermoelectric systems, including the Carnot efficiency and thermoelectric properties such as the Seebeck coefficient, thermal conductivity, and electrical resistivity. The average energy conversion efficiency of roughly 1 ZT is 10%. A 20% rise occurs when the ZT value is about 3. When the thermoelectric figure of merit (ZT) hits around 4, the conversion efficiency might reach 30%. We have now successfully achieved a ZT value larger than 2, allowing its practical application. To achieve high-efficiency thermoelectric materials, the materials must have a high Seebeck coefficient and electrical conductivity while also having a high coefficient of thermal conductivity. **Figure 18.2** depicts a graphical representation of the thermoelectric module.

As a result of their exceptionally high thermoelectric figure of merit (ZT), chalcogenides, half-Heusler compounds, clathrates, Zintl phases, and skutterudites have attracted a lot of attention in recent years. The vast majority of the aforementioned substances are almost entirely made up of heavy metals and have a restricted availability of the element in their make-up. The development of thermoelectric materials therefore requires not only the achievement of a high figure of merit (ZT), but also the ability to function well throughout a wide temperature range while making use of materials that are both non-toxic and commercially feasible. $PbTe$, Bi_2Te_3, $NaPb_xSbTe_{x+2}$, and $AgSbTe_2$ are some famous examples of the telluride compounds that form the basis of the bulk of thermoelectric materials that are currently available on the market. These materials provide a number of benefits, including a low thermal conductivity (2.3 W/m.K for PbTe and 1.7 W/m.K for Bi_2Te_3), a high Seebeck coefficient of 500 V/K, and the capacity to be easily doped with either p-type or n-type atoms. Additionally, these materials have a high Seebeck coefficient of 500 V/K. Because of the prohibitively expensive nature of tellurium-based thermoelectric materials, researchers have been looking at other potential solutions. Because of their low thermal conductivity and high Seebeck coefficients, sulfur (S; Bi_2S_3) and selenium (Se; PbSe) have been identified as two of the most promising options among these different alternatives.

18.5.1.3 *Approaches to Improve the Thermoelectric Efficiency*

(a) *Doping to modify charge carrier concentration*: to modify the electrical properties of chalcogenide materials, the doping process calls for the incorporation of dopants, which are essentially foreign elements, into the lattice structure of the material. The manipulation of dopants makes it possible to regulate the concentration of charge carriers, which in turn has an effect on both the electrical conductivity and the Seebeck coefficient. In the process of n-type doping, extra electrons are introduced into the material, which results in an increase in the electron concentration. Doping of the p-type, on the other hand, entails the creation of holes to increase the engagement of holes. The choice of a dopant, in addition to the concentration of the dopant and its position within the crystal lattice, can have a significant impact on the electrical properties of a material, which in turn can have an effect on the material's thermoelectric efficiency.

(b) *Nanostructuring and grain boundary engineering*: Nanostructuring is modifying a material's nanoscale structure to create new nanoscale features, such as nanoparticles, nanowires, or nanograins. Phonons, vibrations responsible for the transfer of heat and an increase in electron scattering, can be dispersed by these features. The material's thermoelectric efficiency is enhanced while its thermal conductivity is decreased. Grain boundary engineering seeks to improve a material's internal grain size and distribution. The grain boundaries can act as phonon-scattering, reducing the material's thermal conductivity.

(c) *Bandgap engineering*: The term "band engineering" describes the process of modifying the electronic band structure of a material to improve its electronic properties and thermoelectric performance. Optimizing the Seebeck coefficient and electrical conductivity requires shifting the position and size of the energy bands. Energy filtering is a technique used to improve the thermoelectric properties of a material by permitting the passage of some energy levels of charge carriers while blocking the path of others. This method is beneficial for studying materials with complicated band topologies.

(d) *Phonon engineering and reduction of thermal conductivity*: Phonons carry heat from one part of a substance to another. Improved performance in thermoelectric chalcogenides is at the expense of their heat conductivity. In phonon engineering, scattering sites like nanoparticles or point defects are purposefully introduced to a material to reduce heat transport efficiency via phonons. Materials with slow phonon velocities or robust anharmonic features could also facilitate heat transfer.

(e) *Transport phenomena*: For chalcogenides to be useful, they must be able to regulate their thermal conductivity without sacrificing their excellent electrical conductivity. In addition to electrical conductivity and the Seebeck coefficient, phonon-scattering is crucial in lowering thermal conductivity through interactions with grain boundaries, nanostructures, and point defects. The engineering, as mentioned earlier, demonstrates interconnectedness and offers the possibility of synergistic consequences when combined.

Crystal structures of thermoelectric chalcogenide materials are typically quite detailed, with layers and lattices and whatever. They have better thermoelectric properties because of the heavy metals they contain and the inherent anisotropy of their crystal formations. There are several excellent case studies demonstrating the usefulness of thermoelectric chalcogenides **(Figure 18.3)**. The ZT values of lead chalcogenides like PbTe are particularly

TABLE 18.3

A List of Transition Metal Chalcogenide PVs and Their Efficiencies

Material	Maximum ZT	Temperature Range	Reference
$BaZrS_3$	1.75–1.94		[36]
PbTe	~2.2	600–900 K	[37]
Sb_2X_3	1.1–1.8 eV		[38]
Cu2Te	~1.3		[39]
Cu_2Se	~1.5		[39]

outstanding at high temperatures. As a result of their intricate crystal structures and inherent anisotropy, copper chalcogenides (Cu_2Se and Cu_2Te) also show promise as thermoelectric materials. A list of the transition metal chalcogenide PVs and their efficiencies are tabulated in **Table 18.1**.

Challenges and Future Directions: Thermoelectric chalcogenides have made great strides in recent years, but there is still work to be done to improve their efficiency and commercial feasibility. Novel material design solutions are still needed for the challenging task of balancing electrical and thermal properties. For the broad use of these materials, it is crucial to think about the sustainability of their procurement and processing.

Advanced computer modeling, materials synthesis techniques, and device engineering will determine the future of thermoelectric chalcogenides. Exploring the underlying dynamics governing their thermoelectric behavior can pave the way for game-changing applications in energy harvesting and usage. Because of their adaptability, thermoelectric chalcogenides can be used in many different contexts. Sustainable energy generation may be possible with the help of solid-state thermoelectric generators, which can use waste heat from factories or cars. They can be used to power electronics with body heat, and their compatibility with flexible substrates paves the way for wearable energy devices. Combining thermoelectric chalcogenides with other technologies, such as photovoltaics, to maximize the conversion of solar energy is one example of a cutting-edge strategy. The fact that they may be used in a hostile environment like space is a testament to their versatility.

References

1. World Energy Balances Overview, International Energy Agency, Paris. 2021. 38. Available from: www.iea.org/re 39 ports/world-energy-balances-overview. 40 License: CC BY 4.0.
2. F.H. Alharbi, S. Kais, Theoretical Limits of Photovoltaics Efficiency and Possible Improvements by Intuitive Approaches Learned from Photosynthesis and Quantum Coherence, *Renewable and Sustainable Energy Reviews*, 43 (2015) 1073–1089.
3. J.A. Wilson, A.D. Yoffe, The Transition Metal Dichalcogenides Discussion and Interpretation of the Observed Optical, Electrical and Structural Properties, *Advances in Physics*, 18 (1969) 193–335.
4. M. Velický, P.S. Toth, From Two-Dimensional Materials to Their Heterostructures: An Electrochemist's Perspective, *Applied Materials Today*, 8 (2017) 68–103.
5. M.Z. Iqbal, S. Alam, M.M. Faisal, S. Khan, Recent Advancement in the Performance of Solar Cells by Incorporating Transition Metal Dichalcogenides as Counter Electrode and Photo Absorber, *International Journal of Energy Research*, 43 (2019) 3058–3079.

6. F.H. Isikgor, S. Zhumagali, L.V. T. Merino, M. De Bastiani, I. McCulloch, S. De Wolf, Molecular Engineering of Contact Interfaces for High-Performance Perovskite Solar Cells, *Nature Reviews Materials*, 8 (2023) 89–108.

7. Z. Guo, A.K. Jena, G.M. Kim, T. Miyasaka, The High Open-Circuit Voltage of Perovskite Solar Cells: A Review, *Energy & Environmental Science*, 15 (2022) 3171–3222.

8. F. Werlinger, C. Segura, J. Martínez, I. Osorio-Roman, D. Jara, S.J. Yoon, A.F. Gualdrón-Reyes, Current Progress of Efficient Active Layers for Organic, Chalcogenide and Perovskite-Based Solar Cells: A Perspective, *Energies,* 16 (2023) 5868.

9. S.K. Matta, C. Tang, A.P. O'Mullane, A. Du, S.P. Russo, Density Functional Theory Study of Two-Dimensional Post-Transition Metal Chalcogenides and Halides for Interfacial Charge Transport in Perovskite Solar Cells, *ACS Applied Nano Materials*, 5 (2022) 14456–14463.

10. A. Lavín, R. Sivasamy, E. Mosquera, M.J. Morel, High Proportion ZnO/CuO Nanocomposites: Synthesis, Structural and Optical Properties, and Their Photocatalytic Behavior, *Surfaces and Interfaces*, 17 (2019) 100367.

11. R. Sekar, R. Sivasamy, B. Ricardo, P. Manidurai, Ultrasonically Synthesized TiO2/ZnS Nanocomposites to Improve the Efficiency of Dye Sensitized Solar Cells, *Materials Science in Semiconductor Processing*, 132 (2021) 105917.

12. T. Devendra, S.H. Oliver, L. Giulia, Chalcogenide Perovskites for Photovoltaics: Current Status and Prospects, *Journal of Physics: Energy*, 3 (2021) 034010 (1–9).

13. M. Yamaguchi, T. Takamoto, K. Araki, N. Ekins-Daukes, Multi-Junction III–V Solar Cells: Current Status and Future Potential, *Solar Energy*, 79 (2005) 78–85.

14. R. Sivasamy, K. Paredes-Gil, J.V. Ramaclus, E. Mosquera, S. Kaliyamoorthy, K.M. Batoo, Sandwich-Like GaN/MoSe$_2$/GaN Heterostructure Nanosheet: A First-Principle Study of the Structure, Electronic, Optical, and Thermodynamical Properties, *Surfaces and Interfaces*, 34 (2022) 102298.

15. R. Sivasamy, F. Quero, K. Paredes-Gil, K.M. Batoo, M. Hadi, E.H. Raslam, Comparison of the Electronic, Optical and Photocatalytic Properties of MoSe2, InN, and MoSe2/InN Heterostructure Nanosheet-A First-Principle Study, *Materials Science in Semiconductor Processing*, 131 (2021) 105861.

16. I. Celik, R. Hosseinian Ahangharnejhad, Z. Song, M. Heben, D. Apul, Emerging Photovoltaic (PV) Materials for a Low Carbon Economy, *Energies,* 13 (2020) 4131.

17. M.H. Miah, M.B. Rahman, M. Nur-E-Alam, N. Das, N.B. Soin, S.F.W.M. Hatta, M.A. Islam, Understanding the Degradation Factors, Mechanism and Initiatives for Highly Efficient Perovskite Solar Cells, *ChemNanoMat*, 9 (2023) e202200471.

18. M. Verma, G.P. Mishra, Analytical Model of InP QWs for Efficiency Improvement in GaInP/Si Dual Junction Solar Cell, *Physica Status Solidi*, 220 (2023) 2200500.

19. L. Zhu, Y. Wang, X. Pan, H. Akiyama, Theoretical Modeling and Ultra-Thin Design for Multi-Junction Solar Cells with a Light-Trapping Front Surface and Its Application to InGaP/GaAs/InGaAs 3-Junction, *Optics Express*, 30 (2022) 35202–35218.

20. F. Dimroth, M. Grave, P. Beutel, U. Fiedeler, C. Karcher, T.N.D. Tibbits, E. Oliva, G. Siefer, M. Schachtner, A. Wekkeli, A.W. Bett, R. Krause, M. Piccin, N. Blanc, C. Drazek, E. Guiot, B. Ghyselen, T. Salvetat, A. Tauzin, T. Signamarcheix, A. Dobrich, T. Hannappel, K. Schwarzburg, Wafer Bonded Four-Junction GaInP/GaAs//GaInAsP/GaInAs Concentrator Solar Cells with 44.7% Efficiency, *Progress in Photovoltaics: Research and Applications*, 22 (2014) 277–282.

21. M. Feifel, D. Lackner, J. Ohlmann, J. Benick, M. Hermle, F. Dimroth, Direct Growth of a GaInP/GaAs/Si Triple-Junction Solar Cell with 22.3% AM1.5g Efficiency, *Solar RRL*, 3 (2019) 1900313.

22. S.M. McLeod, C.J. Hages, N.J. Carter, R. Agrawal, Synthesis and Characterization of 15% Efficient CIGSSe Solar Cells from Nanoparticle Inks, *Progress in Photovoltaics: Research and Applications*, 23 (2015) 1550–1556.

23. D.L. Lepkowski, T.J. Grassman, J.T. Boyer, D.J. Chmielewski, C. Yi, M.K. Juhl, A.H. Soeriyadi, N. Western, H. Mehrvarz, U. Römer, A. Ho-Baillie, C. Kerestes, D. Derkacs, S.G. Whipple, A.P. Stavrides, S.P. Bremner, S.A. Ringel, 23.4% Monolithic Epitaxial GaAsP/Si Tandem Solar Cells and Quantification of Losses from Threading Dislocations, *Solar Energy Materials and Solar Cells*, 230 (2021) 111299.

24. M.A. Green, E.D. Dunlop, M. Yoshita, N. Kopidakis, K. Bothe, G. Siefer, X. Hao, Solar Cell Efficiency Tables (Version 62), *Progress in Photovoltaics: Research and Applications*, 31 (2023) 651–663.

25. I. García, L. Barrutia, S. Dadgostar, M. Hinojosa, A. Johnson, I. Rey-Stolle, Thinned GaInP/GaInAs/Ge Solar Cells Grown with Reduced Cracking on Ge|Si Virtual Substrates, *Solar Energy Materials and Solar Cells*, 225 (2021) 111034.

26. H. Shu, C. Peng, Q. Chen, Z. Huang, C. Deng, W. Luo, H. Li, W. Zhang, W. Zhang, Y. Huang, Strategy of Enhancing Built-in Field to Promote the Application of C-TiO2/SnO2 Bilayer Electron Transport Layer in High-Efficiency Perovskite Solar Cells (24.3%), *Small*, 18 (2022) 2204446.

27. M. Yao, A. Marzouki, S. Hao, S. Salmanov, M. Otonicar, V. Loyau, B. Dkhil, Grain Size and Piezoelectric Effect on Magnetoelectric Coupling in BFO/PZT Perovskite-Perovskite Composites, *Journal of Alloys and Compounds*, 948 (2023) 169731.

28. Y. Shoji, R. Oshima, K. Makita, A. Ubukata, T. Sugaya, 1.5 eV GaInAsP Solar Cells Grown via Hydride Vapor-Phase Epitaxy for Low-Cost GaInP/GaInAsP//Si Triple-Junction Structures, *Advanced Energy and Sustainability Research*, 4 (2023) 2200198.

29. A. Mehrotra, A. Freundlich, Superior Radiation and Dislocation Tolerance of IMM Space Solar Cells, *2012 38th IEEE Photovoltaic Specialists Conference* (2012) 003150–003154.

30. M. Li, J. Wang, A. Jiang, D. Xia, X. Du, Y. Dong, P. Wang, R. Fan, Y. Yang, Metal Organic Framework Doped Spiro-OMeTAD with Increased Conductivity for Improving Perovskite Solar Cell Performance, *Solar Energy*, 188 (2019) 380–385.

31. S. Ramesh, S. Marutheeswaran, J.V. Ramaclus, D.C. Paul, Electronic Structure Study on 2D Hydrogenated Icosagens Nitride Nanosheets, *Superlattices and Microstructures*, 76 (2014) 213–220.

32. R. Sivasamy, M. Srinivasan, R. Espinoza-González, E. Mosquera, Electronic and Optical Studies on Two-Dimensional Hydrogenated Stirrup Triels Nitride Nanosheets: A First-Principle Investigation, *Materials Science and Engineering: B*, 264 (2021) 114978.

33. F. Dimrot, *World's Most Efficient Solar Cell with 47.6 Percent Efficiency.* Press Release Fraunhofer Institute for Solar Energy Systems ISE (2022) 1–3.

34. J.-H. Lee, J. Kim, T.Y. Kim, M.S. Al Hossain, S.-W. Kim, J.H. Kim, All-in-One Energy Harvesting and Storage Devices, *Journal of Materials Chemistry A*, 4 (2016) 7983–7999.

35. P.V. Akshay Thalkar, Sagar Nikam, Swapnil Patil, Lalit Shendre, Study of Thermoelectric Air Conditioning for Automobiles, *International Research Journal of Engineering and Technology*, 5 (2018) 1084.

36. K.V. Sopiha, C. Comparotto, J.A. Márquez, J.J.S. Scragg, Chalcogenide Perovskites: Tantalizing Prospects, *Challenging Materials*, 10 (2022) 2101704.

37. O. Khshanovska, T. Parashchuk, I. Horichok, Estimating the Upper Limit of the Thermoelectric Figure of Merit in n- and p-Type PbTe, *Materials Science in Semiconductor Processing*, 160 (2023) 107428.

38. S. Barthwal, R. Kumar, S. Pathak, Present Status and Future Perspective of Antimony Chalcogenide (Sb2X3) Photovoltaics, *ACS Applied Energy Materials*, 5 (2022) 6545–6585.

39. C. Xing, Y. Zhang, K. Xiao, X. Han, Y. Liu, B. Nan, M.G. Ramon, K.H. Lim, J. Li, J. Arbiol, B. Poudel, A. Nozariasbmarz, W. Li, M. Ibáñez, A. Cabot, Thermoelectric Performance of Surface-Engineered Cu1.5-xTe-Cu2Se Nanocomposites, *ACS Nano*, 17 (2023) 8442–8452.

19

Electrical Energy Storage Devices Based on 2D Semiconducting Materials

Prakash Chandra*

Correspondence: prakash.chandra@sot.pdpu.ac.in

19.1 Introduction

Two-dimensional semiconducting materials (2D-SCMs) have captivated the attention of the scientific community recently owing to their unusual properties and potential for miscellaneous utilities. As compared to the conventional three-dimensional structure, 2D-SCMs are made up of atomically thin layers, characteristically only a few atoms thick and the 2D morphology bestows unexpected features. Graphene is an outstanding illustration for these 2D-SCMs. The graphene is composed of a single layer of carbon atoms organized in a hexagonal lattice [1,2]. Graphene exhibits exceptional mechanical strength, high electrical and thermal conductivity, and a large surface area, making it suitable for a wide range of applications, including electronics, energy storage, and sensors [3,4].

Beyond graphene, there is a diverse family of 2D-SCMs that includes TMDCs [5], black phosphorus (BP) [6,7], and MXenes [8,9]. TMDCs like MoS_2 [10] and tungsten WSe_2 [11], possess unique optical and electronic properties due to their layered structure [12–15]. These materials exhibit a direct bandgap in the monolayer form, making them attractive for optoelectronic devices like photovoltaics and light-emitting diodes. BP, MXenes, and layered metal oxides are significant 2D-SCMs. Their distinct features-expansive surface area, adaptable bandgap, mechanical flexibility, and superior charge mobility render them sought-after across diverse applications. Ongoing research optimizes these materials, crafting novel device structures, poised to revolutionize electronics, energy storage, catalysis, and optoelectronics [16].

2D-SCMs have shown immense potential in various applications due to their unique properties and versatility [17,18]. These 2D-SCMs have gained attention in the field of electronics for applications such as transistors, integrated circuits, and flexible electronics. Graphene, with its high electrical conductivity and mechanical flexibility, has been explored for high-speed transistors and transparent conductive electrodes [19,20]. TMDCs like MoS_2 and WSe_2 are under study for advanced transistors and optoelectronic tools. Their light absorption and charge transport make them appealing for catalyzing hydrogen evolution and oxygen reduction in fuel cells. With high surface areas and sensitivity, they detect various analytes and excel in optoelectronics, such as LEDs, photodetectors, and lasers. Graphene and similar materials are explored for membrane and filtration applications [21]. Their atomically thin structure allows for precise control of permeability, making them useful for water filtration, gas separation, and desalination. 2D-SCMs have

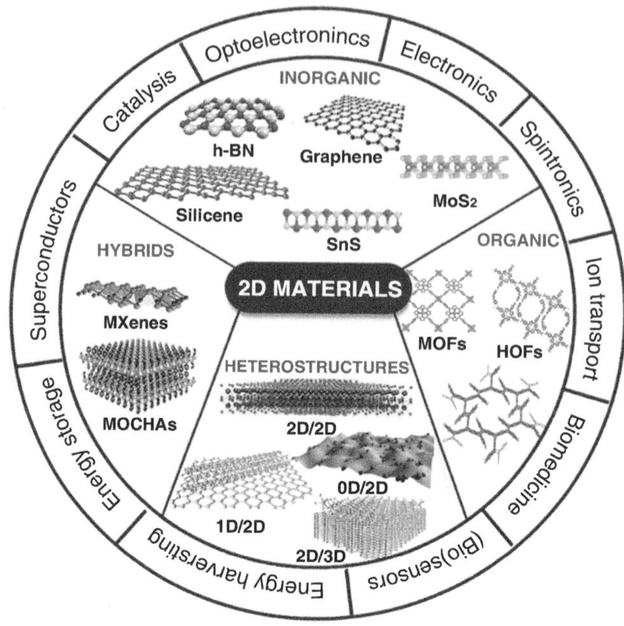

FIGURE 19.1
Miscellaneous applications of advanced 2D materials. Adapted with permission [23], Copyright (2022), MDPI.

shown promise in biomedical applications, including drug delivery systems, bioimaging, and tissue engineering. Their biocompatibility, large surface area, and tunable properties make them attractive for various biomedical applications.

The evolving field of 2D-SCMs continues to expand through ongoing research, unveiling new materials and enhancing their potential across diverse applications. In energy storage and conversion devices (EESDs), graphene, MXenes, TMDCs, and BP show promise, offering high energy/power densities and rapid charge capabilities in batteries and supercapacitors [22]. These 2D-SCMs can serve as excellent electrode materials due to their large surface area and high electrical conductivity (see Figure 19.1).

19.2 Unique Properties of 2D-SCMs

The unique optical and electronic properties of 2D-SCMs contribute to improved light absorption and charge transport within EESDs [24,25]. Uniqueness likes in the high surface area, tunable band gap, mechanical flexibility, and excellent charge transportation properties make them suitable for wide range of applications [26]. The outstanding and unique properties are discussed in the present section.

19.2.1 High Surface Area

The atomic thinness of 2D-SCMs yields an extraordinary surface-to-volume ratio, a boon for electrical energy storage devices (EESDs). This vast surface area accommodates numerous active sites pivotal for electrochemical reactions in batteries and supercapacitors.

These sites engage in ion/electron storage and release, augmenting energy capacity and performance.

The increment in the active sites expedites electrochemical processes, fostering rapid ion/electron transport and enabling swift charge/discharge rates and high power density. This reduces internal resistance, enhancing overall electrochemical kinetics. Moreover, the expansive surface area enhances electrolyte ion access, promoting efficient ion adsorption/intercalation, thus elevating charge storage capacity and mitigating diffusion limitations that can impede device performance. Surface functionalization, leveraging the abundant surface area, allows tailored chemistry through attachments like functional groups or nanoparticles. This enhances electrochemical activity, augmenting interactions between electrode material and electrolyte, ultimately improving energy storage performance. 2D-SCMs fundamentally elevate energy storage by facilitating active sites, efficient diffusion, enhanced electrolyte access, and surface customization for optimized device efficiency and capacity.

19.2.2 Tunable Bandgap

The bandgap of 2D-SCMs can be tuned by altering their thickness or composition, offering versatility in designing materials with specific energy levels. This tunability of bandgap is advantageous for various energy storage applications. 2D-SCMs with narrow bandgaps, such as certain TMDCs, exhibit excellent light absorption properties. These materials can efficiently capture a broader range of solar energy, including visible and near-infrared wavelengths. By designing heterostructures or hybrid devices using 2D-SCMs with narrow bandgaps, it is possible to enhance the light absorption and conversion efficiency of photovoltaic devices [27, 28].

In battery applications, 2D-SCMs with wider bandgaps are typically preferred. These materials exhibit better chemical and electrochemical stability and lower self-discharge rates. The wider bandgap helps mitigate unwanted side reactions, improving the overall cycle life and energy density of the battery. Additionally, wide bandgap materials often have higher open-circuit voltages, which can enhance the overall energy storage capacity of the battery. By tailoring the bandgap of 2D-SCMs, it is possible to optimize their energy storage properties for specific applications. For instance, in addition to photovoltaics and batteries, narrower bandgap 2D-SCMs may find use in thermoelectric devices, where they can efficiently convert waste heat into electrical energy. Similarly, wider bandgap 2D-SCMs can be suitable for electrochemical capacitors or other energy storage systems that require high power density and fast charge-discharge rates. The ability to tune the bandgap of 2D-SCMs is a valuable feature that allows researchers to design and engineer materials with specific energy levels, tailoring them for different energy storage applications and optimizing their performance.

19.2.3 Mechanical Flexibility

The mechanical flexibility of 2D-SCMs is a highly advantageous property for the integration of EESDs into flexible or wearable electronics. These mechanically flexible 2D-SCMs are comfortable, durable, lightweight, and portable materials for energy storage portable devices. Some salient features of these 2D-SCMs are discussed in this section [29–31].

The exceptional thinness of 2D-SCMs grants them remarkable flexibility, allowing bending, twisting, or stretching without compromising their structural integrity. This flexibility revolutionizes energy storage, enabling systems that conform to diverse shapes like curved surfaces or irregular geometries, crucial for applications where rigid EESDs would prove impractical. Their key utility lies in seamlessly integrating into flexible or wearable electronic devices like smart textiles, electronic skins, or flexible displays. These devices necessitate energy storage components enduring repeated mechanical deformations. The mechanical flexibility of 2D-SCMs ensures systems that endure stress while maintaining peak performance, enhancing durability in demanding environments, and reducing structural risks or failures. Moreover, their ultrathin and lightweight nature make 2D-SCMs ideal for portable electronics. Their incorporation into thin, compact systems reduces device weight and size significantly, pivotal for mobile and wearable applications prioritizing lightweight and portable energy storage. The mechanical flexibility of 2D-SCMs opens up possibilities for the development of EESDs that can conform to various shapes, integrate into flexible or wearable electronics, withstand mechanical stress, and enhance the durability and portability of the devices. This property has the potential to revolutionize the design and functionality of energy storage systems in the context of emerging technologies and applications.

19.2.4 Excellent Charge Transport Properties

The high carrier mobility of 2D-SCMs is a crucial advantage that contributes to efficient charge transport within EESDs. The high carrier mobility of 2D-SCMs enables rapid movement of charges (e.g., ions or electrons) within the material. This facilitates fast charge–discharge rates in EESDs. The efficient charge transport allows for quick charging and discharging, reducing the time required to store or release energy. This is particularly important for applications where rapid energy delivery is necessary, such as electric vehicles or high-power electronics. The efficient charge transport in 2D-SCMs also contributes to high power density in energy storage systems. The ability to rapidly move charges ensures that a large amount of power can be delivered or extracted from the device within a short period. This is essential for applications that require high bursts of power, such as energy harvesting, grid stabilization, or high-performance electronics. Efficient charge transport in 2D-SCMs helps minimize energy loss during operation. When charges can move quickly and smoothly through the material, there is a reduced likelihood of encountering obstacles or experiencing resistive losses. This results in higher energy efficiency and improved overall performance of the electrical energy storage device [32–34].

The combination of high carrier mobility, fast charge-discharge rates, and high power density makes 2D-SCMs attractive for a wide range of energy storage applications, including batteries, supercapacitors, fuel cells, and photovoltaics. Researchers are actively exploring different 2D-SCMs and optimizing their properties to further enhance energy storage performance. They are also developing innovative device architectures to maximize the benefits of efficient charge transport and address any remaining challenges, such as stability, scalability, and interface engineering.

19.3 Electrochemical Properties of 2D-SCMs in Energy Storage

The key electrochemical properties of these 2D-SCMs include their charge storage, redox properties, and catalytic properties of these 2D-SCMs. The electrochemical properties of 2D-SCMs play a crucial role in their application for energy storage.

19.3.1 Charge Storage Capabilities of the 2D-SCMs

Two-dimensional semiconducting materials (2D-SCMs), comprising single-layered structures, tout a significant advantage in their extensive surface area, stemming from their atomically thin composition. This characteristic yields a surplus of active sites for diverse electrochemical reactions, intensifying charge storage capabilities notably in batteries and supercapacitors.

Their amplified surface area fosters a dense landscape for electrochemical reactions, augmenting the storage potential for ions/electrons during charging and elevating overall energy storage capabilities. The expansive interface facilitates crucial ion/electron exchange between material and electrolyte, pivotal in electrochemical devices. Abundant active sites on 2D-SCMs enable efficient adsorption/desorption or intercalation/deintercalation during charge cycles, boosting charge storage potential significantly. Their vast surface area enhances interior ion/electron accessibility, streamlining diffusion and enabling swifter charge/discharge rates. In supercapacitors, higher surface area correlates directly with capacitance, elevating energy storage potential. Similarly, in batteries, surface-based ion/electron storage on 2D-SCMs bolsters overall storage capacity, further enhancing their energy storage capabilities [35–38].

19.3.2 Redox Properties of SCMs

Numerous 2D-SCMs showcase redox reactions, pivotal for EESDs, enabling reversible changes in oxidation states—crucial for storing and releasing electrical energy without significant degradation over multiple cycles. Tailoring these reactions through composition, defects, and surface chemistry ensures device longevity and stability.

Customizing 2D-SCMs via dopants or defects influences their redox behavior, altering electronic structures and enhancing electrochemical activity, thus boosting energy storage capabilities. Surface modifications impact interactions with electrolytes, optimizing charge transfer processes and redox kinetics for specific storage applications. In diverse EESDs like batteries or supercapacitors, tailored redox reactions, particularly in lithium-ion batteries, involving reversible ion intercalation, are fundamental for energy storage. High carrier mobility and superior ion diffusion in 2D-SCMs facilitate swift ion/electron transport, vital for rapid charge/discharge rates, augmenting power density within the system. Electrochemical stability, dependent on structural resilience, resistance to side reactions, and tolerance to volume changes, influences device durability, ensuring the prolonged cycle life of EESDs.

19.3.3 Catalytic/Electrocatalytic Properties of 2D-SCMs

2D-SCMs exhibit catalytic prowess, aiding electrochemical reactions by reducing activation energy or enhancing kinetics, a crucial attribute in EESDs like fuel cells. Their catalytic activity boosts reaction kinetics within EESDs, providing active surfaces that lower activation energy, foster efficient charge transfer, and escalate reaction rates, yielding faster charge/discharge rates and heightened energy conversion efficiency.

These materials diminish overpotentials, reducing additional energy needs for electrochemical reactions, thus enabling more efficient energy conversion and storage and elevating device performance and overall efficiency. Tailoring 2D-SCMs' catalytic activity to specific electrochemical reactions—like oxygen reduction in fuel cells or oxygen evolution in water electrolysis—ensures high efficiency and system performance. Moreover, catalysts in 2D-SCMs fortify EESD durability and stability by curbing electrode degradation, limiting side reactions, and sustaining activity across extended cycles. They also serve as ideal supports for catalytic materials, enhancing dispersion, stability, and activity, further improving overall performance. This catalytic prowess of 2D-SCMs significantly elevates energy storage device efficiency by lowering activation energy, enhancing kinetics, and improving selectivity, holding promise for fuel cells, batteries, and other storage systems. Ongoing research optimizes compositions, surface modifications, and device architectures to bolster electrochemical performance, enhance energy storage efficiency, and tackle stability and scalability challenges, fueling advancements in sustainable energy storage technologies.

19.4 Classification of 2D-SCMs for Energy Storage Applications

2D-SCMs can be classified into different categories based on their chemical composition and structure. Here are some common classifications of 2D-SCMs used in energy storage applications:

19.4.1 Graphene

Graphene is a single layer of carbon atoms arranged in a 2D honeycomb lattice. It exhibits excellent electrical conductivity, mechanical strength, and high surface area. Graphene has been extensively investigated for energy storage applications, including batteries, supercapacitors, and fuel cells. Its high conductivity makes it suitable for fast charge transport, while its large surface area enhances energy storage capacity.

19.4.2 Transition Metal Dichalcogenides

TMDCs are 2D materials with a layered structure that have garnered attention in energy storage applications. They possess unique properties that make them desirable for electrode materials. TMDCs like MoS_2 or WS_2 possess tunable direct bandgap, permitting tailored energy levels appropriate for definite utilities. Narrow bandgap TMDCs work in photovoltaics to capture a diversity of solar energy, while wider bandgap materials find usefulness in batteries for higher energy density. TMDCs display outstanding charge carrier mobility, enabling efficient charge transport and resulting in fast charge–discharge rates and high power density in energy storage systems. These characteristics minimize energy loss during operation. TMDCs have been extensively explored as electrode materials in batteries, supercapacitors, and photovoltaics. In batteries, they offer high capacity and fast charge–discharge rates. In supercapacitors, TMDCs enhance energy storage capabilities. TMDCs are also utilized in photovoltaics to improve solar cell efficiency through heterostructures or hybrid devices. Ongoing research aims to optimize TMDC properties for improved performance in energy storage applications, leveraging their tunable bandgap and high charge carrier mobility.

19.4.3 MXenes

MXenes, a class of 2D materials, have garnered attention for their unique properties and potential in energy storage applications. Synthesized by selectively etching layered MAX phases, MXenes exhibit advantageous characteristics. They possess metallic conductivity, enabling effective electron transmission, fast charge/discharge rates, and high power density in EESDs. MXenes are hydrophilic, augmenting communication with aqueous electrolytes in supercapacitors and improvising ion adsorption and capacitance. The layered structure of MXenes, with significant interlayer spacing, permits facile ion diffusion and transport, promoting rapid ion intercalation/deintercalation in batteries and fast ion adsorption/desorption in supercapacitors. The large interlayer spacing also contributes to a high specific capacitance. MXene surfaces can be functionalized or modified, offering further tunability. Surface modifications enhance stability, conductivity, and specific capacitance, making MXenes well-suited for energy storage. MXenes, like titanium carbide ($Ti_3C_2T_x$), show promise as electrode materials in batteries and supercapacitors due to their high electrical conductivity, large interlayer spacing, and hydrophilicity. Ongoing research explores MXenes' potential and optimizes their properties for diverse energy storage applications.

19.4.4 Black Phosphorus

BP, a 2D material with a layered structure, shows promise for energy storage applications, particularly as an electrode material in batteries. BP's layered structure allows for efficient ion diffusion and transport, enabling rapid charge-discharge cycles. Its large surface area facilitates electrochemical reactions, contributing to high energy storage capacity. BP exhibits high carrier mobility, facilitating fast charge and discharge rates, making it suitable for high-power applications. The tunable bandgap of BP, achieved through layer control or strain engineering, allows for the optimization of its energy levels for various energy storage applications. BP has a high energy density, enabling the storage of a significant amount of energy in a compact space. It also maintains good electrochemical performance at high charge-discharge rates, crucial for applications requiring fast charging or high-power output. The combination of high carrier mobility, tunable bandgap, high energy density, and stability positions BP as a promising electrode material, especially as a cathode, in energy storage. Ongoing research aims to explore and optimize BP's properties and develop innovative battery architectures to enhance its performance and overcome associated challenges.

19.4.5 Layered Transition Metal Oxides

Layered transition metal oxides are promising materials for energy storage applications. Their composition consists of transition metal cations sandwiched between oxide layers, providing unique properties. The layered structure enables efficient ion intercalation and deintercalation, facilitating ion transport and contributing to high energy storage capacity. These oxides often exhibit high theoretical capacity due to the ample space available for ion storage. Their structural integrity is maintained during charge–discharge cycles, ensuring long-term stability. The properties of layered transition metal oxides can be tuned through composition, doping, and selection of transition metal elements, allowing for optimization of capacity, voltage, and cycling stability. Some oxides can store multivalent ions, offering the potential for high-capacity and high-energy-density battery systems.

Common examples of layered transition metal oxides used in energy storage applications include lithium cobalt oxide ($LiCoO_2$), lithium nickel manganese cobalt oxide ($LiNiMnCoO_2$ or NMC), and sodium nickel cobalt aluminum oxide (Na-NCA). These materials have been extensively investigated for their use as cathode materials in lithium-ion batteries due to their high capacity and stability. These layered transition metal oxides offer attractive properties for energy storage applications, including high capacity, intercalation chemistry, structural stability, and tunability. Ongoing research and development aim to further optimize these materials and explore their potential for next-generation energy storage systems.

19.4.6 Boron Nitride

Boron nitride (BN) is a 2D material with excellent thermal and electrical insulation properties. Although primarily known for these characteristics, BN has also shown promise in energy storage applications. It can enhance battery performance and safety when used as an additive in electrolytes. BN improves stability, suppresses undesirable side reactions, and enhances overall cycling performance. BN-based materials have been investigated for supercapacitor electrodes, demonstrating high capacitance and fast charge-discharge rates. BN incorporation in anode materials, such as BN-graphene composites, enhances ion adsorption and diffusion for improved lithium-ion storage capacity and cycling stability. BN has also been explored for fuel cells, acting as catalyst supports to enhance efficiency and durability. In photovoltaics, BN can serve as a protective layer or component in heterostructures, improving solar cell performance and stability. Further research is needed to optimize BN's properties for specific energy storage systems, but its development holds the potential for advancing the performance and safety of energy storage technologies.

19.4.7　2D–2D Semiconductor Hetero Junctions for Energy Storage

2D–2D semiconductor heterojunctions have emerged as a promising approach to enhance energy storage performance by combining two different 2D-SCMs. These heterojunctions possess unique electronic, optical, and electrochemical properties that enable improved charge separation, expanded absorption spectra, tailored band alignments, and synergistic effects. Ongoing research aims to optimize their design, fabrication, and integration into EESDs to achieve enhanced efficiency, stability, and functionality. Efficient charge separation and transport in heterojunctions are achieved by controlling the band offsets and alignments at the interface. This enables the reduction of recombination and improvement of charge storage kinetics. Additionally, the combination of complementary bandgap materials allows for broader absorption of the electromagnetic spectrum, leading to enhanced light harvesting capabilities.

In the context of EESDs such as batteries and supercapacitors, 2D–2D heterojunctions enhance crucial properties like ion diffusion, surface reactivity, and capacitance, resulting in improved overall performance. The precise engineering of band alignments enables the creation of desired device characteristics and the optimization of energy storage performance. Several specific types of 2D–2D heterojunctions have been investigated in recent research. TMDC heterojunctions exhibit unique electronic properties that are determined by the specific combination of TMDC materials involved. The band offsets at the interface can be type I or type II, influencing the populations of electrons and holes. Interface states, arising from lattice mismatch or charge transfer effects, significantly impact charge carrier dynamics. TMDC heterojunctions offer the possibility of tuning the effective bandgap by combining materials with different bandgaps, which has implications for applications like photovoltaics. Quantum confinement effects resulting from charge carrier confinement modify the band structure and enhance optical and electrical properties. However, the band structure is influenced by factors such as the types of TMDCs, stacking order, and interlayer interactions or defects. To understand these heterojunctions in detail, researchers employ theoretical modeling and experimental techniques such as spectroscopy and transport measurements.

Graphene-based heterojunctions, such as graphene–MoS_2, graphene–WS_2, and graphene–BP, combine the unique properties of graphene with those of respective 2D materials, resulting in intriguing band structures. Graphene, with its linear dispersion relation near the Fermi level and high charge carrier mobility, forms a Dirac cone structure with massless Dirac fermions. The band alignment at the interface between graphene and the other 2D material plays a crucial role in determining the electronic properties, and it can be of type I or type II. Although graphene lacks an intrinsic energy bandgap, when combined with 2D materials, the resulting heterojunctions exhibit a modified bandgap, enabling control over conductivity and optoelectronic properties. The interface between the two materials introduces interfacial states that significantly impact charge transport and device performance. Heterojunctions induce quantum confinement effects in the 2D material, leading to enhanced optical absorption and improved carrier mobility. Researchers are actively exploring the engineering of graphene-based heterojunctions to create devices with tailored properties for applications in electrical, optical, and energy storage domains. MXene-based heterostructures involve the stacking or integration of MXene layers with other 2D materials like graphene, TMDCs, or BN. The band alignment between MXene and the other 2D material is vital in determining the electronic properties, charge transfer, and device performance. MXene-based heterostructures can exhibit modified bandgaps, offering tunability for controlling electrical and optical properties, which is crucial for applications like photovoltaics and optoelectronics. The metallic conductivity of MXenes enables efficient charge

19.5 Electroanalytical Tools Applicable to Understand the Energy Storage Capability of 2D-SCMs

Electroanalytical techniques are commonly employed to understand the electrochemical properties of 2D-SCMs. These techniques provide valuable insights into the material's charge storage behavior, electrochemical kinetics, and stability. Here are some key electroanalytical tools used for studying the properties of 2D-SCMs.

19.5.1 Cyclic Voltammetry

Cyclic voltammetry (CV) is indeed a widely used electroanalytical technique for characterizing the redox behavior and charge storage capacity of 2D-SCMs. During CV measurements, a potential sweep is applied to the material over a specified voltage range, typically in a repetitive cycle. The potential is swept linearly between a starting potential and an endpoint and then reversed back to the starting potential. The resulting current response is recorded as a function of the applied potential. During the potential sweep, electrochemical reactions occur at the surface of the 2D-SCM material, involving the transfer of electrons and ions. These reactions can be reversible or irreversible, depending on the material and the specific redox processes involved. CV provides valuable information about the redox behavior of the 2D-SCM. It helps identify redox peaks, which are characteristic features in the current response curve. The position and shape of these peaks can provide insights into the thermodynamics and kinetics of the redox processes, such as the potential at which the redox reactions occur and the rate at which they take place. Additionally, CV can indicate the reversibility of the redox reactions. Reversible redox reactions exhibit symmetric anodic and cathodic peaks, indicating efficient charge storage and release. On the other hand, irreversible reactions may show asymmetric peaks or broad current responses, suggesting limitations in charge storage capacity or kinetic processes. The magnitude of the current response in CV is proportional to the charge storage capacity of the 2D-SCM material. By integrating the current response over the voltage range, the total charge associated with the redox processes can be determined. This provides an estimation of the material's charge storage capacity, which is crucial for assessing its suitability for energy storage applications. Overall, CV is a powerful technique that provides valuable information about the redox behavior, reversibility, and charge storage capacity of 2D-SCMs. It helps in understanding the electrochemical properties of these materials and optimizing their performance for various energy storage applications, such as batteries and supercapacitors.

19.5.2 Chronoamperometry

Chronoamperometry (CA) is an electroanalytical technique used to study the charge storage dynamics and kinetics of 2D-SCMs. In CA, a constant potential is applied to the 2D-SCM material, and the resulting current is measured as a function of time. By applying a constant potential, the charge storage processes within the material can be monitored over a specific time interval. CA provides valuable information about the charge storage and diffusion processes occurring in the 2D-SCM. The current response observed during CA reflects the movement of charge carriers (electrons or ions) within the material. It allows for the characterization of charge storage and release kinetics, as well as the determination of the charge storage capacity and rate capabilities of the material. The measured current response in CA can be analyzed to understand the charge transfer reactions taking place at the material–electrolyte interface. It provides insights into the kinetics of these charge transfer processes, including information about the rate of electron or ion transfer and any associated limitations or resistances. Furthermore, CA can be used to study the charge storage behavior under various conditions, such as different applied potentials or electrolyte compositions. By varying the experimental parameters, researchers can gain a better understanding of the factors influencing the charge storage dynamics and optimize the performance of the 2D-SCMs for energy storage applications. In summary, CA is a valuable electroanalytical technique that enables the study of charge storage dynamics, diffusion processes, and charge transfer kinetics in 2D-SCMs. It provides essential information for understanding and optimizing the energy storage properties of these materials, contributing to the development of improved EESDs such as batteries and supercapacitors.

19.5.3 Electrochemical Impedance Spectroscopy

Electrochemical impedance spectroscopy (EIS) is a powerful electroanalytical technique used to study the electrical properties and charge transfer kinetics of 2D-SCMs. Here's a more detailed explanation of the technique: EIS involves applying a small amplitude AC signal to the 2D-SCM material over a range of frequencies and measuring the resulting impedance response. The impedance is the ratio of the applied AC voltage to the resulting AC current and is represented by a complex number. EIS provides valuable information about the electrical behavior and charge transfer kinetics within the 2D-SCM material.

The impedance response obtained from EIS can be analyzed using different models and mathematical techniques to extract various electrochemical parameters.

Charge transfer resistance (Rct): The charge transfer resistance represents the resistance encountered by the charge carriers during the charge transfer processes at the material-electrolyte interface. It provides insights into the kinetics and efficiency of charge transfer reactions.

Double-layer capacitance (Cdl): The double-layer capacitance is related to the capacitance of the electrical double layer formed at the material-electrolyte interface. It provides information about the surface area and electrochemical activity of the material.

Warburg impedance (Zw): The Warburg impedance is associated with the diffusion of charge carriers within the material. It provides information about the diffusivity of ions or electrons and can reveal the presence of mass transport limitations. Frequency-dependent behavior: By analyzing the impedance response at different frequencies, information about the electrode processes, adsorption/desorption phenomena, and diffusion processes can be obtained.

EIS is particularly useful for understanding the charge transfer kinetics, capacitive behavior, and diffusivity of ions or electrons within the 2D-SCM material. It provides insights into the electrochemical performance and can be used to optimize the material for energy storage applications. By comparing EIS measurements before and after cycling or under different experimental conditions, researchers can gain insights into the stability, degradation, and electrochemical behavior of the 2D-SCMs over time. In summary, EIS is a powerful technique for studying the electrical properties, charge transfer kinetics, and diffusion processes of 2D-SCMs. It provides valuable information for understanding and optimizing the performance of these materials in energy storage applications such as batteries, supercapacitors, and fuel cells.

19.5.4 Galvanostatic/Galvanodynamic Techniques

Galvanostatic and galvanodynamic techniques are commonly used in electrochemical characterization to study the charge storage capacity, ion diffusion, and kinetic limitations of 2D-SCMs. In galvanostatic techniques, a constant current is applied to the 2D-SCM material, and the resulting potential response is monitored over time. This technique allows for the characterization of the charge storage capacity and electrochemical behavior of the material. By measuring the potential response as a function of time, researchers can obtain information about the capacity of the material to store charge. The charge storage capacity is typically evaluated in terms of the specific capacitance, which represents the amount of charge stored per unit mass or surface area of the material. Galvanostatic techniques also provide insights into the kinetics of charge transfer reactions and ion diffusion within the 2D-SCM. By analyzing the potential-time curves, information about the rate of charge storage or release, as well as any potential limitations or resistances, can be obtained.

Galvanodynamic techniques involve applying a controlled current profile to the 2D-SCM material, where the current is varied over time according to a predefined profile. This technique allows for the investigation of dynamic processes and can provide more detailed information about the charge storage and ion diffusion behavior. By applying specific current profiles, such as step changes or sinusoidal waveforms, researchers can gain insights into the charge storage dynamics, reaction kinetics, and diffusion processes within the material. The resulting potential response is analyzed to understand the behavior of the 2D-SCM under different current profiles and to extract relevant electrochemical parameters. Both galvanostatic and galvanodynamic techniques provide important information about the charge storage capacity, ion diffusion behavior, and kinetic limitations of 2D-SCMs. These techniques are crucial for characterizing the electrochemical performance of the materials and optimizing their use in energy storage applications. By comparing measurements obtained under different experimental conditions, such as varying current densities or cycling rates, researchers can gain a deeper understanding of the electrochemical behavior and stability of the 2D-SCMs over time. The galvanostatic and galvanodynamic techniques are valuable tools for studying the charge storage capacity, ion diffusion, and kinetic limitations of 2D-SCMs. These techniques provide important insights into the electrochemical behavior of the materials and contribute to the development of high-performance EESDs.

19.5.5 Scanning Electrochemical Microscopy

Scanning electrochemical microscopy (SECM) is a powerful electrochemical technique that allows for spatially resolved imaging of the electrochemical activity and charge transfer processes of 2D-SCMs. SECM combines principles of scanning probe microscopy with electrochemical measurements to provide information about the local electrochemical activity and charge transfer processes on the surface of the 2D-SCM material. The technique involves scanning a microelectrode, typically a small ultramicroelectrode or a nanoelectrode, in close proximity to the surface of the 2D-SCM while monitoring the resulting electrochemical currents or potentials. SECM operated primarily in two modes, that is, feedback mode and imaging mode.

Feedback mode: In this mode, the microelectrode is held at a constant distance from the surface, and the electrochemical current or potential is measured as a function of the position of the microelectrode. By mapping the local electrochemical activity, researchers can obtain information about the charge storage capacity and electrochemical reactivity of the 2D-SCM material.

Imaging mode: SECM can also be used to generate spatially resolved images of the electrochemical activity on the surface of the 2D-SCM. By scanning the microelectrode in a raster pattern over the surface, an image of the electrochemical activity distribution can be obtained. This imaging capability allows for the visualization of localized redox reactions and the mapping of reactive species on the surface.

SECM provides valuable information about the charge transfer processes, electrochemical reactivity, and spatial distribution of reactive species on the surface of 2D-SCMs. It can be used to investigate the influence of defects, surface modifications, or environmental conditions on the electrochemical properties of the material. Additionally, SECM can be combined with other imaging techniques, such as scanning tunneling microscopy or atomic force microscopy, to correlate the electrochemical activity with the surface morphology or topography of the 2D-SCM material. In summary, SECM is a powerful electrochemical technique that enables spatially resolved imaging of the electrochemical activity and charge transfer processes of 2D-SCMs. It provides valuable information about the local charge storage capacity, electrochemical reactivity, and distribution of reactive species on the material's surface. SECM is a valuable tool for understanding the electrochemical behavior of 2D-SCMs and optimizing their performance in energy storage applications. These electroanalytical techniques, among others, provide valuable information about the electrochemical behavior, charge storage mechanisms, and stability of 2D-SCMs. By combining multiple techniques, researchers can gain comprehensive insights into the electrochemical properties of these materials and optimize their performance for energy storage applications.

19.6 Conclusion and Outlook

To conclude, EESDs based on 2D semiconductors (2D-SCs) have gained significant attention due to their exceptional properties, including high surface area, tunable electronic properties, and fast ion diffusion. 2D-SCs find applications in various EESDs such as supercapacitors, batteries, solar cells, and hydrogen storage materials. Supercapacitors based on 2D-SCs, such as graphene, TMDCs, and BP, exhibit high specific capacitance, fast charge/discharge rates, and long cycle life, making them promising for applications requiring high power output and fast charging/discharging. 2D-SCs have been utilized as anode or cathode materials in batteries. For example, MXene has been used as an anode material in lithium-ion batteries, showing high specific capacity and cycling stability. TMDCs have been used as cathode materials in sodium-ion batteries, demonstrating high specific capacity and excellent rate performance. 2D-SCs have been explored in solar cells. Perovskite solar cells based on 2D perovskites exhibit high efficiency and stability. TMDCs have been employed as electron transport layers in organic solar cells, leading to improved device performance. Hydrogen storage is an important technology for hydrogen fuel cells, and 2D materials like graphene and BN show promise as hydrogen storage materials due to their high surface area and hydrogen adsorption capacity. The chapter discusses the challenges associated with the scalability and cost-effectiveness of producing 2D-SCs in large quantities, highlighting the need for addressing these issues. Ongoing research and development in the field of 2D-SCs based devices are mentioned, indicating the continuous efforts to advance energy storage technology. 2D-SCM has a wide range of applications for 2D-SCs in EESDs, emphasizing their unique properties and promising performance. It also addresses the challenges and ongoing research in the field, providing insights into the future developments of 2D-SCs in energy storage technology.

References

1. A. Verma, A. Parashar, Characterization of 2D nanomaterials for energy storage, *Recent Advances in Theoretical, Applied, Computational and Experimental Mechanics: Proceedings of ICTACEM 2017*, Springer (2020) 221–226.
2. X. Zhang, L. Hou, A. Ciesielski, P. Samorì, 2D materials beyond graphene for high-performance energy storage applications, *Adv. Energy Mater.* 6 (2016) 1600671.
3. R. Sahoo, A. Pal, T. Pal, 2D materials for renewable Electrical energy storage devices: Outlook and challenges, *Chem. Commun.* 52 (2016) 13528–13542.
4. Z. Wu, J. Qi, W. Wang, Z. Zeng, Q. He, Emerging elemental two-dimensional materials for energy applications, *J Mater. Chem. A.* 9 (2021) 18793–18817.
5. Q. Yun, L. Li, Z. Hu, Q. Lu, B. Chen, H. Zhang, Layered transition metal dichalcogenide-based nanomaterials for electrochemical energy storage, *Adv. Mater.* 32 (2020) 1903826.
6. X. Ren, P. Lian, D. Xie, Y. Yang, Y. Mei, X. Huang, Z. Wang, X. Yin, Properties, preparation and application of black phosphorus/phosphorene for energy storage: a review, *J. Mater. Sci.* 52 (2017) 10364–10386.
7. J. Cheng, L. Gao, T. Li, S. Mei, C. Wang, B. Wen, W. Huang, C. Li, G. Zheng, H. Wang, Two-dimensional black phosphorus nanomaterials: emerging advances in electrochemical energy storage science, *Nano-Micro Lett.* 12 (2020) 1–34.
8. T. Najam, S.S.A. Shah, L. Peng, M.S. Javed, M. Imran, M.-Q. Zhao, P. Tsiakaras, Synthesis and nano-engineering of MXenes for energy conversion and storage applications: recent advances and perspectives, *Coord. Chem. Rev.* 454 (2022) 214339.

9. P. Lokhande, A. Pakdel, H. Pathan, D. Kumar, D.-V.N. Vo, A. Al-Gheethi, A. Sharma, S. Goel, P.P. Singh, B.-K. Lee, Prospects of MXenes in energy storage applications, *Chemosphere*, 297 (2022) 134225.

10. S. Ghosh, S.S. Withanage, B. Chamlagain, S.I. Khondaker, S. Harish, B.B. Saha, Low pressure sulfurization and characterization of multilayer MoS_2 for potential applications in supercapacitors, *Energy*, 203 (2020) 117918.

11. T. Saisopa, A. Bunpheng, T. Wechprasit, P. Kidkhunthod, P. Songsiriritthigul, A. Jiamprasertboon, A. Bootchanont, W. Sailuam, Y. Rattanachai, C. Nualchimplee, A structural study of size selected WSe_2 nanoflakes prepared via liquid phase exfoliation: x-ray absorption to electrochemical application, *Radiat. Phys. Chem.* 206 (2023) 110788.

12. X. Cao, C. Tan, X. Zhang, W. Zhao, H. Zhang, Solution-processed two-dimensional metal dichalcogenide-based nanomaterials for energy storage and conversion, *Adv. Mater.* 28 (2016) 6167–6196.

13. X. Yu, S. Yun, J.S. Yeon, P. Bhattacharya, L. Wang, S.W. Lee, X. Hu, H.S. Park, Emergent pseudo-capacitance of 2D nanomaterials, *Adv. Energy Mater.* 8 (2018) 1702930.

14. L. Peng, Z. Fang, Y. Zhu, C. Yan, G. Yu, Holey 2D nanomaterials for electrochemical energy storage, *Adv. Energy Mater.* 8 (2018) 1702179.

15. W. Liu, B. Ullah, C.-C. Kuo, X. Cai, Two-dimensional nanomaterials-based polymer composites: fabrication and energy storage applications, *Adv. Polym. Technol.* 2019 (2019) 1–15.

16. Y. Yang, X. Liu, Z. Zhu, Y. Zhong, Y. Bando, D. Golberg, J. Yao, X. Wang, The role of geometric sites in 2D materials for energy storage, *Joule.* 2 (2018) 1075–1094.

17. Q. Guo, N. Chen, L. Qu, Two-dimensional materials of group-IVA boosting the development of energy storage and conversion, *Carbon Energy.* 2 (2020) 54–71.

18. T.A. Shifa, F. Wang, Y. Liu, J. He, Heterostructures based on 2D materials: a versatile platform for efficient catalysis, *Adv. Mater.* 31 (2019) 1804828.

19. T. Palaniselvam, J.-B. Baek, Graphene based 2D-materials for supercapacitors, *2D Mater.* 2 (2015) 032002.

20. Y. Dong, Z.-S. Wu, W. Ren, H.-M. Cheng, X. Bao, Graphene: a promising 2D material for electrochemical energy storage, *Sci. Bull.* 62 (2017) 724–740.

21. G.F. Smaisim, A.M. Abed, H. Al-Madhhachi, S.K. Hadrawi, H.M.M. Al-Khateeb, E. Kianfar, Graphene-based important carbon structures and nanomaterials for energy storage applications as chemical capacitors and supercapacitor electrodes: a review, *BioNanoScience.* 13 (2023) 219–248.

22. J. Nan, X. Guo, J. Xiao, X. Li, W. Chen, W. Wu, H. Liu, Y. Wang, M. Wu, G. Wang, Nanoengineering of 2D MXene-based materials for energy storage applications, *Small.* 17 (2021) 1902085.

23. D. Scarano, F. Cesano, Graphene and other 2D layered nanomaterials and hybrid structures: synthesis, properties and applications, *Mater.* 14 (2021) 7108.

24. H. Yoo, K. Heo, M.H.R. Ansari, S. Cho, Recent advances in electrical doping of 2D semiconductor materials: methods, analyses, and applications, *Nanomater.* 11 (2021) 832.

25. K.F. Mak, J. Shan, Photonics and optoelectronics of 2D semiconductor TMDs, *Nat. Photonics.* 10 (2016) 216–226.

26. C. Cong, J. Shang, Y. Wang, T. Yu, Optical properties of 2D semiconductor WS_2, *Advanced Opt. Mater.* 6 (2018) 1700767.

27. Y. Lee, S.B. Cho, Y.-C. Chung, Tunable indirect to direct band gap transition of monolayer Sc2CO2 by the strain effect, *ACS Appl. Mater. Interfaces.* 6 (2014) 14724–14728.

28. X. Wang, J. He, B. Zhou, Y. Zhang, J. Wu, R. Hu, L. Liu, J. Song, J. Qu, Bandgap-tunable preparation of smooth and large two-dimensional antimonene, *Angew. Chem.* 130 (2018) 8804–8809.

29. D. Zhou, L. Zhao, B. Li, Recent progress in solution assembly of 2D materials for wearable energy storage applications, *J. Energy Chem.* 62 (2021) 27–42.

30. K. Jiang, Q. Weng, Miniaturized EESDs based on two-dimensional materials, *ChemSusChem.* 13 (2020) 1420–1446.

31. F. Yi, H. Ren, J. Shan, X. Sun, D. Wei, Z. Liu, Wearable energy sources based on 2D materials, *Chem. Soc. Rev.* 47 (2018) 3152–3188.

32. Z. Wu, J. Hao, Electrical transport properties in group-V elemental ultrathin 2D layers, *NPJ 2D Mater. Appl.* 4 (2020) 4.

33. J.Y. Park, S. Kwon, J.H. Kim, Nanomechanical and charge transport properties of two-dimensional atomic sheets, *Adv. Mater. Interfaces.* 1 (2014) 1300089.

34. S.M. Gali, A. Pershin, A.L. Lherbier, J.-C. Charlier, D. Beljonne, Electronic and transport properties in defective MoS_2: impact of sulfur vacancies, *J. Phys. Chem. C.* 124 (2020) 15076–15084.

35. X. Huang, C. Liu, P. Zhou, 2D semiconductors for specific electronic applications: from device to system, *NPJ 2D Mater. Appl.* 6 (2022) 51.

36. S. Kang, D. Lee, J. Kim, A. Capasso, H.S. Kang, J.-W. Park, C.-H. Lee, G.-H. Lee, 2D semiconducting materials for electronic and optoelectronic applications: potential and challenge, *2D Mater.* 7 (2020) 022003.

37. H. Qiu, Y. Zhao, Z. Liu, M. Herder, S. Hecht, P. Samorì, Modulating the charge transport in 2D semiconductors via energy-level phototuning, *Adv. Mater.* 31 (2019) 1903402.

38. H. Duan, P. Lyu, J. Liu, Y. Zhao, Y. Xu, Semiconducting crystalline two-dimensional polyimide nanosheets with superior sodium storage properties, *ACS Nano.* 13 (2019) 2473–2480.

Two-Dimensional (2D) Semiconductors for Solar to Hydrogen Fuel

Spandana Gonuguntla, Aparna Jamma, Bhavya Jaksani, and Ujjwal Pal*

Correspondence: uapl03@gmail.com; ujjwalpal@iict.res.in

20.1 Introduction

The increasing demands of global energy have continuously triggered extensive research innovations for energy generation which leads to zero emission of effluent or eradicating the release of greenhouse gases. Hydrogen is found to be the most viable alternative source of energy, which can be produced through visible-light-induced water splitting. Water splitting is an efficient technique with net-zero emission of any other harmful side products wherein the energy required to break the water molecule is +1.23 eV [1]. Semiconductor photocatalysts have the potential to revolutionize the field of renewable hydrogen production by utilizing abundant sunlight and water resources [2]. Subsequently, numerous photocatalysts have been engineered to facilitate the photocatalytic generation of hydrogen from water. This photocatalytic process entails two distinct half-reactions: the photo-oxidation of H_2O in Equation (1) and the reduction of H^+ as depicted in Equation (2) and the final reaction in Equation (3).

$$H_2O + 2h^+ \rightarrow 2H^+ + \frac{1}{2}O_2 \quad E^0\text{oxidation} = -1.23V \quad (1)$$

$$2H^+ + 2e^- \rightarrow H2 \quad\quad E^0\text{reduction} = 0.00V \quad (2)$$

$$H_2O \rightarrow H_2 + \frac{1}{2}O_2 \quad\quad\quad (3)$$

The photocatalytic mechanism of the 2D layered materials toward hydrogen generation involves light absorption by the layered 2D materials which govern the photon of energy greater than that of the material's bandgap and when these photons get absorbed leads to the generation of electron–hole pairs. The thus-created electron–hole pair gets rapidly separated within the material surface due to its layered structures. The electrons migrate to the conduction band (CB) leaving behind the holes in the valence band (VB). These electrons lead to the involvement in the redox reactions over their surfaces as shown in Figure 20.1, electrons are involved in the reduction of water molecules, converting them into hydrogen ions (H^+) and ultimately, molecular hydrogen (H_2).

For efficient water splitting, an ideal photocatalyst must possess specific fundamental characteristics. The energy gap between the CB and VB should exceed 1.23 eV. Additionally, the CB should exhibit a more negative potential than the redox couple H^+/H_2 vs. reversible

DOI: 10.1201/9781003439448-20

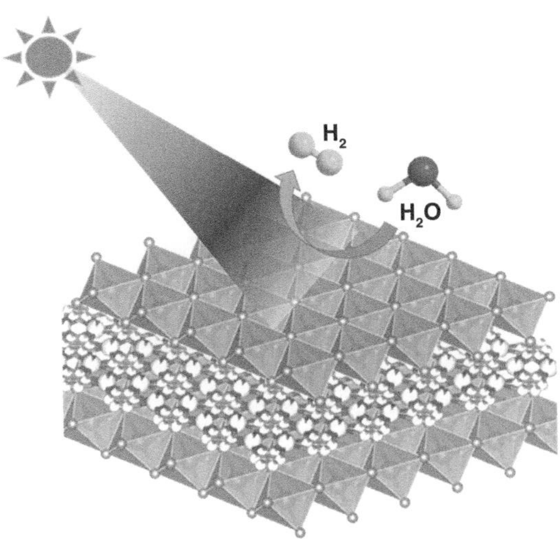

FIGURE 20.1
Photocatalyst: Layered 2D materials for H_2 generation from water.

hydrogen electrode (RHE), while the VB should maintain a more positive potential than O_2/H_2O vs. RHE.

However, ongoing research and development are essential to improve their efficiency, stability, and scalability for practical applications in a sustainable hydrogen economy. Among all the photocatalytic materials, Two-dimensional (2D) materials have garnered significant attention in the field of photocatalytic hydrogen generation through water splitting due to their unique properties and potential advantages such as high surface area, effective light absorption, charge separation, catalytic active sites and their charge separation activity, tunable bandgap, durability, quantum confinement effects, flexibility over their design strategy, and scalability [3]. The additional benefit of the 2D materials is their robust covalent bonding favoring their in-plane orientation toward the heterojunction and heterostructure formations [4]. Notably, graphene stands as the initial example of a 2D material, showcasing unparalleled electronic, mechanical, and optical properties. The advent of the graphene revolution instigated an unprecedented paradigm shift, resulting in an entirely novel realm of investigation concerning analogous architectures. The emergence of 2D TMDs has garnered substantial and growing interest in the contemporary scientific field/era [5]. TMDs, comprised of hexagonally arranged layers of metal atoms (M) intercalated between two layers of chalcogen atoms (X) according to the stoichiometric composition MX2, have emerged as focal points within this domain. The unified confinement of the covalent interactions over the adjacent layers which are majorly bonded through van der Waals forces, collectively lead to the formation of 3-D crystalline lattice planes and structures. In continuation to these catalysts, various other materials have been developed such as the layered construction of the 2D graphitic carbon nitride (g-C_3N_4), layered double hydroxides (LDH), perovskites, MXenes, and metal oxides. The MoS_2, possessing their unified optical as well as electronic properties leads to the reduction over the layer counts. The studies demonstrated that the bulk MoS_2 possesses indirect

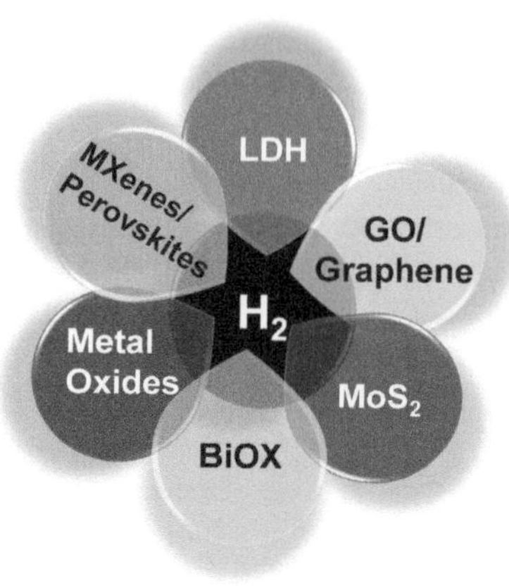

FIGURE 20.2
Various 2D materials for photocatalytic hydrogen generation.

band gaps which transform into direct band gaps through the arrangement of layers and their substrate criteria boosting their efficiencies. Graphitic carbon nitride (g-C_3N_4) is found to be another conjugated semiconductor photocatalyst, with planar arrangements of the carbon and nitrogen atoms [6]. The predominant electronic function assumed by graphene at the interface with semiconductor (SC) materials demonstrates its roles as an electron acceptor, transporter, and mediator within the framework of graphene-incorporated composites. This multifaceted engagement emerges from the inherent conductive two-dimensional (2D) architecture intrinsic to graphene, as substantiated across the scientific literature [7].

Distinct in its constitution as a 2D conductive medium characterized by a diminutive Fermi level, graphene engenders the promotion of both electron–hole delocalization and facilitated electron transfer across disparate photoelectrode architectures. Moreover, the 2D morphology inherent to these materials engenders an elongated trajectory of charge separation, effectively curbing the transverse distance that photoexcited electrons essential in achieving the solid/water interphase, thus favoring the charge recombination dynamics [8]. Different photocatalysts have been considered over the 2D-based semiconductor materials and they have been shown in Figure 20.2.

It is pertinent to note that frequently, the 2D materials do not function as primary photocatalysts or photoelectrodes. Nevertheless, these entities have been judiciously harnessed as sensitizer species, mediators of electron flux, co-catalytic agents, and protective overlays in tandem with other SC materials. Through these orchestrated collaborations, the amalgamation of 2D and SC constituents in hybrid architectures precipitates synergistic impacts that conduce to the enhancement of the electronic, optical, and photoelectrochemical (PEC) characteristics innate to 2D/SC electrodes.

Additionally, there is a growing interest in the investigation of emerging 2D materials such as black phosphorus (BP), MXenes, and organic/inorganic perovskites, which have recently attracted significant attention. The morphological configuration, photocatalytic properties, band gap tuning, characteristics of various electron transfer paths, and explain how they regulate the interfacial charge transfer and redox kinetics. Significant attention has also been given to highlighting the atomically thin two-dimensional materials with large lateral sizes, and precise and uniform thicknesses with minimal defects, discussing their unusual physical and chemical properties due to their special dimension effects.

The advantages of a 2D structure in photocatalysis can be attributed to four key factors:

(i) *Ample specific surface area*: 2D photocatalysts offer a substantial specific surface area, providing an abundance of highly exposed active sites. This increased surface area enhances the opportunities for catalytic reactions.

(ii) *Short transport paths*: Photoexcited holes and electrons have shorter transport paths to reach the photocatalyst surface for reactions. This minimizes the undesirable recombination of electron–hole pairs within the bulk material, a phenomenon that often limits the efficiency of traditional 3D photocatalysts.

iii. *Enhanced conductivity*: 2D photocatalysts, owing to their abundant surface defects, exhibit improved conductivity. This enhanced conductivity facilitates efficient charge transfer to adsorbates, contributing to better catalytic performance.

iv. *Superior mechanical properties*: 2D photocatalysts often demonstrate excellent durability, especially when they are combined with other materials in composites. This durability makes them suitable for long-term and practical applications in various catalytic processes.

In conclusion, drawing on the research landscape, we put forward a set of current issues and obstacles within semiconducting 2D systems. These challenges can serve as valuable sources of inspiration for the future advancement of highly efficient tailored semiconducting 2-D materials for solar to fuel generation.

20.2 Graphene-Based Materials

The 2D material possesses novel properties with strong covalent bonds forming the in-plane stability along with the weak van der Waals bonds. Graphene is a hydrophobic material that limits its efficacy in water-splitting reactions [9]. The cardinal electronic functions assumed by graphene at the interface with semiconductor (SC) materials entail its tripartite roles as an electron acceptor, transporter, and mediator within the framework of graphene-incorporated composites [10]. This multifaceted engagement emanates from the inherent conductive two-dimensional (2D) architecture intrinsic to graphene. Graphene oxide and reduced graphene oxide have unique features suitable for water splitting. Polymeric graphitic carbon nitride (g-C_3N_4 or GCN) has gained significant scientific attention due to its notable thermal stability, commendable chemical durability, and favorable electronic structure. It possesses advantageous features such as appropriate band energy, non-toxicity, and abundance, as well as a moderate band gap of approximately 2.7 to 2.9 eV [11]. g-C_3N_4 does exhibit certain limitations, including limited utilization of visible light (absorption restricted to wavelengths <460 nm) and swift recombination of photogenerated charges. The incorporation of graphene into a nanoparticulate semiconductor can address the absence of a bandgap and significantly augment photocatalytic efficiency. Exemplifying this mode of synthesis with paramount proficiency is the fabrication of graphene manifested as reduced graphene oxide (rGO). An alternative technique involves intercalation, wherein the interstitial interstices amid stratified configurations are

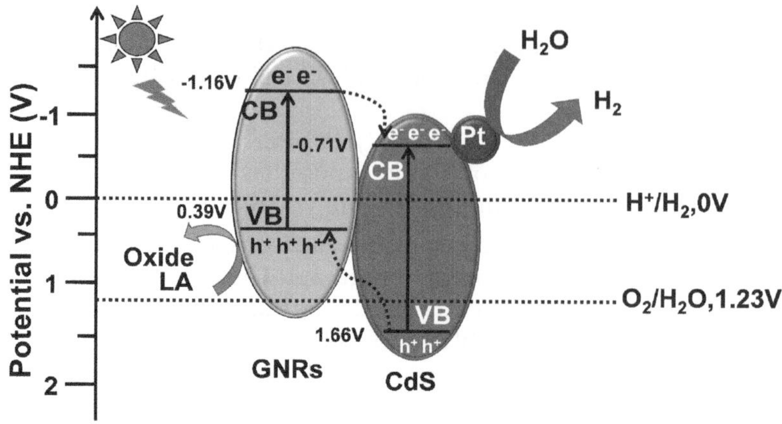

FIGURE 20.3
Schematic diagram of the band arrangements and charge transfer mechanism for the CdS/GNR heterojunction [13].

harnessed to introduce exogenous molecular entities, thereby effectuating the disengagement of sheets via mitigation of their interlayer cohesive interactions [12].

The enhancement observed in this context is underpinned by two pivotal factors: firstly, graphene introduces augmented porosity, thereby resulting in an enhancement in both surface area and the populace of active sites. Secondly, graphene operates as an electron acceptor, adsorbed by the electrons triggered within the CdS substrate. Moreover, graphene substantially decreases the recombination tendencies of photoexcited electron–hole pairs, thereby substantiating its utility in this domain. This phenomenon is effectively elucidated by a schematic representation in Figure 20.3, which expounds upon the intricacies of the photocatalytic reaction. The diagram delineates the fundamental process of generating photoinduced charge carriers, hinging upon the capacity to promote electrons from the VB to the CB of CdS. The construction of a hybrid heterojunction possessing CdS nanoparticles over the graphene nanoribbons (GNRs) led to the remarkable electron–hole recombination criteria. The CdS/GNR with 10wt% of GNR with higher efficiency of 22.4 mmolh^{-1}g^{-1} [13]. This phenomenon is viable through three distinct conduits: firstly, the carbon moieties residing on the graphene surface; secondly, the platinum species adorning the CdS substrate; and thirdly, the strategic localization of platinum upon the graphene matrix. The genesis of H_2 is attributed to the orchestrated interplay of electrons and adsorbed H^+ ions. Notably, graphene additionally augments the temporal longevity of charge carriers, thereby expediting the efficacious dissociation of photoinduced electron–hole pairs.

20.3 Boron Nitride

Boron nitride (BN) is one of the elements of the class of 2D materials. These BN materials possess various structural variations such as hexagonal BN, amorphous BN, rare wurtzite BN, and cubic BN. Hexagonal boron nitride (h-BN) exhibits remarkable attributes including elevated thermal durability and conductivity, notable mechanical robustness, and superb lubrication characteristics. These distinctive traits hold significant promise for multifarious applications, notably in the realms of thermoelectric apparatuses and micro-electronics. h-BN also manifests itself as a semiconductor featuring a considerable band-gap exceeding 5 eV, a fact that renders it unsuitable for deployment as a photocatalyst [14]. Nevertheless, the extensive utility of h-BN-based materials within PEC and photocatalytic (PC) systems is feasible through strategic manipulation of their characteristics, achieved by means such as elemental doping and surface functionalization, synergized with other semiconductor materials like titanium dioxide (TiO_2) and tungsten trioxide (WO_3) [15]. Among all the BN-based composites h-BN is found to be one of the most crystalline and stable composites, their arrangements of the boron and nitrogen atoms over 2-dimensional planes, possess a honey-comb structure, like the graphene structure.

20.4 Transition Metal Dichalcogenides

TMDs have emerged as a novel subclass of two-dimensional materials, renowned for their extraordinary attributes that render them potential noble metal substitutes in catalytic applications [16]. The chemical composition of TMDs follows a generic formula MX_2, wherein M denotes a transition metal, and X signifies a chalcogen. Notably, these materials exhibit planar X–M–X layers, stacked exclusively via van der Waals (vdW) interactions. The group of transition metal dichalcogenides (TMDs) includes molybdenum diselenide ($MoSe_2$), molybdenum disulfide (MoS_2), silicene (two-dimensional silicon), borophene (two-dimensional boron), and tungsten diselenide (WSe_2), alongside hexagonal h-BN, tungsten disulfide (WS_2) and germanene (two-dimensional germanium).

TMDs possess a distinctive characteristic wherein their indirect bandgap transitions into a direct bandgap when transitioning from bulk materials to monolayers. The bulk crystals of TMDs, comprising layers held together by van der Waals bonding, can be mechanically exfoliated to obtain bilayers or monolayers. This intriguing alteration in their electrical and optical properties introduces quantum confinement and surface effects, rendering them highly versatile for various applications. Within TMDs, the magnitude of spin-orbit coupling is substantial, yielding spin-split bands and thereby affording avenues for electrical modulation of electron spins. This property endows TMDs with marked utility across relevant domains [17]. Notably, TMDs have garnered considerable interest within the catalytic realm. Exploiting their layered architectures, TMDs offer the potential for deliberate adjustment of their crystal lattices and associated strain landscapes, thereby enabling augmentation of the catalytic efficacy exhibited by catalysts based on TMDs [18].

TMD nanostructures, encompassing MoS_2 [19], $MoSe_2$, WS_2, $MoTe_2$, and WSe_2, have garnered frequent utilization in catalytic applications due to their remarkable properties. The prominent features of their enhanced surface areas, huge numbers of excellent electronic properties, and along with high chemical stabilities. WS_2 and MoS_2 encompass two primary crystalline phases denoted as 1T (metallic) and 2H (semiconducting), each characterized by distinctive material attributes [20]. Nevertheless, strategies exist to induce interconversion between these phases, contingent upon the intricacies of design, structural arrangement, and specific quantum states. The crystallographic configurations of the 1T and 2H phases of MoS_2 are visually expounded upon in Figure 20.4.

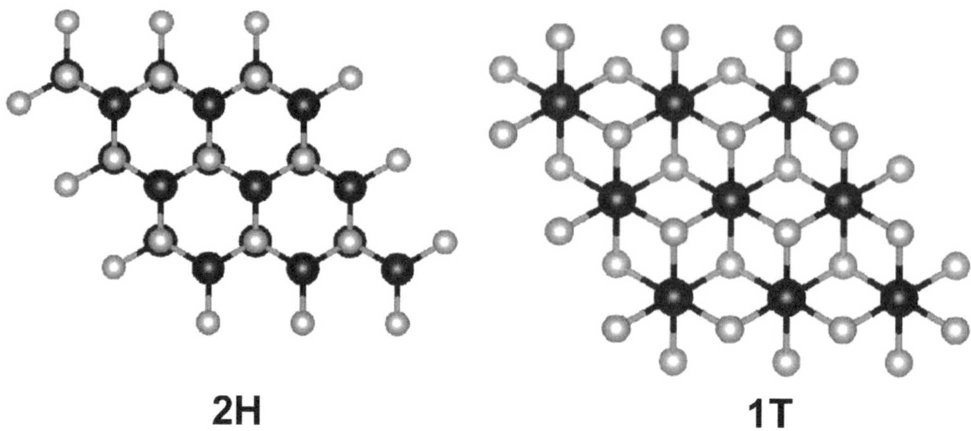

FIGURE 20.4
Structures of two phases (1T and 2H) of MoS_2.

Zhang et al. demonstrated a prominent strategy toward the synthesis of $1\%MoS_2/Fe_2O_3/$ g-C_3N_4, through hydrothermal synthesis, which possesses superior hydrogen generation efficiency of 7.82 mmolg^{-1}h^{-1} in comparison to that of the pristine g-C_3N_4 (1.56 mmolg^{-1}h^{-1}). The effective fabrication of a composite catalyst has yielded additional conduits for charge carrier migration, consequently mitigating the recombination of photogenerated electron–hole pairs [21]. In the presence of illumination, VBs of CN, Fe_2O_3, and MoS_2 engendered electron excitation, ensuring the migration of electrons from valence to conduction bands, with the creation of holes within the valence band.

20.5 LDH Materials

LDHs, also referred to as bimetallic hydroxides, are characterized by the overarching formulation $[M_{1-x}^{2+}M_x^{3+}(OH)_2][A_{x/n}].mH_2O$, manifest as two-dimensional substrates instrumental in co-catalytic/catalytic capacities within PEC/photocatalytic water-splitting frameworks [22,23]. Diverse iterations of LDHs can be synthesized by deliberate manipulation of the parameter x, affording varied chemical compositions and structural configurations [24].

Modifications involving the introduction of metal cations and anionic species into the LDH interlayers exert a significant influence on the electronic and optical characteristics. The incorporation of metal cations with a higher work function can enhance the optical attributes of LDH materials while facilitating the effective separation of photogenerated carriers. The energy conversion field shows great potential for LDHs, due to their unique 2D structures and varying bandgaps which favor the photocatalytic hydrogen generation efficiency [25]. The narrow photo-excitation range, as well as low efficiency toward the charge separation, is the major obstacle for many photocatalysts for industrial applications in energy conversion, through photocatalysis. LDHs have become a popular option for photocatalysts, partly due to their ability to adjust their chemical composition and their layered structure. The manipulation of metal cations allows for precise adjustment of light absorption to match specific visible wavelengths. The manipulation of cationic metals and the introduction of intercalating anions contribute to the broadening of LDH-based materials. The synergistic combination of various components can yield photocatalysts with increased activity [26].

Modifying the band structures and expanding the range of visible-light absorption represents an efficient strategy to augment optical reactivity, as well as to enhance the separation and mobility of charge carriers. Utilizing diffuse reflectance ultraviolet–visible (UV–vis) spectroscopy proves to be a valuable method for visualizing the optical characteristics of LDH. The absorption spectrum of LDH typically presents three discernible absorption bands, spanning the regions of 200–300 nm, 300–500 nm, and 600–800 nm, thereby encompassing both the UV and visible spectra. The absorption peak observed in the 200–300 nm range of the UV-active region is typically attributed to charge transfer between the ligand and the metal ion, specifically referred to as ligand-to-metal charge-transfer (LMCT). The absorption bands observed between 380 and 740 nm are assigned to transitions that are allowed by the spin, known as spin-allowed transitions [27]. Numerous distinct active LDHs have been successfully synthesized, each comprising unique combinations of divalent (M^{2+}) and trivalent (M^{3+}) or tetravalent (M^{4+}) metal cations, such as ZnCr, NiCo, CoFe, ZnFe, CuCr, NiFe, among others. These LDHs have been engineered to exhibit diverse photoactive properties that align with their compositional, structural, and electronic versatility [28]. The initial documentation of their photocatalytic performance can be attributed to the work of Silva et al. in 2009, wherein a series of LDHs, including ZnTi, ZnCe, and ZnCr, was investigated for their ability to facilitate the oxidation of water into O2 gas when subjected to visible light irradiation [29]. Spectroscopic analysis of these LDHs, as elucidated through diffuse reflectance spectra, revealed unique optical characteristics. Notably, ZnCe LDH displayed a prominent absorption peak around 280 nm, while ZnTi exhibited a more pronounced absorption peak at approximately 380 nm. In contrast, ZnCr LDH exhibited dual absorption maxima at 410 and 570 nm. ZnCr LDH emerged as the most proficient photocatalyst primarily attributed to its pronounced absorption of visible light. Additionally, these LDHs were characterized as "doped semiconductors," where the higher-valent metal served as a dopant. Among the extensive body of literature, the CdS/NiFe nanocomposite stood out, demonstrating an impressive hydrogen evolution rate of 72 mmol g^{-1} h^{-1}. The incorporation of noble-metal-free CdS-based cocatalyst nanostructures in photocatalytic hydrogen evolution processes presents innovative avenues for the development of economically viable catalysts for hydrogen production [30].

20.6 Metal Oxides

In recent trends, several sets of metal oxides, including TiO_2 [31], Fe_2O_3, ZnO, SnO_2, and WO_3, have garnered extensive employment as photocatalysts in the context of solar-driven water splitting [32]. While the allure of these 2D materials has augmented considerably, the non-layered oxide nanosheets, notably TiO_2 and Fe_2O_3, have undergone thorough scrutiny and been one of the prominent photocatalysts against their stratified counterparts such as graphene and TMDs [33]. The synthesis of non-layered 2D nanomaterials remains an intricate endeavor, attributable to the robust affinity existing between metal cations and oxygen anions. Despite this challenge, diverse non-layered 2D nanosheets, exemplified by TiO_2, WO_3, and SnO_2, have been successfully engineered and subsequently deployed in diverse applications [34–39]. Among these non-layered oxide nanosheets, those grounded in TiO_2 have garnered comprehensive investigation, predicated upon their commendable stability, non-toxicity, cost-effectiveness, and natural abundance [40–43]. The synthesis of MoS_2-loaded TiO_2 photocatalysts has been thoroughly demonstrated in one of our works wherein various morphologies of TiO_2 have been systematically considered such as TiO_2 (commercial, nanosheets, hierarchical, etc.) when loaded over with the MoS_2 nanosheets it is evident that the TiO_2 nanosheets over MoS_2 resulted in an efficiency of 77.41 $\mu molh^{-1}g^{-1}$ [43].

20.7 2D Metal Oxyhalide-Based Photocatalysts

Bismuth oxyhalides (BiOX, where X = Cl, Br, or I) have elicited significant intrigue due to their discrete architecture coupled with commendable optical and electrical attributes [44–46]. The configuration of BiOX encompasses $[Bi_2O_2]^{2+}$ layers interposed amid bilayers of halogen ions, yielding inherent, internally generated electrostatic fields, concomitant with remarkable photocatalytic proficiencies. Leveraging these distinctive characteristics, BiO_X species emerge as viable candidates for photocatalytic applications, particularly in the domain of water splitting.

20.8 MXenes and MXenes-Based Photocatalysts

Over the past few years, MXenes have emerged as highly promising next-generation materials for photocatalytic hydrogen generation applications, and the compound with Ti_3C_2 composite has been widely studied. MXenes, a rapidly advancing category of 2D materials encompassing transition metal nitrides, carbides, and carbonitrides, manifest substantial promise in their role as photocatalysts for hydrogen production [47]. Their growing recognition in this arena can be credited to a range of exceptional attributes, which include remarkable photophysical properties, greatly reduced thickness, distinctive morphology, structural and chemical stability, photostability, and a substantial active surface area.

MXene-semiconductor hybrids have proven to be highly effective photocatalysts due to their unique interface characteristics. In the context of photocatalysis, Schottky heterojunctions offer distinct advantages, such as faster charge separation and a lower Schottky barrier. This phenomenon finds its illustration in MXene-derived TiO_2 and MXene/TiO_2-based nanocomposites, where the mechanism underlying charge carrier separation is ascribed to the creation of a Schottky junction at the boundary between the electrically conductive Ti_3C_2 (MXene) and the optically responsive TiO_2, leading to the remarkable enhancement of photocatalytic hydrogen (H_2) evolution. These alterations are designed with the intention of promoting the effective direction of electrons, capitalizing on the metal–semiconductor interaction in cases where MXene functions as a cocatalyst or substrate, ultimately amplifying the comprehensive photocatalytic efficacy. In contemporary research, computational methodologies have unveiled a plethora of prospective MXene configurations and their precursor derivatives. This has led to the identification of a vast array of non-stoichiometric MXenes, characterized by finely adjusted properties and combinations of transition metals or carbon nitrides. These innovative materials are synthesized through the creation of solid solutions at the M and X sites. Presently, scientists are actively exploring strategies to introduce extra X elements into 2D boride systems [48]. The etching parameters are crucial for achieving a full conversion from MAX phases to MXenes. Ongoing efforts are dedicated to enhancing this process, with one notable approach being the suggested substitution of harsh HF with a milder etchant, ammonium bifluoride (NH_4HF_2).

Perovskite-based mixed metal oxide materials with a general formula of ABO_3—here A,B represent two diverse metal ions and oxygen for O—have attained huge attention over the past decades due to their sacred features such as their enhanced photostability, low toxicity, and tunable photophysical properties over various inorganic based semiconductor photocatalysts. Some of the perovskites involve alkaline titanates, alkaline earth tantalates—BTa_2O_6, $B_5Ta_4O_{15}$, $B_2Ta_2O_7$ (B = Ca, Sr, and Ba), $AgNbO_3$, and so on, and alkali tantalates ($MTaO_3$; M = Li, Na, and K). According to the literature and synthetic strategies, it is evident that the doping of the semiconductor materials with other elements is an efficient method for controlling the band gap alignment of the specific materials [49]. Rhodium-doped $SrTiO_3$ is most widely utilized for both photocatalytic and photoelectrochemical water splitting. Furthermore, rhodium-doped $BaTiO_3$ has demonstrated its efficiency and stability as a p-type semiconductor photocatalyst for generating hydrogen using visible light. However, there is still a need to investigate how precisely engineered dopants influence local structures to enhance photocatalytic activity. Niobium oxide semiconductors, known for their remarkable stability under solar radiation and non-toxic nature, have garnered significant attention due to their intriguing and potentially valuable properties, such as photocatalytic, piezoelectric, ferroelectric, and various optical characteristics. Potassium niobate ($KNbO_3$), with its appropriate CB potential in relation to H^+/H_2 potential, has been studied for photocatalytic water splitting to produce hydrogen. In recent years, numerous efforts have been made to enhance the photoactivity of $KNbO_3$. Nevertheless, exploration of the photocatalytic properties of these perovskite materials remains an ongoing area of investigation [50]. A comprehensive comparison of all these materials is presented in Table 20.1.

20.9 Scope and Future Perspectives

2D-based nanomaterials offer a path to cleaner and sustainable hydrogen production. Future developments may involve enhancing the efficiency through novel materials, tailored band structures, and co-catalysts. Researchers are also focusing on scalability, stability, and integration with water-splitting systems further ensuring environmental sustainability and transitioning to real-world applications. As future prospects, 2D-based photocatalysts have the potential to revolutionize hydrogen production and contribute to greener energy. Integrating 2D photocatalysts into sustainable energy systems represents a critical advancement. These photocatalysts can seamlessly join a larger network of clean energy technologies, encompassing solar panels, energy storage solutions, and efficient hydrogen generation. This synergistic integration will serve as a cornerstone in the establishment of holistic, environmentally friendly energy systems. A more profound comprehension of photocatalytic reactions, combined with the progression of ultrafast characterization methods and innovative dynamic simulation algorithms, has the potential to bring about significant enhancements in photocatalyst performance. In this context,

Table 20.1 Comparison of Photocatalytic Efficiencies of Various 2D-Based Photocatalysts

S. No	Composite Name	Light Source	H_2 Evolution Rate $(mmolg^{-1}h^{-1})$	Sacrificial Agent	References
1	CuO/rGO	300 W Xe lamp	19.2 mmol $g^{-1}h^{-1}$	Methanol	12
2	CdS/GNR with 10wt% of GNR	300 W Xe lamp	22.4 mmol $g^{-1}h^{-1}$	Lactic acid	13
3	$Cu(OH)_2$/P(g-C_3N_4)/MoS_2	400 W Xe lamp	12.01 mmol g^{-1} h^{-1}	Na_2S/Na_2SO_3	19
4	MoS_2/Fe_2O_3/g-C_3N_4	300 W Xe lamp	7.82 mmol g^{-1} h^{-1}	Lactic acid	21
5	NiFeLDH-Rh	420 W Xe lamp	2.09 mmol$g^{-1}h^{-1}$	Lactic acid	22
6	CuCdCe-LDH/g-$C_3N_4$3:1	420 W Xe lamp	3.5 mmol$g^{-1}h^{-1}$	Na_2S/Na_2SO_3	23
7	ML200 hybrid composite	300 W Xe lamp	127.6 mmol $g^{-1}h^{-1}$	TEOA	26
8	T-LDH/PbI_2 NC	500 W Mercury-Xenon	107.53 mmol h^{-1} cm^{-2}	KOH	28
9	CdS/NiFe	150 W Xe lamp	72 mmol $g^{-1}h^{-1}$	Lactic acid	30
10	α-Fe_2O_3/2D g-C_3N_4	300 W Xe lamp	>3 × 10^4 μmol g^{-1} h^{-1}	TEOA	33
11	20% CZS10/ZnO	300 W Xe lamp	40.14 mmol g^{-1} h^{-1}	Na_2S/Na_2SO_3	34
12	WO_3/TiO_2/Fe_2O_3	AAA solar light line A1, with Xe lamp	10.2 mL h^{-1}	Methanol	37
13	B:SnO_2+Pd	250 W mercury lamp	63.6184 μmol g^{-1} h^{-1}	EDTA + methanol	38
14	SnC-3	300 W Xe lamp	46.72 mmol$g^{-1}h^{-1}$	Lactic acid	39
15	2 nm Cu/TiO_2	150 W Xe lamp	334 μmol g^{-1} h^{-1}	TEOA	40
16	h-BN/TiO_2(HPT)/g-C_3N_4	450 W Xe lamp	2.02 mmolg^{-1} h^{-1}	Methanol	41
17	RGNPT-7.5Wt%	450 W Xe lamp	8,215 μmol$g^{-1}$for 4 h	Methanol	42
18	Sn-doped ZnO/BiOCl (ZBC-S)		4146.77 μmol g^{-1} h^{-1}		46
19	CeO_2/MXene	300 W Xe lamp	454.32 μmol·g^{-1}·h^{-1}	TEOA	48
20	$MAPbI_3$/Pt/3S-TiO_2HoMSs	300 W Xe lamp	6856.2 μmol $g^{-1}h^{-1}$	8 mL HI and 2 mL H_3PO_2	49

2D photocatalysts are anticipated to assume a central role in the progression toward practical photocatalyst systems for industrial applications.

References

1. Q. Mo, L. Zhang, S. Li, H. Song, Y. Fan, C. Y. Su, Engineering Single-Atom Sites into Pore-Confined Nanospaces of Porphyrinic Metal-Organic Frameworks for the Highly Efficient Photocatalytic Hydrogen Evolution Reaction, *J. Am. Chem. Soc.* 144 (2022) 22747–22758.
2. Q. Wang, K. Domen, Particulate Photocatalysts for Light-Driven Water Splitting: Mechanisms, Challenges, and Design Strategies, *Chem. Rev.* 120 (2020) 919–985.
3. P. Ganguly, M. Harb, Z. Cao, L. Cavallo, A. Breen, S. Dervin, D. D. Dionysiou, S. C. Pillai, 2D Nanomaterials for Photocatalytic Hydrogen Production, *ACS Energy Lett.* 4 (2019) 1687–1709.

4. X. Chen, S. Shen, L. Guo, S. S. Mao, Semiconductor-Based Photocatalytic Hydrogen Generation, *Chem. Rev.* 110 (2010) 6503–6570.

5. S. C. Zhu, F. X. Xiao, Transition Metal Chalcogenides Quantum Dots: Emerging Building Blocks Toward Solar-to-Hydrogen Conversion, *ACS Catal.* 13 (2023) 7269–7309.

6. D. K. Lee, K. S. Han, W. H. Shin, J. W. Lee, J. H. Chi, K. M. Choi, Y. Lee, H. Kim, W. Choi, J. K. Kang, Graphitic Domain Layered Titania Nanotube Arrays for Separation and Shuttling of Solar- Driven Electrons, *J. Mater. Chem. A.* 1 (2013) 203–207.

7. T. Su, Q. Shao, Z. Qin, Z. Guo, Z. Wu, Role of Interfaces in Two-Dimensional Photocatalyst for Water Splitting, *ACS Catal.* 8 (2018) 2253–2276.

8. J. Kosco, S. G. Carrero, C. T. Howells, T. Fei, Y. Dong, R. Sougrat, G. T. Harrison, Y. Firdaus, R. Sheelamanthula, B. Purushothaman, F. Moruzzi, W. Xu, L. Zhao, A. Basu, S. D. Wolf, T. D. Anthopoulos, J. R. Durrant, I. McCulloch, Generation of Long-Lived Charges in Organic Semiconductor Heterojunction Nanoparticles for Efficient Photocatalytic Hydrogen Evolution, *Nature Energy,* 7 (2022) 340–351.

9. A. T. Hoang, A. Pandey, W. H. Chen, S. F. Ahmed, S. Nizetic, K. H. Ng, Z. Said, X. Q. Duong, U. Agbulut, H. Hadiyanto, X. P. Nguyen, Hydrogen Production by Water Splitting with Support of Metal and Carbon-Based Photocatalysts, *ACS Sustain. Chem. Eng.* 11 (2023) 1221–1252.

10. N. Gurbani, C. P. Han, K. Marumoto, R. S. Liu, R. J. Choudhary, N. Chouhan, Biogenic Reduction of Graphene Oxide: An Efficient Superparamagnetic Material for Photocatalytic Hydrogen Production, *ACS Appl. Energy Mater.* 1 (2018) 5907–5918.

11. Y. Huang, T. Yang, H. Yu, X. Li, J. Zhao, G. Zhang, X. Li, L. Yang, J. Jiang, Theoretical Calculation of Hydrogen Generation and Delivery via Photocatalytic Water Splitting in Boron-Carbon-Nitride Nanotube/Metal Cluster Hybrid, *ACS Appl. Mater. Interfaces.* 12 (2020) 48684–48690.

12. A. A. Yadav, Y. M. Hunge, S. W. Kang, Spongy Ball-Like Copper Oxide Nanostructure Modified by Reduced Graphene Oxide for Enhanced Photocatalytic Hydrogen Production, *Mater. Res. Bull.* 133 (2021) 111026.

13. X. Luan, H. Dai, Q. Li, F. Xu, Y. Mai, A Hybrid Photocatalyst Composed of CdS Nanoparticles and Graphene Nanoribbons for Visible-Light-Driven Hydrogen Production, *ACS Appl. Energy Mater.* 5 (2022) 8621–8628.

14. Q. Weng, X. Wang, X. Wang, Y. Bando, D. Golberg, Functionalized Hexagonal Boron Nitride Nanomaterials: Emerging Properties and Applications, *Chem. Soc. Rev.* 45 (2016) 3989–4012.

15. J. Zhao, Z. Chen, Carbon-Doped Boron Nitride Nanosheet: An Efficient Metal-Free Electrocatalyst for the Oxygen Reduction Reaction, *J. Phys. Chem. C* 119 (2015) 26348–26354.

16. A. Gautam, S. Sk, U. Pal, Recent Advances in Solution Assisted Synthesis of Transition Metal Chalcogenides for Photo-Electrocatalytic Hydrogen Evolution, *Phys. Chem. Chem. Phys.* 24 (2022) 20638–20673.

17. M. Z. Rahman, F. Raziq, H. Zhang, J. Gascon, Key Strategies for Enhancing H_2 Production in Transition Metal Oxide Based Photocatalysts, *Angew. Chem. Int. Ed.* (2023) e202305385.

18. S. Chandrappa, D. H. K. Murthy, N. L. Reddy, S. J. Babu, D. Rangappa, U. Bhargav, V. Preethi, M. M. Kumari, M. V. Shankar, Utilizing 2D Materials to Enhance H_2 Generation Efficiency via Photocatalytic Reforming Industrial and Solid Waste, *Environ. Res.* 200 (2021) 111239.

19. C. S. Vennapoosa, S. Gonuguntla, S. Sk, B. M. Abraham, U. Pal, Ternary $Cu(OH)_2$/$P(g\text{-}C_3N_4)$/MoS_2 Nanostructures for Photocatalytic Hydrogen Production, *ACS Appl. Nano Mater.* 5 (2022) 4848–4859.

20. R. Kappera, D. Voiry, S. E. Yalcin, B. Branch, G. Gupta, A. D. Mohite, M. Chhowalla, Phase-Engineered Low-Resistance Contacts for Ultrathin MoS_2 Transistors, *Nat. Mater.* 13 (2014) 1128–1134.

21. Y. Zhang, J. Wan, C. Zhang, X. Cao, MoS_2 and Fe_2O_3 co-Modify $g\text{-}C_3N_4$ to Improve the Performance of Photocatalytic Hydrogen Production, *Sci. Rep.* 12 (2022) 3261.

22. C. S. Vennapoosa, A. Karmakar, Y. T. Prabhu, B. M. Abraham, S. Kundu, U. Pal, Rh-Doped Ultrathin NiFeLDH Nanosheets Drive Efficient Photocatalytic Water Splitting, *Int. J. Hydrogen Energy* 52 (2023). https://doi.org/10.1016/j.ijhydene.2023.02.025

23. C. S. Vennapoosa, S. Varangane, B. M. Abraham, V. Perupogu, S. Bojja, U. Pal, Controlled Photoinduced Electron Transfer from $g\text{-}C_3N_4$ to CuCdCe-LDH for Efficient Visible Light Hydrogen Evolution Reaction, *Int. J. Hydrogen Energy* 47 (2022) 40227–40241.

24. C. S. Vennapoosa, V. Tejavath, Y. T. Prabhu, A. Tiwari, B. M. Abraham, V. S. Upadhyayula, U. Pal, Sulphur Vacancy-Rich PCdS/NiCoLDH Promotes Highly Selective and Efficient Photocatalytic CO_2 Reduction to MeOH, *J. CO2 Util.* 67 (2023) 102332.

25. H. Boumeriame, E. S. D. Silva, A. S. Cherevan, T. Chafik, J. L. Faria, D. Eder, Layered Double Hydroxide (LDH)-Based Materials: A Mini-Review on Strategies to Improve the Performance for Photocatalytic Water Splitting, *J. Energy Chem.* 64 (2022) 406–431.

26. M. Sohail, H. Kim, T. W. Kim, Enhanced Photocatalytic Performance of a Ti-Based Metal-Organic Framework for Hydrogen Production: Hybridization with ZnCr-LDH Nanosheets, *Sci. Rep.* 9 (2019) 7584.

27. A. Sherryna, M. Tahir, Recent Developments in Layered Double Hydroxide Structures with Their Role in Promoting Photocatalytic Hydrogen Production: A Comprehensive Review, *Int. J. Energy Res.* 46 (2021) 2093–2140.

28. F. Mohamed, N. Bhnsawy, M. Shaban, Reusability and Stability of a Novel Ternary (Co–Cd–Fe)-LDH/PbI_2 Photoelectrocatalytst for Solar Hydrogen Production, *Sci. Rep.* 11 (2021) 5618.

29. C. G. Silva, Y. Bouizi, V. Fornés, H. García, Layered Double Hydroxides as Highly Efficient Photocatalysts for Visible Light Oxygen Generation from Water, *J. Am. Chem. Soc.* 131 (2009) 13833–13839.

30. H. Lee, D. A. Reddy, Y. Kim, S. Y. Chun, R. Ma, D. P. Kumar, J. K. Song, T. K. Kim, Drastic Improvement of 1D-CdS Solar-Driven Photocatalytic Hydrogen Evolution Rate by Integrating with NiFe Layered Double Hydroxide Nanosheets Synthesized by Liquid-Phase Pulsed-Laser Ablation, *ACS Sustain. Chem. Eng.* 6 (2018) 16734–16743.

31. S. Gonuguntla, R. Kamesh, U. Pal, D. Chatterjee, Dye sensitization of TiO_2 Relevant to Photocatalytic Hydrogen Generation: Current Research Trends and Prospects, *J. Photochem. Photobiol. C: Photochem. Rev.* 57 (2023) 100621.

32. Y. J. Yuan, Z. J. Ye, H. W. Lu, B. Hu, Y. H. Li, D. Q. Chen, J. S Zhong, Z. T. Yu, Z. G Zou, Constructing Anatase TiO_2 Nanosheets with Exposed (001) Facets/Layered MoS_2 Two-Dimensional Nanojunctions for Enhanced Solar Hydrogen Generation, *ACS Catal.* 6 (2016) 532–541.

33. X. She, J. Wu, H. Xu, J. Zhong, Y. Wang, Y. Song, K. Nie, Y. Liu, Y. Yang, M. T. F. Rodrigues, R. Vajtai, J. Lou, D. Du, H. Li, P. M. Ajayan, High Efficiency Photocatalytic Water Splitting Using 2D α-Fe_2O_3/g-C_3N_4 Z-Scheme Catalysts, *Adv. Energy Mater.* 7 (2017) 1700025.

34. G. Liu, J. Chen, Z. Xie, S. Lin, L. Xie, Y. Deng, C. Z. Lu, Hydrothermal Synthesis of $Cd_{0.5}Zn_{0.5}S$/ZnO Heterojunctions with Controlled pH and Enhanced Photocatalytic Hydrogen Production Activity, *ACS Appl. Energy Mater.* 5 (2022) 3502–3513.

35. S. D. Balgude, S. S. Barkade, S. P. Mardikar, Metal Oxides for High-Performance Hydrogen Generation by Water Splitting, *Multifunctional Nanostructured Metal Oxides for Energy Harvesting and Storage Devices*, Taylor & Francis/CRC Press, vol. 1 (2020) 26.

36. S. Ali, J. A. Nasir, R. N. Dara, Z. Rehman, Modification Strategies of Metal Oxide Photocatalysts for Clean Energy and Environmental Applications: A Review, *Inorg. Chem. Commun.* 145 (2022) 110011.

37. P. Jineesh, T.C. Bhagya, R. Remya, S.M.A. Shibli, Photocatalytic Hydrogen Generation by WO_3 in Synergism with Hematite-Anatase Heterojunction, *Int. J. Hydrogen Energy.* 45 (2020) 18946–18960.

38. S. Kumar, Bhawna, S. K. Yadav, A. Gupta, R. Kumar, J. Ahmed, M. Chaudhary, Suhas, V. Kumar, B-Doped SnO_2 Nanoparticles: A New Insight into the Photocatalytic Hydrogen Generation by Water Splitting and Degradation of Dyes, *Environ. Sci. Pollut. Res.* 29 (2022) 47448–47461.

39. K. Mallikarjuna, G.A.K.M. R. Bari, S.V.P. Vattikuti, H. Kim, Synthesis of Carbon-Doped SnO_2 Nanostructures for Visible-Light-Driven Photocatalytic Hydrogen Production from Water Splitting, *Int. J. Hydrogen Energy.* 45 (2020) 32789–32796.

40. I. Mondal, S. Gonuguntla, U. Pal, Photoinduced Fabrication of Cu/TiO_2 Core–Shell Heterostructures Derived from Cu-MOF for Solar Hydrogen Generation: The Size of the Cu Nanoparticle Matters, *J. Phys. Chem.* 123 (2019) 26073–26081.

41. S. Gonuguntla, S. Sk, A. Tiwari, H. Mandal, P. N. Lakavath, V. Perupoga, U. Pal, Regulating Surface Structures for Efficient Electron Transfer Across h-BN/TiO_2/g-C_3N_4 Photocatalyst for Remarkably Enhanced Hydrogen Evolution, *J. Mater. Sci.: Mater. Electron.* 32 (2021) 12191–12207.

42. K. Santhosh, S. Sk, S. Chouti, S. Gonuguntla, S. P. Ega, A. Tiwari, U. Pal, Tailoring Hierarchical Porous TiO_2 Based Ternary rGO/NiO/TiO_2 Photocatalyst for Efficient Hydrogen Production and Degradation of Rhodamine B, *J. Mol. Struct.* 1235 (2021) 13022.
43. A. Gautam, Y. T. Prabhu, U. Pal, Efficient Charge Transfer on the Tunable Morphology of TiO_2/MoS_2 Photocatalyst for an Enhanced Hydrogen Production, *New J. Chem.* 45 (2021) 10257–10267.
44. M. Shi, G. Li, J. Li, X. Jin, X. Tao, B. Zeng, P. E. A. Pidko, P. R. Li, P. C. Li, Intrinsic Facet-Dependent Reactivity of Well-Defined BiOBr Nanosheets on Photocatalytic Water Splitting, *Angew. Chem.* 132 (2020) 6652–6657.
45. X. Wei, M. U. Akbar, A. Raza, G. Li, A Review on Bismuth Oxyhalide Based Materials for Photocatalysis, *Nanoscale Adv.* 3(2021) 3353–3372.
46. Y. Guo, C. Qi, B. Lu, P. Li, Enhanced Hydrogen Production from Water Splitting by Sn-Doped ZnO/BiOCl Photocatalysts and Eosin Y Sensitization, *Int. J. Hydrogen Energy.* 47 (2022) 228–241.
47. X. Pan, X. Yang, M. Yu, H. Kang, M. Q. Yang, Q. Qian, X. Zhao, S. Liang, Z. Bian, 2D MXenes Polar Catalysts for Multi-Renewable Energy Harvesting Applications, *Nat Commun.* 14 (2023) 4183–4193.
48. H. Zhu, X. Fu, Z. Zhou, 3D/2D Heterojunction of CeO_2/Ultrathin MXene Nanosheets for Photocatalytic Hydrogen Production, *ACS Omega.* 7 (2022) 21684–21693.
49. W. Han, Y. Wei, J. Wan, N. Nakagawa, D. Wang, Hollow Multishell-Structured TiO_2/$MAPbI_3$ Composite Improves Charge Utilization for Visible-Light Photocatalytic Hydrogen Evolution, *Inorg. Chem.* 61 (2022) 5397–5404.
50. I. Mondal, M. Srikanth, K. Srinivasu, Y. Soujanya, U. Pal, Understanding the Structural and Electronic Effect of Zr^{4+}-Doped $KNb(Zr)O_3$ Perovskite for Enhanced Photoactivity: A Combined Experimental and Computational Study, *J. Phys. Chem. C.* 121 (2017) 2597–2604.

21

2D Semiconductors for Next-Generation Thermoelectric Materials

Abu B. Siddique, Sergio O. Martinez, and Mallar Ray*
Correspondence: mallar.ray@tec.mx

21.1 Background

The efficiency of a thermoelectric material is measured by a dimensionless parameter, $ZT = \dfrac{S^2 \sigma}{\kappa} T$ —the thermoelectric figure of merit, where S, σ, κ and T are Seebeck coefficient, electrical conductivity, thermal conductivity, and temperature, respectively [1]. Both TE generator (TEG—utilizes Seebeck effect to generate electricity from thermal gradient) and a TE cooler (TEC—utilizes Peltier effect to cool a junction by passing current) demand materials with high values of ZT (> 1 for TEGs and ~1 for TECs) for efficient design of a working device. It is easy to realize why maximizing ZT is a big challenge; the parameters S, σ and κ of a material are strongly correlated and optimizing one adversely affects another. It is well known that σ and κ scale proportionately in most materials following the Wiedemann–Franz law: $\dfrac{k}{\sigma} = LT$, where L is the Lorentz number. The Seebeck coefficient on the other hand decreases with increasing carrier density, n ($S \sim n^{-\frac{2}{3}}$) [2], implying that attempts to increase electrical conduction by manipulating charge carrier density results in decreasing S. The situation is therefore conflicting in all respect. The trade-off between S, σ and κ in typical materials is shown in the Figure 21.1(a). Though, several materials have been developed with ZT values >1, the values of ZT for most materials are restricted to ≪1. So far, layered semiconductors like bismuth telluride (Bi_2Te_3) based alloys, have been the materials of choice for TE applications. These materials have very narrow bandgaps (0.15–0.3 eV) and large band degeneracy, which helps in the enhancement of σ, while maintaining high S [3]. They also have inherently low lattice thermal conductivity (due to heavy Bi and Te atoms) and high weighted electronic mobility [4]. The small bandgap, however, also promote the generation of minority carriers which in turn restricts the TE performance of Bi_2Te_3 alloys.

The advent of nanomaterials has opened up a new possibility for developing high ZT materials due to the non-conventional charge and heat transport phenomena observed in low-dimensional systems which sometimes allow independent tuning of the strongly interrelated parameters. Nanoscale phenomena have been used to boost the power factor ($S^2\sigma$) by exploiting quantum size effect [5–7], interface energy filtering [8–10], band-structure engineering [11], and so on, while nanoscale inclusions in nanocomposites have been utilized to decrease thermal conductivity [6, 12], in attempts to develop high ZT materials. The most common strategy to enhance ZT in nanoscale materials involve

DOI: 10.1201/9781003439448-21

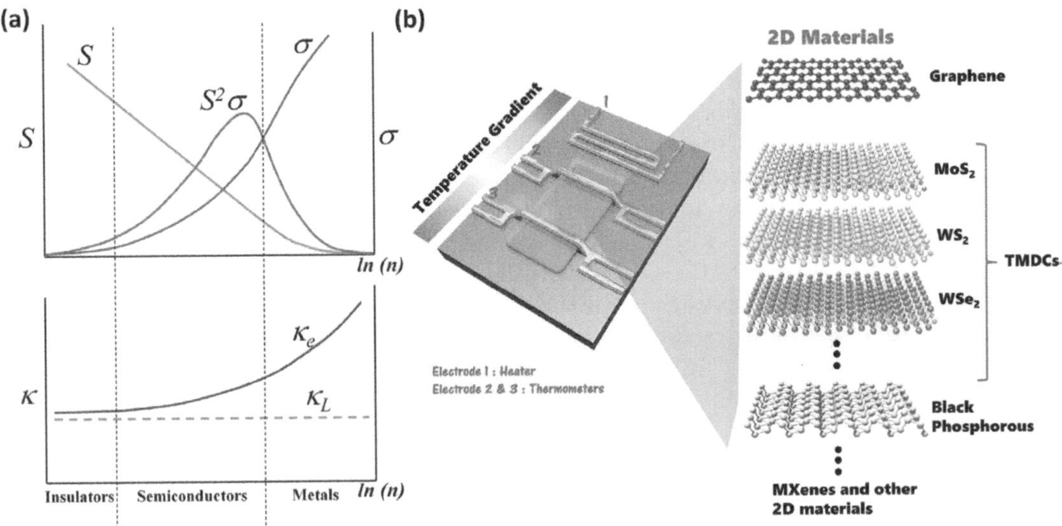

FIGURE 21.1
(a) For various charge carrier concentrations, the relationship between the Seebeck coefficient (S), electrical conductivity (σ), and the electronic (k_e) and lattice (k_l) thermal conductivity. b) Thermoelectric measurement technique employed for the characterization of two-dimensional (2D) materials along with the different 2D materials. Adapted with permission [13], © 2018 WILEY-VCH.

reduction of lattice thermal conduction by enhanced phonon scattering. Thermal conductivity of a material depends on heat transport by lattice phonons and by free charge carriers, implying that the total thermal conductivity can be expressed as: $k = k_l + k_e$, where, k_l is the lattice thermal conductivity and k_e is the electronic thermal conductivity (assuming electrons to the free carriers). In principle, we can reduce k_l by reducing lattice transport without affecting charge transport by the free carriers, which allows maintaining high σ, while reducing k. However, point scattering only works for short-wave phonons, and the coupling relationship between S and σ inhibits ZT reinforcement, therefore such a strategy cannot lower lattice conduction beyond a certain limit.

21.2 The Potential of 2D Materials for TE Applications

In this chapter we will focus on the potential of 2D semiconducting materials (2D-SCM) for development of next-generation thermoelectric materials and devices. As material dimensions diminish from 3D (conventional bulk materials) to 2D (thin films) to 1D (nanowire) to 0D (nanodot), the density of states (DOS) change significantly. Lower-dimensional materials have a steeper DOS near the Fermi level than bulk materials, making it easier to manipulate S and σ. Layered materials exhibit intriguing electrical, mechanical, thermal, and optoelectronic properties. In fact, one of the reasons why Bi_2Te_3 alloys have remained the most attractive TE material is due to their layered morphology. In a typical layered material, van der Waals interactions hold atomically thin layers together, allowing them to be isolated and investigated as quasi 2D materials. New theoretical and experimental methods have advanced thermoelectric physics and 2D material performance. In a 2D system, dimensional confinement and different interface scattering/reflection would decrease phonon transmission. This creates more effective suppression of lattice thermal conduction (k_l), one of the fundamental forms of phonon activity, by blocking phonons at varying frequencies in a hierarchical fashion. In addition, the microstructure may be used to actively influence the behavior of the carrier density, n and electronic conductivity, k_e to affect an overall reduction in k. This opens up new possibilities for the continued development of TE materials through the synthesis of 2D materials specifically for this application. 2D-SCM are therefore interesting candidates for next-generation TE material because they not only improve the power factor but also reduce thermal conductivity. Graphene, transition metal dichalcogenides (TMDCs), Mxenes, silicene, and phosphorene compounds are the most studied materials as shown in the schematic in the Figure 21.1(b) [13].

21.2.1 Graphene

Isolation of graphene, the ideal 2D honeycomb crystal of carbon atoms, triggered the massive recent interest in 2D materials [14]. Strong in-plane covalent bonds hold each carbon atom in graphene's three equidistant neighbors together. The unassociated p_z-orbital electron generates weak π-bonds with surrounding orbitals. Thus, delocalized π electrons can freely move above and below graphene sheets like a 2D electron gas. With linear energy dispersion for electrons and holes in the conduction and valence bands, intrinsic graphene is a zero-band-gap semiconductor. At two inequivalent high-symmetry positions (Dirac points) at the Brillouin zone boundaries, K and K', the conduction and valence bands produce Dirac cones. Instead of obeying the nonrelativistic Schrodinger equation, electrons and holes follow the relativistic Dirac equation for massless fermions. It is possible to evaluate the inherent electron and hole

concentrations as: $n = p = \dfrac{\pi}{6}\left(\dfrac{k_B T}{h v_F}\right)^2$; graphene carrier Fermi velocity is $v_F{\sim}10^8$ cm/s [15].

At 10 K, intrinsic carrier concentrations are linearly distributed between 10^8 cm^{-2} and 10^{11} cm^{-2} at room temperature. Graphene has high inherent mobility values without extrinsic disorder given by $\mu = 2*10^5 * \dfrac{cm^2}{Vs}$ at ambient temperature [16]. Many transport tests involve graphene samples on substrates. According to current knowledge, just one study has quantified suspended graphene's Seebeck coefficient. A single effort found a relatively low value of 9 μV/K at ordinary ambient temperature. This shows that graphene's Fermi level is near the charge neutrality point [17]. This experiment found no carrier concentration or gate voltage dependence. The substrate greatly impacts graphene transport. Gate voltage and layer number affect sheet resistance, Seebeck coefficient, and power factor of single- and multilayer graphene films on SiO_2 and hBN substrates, as illustrated in Figure 21.2(a) and (b). Adjusting gate voltage is equivalent to scanning chemical potential (μ) or doping concentration. Intrinsic graphene has maximal resistance at zero voltage [18]. However, in practical systems, this maximum is usually pushed closer to the average impurity potential's zero point [19]. Quantum mechanical disorder promotes a finite DOS at the Dirac point. Graphene sheet resistance in testing depends on sample size, quality, environment, and substrate. Outside the Dirac point, sheet conductance is linear and sublinear with voltage [16].

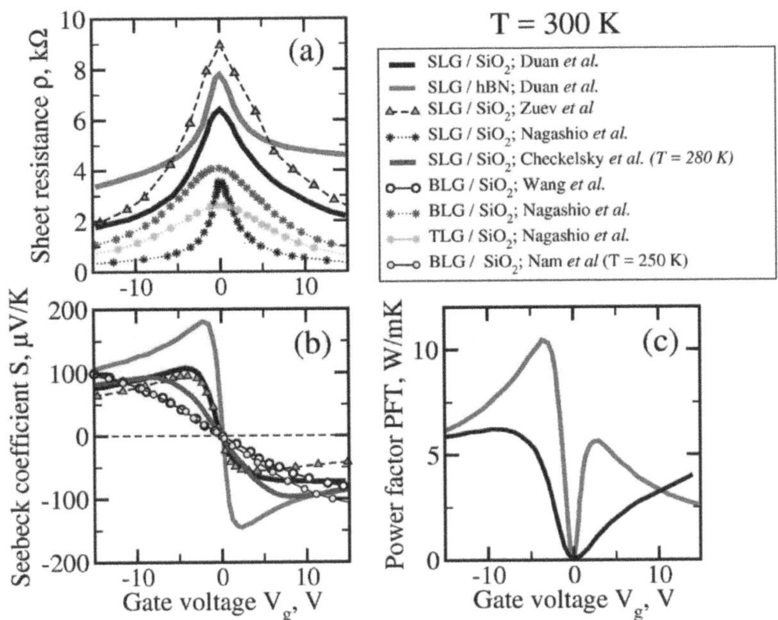

FIGURE 21.2

(a) Sheet resistance, (b) Seebeck coefficient, and (c) thermoelectric power factor as a function of gate voltage for single-layer graphene sheets produced on SiO$_2$ and hexagonal BN substrates and bilayer and trilayer graphene on SiO$_2$ substrate at ambient temperature. Adapted with permission [21], © 2018 Taylor & Francis.

Most graphene is grown on SiO$_2$ substrates. Graphene has electron–hole charge fluctuations (sometimes known as "puddles") due to its roughness, strong chemical reactivity, and high charged impurity content [20]. These issues do not affect chemically inert graphene monolayers with alkyl terminations or hBN. The solid curves in Figure 21.2(a) show that graphene on hBN has a larger resistance, although this is misleading because the resistance also relies on the flakes' form. Figure 21.2(a) shows that graphene/hBN conducts much better than graphene/SiO$_2$ [21]. Similarly, variations due to the number of layers in graphene on different substrates in Seebeck coefficient and thermoelectric power factor in relation to gate voltage are exhibited in Figure 21.2(b) and (c), respectively. Graphene has the highest heat conductivity, surpassing graphite, diamond, and carbon nanotubes. Several groups have assessed suspended graphene's conductivity using assumptions. The values range from 600 W/mK to several thousand W/mK at ambient temperature. Li Shi's critique examines these measurements' discrepancies [22].

Graphene, a semimetal, has "lattice" and "electronic" contributions from phonons and charged carriers. Recent ab initio calculations showed that the electronic contribution to thermal conductivity ranges from 2% to 10% (~80–300 W/mK) and increases with doping density. At low temperatures T < 200 K, electron-impurity scattering dominates the

Lorenz number $\dfrac{k(T)}{\sigma(T)} = L(T)T$, whereas at normal temperature, inelastic electron–phonon

scattering dominates. A method of Joule self-heating at low temperatures between 50 and 160 K and Johnson noise thermometry studies at T < 150 K validated the Wiedemann–Franz law [23].

In suspended graphene at ambient temperature, electron mean-free pathways are 100 nm, while phonon mean-free paths are 1 mm. Sample size greatly affects graphene thermal conductivity. Simulations indicate that samples with L < 2 µm can reduce heat conductivity by 50%. Phonon-boundary scattering increases with sample size, lowering thermal conductivity. Graphene's thermal conductivity makes thermoelectric uses problematic. There are many reports that discuss defect engineering, band engineering, nanostructuring, isotopic superlattice structure, grain boundary reduction of thermal conductivity in oxidized graphene, and functionalization to counter this problem [24].

The thermoelectric power factor decreases as defect density grows due to this process. When defect density increases, thermal conductivity decreases more than electrical conductivity, hence possessing ZT values up to three times higher than graphene in its original condition. Adding a determined amount of flaws to graphene lowers its thermal conductivity, improving the performance of graphene-based thermoelectric devices [25]. The promise of graphene is even better for active hotspot cooling, because it has a high power factor and conducts heat well. In such applications besides efficient cooling it is also important to remove the heat from the hotspot efficiently. Hence, graphene-based thermoelectric has significant potential in next-generation hotspot cooling in nanoelectronic devices, where heat must be quickly removed to prevent thermal breakdown.

21.2.1.1 Graphene-Based TE Devices

Graphene-based TE materials have already been integrated into devices with a wide range of intriguing applications beyond cooling and electricity harvesting, demonstrating their commercial viability. This is because several different graphene-based TE hybrids have been developed and graphene's own TE performance has been significantly improved. Graphene has been shown to enhance the TE performance of these hybrids [26].

(a) *TE generator*: A graphene-based flexible TE generator with a wristband-type structure has been developed as shown in Figure 21.3(a) and (b) and manufactured by building rGO film on a 3D-printed poly-dimethylsiloxane (PDMS) grid with a controlled porosity structure. The graphene based flexible TE generator was designed to be stretchable [27]. For a temperature difference of 50 K, it generates 57.33 mV g^{-1} and the power density reaches 4.19 µW g^{-1} when the temperature is set to 15°C. The graphene-based TE generator is a promising candidate for self-powered wearable microelectronics due to the decreased Young's modulus and higher flexibility of PDMS.

(b) *TE supercapacitor*: Supercapacitors made of traditional inorganic materials are unable to charge via heat and are too rigid for use in wearable applications. A new planar-type "thermally" chargeable supercapacitor (pTCSC) was developed by Kim et al. [28] that converts thermal energy into electricity without an external power supply, as shown in Figure 21.3c. Simple and scalable laser irradiation on graphene oxide films intercalated by sulfate ions (SGO) controlled the mobility and concentration of dissociated protons without liquid-phase electrolytes in the pTCSC. It generated a high thermally induced voltage of up to $S = 9$ mV K^{-1} and has surpassed classic inorganic and organic thermoelectric materials' Seebeck effect-based voltages. The enhanced proton mobility and concentration were found to produce high thermopower with good electrical conductivity. Meanwhile, lowered SGO electrodes stored harvested electrical energy. Their "single" pTCSC gadget

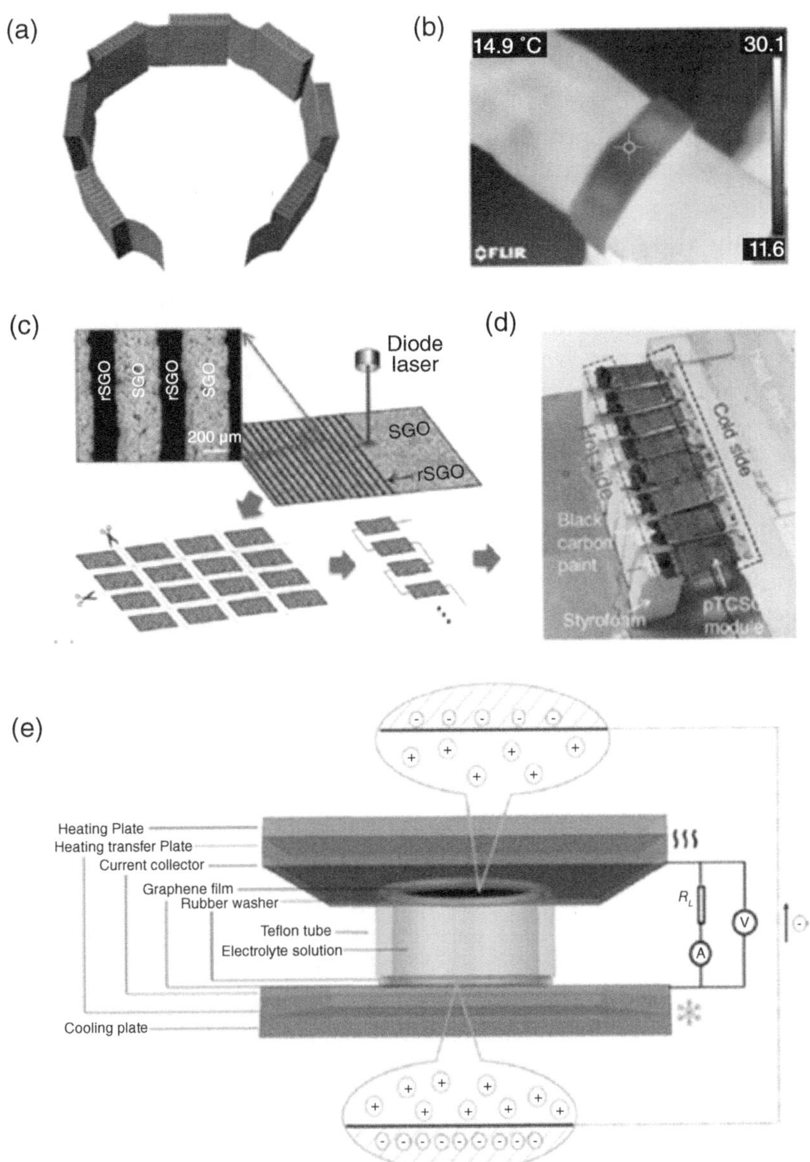

FIGURE 21.3
(a) Graphene-based TE generator schematic. (b) Wristband generator thermal infrared spectrum. Adapted with permission [27], © 2018 Elsevier Ltd.; (c) SGO/rSGO patterns by laser reduction, small module tailoring, and series connection. (d) Module-connected generator. Adapted with permission [28], © 2018 Elsevier Ltd.; (e) A diagrammatic representation of a graphene-based closed TE cell. Adapted with permission [29], Copyright © 2019, American Chemical Society.

generated 58 mV thermally charged voltage without power supply at 10.5 K temperature gradient. They showed that the batch fabricated with pTCSC can run an electrochromic device like a smart window that darkens with power supply when a temperature gradient is introduced (Figure 21.3d).

(c) *TE cell*: Using nanoporous graphene as the foundation, Yang et al. developed a TE cell (Figure 21.3e) [29]. The closed TE cell is contacted by two graphene electrodes, one on the hot side and one on the cold side. The cell might collect energy through the redistribution of the electrodes using a certain T. Under, electrodes based on nanoporous graphene produced 168.91 mV obtained under $\Delta T = 35$ K with an S of 4.54 mVK^{-1}, which is significantly greater than the output of typical TE generators.

21.2.2 Transition Metal Dichalcogenides

TMDCs, a novel class of layered materials with exotic characteristics, have garnered scientific attention. Single-layer and few-layer samples can be isolated by covalent in-plane bonding and weak van der Waals out-of-plane bonding. These materials were initially studied for their optical and optoelectronic capabilities. Researchers focused on the shift from an indirect to a direct bandgap when the number of layers reduced from n to 1 [30]. The absence of inversion symmetry in a TMDC monolayer allowed study of spin–valley locking, valleytronics, and piezoelectricity. The usual chemical formula for this class of compounds is MX_2, where M is a transition metal (Mo, Hf, W, Ti, Nb, V, etc.) and X is a chalcogen. These 2D TMDCs have also shown a wide range of complicated electronic behavior, including transitions from metal to insulator and electronic correlations that drive charge density waves, which were observed in single atomic layer and heterostructures with bandgap tuning and renormalization capabilities [31, 32]. How a material behaves as a metal, semiconductor, or insulator depends on how much transition metal coordination fills nonbonding d-states compared to σ and σ^* bonding states, and whether the E_F is inside a half-filled band or between the d-bands due to M atoms.

Hicks and Dresselhaus' original theory was troublesome in experimental application [33]. Increasing the thermoelectric power factor $S^2\sigma$ is expected due to a discretized electronic DOS, which should also raise the ZT. Experimentally testing this notion has proven difficult. Due to their atomically clean surfaces, lack of dangling bonds, and little quantum well thickness variation, 2D TMDCs are ideal for studying how discretized DOS affects the thermoelectric power factor. Back gate or ionic liquid-controlled field-effect doping can change carrier concentrations, reducing ionized impurity scattering on electronic and thermoelectric transport. This is conceivable because field-effect transistor (FET) devices can synthesize or exfoliate high-quality single and few-layer TMDCs. Phase transitions and dominant scattering mechanisms can be studied with the transverse Nernst coefficient.

Semiconducting TMDCs can use their 2D DOS to increase the Seebeck coefficient. This is shown by the band structure of group VI B semiconducting compounds such as MoS_2, $MoSe_2$, WS_2, and WSe_2 [34, 35]. It is generally known that transit effective mass increases the Seebeck coefficient [36]. Exfoliated MoS_2 and $MoSe_2$ on SiO_2/h-BN are n-type, whereas WS_2 and WSe_2 are p-type, presumably due to natural traps that preserve the Fermi level near the conduction or valence band. While full theoretical studies like these are still in their infancy for several TMDC materials, a considerable band degeneracy will lead to a high power factor in the thin layer limit. So far, 2D TMDC research has focused on MoS_2 (n-type) and WSe_2 (p-type). The first photothermoelectric effect measurement on MoS_2 showed that light absorption creates a temperature gradient in a monolayer device, generating thermovoltage that drives photocurrent [37]. The spatial resolution of the photocurrent caused an exceptionally large Seebeck, which scanning photocurrent microscopy detected. The Seebeck coefficient of monolayer MoS_2 built using chemical vapor deposition (CVD) was up to 30 mV K^{-1} in the OFF state [38]. Seebeck experiments at different temperatures ($S \propto T^{1/3}$ in 2D) showed that electron transport in the MoS_2 channel reduced charged particle mobility due to range hopping. CVD MoS_2 has higher sulfur vacancy densities than manually exfoliated MoS_2, reducing mobility and electrical conductivity. This is unsuitable for thermoelectric applications because high power factor requires high electrical conductivity and Seebeck. Recent, extensive studies on exfoliated monolayer and few-layer MoS_2 have revealed significant valley degeneracies, high power factor, big Seebeck, and strong mobility in the ON state, especially for bilayer MoS_2 [39]. Interestingly, gate-voltage-dependent Seebeck experiments can illuminate conduction electron scattering in the 2D MoS_2 channel.

It is also shown that for similar chemical doping, 2D TMDCs have ten times greater power factors than 3D, largely due to their high mobilities at the same carrier concentration [40]. This highlights the importance of 2D TMDCs as unique systems with controlled dispersion on electronic properties and high thermoelectric power factor. This challenges the idea that efficient bulk thermoelectrics with lower effective masses are better due to their inverse connection to phonon-limited mobility. Since mobilities and carrier concentration can be tailored creatively, these notions may not hold in two dimensions. However, a detailed theoretical description of the discretized 2D DOS's benefits for electrical power factors is needed to better understand and tune the thermoelectric feature.

Phonon transport is particularly anisotropic in multilayer materials like MoS_2, WSe_2, and other TMDCs, with higher thermal conductivity along the a- and b-axes than along the c-axis. The out-of-plane ZA mode contributes greatly to graphene's in-plane thermal conductivity, but it's less active in TMDCs' triatom layers. Due to strong covalent bonding, 2D MoS_2 has a high in-plane phonon conductivity between 36 and 52 W m^{-1} K^{-1}. Thus, even with a high TE power factor, the in-plane device ZT can be 0.05–0.1 [41]. The real challenge is using novel nanostructuring techniques to leverage the high power factor while reducing phonon conductivity.

21.2.3 MXene

Discovered in 2011, MXene possesses a distinctive composition of $M_{n+1}X_n$ (n = 1, 2, or 3), wherein M and X represent transition metals and carbon or nitrogen, respectively. MXene, a new 2D material with unique electrical and thermal properties, has garnered attention in thermal energy conversion and management. However, 2D MXenes need customized surface modification to suit application requirements or overcome specific restrictions [42]. The high electrical conductivity of 2D MXene made from titanium carbide is especially helpful for Joule heating and it can be a good material for heat removal. 2D MXenes that are based on molybdenum (Mo), like $(Mo_2Ti_2)C_3T_x$, Mo_2CT_x, and $(Mo_2Ti)C_2T_x$, constitute a special type of 2D-SCM that possess high thermopower (>100 V K^{-1}) [43]. $(Mo_2Ti)C_2T_x$ has been reported to have electrical conductivity, $\sigma = 1380$ S cm^{-1}, power factor ($S^2\sigma = 309$ WmK^{-2}), and thermopower, $S = 47.3$ V K^{-1} [44], suggesting that Mo-based 2D MXenes are a good choice for TE applications. Theoretical calculations using the Boltzmann theory and first-principles electronic structure calculations have been performed to evaluate the TE parameters of $Sc_2C(OH)_2$, Nb_2CF_2, and Mo_2CF_2. Figure 21.4 shows the variation of DOS, carrier density (n), S, σ, and the power factor with chemical potential (m) for multilayers of $Sc_2C(OH)_2$ (Figure 21.4(a)–(c)), Nb_2CF_2 (Figure 21.4(f)–(j)), and Mo_2CF_2 (Figure 21.4(k)–(o)), respectively [43]. The theoretical calculations suggest promising characteristics of MXene as a next-generation 2D-SCM for thermoelectric. According to these findings, Mo-based MXene exhibits the greatest potential as an effective thermoelectric material. In addition, thermoelectric conversion can be accomplished with the help of Nb-based MXenes. $(Mo_2Ti)C_2T_x$, Nb_2CTx, and $Ti_3C_2T_x$ were the constituent elements of the thermoelectric nanogenerator that was developed by Huang and Kim et al [45]. Niobium-based 2D MXene, such as Nb_2CT_x, is also a type of 2D MXene that can be used to store energy and convert thermal

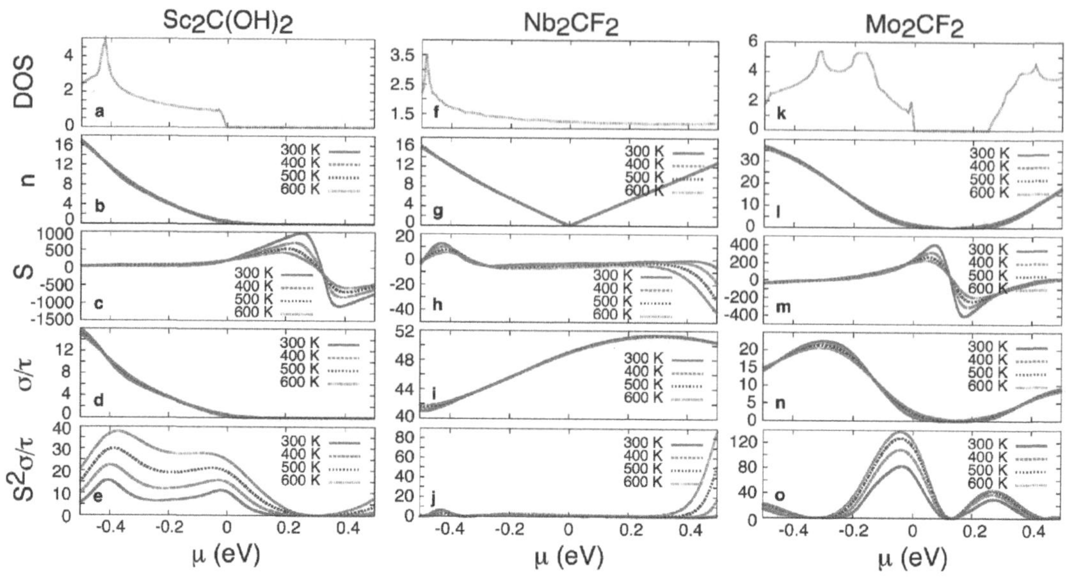

FIGURE 21.4

Density of states (DOS), carrier density (n), the Seebeck coefficient (S), and electrical conductivity (σ); τ is the relaxation time and the power factor for multilayers of $Sc_2C(OH)_2$ (a–e), Nb_2CF_2 (f–j), and Mo_2CF_2 (k–o) plotted against chemical potential, m. Reproduced with permission [43], © 2014 Royal Society of Chemistry.

energy [45]. $Sc_2C(OH)_2$, Ti_2CO_2, Zr_2CO_2, and Hf_2CO_2 are all possible 2D MXenes that could be efficient thermoelectric materials. Experimental investigation showed that Nb_2CT_x has the potential to be one of the thermoelectric terminals. Electrical conductivity, thermo-power, and power factor were measured to be 121.3 S cm^{-1}, 30 V K^{-1}, and 11.06 WmK^{-2}, respectively, for Nb_2CT_x that was annealed at 573 K. In comparison, those values were 169 S cm^{-1}, 27 V K^{-1}, and 14.7 W mK^{-2}, respectively, for $(Mo_2Ti)C_2T_x$ after being annealed at 633 K. Notably, because 2D MXenes can be processed in solutions, a thermoelectric nanogenerator could be made by drop-casting. This was made possible by the solution processability of the material. At a temperature difference of 30 K, the output power of a thermoelectric nanogenerator comprised of 20 n–p couples was measured to be 33.5 nW [46].

21.2.4 Silicene

Silicene has garnered attention in recent years. There was a little investigation on silicene until 2009, when it was theoretically proven that its low buckled structure was dynamically stable. Despite its low buckling geometry, silicene has most of planar graphene's electrical properties. Graphene and silicene share the "Dirac cone," high Fermi velocity, and carrier mobility. Unlike graphene, silicene has many advantages, including (1) simpler valley polarization and greater applicability for valleytronics research, (2) the possibility of realizing a quantum spin Hall effect at temperatures amenable to experimentation due to enhanced spin-orbit coupling, and (3) improved band gap tunability, essential for room-temperature FET operation. Since 2012, silicene sheets of various superstructures have been synthesized on MoS_2, ZrB_2, Ir, Ag, and ZrC substrates. Multilayer silicene films have also developed on Ag. The favorable results of the trials have prompted several researchers to study silicene's fundamental properties and device technology applications. Research on silicene includes thermoelectric effect, superconductivity, tunneling FET, Hall effect, spin filter, quantum anomalous Hall effect, quantum spin Hall effect, quantum valley Hall effect, and gas sensor [47].

Because of its near-zero bandgap, silicene has the potential to exhibit high electrical conductivity tailored with doping, similar to that of its organic counterpart, graphene. However, this property is dependent on the material being doped. However, silicene is forecasted to come up with significantly lower thermal conductivity compared to graphene due to the larger atomic mass it possesses and the softer interatomic bonding it possesses thereby rendering it more promising for TE applications.

A high TE figure of merit is predicted for silicene nanoribbons with hydrogen passivated edges using nonequilibrium green's functions and nonequilibrium molecular dynamics simulations. Additionally, it has been found that silicene nanoribbons with armchairs are more efficient than zigzag. The figure of merit of armchair silicene nanoribbons at ambient temperature demonstrates a width-dependent oscillation, whereas the ZT value of ZZ silicene nanoribbons gradually decreases with increasing ribbon width. It was also discovered that the TE performance had a significant relationship with temperature. According to the results of the calculations, the ZT values of the substance might be increased to as high as 4.9 if the doping level and temperature were optimized [48]. In the assessment, ab initio approaches have been performed to silicene nanoribbons of varying widths. These techniques have shown that the ZT at ambient conditions can reach a maximum of 2.5 [49]. Theoretically, the S of intrinsic silicene at room temperature is projected to be approximately 80 $\mu V\ K^{-1}$, and it is insensitive to changes in temperature. It is possible to open up a large bandgap by providing an electric field that is perpendicular to the direction of the bandgap, which results in a significantly increased Seebeck coefficient [50]. In addition to this, it was discovered that the ZT values of silicene nanoribbons may be significantly improved through the manipulation of both the kind and amount of edge flaws [51]. In conclusion, valley and spin-resolved TE transport are both conceivable in silicene. This is because the band structure of silicene is spin-polarized at distinct valleys. Because there is a link between the valley degree of freedom and the spin degree of freedom, it is possible to inject pure spin and valley currents by applying a thermal bias [52].

Considering the requirements for efficient cooling applications using TECs, there is a need for TE materials that operate at near-room temperatures and possess a high figure of merit. It is indisputable that Bi_2Te_3-based thermoelectric materials continue to exhibit superior performance in the vicinity of room temperature. Nevertheless, the element tellurium is characterized by its limited abundance on the earth's surface and is expensive. The topic of identifying a cost-effective thermoelectric material that exhibits high efficiency at room temperature remains unresolved. In recent times, significant advancements have been achieved by researchers in this particular domain. Recently, penta-silicene and its nanoribbon have been experimentally produced on an Ag (110) surface with unusual electrical characteristics. Both of these compounds are new members of the silicene family. According to the Boltzmann transport theory and the results of ab initio calculations, Gao et al. discovered that penta-silicene exhibits outstanding room-temperature figures of merit ZT of 3.4 and 3.0 at the achievable hole and electron concentrations, respectively, as shown in Figure 21.5.

21.2.5 Phosphorenes

Phosphorene has been the focus of an ever-increasing number of research activities ever since its first fabrication by exfoliation in 2014. The number of papers related to phosphorene has been growing at a rate that is significantly faster than that of publications

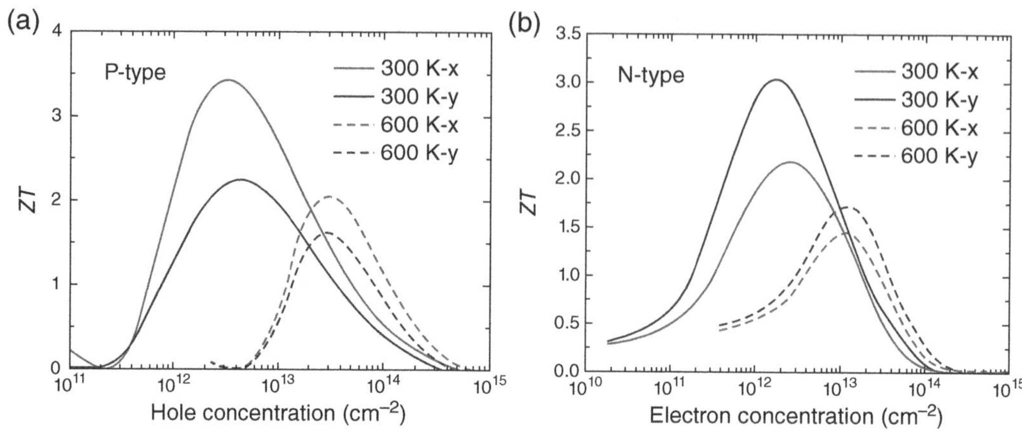

FIGURE 21.5

The figure of merit ZT for both (a) p-type and (b) n-type penta-silicene is determined as a function of the carrier concentration along the x- and y-axes at temperatures of 300 and 600 K. Adapted with permission [53], Copyright © 2020 American Chemical Society.

related to other two-dimensional materials. Phosphorene possesses a variety of one-of-a-kind qualities, including a puckered layer structure, which has generated specific interest in this material. Due to its broadly tunable band gap, significant in-plane anisotropy, and high carrier mobility, phosphorene is being studied extensively. Electronic, optoelectronic, and spintronic devices, sensors, actuators, and thermoelectric, as well as energy conversion and storage devices, all fall under the purview of these studies and applications [54].

Using first principle-based ballistic electron and phonon-transport simulations with hybrid functionals, the TE transport properties of single-layer black and blue-phosphorene structures have been investigated. The highest values of ZT, corresponding to armchair and zigzag pattern of black phosphorene are computed as being less than those found with ab initio calculations by implementing semiclassical Boltzmann transport theory. These values are 0.5 and 0.25, respectively [55]. On the other hand, it is projected that the maximum value of blue-phosphorene's ZT at ambient temperature will be quite high, and surprising values reaching as high as 2.5 were obtained for increased temperatures. In addition to the fact that these results were obtained near the ballistic limit, other calculations demonstrated that there is a significant likelihood of good TE performance of blue-phosphorene in applications belonging to the next generation of TE technologies [55].

21.3 Limitations of 2D SCM

- 2D material's electronic and phonon-transport mechanisms are still not fully understood. Even though a number of potential mechanisms have been put out and the power factors of many single- and few-layer 2D materials have been researched. However, there is still little understanding of how the interfacial electronic and phonon-transport affect thermoelectric performance. Their greater understanding ought to direct the creation of innovative thermoelectric composites and enhance their thermoelectric capabilities.

- Two-dimensional materials have substantially lower thermoelectric performance than bulk materials. A high ZT value requires precision chemical doping during composite production and functionalization, which remains a challenging task with regard to controlled doping.

- 2D materials usually have a very thermal conductivity which limits their performance as high-efficiency TE materials. Graphene, TMDCs, and hBN are examples of popular 2D materials, and their intrinsic heat conductivity is significantly higher than that of conventional semiconductor thermoelectric materials. However, large thermal conductivity makes these materials ideal for TE-based hot spot cooling applications.

- Demand for wearable and portable flexible electronics has increased interest in thin-film TE materials. Due to their superior electrical and mechanical capabilities, 2D materials and their composites present huge opportunities for thin-film TE components and devices. However, measuring the thermal conductivity of thin films remains a big challenge.

- The simple and scalable creation of high-quality 2D material is crucial. This is a topic that is being actively pursued right now. CVD, mechanical exfoliation, atomic layer deposition, molecular beam epitaxy, physical vapor deposition liquid exfoliation, and mechanical exfoliation are only some of the technologies that have been used to manufacture 2D materials thus far. However, producing such materials with a regulated structure via scale-up processes is still a big challenge.

21.4 Conclusion

The rich physics of multilayer and 2D materials has inspired numerous novel TE concepts, both from the standpoint of fundamental understanding and for developing new devices and components. This update reviews the past decade of TE research based on the well-known layered 2D materials and highlights the most relevant findings. In addition to DOS confinement, electron correlations and energy-dependent scattering offer new ways to control TE transport in graphene, as shown by the violation of the Mott relation and increased Seebeck and power factor. Band convergence and degeneracy in phosphorene and TMDCs result in high electronic power factors, suggesting more possibility for high-performance TE materials. Since native imperfections in 2D thermoelectric materials enhance the Seebeck coefficient and electrical conductivity, understanding and managing point defects and their distribution seems to be crucial for developing high-efficiency materials. The many ways to improve TE performance in 2D-layered materials require further research. Applications aside, 2D-like bulk materials with 2D DOS and poor heat conductivity are intriguing. The atomically clean surfaces of layered compounds provide a nanoscale platform for studying electron–phonon, photon–phonon, and excitonic processes in 2D heterostructures using TE signatures. Finally, coupled and magnetic effects may be studied in 2D due to the low atomic density. High-throughput machine learning tools are uncovering a massive library of materials and unique physics. Many exciting discoveries are sure to follow.

References

1. H. Lai, Y. Peng, J. Gao, M. Kurosawa, O. Nakatsuka, T. Takeuchi, *et al.*, "Silicon-based low-dimensional materials for thermal conductivity suppression: recent advances and new strategies to high thermoelectric efficiency," *Japanese Journal of Applied Physics*, vol. 60, p. SA0803, 2020.
2. E. Levin, "Charge carrier effective mass and concentration derived from combination of Seebeck coefficient and Te 125 NMR measurements in complex tellurides," *Physical Review B*, vol. 93, p. 245202, 2016.
3. B. Y. Yavorsky, N. Hinsche, I. Mertig, and P. Zahn, "Electronic structure and transport anisotropy of Bi 2 Te 3 and Sb 2 Te 3," *Physical Review B*, vol. 84, p. 165208, 2011.
4. I. T. Witting, T. C. Chasapis, F. Ricci, M. Peters, N. A. Heinz, G. Hautier, *et al.*, "The thermoelectric properties of bismuth telluride," *Advanced Electronic Materials*, vol. 5, p. 1800904, 2019.
5. L. D. Hicks and M. S. Dresselhaus, "Effect of quantum-well structures on the thermoelectric figure of merit," *Physical Review B*, vol. 47, p. 12727, 1993.
6. G. Chen, "Thermal conductivity and ballistic-phonon transport in the cross-plane direction of superlattices," *Physical Review B*, vol. 57, p. 14958, 1998.
7. R. Venkatasubramanian, E. Siivola, T. Colpitts, and B. O'Quinn, "Thin-film thermoelectric devices with high room-temperature figures of merit," *Nature*, vol. 413, pp. 597–602, 2001.
8. D. Vashaee, Y. Zhang, A. Shakouri, G. Zeng, and Y.-J. Chiu, "Cross-plane Seebeck coefficient in superlattice structures in the miniband conduction regime," *Physical Review B*, vol. 74, p. 195315, 2006.
9. D. Vashaee and A. Shakouri, "Electronic and thermoelectric transport in semiconductor and metallic superlattices," *Journal of Applied Physics*, vol. 95, pp. 1233–1245, 2004.
10. G. Zeng, X. Fan, C. LaBounty, E. Croke, Y. Zhang, J. Christofferson, *et al.*, "Cooling power density of SiGe/Si superlattice micro refrigerators," *MRS Online Proceedings Library (OPL)*, vol. 793, 2003.
11. Y. Pei, G. Tan, D. Feng, L. Zheng, Q. Tan, X. Xie, *et al.*, "Integrating band structure engineering with all-scale hierarchical structuring for high thermoelectric performance in PbTe system," *Advanced Energy Materials*, vol. 7, p. 1601450, 2017.
12. T. Harman, P. Taylor, M. Walsh, and B. LaForge, "Quantum dot superlattice thermoelectric materials and devices," *Science*, vol. 297, pp. 2229–2232, 2002.
13. J. Wu, Y. Chen, J. Wu, and K. Hippalgaonkar, "Perspectives on thermoelectricity in layered and 2D materials," *Advanced Electronic Materials*, vol. 4, p. 1800248, 2018.
14. K. S. Novoselov, A. K. Geim, S. V. Morozov, D.-e. Jiang, Y. Zhang, S. V. Dubonos, *et al.*, "Electric field effect in atomically thin carbon films," *Science*, vol. 306, pp. 666–669, 2004.
15. T. Fang, A. Konar, H. Xing, and D. Jena, "Carrier statistics and quantum capacitance of graphene sheets and ribbons," *Applied Physics Letters*, vol. 91, p. 092109, 2007.
16. S. á Morozov, K. á Novoselov, M. á Katsnelson, and F. á Schedin, "DC áElias, JA áJaszczak, AK áGeim," *Physical Review Letters*, vol. 100, p. 016602, 2008.
17. X. Xu, Y. Wang, K. Zhang, X. Zhao, S. Bae, M. Heinrich, *et al.*, "Phonon transport in suspended single layer graphene," *arXiv preprint arXiv:1012.2937*, 2010.
18. L. Chen, A. Bhattacharjee, P. Puhl-Quinn, H. Yang, N. Bessho, S. Imada, *et al.*, "Observation of energetic electrons within magnetic islands," *Nature Physics*, vol. 4, pp. 19–23, 2008.
19. N. M. Peres, "Colloquium: The transport properties of graphene: An introduction," *Reviews of modern physics*, vol. 82, p. 2673, 2010.
20. Q. H. Wang, Z. Jin, K. K. Kim, A. J. Hilmer, G. L. Paulus, C.-J. Shih, *et al.*, "Understanding and controlling the substrate effect on graphene electron-transfer chemistry via reactivity imprint lithography," *Nature Chemistry*, vol. 4, pp. 724–732, 2012.
21. M. Markov and M. Zebarjadi, "Thermoelectric transport in graphene and 2D layered materials," *Nanoscale and Microscale Thermophysical Engineering*, vol. 23, pp. 117–127, 2019.
22. Y. M. Zuev, J. S. Lee, C. Galloy, H. Park, and P. Kim, "Diameter dependence of the transport properties of antimony telluride nanowires," *Nano Letters*, vol. 10, pp. 3037–3040, 2010.
23. T. Y. Kim, C.-H. Park, and N. Marzari, "The electronic thermal conductivity of graphene," *Nano Letters*, vol. 16, pp. 2439–2443, 2016.

24. T. A. Amollo, G. T. Mola, M. Kirui, and V. O. Nyamori, "Graphene for thermoelectric applications: prospects and challenges," *Critical Reviews in Solid State and Materials Sciences*, vol. 43, pp. 133–157, 2018.

25. Y. Anno, Y. Imakita, K. Takei, S. Akita, and T. Arie, "Enhancement of graphene thermoelectric performance through defect engineering," *2D Materials*, vol. 4, p. 025019, 2017.

26. P.-A. Zong, J. Liang, P. Zhang, C. Wan, Y. Wang, and K. Koumoto, "Graphene-based thermoelectrics," *ACS Applied Energy Materials*, vol. 3, pp. 2224–2239, 2020.

27. W. Zeng, X.-M. Tao, S. Lin, C. Lee, D. Shi, K.-h. Lam, *et al.*, "Defect-engineered reduced graphene oxide sheets with high electric conductivity and controlled thermal conductivity for soft and flexible wearable thermoelectric generators," *Nano Energy*, vol. 54, pp. 163–174, 2018.

28. S. L. Kim, J.-H. Hsu, and C. Yu, "Intercalated graphene oxide for flexible and practically large thermoelectric voltage generation and simultaneous energy storage," *Nano Energy*, vol. 48, pp. 582–589, 2018.

29. Z. Yang, F. Dang, C. Zhang, S. Sun, W. Zhao, X. Li, *et al.*, "Harvesting Low-Grade Heat via Thermal-induced electric double layer redistribution of nanoporous graphene films," *Langmuir*, vol. 35, pp. 7713–7719, 2019.

30. K. F. Mak, C. Lee, J. Hone, J. Shan, and T. F. Heinz, "Atomically thin MoS 2: A new direct-gap semiconductor," *Physical Review Letters*, vol. 105, p. 136805, 2010.

31. B. Radisavljevic and A. Kis, "Mobility engineering and a metal–insulator transition in monolayer MoS2," *Nature Materials*, vol. 12, pp. 815–820, 2013.

32. B. W. Baugher, H. O. Churchill, Y. Yang, and P. Jarillo-Herrero, "Intrinsic electronic transport properties of high-quality monolayer and bilayer MoS2," *Nano Letters*, vol. 13, pp. 4212–4216, 2013.

33. M. Dresselhaus, G. Dresselhaus, X. Sun, Z. Zhang, S. Cronin, and T. Koga, "Low-dimensional thermoelectric materials," *Physics of the Solid State*, vol. 41, pp. 679–682, 1999.

34. W. Lee, H. Li, A. B. Wong, D. Zhang, M. Lai, Y. Yu, *et al.*, "Ultralow thermal conductivity in all-inorganic halide perovskites," *Proceedings of the National Academy of Sciences*, vol. 114, pp. 8693–8697, 2017.

35. K. Hippalgaonkar, Y. Wang, Y. Ye, D. Y. Qiu, H. Zhu, Y. Wang, *et al.*, "High thermoelectric power factor in two-dimensional crystals of Mo S 2," *Physical Review B*, vol. 95, p. 115407, 2017.

36. H. Ng, D. Chi, and K. Hippalgaonkar, "Effect of dimensionality on thermoelectric powerfactor of molybdenum disulfide," *Journal of Applied Physics*, vol. 121, p. 204303, 2017.

37. M. Buscema, M. Barkelid, V. Zwiller, H. S. van der Zant, G. A. Steele, and A. Castellanos-Gomez, "Large and tunable photothermoelectric effect in single-layer MoS2," *Nano Letters*, vol. 13, pp. 358–363, 2013.

38. J. Wu, H. Schmidt, K. K. Amara, X. Xu, G. Eda, and B. Özyilmaz, "Large thermoelectricity via variable range hopping in chemical vapor deposition grown single-layer MoS2," *Nano Letters*, vol. 14, pp. 2730–2734, 2014.

39. M. Kayyalha, J. Maassen, M. Lundstrom, L. Shi, and Y. P. Chen, "Gate-tunable and thickness-dependent electronic and thermoelectric transport in few-layer MoS2," *Journal of Applied Physics*, vol. 120, p. 134305, 2016.

40. M. Samanta, T. Ghosh, S. Chandra, and K. Biswas, "Layered materials with 2D connectivity for thermoelectric energy conversion," *Journal of Materials Chemistry A*, vol. 8, pp. 12226–12261, 2020.

41. X. Wang and A. Tabarraei, "Phonon thermal conductivity of monolayer MoS2," *Applied Physics Letters*, vol. 108, p. 191905, 2016.

42. M. Naguib, M. Kurtoglu, V. Presser, J. Lu, J. Niu, M. Heon, *et al.*, "Two-dimensional nanocrystals produced by exfoliation of Ti3AlC2," *Advanced Materials*, vol. 23, pp. 4248–4253, 2011.

43. M. Khazaei, M. Arai, T. Sasaki, M. Estili, and Y. Sakka, "Two-dimensional molybdenum carbides: potential thermoelectric materials of the MXene family," *Physical Chemistry Chemical Physics*, vol. 16, pp. 7841–7849, 2014.

44. H. Kim, B. Anasori, Y. Gogotsi, and H. N. Alshareef, "Thermoelectric properties of two-dimensional molybdenum-based MXenes," *Chemistry of Materials*, vol. 29, pp. 6472–6479, 2017.

45. D. Huang, H. Kim, G. Zou, X. Xu, Y. Zhu, K. Ahmad, *et al.*, "All-MXene thermoelectric nanogenerator," *Materials Today Energy*, vol. 29, p. 101129, 2022.

46. A. N. Gandi, H. N. Alshareef, and U. Schwingenschlögl, "Thermoelectric performance of the mxenes M2Co2 (M= Ti, Zr, or Hf)," *Chemistry of Materials*, vol. 28, pp. 1647–1652, 2016.
47. J. Zhao, H. Liu, Z. Yu, R. Quhe, S. Zhou, Y. Wang, *et al.*, "Rise of silicene: A competitive 2D material," *Progress in Materials Science*, vol. 83, pp. 24–151, 2016.
48. L. Pan, H. Liu, X. Tan, H. Lv, J. Shi, X. Tang, *et al.*, "Thermoelectric properties of armchair and zigzag silicene nanoribbons," *Physical Chemistry Chemical Physics*, vol. 14, pp. 13588–13593, 2012.
49. K. Yang, S. Cahangirov, A. Cantarero, A. Rubio, and R. D'Agosta, "Thermoelectric properties of atomically thin silicene and germanene nanostructures," *Physical Review B*, vol. 89, p. 125403, 2014.
50. Y. Yan, H. Wu, F. Jiang, and H. Zhao, "Enhanced thermopower of gated silicene," *The European Physical Journal B*, vol. 86, pp. 1–5, 2013.
51. W. Zhao, Z. Guo, Y. Zhang, J. Ding, and X. Zheng, "Enhanced thermoelectric performance of defected silicene nanoribbons," *Solid State Communications*, vol. 227, pp. 1–8, 2016.
52. Z. Ping Niu and S. Dong, "Valley and spin thermoelectric transport in ferromagnetic silicene junctions," *Applied Physics Letters*, vol. 104, p. 202401, 2014.
53. Z. Gao and J.-S. Wang, "Thermoelectric penta-silicene with a high room-temperature figure of merit," *ACS Applied Materials & Interfaces*, vol. 12, pp. 14298–14307, 2020.
54. M. Akhtar, G. Anderson, R. Zhao, A. Alruqi, J. E. Mroczkowska, G. Sumanasekera, *et al.*, "Recent advances in synthesis, properties, and applications of phosphorene," *NPJ 2D Materials and Applications*, vol. 1, p. 5, 2017.
55. C. Sevik and H. Sevinçli, "Promising thermoelectric properties of phosphorenes," *Nanotechnology*, vol. 27, p. 355705, 2016.